a LANGE medical book

Histology & Cell Biology

Examination & Board Review

fourth edition

Douglas F. Paulsen, PhD
Professor of Anatomy
Director of Graduate Studies
Morehouse School of Medicine
Atlanta, Georgia

Lange Medical Books/McGraw-Hill
Medical Publishing Division

New York St. Louis San Francisco Auckland Bogotá Caracas Lisbon London
Madrid Mexico City Milan Montreal New Delhi San Juan
Singapore Sydney Tokyo Toronto

McGraw-Hill

A Division of The McGraw·Hill Companies

Histology & Cell Biology: Examination & Board Review, Fourth Edition

1234567890 DOWDOW 09876543210

ISBN: 0-8385-0593-7
ISSN: 1045-4586

Notice

This book was set in Times Roman by Rainbow Graphics, LLC.
The editors were Janet Foltin, Harriet Lebowitz, and Lester A. Sheinis.
The production supervisor was Richard C. Ruzycka.
The production service was Rainbow Graphics, LLC.
The cover designer was Mary Skudlarek.
The art managers were Eve Siegel and Charissa Baker.
The indexer was Kathleen Garcia.

R.R. Donnelley & Sons Company was printer and binder.

This book is printed on acid-free paper.

*This book is dedicated to my students at the Morehouse School of
Medicine for their commitment to providing quality primary care
to the economically disadvantaged and the medically underserved.
This edition is dedicated to the memory of Professor Curtis L. Parker,
a trusted friend, honored mentor, and talented colleague.*

Contents

Preface

As with all Lange Medical Books, this book is intended to meet the needs of students. The first through third editions were written as companions to Lange's popular *Basic Histology,* Junqueira, Carneiro, and Kelley, now in its ninth edition. In the fourth edition of this review, the correspondence with the text's chapter layout has been retained to continue that relationship. However, the title of this review has been changed to reflect the increasing importance of cell biology to our understanding of tissue structure and function, and thus the increasing coverage this topic has been accorded with each edition of this review. This fourth edition contains substantial revisions of the first four chapters to include additional cell biology methods and concepts that have become fundamental to medical education. The reader is still reminded that although this review can be used as an adjunct to any standard textbook of histology, it is not intended as a substitute.

All of the synopses have been edited to make them more concise. This will be a constant goal in developing this book through future editions. A significant improvement is the inclusion of questions with longer, case-related stems in the Diagnostic Final Examination. The USMLE Part I examinations include more of these every year. Questions in the individual chapters remain similar to those you encounter in your course examinations because this format, in addition to being beneficial in preparing for course and subject board (miniboard) examinations, is more useful in identifying specific knowledge deficits. Readers are encouraged to explore the How to Use This Book section *before* beginning their study or review of histology.

My greatest wish is that you find this book helpful in gaining the knowledge you need for the effective practice of medicine and in preparing for examinations designed to test that knowledge. Readers' comments and suggestions are welcome and have again been helpful in improving the book. Your contributions may assist many other students in the coming years.

I would like to acknowledge the contributions of several people who were instrumental in the publication of this book. Drs. Shanda Blackmon, Gale Newman, Mary Scanlon, Mary Saltarelli, and Nina Zanetti provided helpful advice and information. The encouragement and insight of Lange editors David Barnes and Harriet Lebowitz were invaluable in making improvements in this edition and getting it into print. All readers owe a debt of gratitude to Susan Worley for her efforts to clarify the meaning of the text and make it more readable. Finally, the improvements in the appearance and usefulness of the figures in this edition reflect the artistic efforts and attention to detail of Becky Hainz-Baxter. Many thanks to everyone for their hard work.

May, 2000

Douglas F. Paulsen, PhD
Atlanta, Georgia
paulsen@msm.edu

How to Use This Book

This book will help you use your study time more effectively, both as you study histology and cell biology for the first time and as you review for course, subject board (miniboard), or U.S. Medical Licensing Examinations (USMLE). More specifically, it will aid in quickly finding and filling the gaps in your knowledge of histology and cell biology and allow you to spend less of your valuable time figuring out what you need to learn, so that you can focus your efforts on actually learning. Because histology and cell biology cover all of the body's systems and because they focus not only on microscopic structure, but also on structure–function relationships at the cell and tissue levels, they incorporate many fundamental concepts of importance in anatomy, biochemistry, embryology, immunology, neurobiology, pathology, and physiology. Thus, this book, in addition to serving as a review and study guide for histology and cell biology, provides a foundation for any comprehensive system-by-system review of the basic medical sciences.

Each chapter covers one topic and is divided into four sections. The first is a list of **objectives** for you to bear in mind as you study the material. They describe, in general terms, the most significant types of facts and concepts you will need to learn about the topic. Besides assisting you during your first encounter with the material, this section will help you, as you review for an examination, to identify stronger and weaker areas of your knowledge and to tailor your review accordingly.

The second section of each chapter is a set of *MAX-Yield* ™ **study questions** that direct your attention to key facts needed to master the material most often covered on examinations. Again, these questions allow students preparing for examinations to quickly identify gaps in their knowledge. The associated references to the synopsis will help you fill these gaps just as quickly. This section is also useful during your first encounter with histology. If you use these questions to focus your studies and write out the answer to each as you proceed, you will improve your retention and be left with an outstanding personal review for course and USMLE examinations. Many of these questions can be answered with a word, phrase, list, or sketch; others ask you to draw comparisons and are best handled by setting up a chart or table to be filled in with the requested information. Again, this will help you focus your efforts on learning instead of spending time figuring out what to learn. Even if you do not write out the answers, focusing on the questions first will improve your retention of the material you study, making you a more efficient student.

The third section of each chapter is a **synopsis** of the topic, presented in outline form. This section reviews all the basic concepts of histology and related cell biology included in examinations. If any concept reviewed in the synopsis is not immediately clear, you should return to your text and review that concept in more detail. This section is also useful as a prelecture introduction to the material for students enrolled in a histology course. The organization of the information is just as important as the information itself and may be even more important in committing the information to memory.

The fourth and final section of each chapter is a set of **multiple-choice questions** written in formats commonly used by the National Board of Medical Examiners. They will help you become more comfortable with these types of questions and allow you to assess your comprehension of the material covered in the *MAX-Yield* ™ questions before taking your examinations. You will also find several collections of **integrative multiple-choice questions.** These will help you to integrate the information included under separate topic subheadings and avoid confusion between topics you studied separately. The correct answers immediately follow each set of multiple-choice questions and are accompanied by references to sections in the synopsis where explanations of the correct answers may be found.

Finally, the **Diagnostic Final Examination** will help you gauge your preparedness for tests in various topic areas and allow you to schedule your examination preparation time for maximum productivity. Many of these questions have been rewritten for this edition in the so-called long stem format now being adopted in the USMLE to frame questions in clinical vignettes and other problem-solving situations. To gauge your ability to handle such questions in terms of reading speed and comprehension, allow yourself an average of 72 seconds per question, 50 questions per hour. At this rate you should be able to finish the entire examination in approximately one hour. I think you will find it time well spent. The answer key for these questions includes full explanations referenced to the synopses. The explanations cover both the rationale for selecting the correct answer and the reasons for eliminating the incorrect answers. As a result, this chapter will help you to test not only your retention of important facts, but also your ability to apply that knowledge to solving problems.

SUGGESTED PROTOCOL FOR USING THIS BOOK TO REVIEW FOR EXAMINATIONS

For unit examinations

1. For each chapter, read the objectives and the synopsis.
2. Attempt to answer the *MAX-Yield* ™ study questions. Identify and list the gaps in your knowledge. Use the accompanying references to look up the answers you do not know in the synopsis.
3. Take the multiple-choice questions in the chapter as a test, allowing about 45 seconds per question. Again, identify and list the gaps in your knowledge and use the accompanying references to look up the answers in the synopsis.
4. Combine the lists of your knowledge gaps obtained in steps 2 and 3 above. Review the missing information in your text, lecture notes, and written answers to the *MAX-Yield* ™ study questions, if you have them, at least twice before the examination.

For final and USMLE examinations

1. Take the Diagnostic Final Examination, using the answer sheet provided and allowing 72 seconds per question.
2. Score your performance and analyze your relative performance in the various topic areas.
3. Use the explanations and references in the accompanying answer key to resolve your problems. Identify and list the gaps in your knowledge.
4. Follow the directions for unit examinations to further clarify and fill the gaps in your knowledge base.
5. Retake the Diagnostic Final Examination to determine your next step.

SUGGESTED PROTOCOL FOR USING THIS BOOK AS A STUDY GUIDE FOR HISTOLOGY

For each chapter

1. Before the lecture, read the objectives and the *MAX-Yield* ™ study questions. Scan the synopsis twice; the first time examine only the major topic headings to see how the information is organized and the second time read only the boldfaced terms. Use a medical dictionary and look up the meanings of any boldfaced terms you do not already know *before* you attend the

lecture. Each of us normally loses a degree of interest each time an unknown word is encountered in our reading or in lectures. The cumulative effect is boredom, which can cause the mind to wander, or worse, lead to drowsiness. The end result is that a barrage of unfamiliar words in your reading, or in lectures you attend, can substantially reduce the effectiveness of the time you spend in these activities.

2. Attend the lecture. During the lecture, highlight the synopsis to indicate material covered by the professor. Add notes in the margins of the synopsis or in a separate notebook to indicate anything the professor emphasizes or anything mentioned by the professor but missing in the synopsis. Note that the many conflicting responsibilities of professors rarely leave them time to leaf through textbooks or lecture notes as they write examination questions. Typically they use questions previously written by themselves and others as a template or write entirely new questions from their existing knowledge of the subject. Attending the lecture and consulting directly with your professors can be useful in gauging the depth to which a particular topic will be covered on a course examination.

3. Review the lecture notes and highlighted synopsis the day of the lecture. This reinforces the material you learned that day and helps to move it from short-term to long-term memory.

4. Skim the chapter in your textbook, focusing on organization and boldfaced terms rather than detailed content.

5. Use the *MAX-Yield* ™ study questions and lecture notes to direct your in-depth study of the material in the text. Approaching your text with questions in your mind instead of trying to read it as a novel will be a more efficient use of your time and improve your retention.

6. Write out the answers to the *MAX-Yield* ™ study questions for future review. This may be too much for one person to do alone given the time constraints of a full course load. However, a study group of six to eight people can usually split up the questions and get them all done quite easily. Naturally you will know best the ones you do yourself, so pick the hardest ones if you have a choice, and do as many as you can.

7. In preparing for unit and subject board (miniboard) examinations, use the protocol described above.

ABBREVIATIONS

ABP	androgen-binding protein
ACTH	adrenocorticotropic hormone
ADH	antidiuretic hormone
ADP	adenosine diphosphate
ANS	autonomic nervous system
ATP	adenosine triphosphate
ATPase	adenosine triphosphatase
AVP	arginine vasopressin
cAMP	cyclic adenosine monophosphate
CAT	computerized axial tomography
Cdk	cyclin-dependent kinase
CFCs	colony-forming cells
CFU	colony-forming unit
CNS	central nervous system
CRH	corticotropin-releasing hormone
CSF	colony-stimulating factor
DAB	diaminobenzidine
DI	diabetes insipidus
DIC	differential interference contrast (microscopes)
DIT	diiodotyrosine
DNES	diffuse neuroendocrine system
ECF-A	eosinophilic chemotactic factor of anaphylaxis
EM	electron micrograph
EM	electron microscopy
ER	endoplasmic reticulum
FSH	follicle-stimulating hormone
GAG	glycosaminoglycan
GDP	guanosine diphosphate
GH	growth hormone
GHIH	growth hormone inhibiting hormone
GnRH	gonadotropin-releasing hormone
GTP	guanosine triphosphate
H&E	hematoxylin and eosin (stain)
HEV	high-endothelial venule
ICSH	interstitial cell-stimulating hormone
IEF	isoelectric focusing

Ig	immunoglobulin
JG	juxtaglomerular (cells)
LH	luteinizing hormone
LHRH	luteinizing hormone-releasing hormone
MAP	microtubule-associated protein
MHC	major histocompatibility complex
MHC	myosin heavy chain
MIT	monoiodotyrosine
MPF	mitotic phase promoting factor
mRNA	messenger RNA
MSH	melanocyte-stimulating hormone
NA	numerical aperture
NK	natural killer (cells)
PAGE	polyacrylamide gel electrophoresis
PALS	periarterial lymphatic sheath
PAS	periodic acid–Schiff (reaction)
PCR	polymerase chain reaction
PIH	prolactin-inhibiting hormone
PNS	peripheral nervous system
PTH	parathyroid hormone or parathormone
PWP	peripheral white pulp
Rb	retinoblastoma
RBC	red blood cell
RER	rough endoplasmic reticulum
SDS	sodium dodecyl sulfate
SER	smooth endoplasmic reticulum
SRP	signal recognition particle
STH	somatotrophic hormone
TEM	transmission electron microscopy or transmission electron micrograph
TGN	*trans* Golgi network
TnC	troponin C
TnI	troponin I
TnT	troponin T
TRH	thyrotropin-releasing hormone
TSH	thyroid-stimulating hormone
VLDL	very low-density lipoprotein

Methods of Study

1

OBJECTIVES

This chapter should help the student to:

- Know the mathematical relationships among the units of measure used for histologic specimens.
- Name the instruments and techniques used to prepare and study histologic specimens.
- Know the basic steps in preparing specimens for light and electron microscopy.
- Know the advantages and limitations of histologic instruments and techniques.
- Select appropriate methods to reveal specific microscopic features of cells and tissues.
- Describe the basic principles of histochemistry.
- Know the substances of biologic interest that can be localized by histochemical techniques.
- Name the classes of histochemical techniques and describe the advantages and limitations of each.
- Choose appropriate techniques to reveal the location of specific substances in cells and tissues.
- Name and describe methods appropriate for the isolation and study of specific cells and tissues from intact organisms.
- Name and describe methods appropriate for the isolation and identification of specific proteins and nucleic acid sequences from cells and tissues.
- Choose appropriate techniques to reveal the location of specific substances in cells and tissues.

MAX-Yield™ STUDY QUESTIONS

1. List, in order, the basic steps in preparing histologic sections for microscopy (Fig. 1–1) and briefly describe the purpose of each step (Table 1–1).
2. Name the type of microscopy that best visualizes unstained living cells and tissues and serves as a basic tool for tissue culture (IV.D.3[1]).
3. Compare light and electron microscopy in terms of:
    a. Methods of fixation, embedding, sectioning, and staining (Tables 1–1, 1–2, and 1–3)
    b. Thickness of sections (Table 1–1)
    c. The support on which sections are mounted (Table 1–1)
    d. The type, source, wavelength, and path of illuminating beams (IV.A and B; V.A and B)
    e. Magnification (V.A)
4. Compare transmission and scanning electron microscopy in terms of:
    a. Methods of preparing tissue for observation (Table 1–1; Fig. 1–1; V.D.2)
    b. Electron path (V.A, B, and D.2)
    c. Image obtained (V.D.1 and 2)
5. What is cryofracture and what are its advantages over conventional methods of preparing tissues for electron microscopy (Table 1–4)?
6. What is radioautography, and what information can it provide (Table 1–4)?

---

[1] The parenthetical references throughout this book (eg, IV.D.3) refer to sections and subsections in the synopsis of each chapter. When chapter numbers precede the roman numerals (eg, 4.II.A.2.B), they refer to the synopsis section in another chapter.

7. List the substances in cells and tissues that can be identified by histochemistry (VII.A–F).
8. The periodic acid–Schiff (PAS) reaction allows the localization of which substances (VII.E.1)?
9. Describe a typical enzyme histochemical reaction (VIII) in terms of:
    a. How the site of activity of a particular enzyme is marked
    b. Any special considerations necessary in preparing tissue for sectioning
10. Name three types of markers used to make antibodies visible with a microscope (IX.B).
11. List the advantages of studying isolated cells, tissues, or organ rudiments in culture rather than in the intact organism (XI).
12. What is the purpose of cell fractionation and how is it accomplished (XII)?
13. Name three types of column chromatography and describe the characteristics of a protein that are best isolated by each (XIII; Table 1–5).
14. Explain how one-dimensional gel electrophoresis may be used to estimate the molecular weight of a protein (XIV.A).
15. What determines a protein's isoelectric point and how can differences in this characteristic be used to separate proteins (XIV.B)?
16. Under what circumstances would two-dimensional gel electrophoresis be preferable to one-dimensional gel electrophoresis (XIV.C)?
17. Name the type of molecule that is identified in Northern, Southern, and Western blotting (XV).
18. What characteristic of a DNA strand determines the site of restriction nuclease activity (XVI.A)?
19. Describe the composition of a plasmid and several characteristics that make it useful as a cloning vector (XVI.C.1).
20. After the 10th cycle, how many copies of a DNA sequence would be present in a PCR reaction mixture for every copy present after the first cycle (XVI.C.2)?

SYNOPSIS

I. GENERAL FEATURES OF HISTOLOGY & ITS METHODS

A. Goals of Histology: Histology, the study of tissues, is largely a visual science that relies on microscopy to reveal cell, tissue, and organ substructure. Its goals include:
1. Understanding tissue structure at levels not visible to the unaided eye, including three-dimensional relationships among the biochemical constituents.
2. Understanding the relationship between tissue structure and function.
3. Establishing a basis for learning histopathology, which involves the relationship between abnormal tissue structure and functional defects.
4. Providing a basis for treating diseased and injured tissues.

B. Histologic Methods: Histology is a subdiscipline of anatomy (from the Greek "cutting apart"). Its methods involve dividing tissues and organs to prepare them for microscopic examination and chemical analyses.
1. **Microscopy.** Tissue analysis by **light microscopy** (IV) and **electron microscopy** (V) is the goal of these methods.
2. **Tissue preparation for microscopy.** The optics of each microscope type make certain demands on tissue preparation (III). Common procedures include **sectioning** to produce thin, translucent slices and **staining** to reveal otherwise transparent substructures.

C. Advantages and Limitations of Histologic Methods: Only by understanding the advantages and limitations of histologic methods can one interpret the information they provide. Advantages result from making small and complex structures and processes observable. Limitations exist because the methods themselves, especially those that require dividing an organism into pieces, often halt the very life processes being examined.

D. Tissue Structure and Function: These are so closely related that neither can be fully understood without an appreciation of the other. Structure–function relationships should be the primary focus of your initial study and your review.

E. **Units of Measure:** Measurements of cell and tissue components provide useful comparisons of relative sizes. The metric system is used exclusively for this purpose. Common units of measure in histology are the millimeter (mm, 10^{-3} m), the micrometer (μm, 10^{-6} m), and the nanometer (nm, 10^{-9} m).

II. GENERAL FEATURES OF CELL BIOLOGY & ITS METHODS

A. **Goal of Cell Biology:** Cell Biology is the study of cell structure and function. Its goal is to reveal the cellular and molecular basis for normal and abnormal biologic functions. Most of its methods involve disassembling intact organisms to examine the functional roles of their individual components.

B. **Methods of Cell Biology:** Cytology, the study of cells, once referred primarily to electron microscopic analyses of cell structure. With the advent of many new and powerful methods for studying living organisms at the cellular and molecular level, the expression "cell biology" currently is more commonly used. No longer the exclusive domain of electron microscopists, cell biology has blurred the borders between traditional basic and clinical biomedical sciences. Its methods have become invaluable to histologists who seek to understand tissue biology, just as histologic methods have become basic tools for cell biologists in every biomedical subdiscipline.

1. **Cell, tissue, and organ cultures** involve isolating components of organisms in controlled conditions and thereby allowing observation of the effects of treatments without interference from the regulatory mechanisms in intact organisms (XI).
2. **Cell fractionation** involves mechanically breaking cells and subsequently separating their components by centrifugation for electron microscopic or biochemical analysis of the smallest functional components of living cells and tissues (XII).
3. **Chromatography** involves separating molecules based on their physical or chemical properties, which enables their purification for further study (XIII).
4. **Electrophoresis** involves separating charged molecules using an electrical field. When gels undergo this process, both molecule size and charge can contribute to the separation process (XIV).
5. **Membrane transfer and blotting methods** allow the identification of molecules separated by electrophoresis (XV).
6. **Genetic technology** (eg, DNA cloning), exploits advances in our understanding of gene structure and function to explore the basis for health and disease in cells and tissues (XVI).

III. PREPARING TISSUES FOR MICROSCOPY

Light and electron microscopy share the basic preparative methods outlined in Figure 1–1 and described in Tables 1–1 through 1–4. Each method has limitations and associated artifacts that must be considered in interpreting histologic images. Particular attention should be paid to the similarities and differences in tissue preparation for light and electron microscopy. Most methods involve preparing thin sections, which can seduce one into thinking of three-dimensional structures in two-dimensional terms. To overcome this problem, tissues and organs are sectioned in several planes or prepared as serial sections to allow conceptual (often computer-assisted) three-dimensional reconstruction. Because even complex mixtures of stains cannot reveal every tissue component, adjacent sections are sometimes treated with different stains.

IV. LIGHT MICROSCOPY

A. **Light Source:** Light microscopes usually are illumined by bulbs that emit white light of varying intensity. White light comprises a limited range of wavelengths (average = 550 nm). Halogen bulbs with tungsten filaments emit intense white light and are commonly used in compound bright-field microscopes.

B. **Microscope Lenses:** Light microscopes have glass lenses. The **condenser** lens lies between the light source and the specimen. It collects light from the source and projects it as a cone through the specimen. The **objective** lens consists of one or more lenses located between the specimen and the ocular lens. It enlarges and resolves the specimen's image and projects it toward the **ocular** lens. Several objectives, each providing a different magnification, typically are

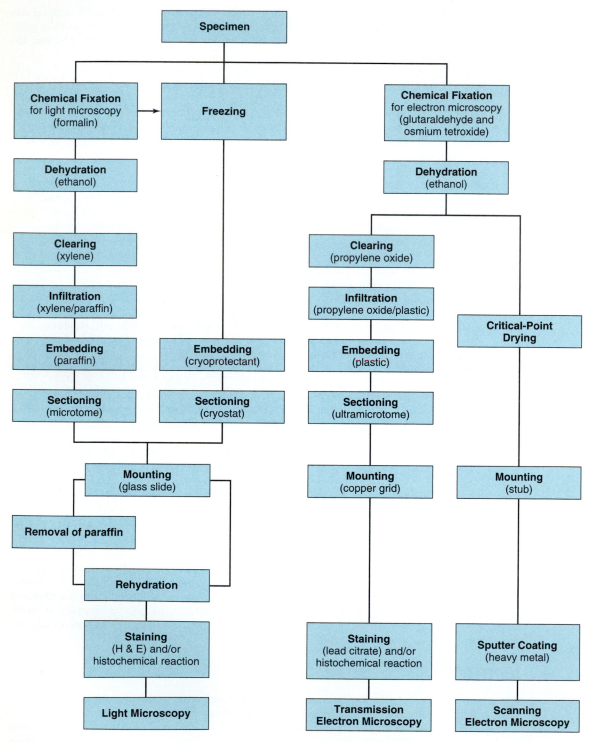

Figure 1–1. Steps in tissue preparation for microscopy. Common techniques and agents are shown in parentheses.

Table 1–1. Basic methods of tissue preparation for microscopy. (Flow chart, Fig. 1–1)

Method and Purpose	Types and Procedures	Limitations and Associated Artifacts
Fixation: Preserves structural organization of cells, tissues, and organs. Prevents bacterial and enzymatic digestion, insolubilizes tissue components to prevent diffusion, and protects against damage from subsequent steps in tissue processing.	**Chemical Fixation:** Common approach. Chemical fixatives are used individually or in mixtures (Table 1–2). Best results are achieved by rapid penetration of living tissue with fixative. Small tissue pieces may be fixed by **immersion.** Entire organs may be fixed by **perfusion** (fixative pumped through vessels serving the tissue of interest).	Fixative-induced changes in chemical composition and fine structure may produce staining artifacts. Structural changes include denaturing and cross-linking proteins.
	Freezing (Physical Fixation): May be used for light or electron microscopy. Tissue is embedded in cryoprotectant (glycerin). Rapid freezing at low temperatures reduces ice crystal formation and associated artifacts. Allows tissue to be sectioned (or fractured [Table 1–4]) without dehydration or clearing (Fig. 1–1). Faster than chemical fixation. Avoids dissolving lipids and denaturing fixative-sensitive proteins (eg, enzymes, antigens).	Not as permanent as chemical fixation. Sections are thicker. Resolution is poor.
Dehydration (substitution): Eases penetration of tissue by clearing agent. Prepares fixed tissue for infiltration with embedding medium.	Replaces water in tissue with organic solvent; commonly, **ethanol.** Fixed tissue is immersed in a series of alcohol–water mixtures with increasing alcohol concentration, to 100% alcohol.	Alcohol may denature proteins of interest. Water loss causes uneven shrinkage of components with different water content. May create unnatural spaces between cells and tissue layers.
Clearing: Prepares fixed tissue for infiltration. Dehydrating agent is replaced with clearing agent.	Dehydrated tissue is immersed in a series of clearing agent–alcohol mixtures with increasing clearing agent concentration, or placed directly into clearing agent. **Xylene** (paraffin solvent) is commonly used for light microscopy. **Propylene oxide** (plastic solvent) is commonly used for EM.	Clearing agents may denature proteins of interest. Some components shrink unevenly as their proteins denature.
Infiltration: Prepares cleared tissue for embedding.	Cleared tissue is immersed in a series of clearing agent–embedding medium mixtures with increasing embedding medium concentrations, at medium-high temperature. Evaporating clearing agent is replaced by embedding medium.	Heat may denature proteins of interest. Bubbles are left behind during poor infiltration.
Embedding: Prepares infiltrated tissue for sectioning. Makes tissue firm and prevents crushing or other tissue disruption during sectioning. Permits thin, uniform sectioning.	Infiltrated tissue is positioned in a mold filled with embedding medium, which hardens into a block. Block is attached to a chuck that holds it in microtome for sectioning. *For light microscopy,* **paraffin** is commonly used; other media are celloidin, plastics, and polyethylene glycol (water soluble) wax. *For EM,* plastics and epoxy resins (eg, **Epon** and **Araldite**) are common. Require a catalyst to harden (polymerize) after infiltration. Harder embedding media allow thinner sectioning, a requirement for EM.	The improved sectioning allowed by embedding has limitations associated with dehydration, clearing, and infiltration.
Sectioning: Most tissues are too thick and opaque for microscopic analysis of internal structure. Thin slices allow light or electrons to penetrate specimen and form image.	*For light microscopy,* standard rotary **microtome** with steel blade cuts 3–8-μm sections of specimens embedded in paraffin, celloidin, or polyethylene glycol. Glass or diamond knives cut 1–5-μm sections of plastic-embedded tissue. Frozen sections, 5–25 μm, are cut with a freezing microtome or in a **cryostat** (standard microtome in a refrigerated chamber). *For EM,* an **ultramicrotome** with a glass or diamond knife cuts very thin sections (0.08–0.1 μm or up to 0.5 μm for high-voltage EM). Ultramicrotomes include stereomicroscope to observe cutting.	Sections typically provide only a two-dimensional image of a three-dimensional structure. Dull knife can crush or pinch tissue. Chatter (wavelike variations in section thickness) results from knife vibration during sectioning. Burr on knife can tear tissue.

(continued)

	Table 1–1. (cont'd.)	
Method and Purpose	**Types and Procedures**	**Limitations and Associated Artifacts**
Mounting: Eases handling and decreases damage to specimen during examination.	*For light microscopy,* sections are placed on **glass slides,** often precoated with thin layer of albumin, gelatin, or polylysine to improve attachment. After staining, sections are covered with glass coverslips to preserve them for repeated examination. *For EM,* specimens are mounted on **copper grids.** Electron beam cannot penetrate glass.	Tissue sections may develop folds, making some regions appear to have more cells and stain darker. For grid-mounted specimens, only portions lying between crossbars are visible.
Staining: Most tissue substructure is indistinguishable even at high magnification. Stains, ligands with specific binding affinities and optical properties, and radiolabels are used to localize and distinguish cell and tissue components. Knowledge of specificities of such substances (Table 1–3) provides additional information about structure and composition.	**For light microscopy:** Once sections are on slide, paraffin is dissolved. Tissue may be rehydrated before staining. Plastic sections stained without removing plastic. Most stain affinities based on reciprocal acid–base characteristics of stain and tissue components. Acidic stains (eg, eosin) bind basic (ie, **acidophilic**) structures and compounds (eg, cytoplasmic proteins). Basic stains (eg, hematoxylin) bind acidic (ie, **basophilic**) tissue components (eg, nucleic acids in ribosomes). Stain mixtures reveal multiple cell components. Hematoxylin and eosin (H & E), most common stain mixture for light microscopy, distinguishes nucleus from cytoplasm.	Acid–base boundaries may not correspond to boundaries between structures. Multiple staining procedures may be needed to characterize a particular cell or tissue component. Because colors are artifacts of staining, it is best to focus more on tissue component structure than on color.
	For TEM: Most stains (contrasting agents) for TEM chosen for electron-absorbing or -scattering ability and affinity for particular cell components. **Heavy metal salts,** such as lead citrate and uranyl acetate, are common. The fixative osmium tetroxide interacts with lipids to form electron-dense precipitate and doubles as a stain for cell membranes.	TEM stains stop electrons from penetrating. TEM images are shadows of heavy metal deposits. Actual tissue structures are not seen.
	For SEM: Specimens not stained per se. First subjected to **critical-point drying,** to prevent surface tension artifacts. After dehydration, specimens soaked in liquid miscible with CO_2 or Freon and put in critical-point chamber. Chamber is heated to a critical temperature (31°C), raising pressure to a critical 73 atm, at which the gas and liquid phases exist without surface tension and liquid escapes specimen without altering structure. Specimen is then mounted on a stub and sputter-coated (sprayed) with fine mist of heavy metal particles (eg, gold) before viewing (Fig. 1–1).	SEM reveals surface architecture in exquisite detail, but heavy metal coating prevents electrons from penetrating to reveal internal structure.

mounted on a rotating turret. The ocular lens is located beyond the objective. It further enlarges the image and projects it onto the observer's retina, a screen, or a photographic emulsion.

C. **Optical Properties of Lenses:**
1. **Magnification** increases the specimen's apparent size and makes it appear closer. It is a property of both objective and ocular lenses. The total magnification value is obtained by multiplying the power of the objective lens by that of the ocular lens.
2. **Resolution** determines a microscopic image's clarity and richness of detail. It measures how close two objects can be while still appearing separate; the smaller the value, the greater the resolution. The resolution of the human eye is 200 μm; of a light microscope, 0.2 μm; and of an electron microscope, 0.002 μm. Increased magnification is useless without improved resolution. Resolution (R) is thus independent of magnification and is calculated from the numerical aperture (NA) of the objective and the wavelength (λ) of illumination:

$$R = \frac{0.61\ \lambda}{NA}$$

Table 1–2. Properties of chemical fixatives and fixative mixtures.

Type	Actions	Examples
Aldehydes	React with amine groups, from cross-links among proteins, and cause coagulation (but not coarse precipitation) of tissue proteins. May interfere with periodic acid–Schiff (PAS) and Feulgen staining specificity.	**Formalin** (formaldehyde gas in water) commonly used for light microscopy. **Glutaraldehyde** commonly used for EM.
Oxidizing agents	Cross-link proteins and precipitate unsaturated lipids.	**Osmium tetroxide.** Often used with glutaraldehyde for EM (see double fixation below). Also, **potassium permanganate** and **potassium dichromate.**
Protein-denaturing agents	Normal protein shape maintained largely by ionic interactions with water molecules. Denature protein by removing associated water, changing the protein's shape. In absence of cross-linking agents, rehydrating tissue may restore protein conformation.	**Acetic acid, methanol, ethanol, acetone.**
Others	Unclear	**Mercuric chloride, picric acid.**
Mixtures	Exploit advantages and minimize disadvantages of various fixatives.	*Light microscopy:* **Bouin's fixative** (picric acid, formalin, acetic acid). *EM:* **Karnovsky's fixative** (paraformaldehyde and glutaraldehyde in buffered saline).
Double fixation	Used for EM. Specimen fixed in buffered glutaraldehyde, washed in phosphate buffer, and postfixed in buffered osmium tetroxide. Osmium reacts with lipids to form black precipitate that stains cell membranes.	

Table 1–3. Examples of common stains and their affinities.

Application	Types	Stains	Affinity
Light microscopy	Basic dyes	Hematoxylin, toluidine blue, methylene blue, Alcian blue	Basophilic tissue components (eg, DNA, RNA, and polyanions such as sulfated glycosaminoglycans)
	Acidic dyes	Eosin, orange G, acid fuchsin	Acidophilic tissue components (eg, basic proteins in cytoplasm)
	Lipid-soluble dyes	Oil red O, Sudan black	Long-chain hydrocarbons (fats, oils, waxes)
	Multicomponent histochemical reaction	Periodic acid–Schiff (PAS) reaction	Complex carbohydrates (glycogen, glycosaminoglycans)
		Feulgen's reaction	Nuclear chromatin (DNA and associated proteins)
Transmission electron microscopy	Heavy metal (electron dense)	Uranyl acetate, lead citrate	Nonspecific; adsorb to surfaces and enhance contrast
		Osmium tetroxide	Actually a fixative, but binds to phosphate groups of membrane phospholipids, enhancing contrast
		Ruthenium red	Polyanions; complex carbohydrates (eg, oligosaccharides of the glycocalyx and glycosaminoglycans of the extracellular matrix)

Table 1–4. Special methods of tissue preparation for microscopy. (See also Chapter 2.)

Method and Purpose	Procedures	Limitations
Cryofracture and Freeze Etching: Permit EM examination of tissues without prior fixation and embedding and thus without related artifacts. Allow verification of results obtained by conventional EM techniques.	Tissue frozen at very low temperatures and fractured with sharp blade (cryofracture). Specimen kept in vacuum while ice sublimates, lowering ice level from specimen surface to reveal more structures (freeze etching). Etched tissue sprayed at an angle with gold particles to give shadowed effect and then coated with a layer of fine carbon particles to form a replica. Tissue and replica returned to atmospheric pressure, tissue dissolved in strong acid, and replica mounted on grid. EM of replica produces shadowed-relief image and provides limited pseudo–three-dimensional view of cell and tissue components.	Image produced is entirely a controlled artifact. None of the original tissue is seen. Resolution is limited by particle size and thickness of coating.
Radioautography (auto-radiography): Allows localization of radioactive elements in cells or tissues. Especially useful in tracking radiolabeled precursors and molecules into which they are incorporated from one part of tissue or cell to another.	Tissue incubated with radiolabeled precursors before fixation (eg, ^{3}H-thymidine for DNA, ^{3}H-uridine for RNA, ^{3}H- or ^{14}C-leucine for protein). Sections mounted on slide, cleared of embedding medium, covered with photographic emulsion, and left in dark. Emulsion exposed (silver bromide reduced to elemental silver) in areas in contact with radiolabel; when developed, black grains (light microscopy) or curling particle tracks (EM) appear over labeled structures. Number of grains or tracks in emulsion proportionate to radiolabel present. For light microscopy, sections often counterstained to reveal tissue architecture.	Radiation exits labeled tissue component in various directions. Some labeled structures may not expose overlying emulsion, and silver grains over one structure may have been exposed by radiation from neighbor.

3. **Numerical aperture (NA)** is related to the width of the lens opening (aperture). The greater the NA, the greater the resolving power.

4. **Refractive index** measures the comparative velocity of light in different media. Owing to the change in refractive index at air–glass interfaces, the air between the lens and the coverslip bends some of the light projected through the specimen. At high magnifications, the accompanying loss of resolution reduces image quality. Using **immersion oil** (with the same refractive index as glass) between the coverslip and an oil immersion objective lens avoids the change in refractive index, thereby improving resolution.

5. **Lens-related artifacts.** Objective lenses comprise a series of glass lenses. The first (frontal) lens is spherical or hemispherical and magnifies the image. Others correct for aberrations or artifacts of lens curvature. Spherical lenses bring light of shorter wavelength into focus closer to the retina than light of longer wavelength, resulting in multiple blurred images. This **chromatic aberration** can be avoided by using **achromatic** or **apochromatic lenses.** Optical properties at the center of a spherical lens differ from those at the periphery. Apochromatic lenses correct for this **spherical aberration.** With spherical lenses either the center or the periphery of the field is out of focus. **Planar lenses** correct this **curvature of field** and provide flat-field focus. The best objective lenses are thus planar apochromatic lenses with a high numerical aperture.

D. **Light Microscope Types:**

1. **Compound bright-field microscopes** are the most common tool of histology and histopathology. The term compound (versus simple) refers to a series of lenses; bright-field means the entire field is illuminated by an ordinary condenser. Specimens must be translucent and stained to provide contrast.

2. **Dark-field microscopes** use a special condenser to provide contrast in unstained material. A disklike shield excludes the center of the light shaft from the condenser so that the specimen is illuminated only from the sides. Objects that deflect light into the objective lens are visible and appear bright on a dark background.

3. **Phase contrast microscopes** use a special lens system to transform invisible differences in phase (light speed) retardation (caused by the different refractive indices of specimen components) into visible differences in light intensity. Fixation and staining are not required. Because phase contrast microscopes allow living specimens to be visualized, they are basic tools for tissue culture. Specimens must be thin and translucent. High resolution is difficult to obtain.

4. **Polarizing microscopes** allow selective visualization of **birefringent** (anisotropic) **structures**—repetitive or crystalline structures such as collagen fibers or myofibrils. Staining is not required. Light traverses a polarizing filter, the condenser projects the polarized light

onto the specimen, and birefringent structures in the specimen rotate the polarized light. The objective lens projects the image through a second polarizing filter, which is oriented so that only light waves oscillating in planes different from that leaving the first polarizing filter can enter the ocular lens and be seen. Birefringent structures appear as bright, often colored, objects on a dark background.

5. **Fluorescence microscopes** allow the localization of substances that are labeled with fluorescing compounds (fluorochromes, such as fluorescein or rhodamine). When stimulated by light of a proper wavelength, fluorochromes emit light of a longer wavelength. An ultraviolet light source is commonly used, and the emitted light is in the visible spectrum. An excitation filter between the light source and the specimen filters out all wavelengths except that needed to stimulate the fluorochrome. A barrier filter between the objective and ocular lenses protects the eyes from ultraviolet rays and projects only the emitted light.

6. **Interference microscopes** combine optical features of phase contrast and polarizing microscopes to provide contrast in unstained material. Relying on differences in refractive index (IV.C.4), they can measure the phase retardation induced by components of a specimen. Unlike standard phase contrast microscopes, they can compare the refracted light with an unimpeded reference beam and provide an electronic readout of the data. Because refractive index and phase retardation are proportionate to mass, these instruments can be used to calculate the mass of cellular components. Modified interference optics, pioneered by Nomarski, are used in **differential interference contrast (DIC)** microscopes.

7. **Confocal scanning microscopes** allow the appreciation of three-dimensional structure in a specimen without cutting sections. Laser optics and computerized imaging methods provide optical sections of thicker specimens. To penetrate the thicker specimen, a narrow beam of concentrated (laser) light is focused at a specific depth (focal plane) and is scanned across the specimen. The imaging system stores the scanned image, point by point (as in a computerized axial tomography [CAT] scan), building up a sharp image of the focal plane while excluding out-of-focus images from other parts of the specimen. A series of optical sections at different depths can be stored digitally, ultimately permitting the reconstruction of a sharp three-dimensional image.

V. ELECTRON MICROSCOPY (EM)

A. **General Principles:** The equation for resolution is the same as for light microscopy (IV.C.2). An electron beam (wavelength = 0.005 nm) is used instead of visible light (wavelength = 397–723 nm), giving electron microscopes much greater resolving power and allowing for magnification as great as 200 times that of light microscopes. Glass lenses are not transparent to wavelengths shorter than 400 nm, but the negatively charged electron beam can be deflected and focused by electromagnets as it travels through a vacuum.

B. **Major Components and Operation of Electron Microscopes:** Together, the cathode and anode are analogous to a light microscope's lamp. The cathode, a metallic filament, emits a spray of electrons when intensely heated in a vacuum by an electric current. The anode is a positively charged metal plate with a small hole at its center. The potential difference between the cathode and anode (60–100 kV) accelerates electrons toward the anode; some pass through the hole to form the electron beam. The condenser electromagnet deflects and focuses a cone of the beam on the specimen. The specimen typically is an ultrathin tissue section stained with electron-absorbing or electron-scattering substances to provide contrast. The objective electromagnet deflects the portion of the electron beam that has passed through the specimen to form and magnify the image. The one or two projector electromagnets are analogous to the light microscope's ocular lenses. They further enlarge the image and project it onto a fluorescent screen or photographic emulsion. The fluorescent screen is a plate coated with material that fluoresces as electrons strike it. Electrons deflected or absorbed by the specimen do not reach the screen, whereas those that pass through the specimen do. The result is a transmission image formed by shadows of the specimen's electron-dense components.

C. **Limitations of Electron Microscopy:** The electron beam must travel in a high vacuum, and thus living tissue cannot be used. Tissue sections must be very thin, or they absorb or deflect the entire beam. The electron beam may alter specimen structure. The image cannot be visualized directly but helps create a fluorescent or photographic image.

D. Types of Electron Microscope:

1. **Transmission electron microscopes (TEMs)** permit visualization of the internal ultrastructure of cells and tissues as well as minute structures in cells or in intercellular spaces (limit of resolution = 0.2 nm). They operate as described earlier (V.B). Specimens are prepared as described in Table 1–1 and Figure 1–1.

2. **Scanning electron microscopes (SEMs)** permit visualization of surface ultrastructure (limit of resolution = 2 nm). After the specimen is coated with a thin layer of heavy metal (Table 1–1), a narrow electron beam is directed across its surface in a point-by-point sequence, generating two major signals. **Secondary electrons** are released from the specimen surface, collected on detectors, and converted electronically into an image that is displayed on a cathode-ray tube. This image provides a three-dimensional representation of the specimen surface. **X-rays** are generated when the electron beam strikes atoms heavier than sodium. Analysis of the x-ray signal can supply information about the concentration and distribution of certain elements in the specimen.

VI. BASIC PRINCIPLES OF HISTOCHEMISTRY

Histochemistry marries histology with chemistry. Its goal is to reveal the chemical composition of tissues and cells beyond the acid–base distribution shown by standard staining methods without disrupting normal distribution. Histochemical analysis is guided by the following criteria:

A. Preservation of Chemical Distribution: The substance must not diffuse away from its original site.

B. Preservation of Chemical Composition: Reactive chemical groups must not be blocked or denatured. Unreactive groups must not become reactive.

C. Specificity: The reaction must be specific to avoid false-positive results.

D. Detectability: The reaction product should be colored or electron-scattering so that it can be visualized with a light or electron microscope.

E. Insolubility: The reaction product should be insoluble so that it remains in close proximity to the substance it marks.

VII. SOME IMPORTANT BIOLOGIC SUBSTANCES & CLASSIC METHODS FOR DETECTING THEM

A. Ions: Most ions are difficult to localize owing to their small size and diffusibility. Some ions, however (eg, iron in red blood cells and phosphate in bone), are immobilized by association with tissue proteins. Incubating iron-containing tissue in potassium ferrocyanide and hydrochloric acid yields a dark blue ferric ferrocyanide precipitate (Perls' reaction). This reaction identifies cells involved in hemoglobin metabolism and diseases involving iron deposits in tissues (hemosiderosis). Tissue phosphates (eg, calcium phosphate in bone) react with silver nitrate to form silver phosphate, which reacts with hydroquinone to form a black precipitate of reduced silver.

B. Lipids: Lipids are dissolved by organic fixatives or clearing agents, leaving gaps in the tissue, but are preserved in frozen sections. For light microscopy, lipids are demonstrated by dyes that are more soluble in lipid than in the dye solvents (eg, Sudan black and oil red O). EM specimens are treated with reagents that react with lipids to form insoluble precipitates (eg, osmium tetroxide). Such methods show normal lipid distribution and disease-related lipid accumulation (eg, fatty change in the liver).

C. Nucleic Acids: DNA and RNA can be localized by specific and nonspecific methods. DNA occurs mainly in nuclei, and its amount per cell varies little. RNA occurs both in nuclei and in cytoplasm, and its amount varies widely, depending on each cell's functional state.

1. **Feulgen's reaction.** Hydrochloric acid partially hydrolyzes DNA, forming free aldehydes.

These react with Schiff's reagent (bleached fuchsin) to form an insoluble magenta precipitate in amounts proportionate to the DNA present.

2. **Acridine orange.** Nucleic acid–acridine orange complexes emit fluorescence in an amount proportionate to the nucleic acid present. The fluorescence is yellow–green for DNA and red–orange for RNA. Neoplastic and other fast-growing cells contain more RNA than slower-growing cells.

3. **Basic dyes.** DNA and RNA stain nonspecifically with basic dyes. RNA's strong affinity for such dyes allows its distribution to be studied by **subtraction.** In this procedure, one of two adjacent sections is treated with ribonuclease (RNase) to remove RNA; subsequently both sections are stained with basic dyes (eg, toluidine blue). RNA-containing structures (eg, ribosomes) are basophilic in the untreated section but absent in the RNase-treated section.

D. **Proteins and Amino Acids:** Older methods of protein identification are specific for particular amino acids. *Examples:* Million reaction for tyrosine, Sakaguchi reaction for arginine, and tetrazotized benzidine reaction for tryptophan. Classes of enzymes can be detected by enzyme histochemistry (VIII). Specific proteins can be localized by immunohistochemistry (IX).

E. **Carbohydrates:** Complex carbohydrates (ie, polysaccharides and oligosaccharides) can be localized by histochemical techniques. Some are immunogenic owing to their large size or their presence as covalently linked components of glycoconjugates (eg, proteoglycans, glycoproteins, glycolipids); these can be analyzed by immunohistochemistry (IX).

1. The **periodic acid–Schiff reaction** is commonly used to demonstrate polysaccharides, particularly **glycogen.** Periodic acid reacts with sugars to form aldehydes. Schiff's reagent then reacts with the aldehydes to form an insoluble magenta pigment (compare C.1). Because this reaction stains many complex carbohydrates, the specific localization of glycogen requires enzymatic subtraction of glycogen from an adjacent section by means of amylase. This method distinguishes among types of glycogen-storage diseases.

2. **Alcian blue** is a nonspecific basic stain at neutral pH, but it is specific for sulfate groups at pH 1. It is used to demonstrate sulfated **glycosaminoglycans** (eg, chondroitin sulfate) that are abundant in the extracellular matrix of cartilage.

3. **Ruthenium red** binds nonspecifically to polyanions and forms an electron-scattering precipitate useful in EM demonstration of polysaccharides.

4. **Lectins** are highly specific sugar-binding proteins. Labeled lectins can localize terminal sugar residues on oligosaccharides, such as those in the glycocalyx of cell membranes. *Examples:* Concanavalin A binds mannose; peanut agglutinin and *Ricinus communis* agglutinin bind galactose; and wheat germ agglutinin binds N-acetyl-D-glucosamine.

F. **Catecholamines:** The catecholamines, including **epinephrine** and **norepinephrine,** fluoresce in the presence of dry formaldehyde vapor at 60°C. This reaction is used to study catecholamine distribution in nerve tissue.

VIII. ENZYME HISTOCHEMISTRY

Enzyme histochemistry relates structure and function. It is used to locate many enzymes, including **acid phosphatase, dehydrogenases,** and **peroxidases.** Because fixation and clearing inactivate enzymes, frozen sections are commonly used. Sections are incubated in solutions containing substrates for the enzymes of interest and reagents that yield insoluble colored or electron-dense precipitates at sites of enzyme activity.

A. **Acid Phosphatase:** In the Gomori method, the tissue is incubated with glycerophosphate and lead nitrate. The enzyme liberates phosphate, which combines with lead to produce lead phosphate, a colorless precipitate. The tissue is then immersed in ammonium sulfide, which reacts with lead phosphate to form lead sulfide, a black precipitate. Owing to their characteristic content of acid phosphatase, lysosomes can be distinguished from other cytoplasmic granules and organelles through the use of enzyme histochemistry.

B. **Dehydrogenases:** Dehydrogenases can be localized by incubating tissue sections with an appropriate substrate and tetrazole. The enzyme transfers hydrogen ions from the substrate to the tetrazole, reducing the tetrazole to formazan, a dark precipitate. Specific dehydrogenases can be targeted by choosing specific substrates.

C. **Peroxidases:** Peroxidases are often demonstrated by incubating tissue with 3,3′ diaminobenzidine (DAB) and hydrogen peroxide. The enzyme transfers hydrogen from DAB to the peroxide, and the oxidized DAB forms an electron-dense dark brown to black precipitate at the site of enzyme activity. This reaction is useful for both light and electron microscopy.

IX. IMMUNOHISTOCHEMISTRY

Immunohistochemistry uses labeled antibodies to localize specific cell and tissue **antigens** and is among the most sensitive and specific histochemical techniques. Because many targeted antigens are proteins, the structure of which might be altered by fixation and clearing, frozen sections are often used. In some cases, water-soluble plastics and waxes can be used for embedding.

A. **Raising Antibodies:** Injection of antigens (proteins, glycoproteins, proteoglycans, and some polysaccharides) causes an injected animal's B lymphocytes to differentiate into plasma cells and produce antibodies. Members of a lymphocyte clone (descendants of a single cell) produce a single type of antibody, which binds to a specific antigenic site, or epitope.
  1. **Polyclonal antibodies.** Large complex antigens may have multiple epitopes and may elicit several antibody types. Mixtures of different antibodies to a single antigen (obtained through fractionation of an injected animal's serum) are called polyclonal antibodies and are commonly raised in rabbits and goats.
  2. **Monoclonal antibodies.** Antibodies specific for a single epitope and produced by a single clone are called monoclonal antibodies and are commonly raised in mice. Lymphocytes from the spleen of an antigen-injected mouse are mixed with myeloma cells (lymphocyte-derived tumor cells) under conditions that cause the lymphocytes and myeloma cells to fuse. Each resulting **hybridoma** cell has the myeloma's capacity for continuous cell division in culture and the lymphocyte's capacity for unique antibody secretion. An isolated hybridoma gives rise to a large clone that produces large quantities of pure antibody.

B. **Labeling Antibodies:** Antibodies are not visible with standard microscopy and must be labeled in a manner that does not interfere with their binding specificity. Common labels include fluorochromes (eg, fluorescein, rhodamine), enzymes demonstrable by enzyme histochemistry (eg, peroxidase, alkaline phosphatase), and electron-scattering compounds for electron microscopy (eg, ferritin, colloidal gold).

C. **Antigen Localization in Tissues:**
  1. **Direct method.** For **direct immunohistochemistry,** antigen-containing tissue is incubated in a solution of labeled antibody. The antibody binds directly to the antigen, and the label appears at the antigenic site.
  2. **Indirect method.** For **indirect immunohistochemistry,** antigen-containing tissue is incubated in a solution containing unlabeled antibody. This **primary antibody,** which binds directly to the antigen in the tissue, is named according to its antibody class and the animal that produced it (eg, mouse IgG). The tissue is subsequently incubated with a labeled **secondary antibody,** which is an antibody to the primary antibody (eg, rabbit antimouse IgG). Several labeled secondary antibodies bind to each primary antibody, and thus more labeled antibodies accumulate at the antigenic site than with the direct method. The indirect method is therefore more sensitive and avoids the risk of altering the primary antibody's binding specificity by attaching a label to it.

X. IN SITU HYBRIDIZATION

In situ hybridization is a method of analyzing the tissue distribution of particular nucleotide sequences in DNA (eg, specific genes) and RNA (eg, specific mRNAs). Hybridization refers to the binding of complementary nucleotide sequences to one another with high specificity. Recombinant DNA technology permits copies of selected single-strand nucleotide sequences to be synthesized in large numbers. Synthetic sequences complementary to the RNA or DNA sequence an investigator wishes to localize are termed **probes** and can be labeled with radioisotopes (eg, ^{32}P), biotin, or digoxigenin. Radiolabeled probes are demonstrated by radioautography. Biotin-labeled probes are demonstrated with enzymes (eg, peroxidase) or fluorochromes covalently linked to avidin, a molecule with high affinity for biotin. Digoxigenin-labeled probes are demonstrated by indirect immunohistochemistry using antidigoxigenin primary antibodies. Labeled probes were first used to analyze nucleic acids isolated from cell or tissue homogenates. The term **in situ** refers to the application of this technique to tissue sections, smears of cells, or even whole embryos; when such specimens are incubated with labeled probes, the probes bind to and reveal the distribution of their complementary sequences.

XI. CELL, TISSUE, & ORGAN CULTURES

These methods are used to study living cells and tissues without the interference of the organism's homeostatic mechanisms. They facilitate the control and manipulation of the environment. Cells and tissue isolated and grown in culture are referred to as **in vitro** (in glass) and those in the intact organism as **in vivo** (in the living). Cells may react differently to in vitro and in vivo treatments.

A. Culture Medium: The medium in which cells and tissues are grown substitutes for the plasma that normally bathes them in vivo. It consists of a buffered isotonic saline solution to which are added nutrients (amino acids, vitamins, hormones, carbohydrates) of rigidly controlled composition. Recent advances have decreased the need for serum and tissue extracts of less well-defined composition in media. Antibacterial and antifungal agents are often added.

B. Culture Types:
 1. **Cell culture.** In **suspension culture,** cells are suspended in culture medium either free or on floating beads. In some cases cells are suspended in semisolid blocks of extracellular matrix or agarose. In **plate culture,** cells attach to plastic or glass tissue culture dishes. The dishes may be coated with substances that improve attachment and cell function, including gelatin, collagen, polylysine, serum albumin, or extracellular matrix extracts. Plate-cultured cells behave differently at different densities. Cells may be cultured in confluent monolayers (ie, the entire culture surface is covered with cells in contact with one another) or at clonal densities (ie, seeded at low densities to avoid cell–cell contact). The latter method allows the growth of individual cell colonies, or **clones.**
 2. **Tissue and organ culture.** Tissues or organ fragments are isolated and grown as explants, usually at the air–medium interface. This method is used to study embryonic differentiation and morphogenesis away from the complex environment of the embryo.

C. Isolation and Study of Pure Cell Strains: Individual cell types may be isolated and studied in vitro to explore their separate contributions to tissue and organ function. Cell suspensions are commonly obtained from tissues by enzymatic digestion (eg, with trypsin or collagenase) of the components that hold the cells together. Cell types may then be separated on the basis of size and mass through specialized forms of centrifugation (eg, elutriation, density gradient centrifugation). Newer methods of isolating a cell type from a heterogeneous population in suspension exploit the differential binding of cells by antibodies attached to a culture surface; others use a **fluorescence-activated cell sorter,** which separates cells labeled with fluorescent antibodies from cells lacking the label.

XII. CELL FRACTIONATION

This procedure is used to isolate and collect cellular components in quantity to study their contributions to cell function. It begins with the mechanical **homogenization** of cells and tissues to break plasma membranes and release the cell components. The components (eg, individual organelles) are then separated by size and density, using either of two centrifugation methods. In **differential centrifugation,** components are separated by their characteristic sedimentation rates, using different amounts of centrifugal force for various periods. In **density gradient centrifugation,** the homogenate is layered on top of a gradient of solute and centrifuged until its components come to rest in portions of the gradient with densities similar to their own.

XIII. COLUMN CHROMATOGRAPHY

Because they are the principal mediators of cell function, proteins have been targeted by many methods to isolate and analyze the content of cell homogenates (XII). Chromatography involves packing a hollow column with a semipermeable material (often tiny resin or agarose beads) and applying a cell or tissue homogenate, or a centrifugal fraction of such a sample, to the top of the column. After the sample percolates into the column, solvent is added and allowed to flow through. Timed fractions of the flow-through material, containing different molecular components of the homogenate, are collected. Component release is retarded by interactions with the packing material and results in separation. Three standard strategies have evolved owing to their proven usefulness (Table 1–5). Almost unlimited variations in packing materials and solvents are available and may be used individually or in combination to isolate molecules of interest.

XIV. ELECTROPHORESIS

A. **One-dimensional Gel Electrophoresis:** An electric field applied to a solution of proteins, or other large molecules, causes them to move in a direction and at rates reflecting their net charge. When the movement occurs in a semipermeable gel (eg, polyacrylamide, agarose), migration in the gel is limited by pore size as well. In **native gel electrophoresis,** proteins migrate in physiologic buffers and retain their native structure and folding; thus the shape of the molecule (eg, globular or filamentous) also affects its migratory rate. More commonly, protein solutions are heated in the presence of a detergent (sodium dodecylsulfate, or **SDS**) and a reducing agent **(mercaptoethanol)** before they are subjected to polyacrylamide gel electrophoresis **(SDS-PAGE).** The reducing agent breaks -S-S- linkages and, together with SDS, promotes unfolding. SDS further coats the protein with negative charges such that each protein's net charge is negative, and all have essentially the same charge density. When applied to a polyacrylamide gel and electrophoresed in buffers containing SDS and mercaptoethanol, proteins in the sample migrate toward the positive pole at rates that depend largely on their size. Smaller molecules are retarded less by the gel and move faster. When proteins of known molecular weight are used as standards, this method allows the molecular weights of individual proteins to be estimated. Typically, SDS-PAGE is carried out in flat or **slab gels,** in which multiple samples are run in parallel **lanes** for comparison. After electrophoresis, the gel is removed from the apparatus and stained (eg, with **Coomassie blue**) to reveal the position of the proteins and thus their molecular weight. Separation of nucleic acids may be carried out in either polyacrylamide or agarose gels. DNA and RNA gels are typically stained with **ethidium bromide,** which complexes with the nucleic acids and causes the bands to fluoresce when exposed to ultraviolet light.

Table 1–5. Column chromatography.

Type	Basis of Separation	Column Packing Material	Solvent(s)
Ion-Exchange Chromatography	**Charge:** Amino acids of proteins confer different net charges. Opposing charges on column packing material retard flow through column to different extents.	Charged resin beads	Buffers of varying ionic strengths (often as a gradient)
Gel-Filtration Chromatography	**Size:** Proteins vary greatly in size. Porous packing materials allow smaller molecules to enter pores and undergo retardation; larger molecules pass between the beads and flow more quickly through the column.	Porous resin or agarose gel beads	Physiologic buffer
Affinity Chromotography	**Binding Affinity:** Most selective chromatography method. Exploits affinity of certain molecules for others. Examples include receptor–ligand, antigen–antibody, and poly A+ mRNA–uridine polymer interactions. Molecules are bound with high affinity until released.	Resin beads coated with covalently bound ligand	Nonspecific molecules washed through with low ionic strength buffer; high ionic strength buffers release specifically bound molecules

B. Isoelectric Focusing: Owing to its amino acid composition, each protein has a characteristic pH (**isoelectric point**) at which it has no net charge and thus will not migrate in an electric field. Using special buffers to imbue a tube of polyacrylamide (**tube gel**) with a stable pH gradient along its length in an electric field, sample proteins can be induced to migrate to, and come to rest at, their own isoelectric points.

C. Two-dimensional Gel Electrophoresis: Because many proteins may have the same molecular weight, complete separation on a one-dimensional gel may not be possible. In such cases, a sample may be subjected first to isoelectric focusing, after which the resulting tube gel may be laid across the top of a slab gel and subjected to SDS-PAGE. As a result of this process, proteins that first separated based on their isoelectric points migrate into the slab gel and are further separated based on their molecular weight. The resulting separation yields spots of smaller amounts of individual proteins. Silver staining of 2-D gels is common owing to its greater sensitivity.

XV. BLOTTING & ELECTROTRANSFER

Exploiting the affinity of antibodies for specific proteins, and that of complementary nucleic acid sequences for one another, additional steps have been developed to analyze molecules first separated by electrophoresis. Gels used for electrophoresis restrict the access of large native molecules, such as antibodies and large nucleic acids, to their target molecules. Thus methods have been developed to remove the separated molecules from these gels and to immobilize them on membranes in the same pattern conferred on them by the initial separation. Transfer from the gels to the membrane may be accomplished by blotting or by electric charge.

In blotting techniques, molecules in the gel are carried by the flow of a buffer from one side of the gel and through the membrane to the other side of the gel. Flow is induced by capillary attraction through absorbent paper layers on the far side of the membrane. As the buffer flows through, the membrane traps the target molecules. In electrotransfer, target molecules migrate from the gel to the membrane in an electric field applied to the assembly. Immobilization is achieved electrostatically for proteins and by heating or exposure to ultraviolet light for RNA and DNA. Subsequently the membranes carrying the more accessible target molecules are incubated with solutions of labeled antibodies or complementary nucleic acid sequences (termed **probes**) to reveal the positions and relative amounts of the molecules of interest. **Western blotting** refers to the use of labeled antibodies to reveal the presence and amount of a specific protein on the membrane. **Northern blotting** refers to the procedure in which labeled probes reveal complementary RNA sequences. **Southern blotting** refers to similar procedures to localize specific DNA sequences. Methods used to label the antibodies and probes determine the detection methods employed. Antibody labeling and detection methods similar to those described for immunohistochemistry (IX) are often used for Western blots. In northern and southern blotting, probes are often labeled by incorporating ^{32}P-labeled nucleotides into the probe during its synthesis, followed by radioautography, which involves exposing x-ray film to the labeled membrane at low temperatures and then developing the film.

XVI. GENETIC TECHNOLOGY

Recent advances in our understanding of DNA structure and function have yielded methods of unprecedented power for unraveling the intricacies of cell and tissue biology. Previously viewed as mysterious agents of heredity, genes have become tools of potentially unlimited diagnostic and therapeutic power.

A. Isolating Genes and DNA Fragments: Studies of how bacteria inactivate artificially introduced foreign DNA led to the discovery of **restriction nucleases** (eg, EcoR1, HindIII), a class of bacterial enzymes that are capable of cutting DNA into fragments by acting at very specific nucleotide sequence target sites. Different bacteria have restriction enzymes that are specific for different sequences, providing molecular biologists with an array of tools for the isolation of very specific DNA fragments, especially when the nucleotide sequence is already known. When large DNA molecules are fragmented by a single restriction enzyme, the fragments generated are of different sizes, reflecting the positions of the target sites in the larger molecule. Because the overall nucleotide sequence is unique to each individual, the banding pattern revealed when the fragments are separated by size, using gel electrophoresis, is equally unique, providing a **genetic fingerprint** of an individual. Comparing banding patterns produced by several enzymes can determine with high confidence whether DNA in multiple samples came from the same individual. Knowing the nucleotide sequence of a gene, and those of its 3′ and 5′ flanking regions, allows individual genes and various upstream and downstream components to be obtained and isolated by electrophoresis. Cutting a band containing a fragment of interest from a gel, and its release from the gel material, allows the fragment to be purified and used for subsequent studies.

B. DNA Sequencing: Unknown nucleotide sequences of genes and other DNA fragments can be determined by several strategies. Commonly, isolated fragments are incubated with a DNA polymerase in conditions that halt assembly of the copy when a particular nucleotide is reached. The result is a mixture of subfragments terminating at each nucleotide in the sequence. Subfragment size depends on which point in the sequence it terminates. Electrophoresis of the mixture arranges the subfragments by size and therefore reveals the order in which each nucleotide occurs in the fragment.

C. Copying Genes and DNA Fragments: Obtaining multiple identical copies of a gene or fragment of interest greatly facilitates further molecular analysis.

  1. DNA cloning. A common method of obtaining such copies involves inserting (cloning) the fragment into bacterial **plasmids** and allowing the DNA replicating machinery in growing bacteria to copy the entire **recombinant DNA** sequence of the plasmid as they copy their own DNA. Plasmids used as carriers (**vectors**) in such procedures are typically small, circular, double-stranded DNA molecules capable of entering living bacteria and maintaining their integrity while they are copied separately from the bacteria's own genome. Cloning vectors typically include genes for antibiotic resistance; thus when bacteria are grown in media containing an antibiotic (eg, ampicillin), only plasmid-containing bacteria survive and continue to grow. Vectors also have cloning sites, with target nucleotide sequences for one or more restriction enzymes that serve as the insertion site for the fragment of interest. Cutting the plasmid with a restriction enzyme linearizes the circular DNA and provides free ends for attaching the fragment. Digesting the fragment and the plasmid with the same enzyme yields matching ends, allowing the fragment to be covalently attached by the enzyme **DNA ligase,** which repairs and recircularizes plasmid. The result is a fragment-containing **recombinant** plasmid that is larger than the original and that can be separated electrophoretically. Often the cloning sites occur or are engineered into sequences for a marking enzyme (eg, beta galactosidase). Inserting the fragment disrupts enzyme expression.

  Bacteria containing recombinant plasmids, when grown on media containing substrate for the enzyme that forms a colored precipitate in the presence of the enzyme, notably lack color and can be easily selected for further growth. Thus bacteria that are capable of growing in media containing the antibiotic, and that remain uncolored on media containing the enzyme substrate, are most likely to contain the desired recombinant plasmids. Colonies of bacteria-containing recombinant plasmids are inoculated into growth media and quickly multiply, making abundant copies of the fragment along with their own DNA and vector DNA. The bacteria can then be lysed, their DNA isolated, the recombinant plasmids separated from the larger bacterial genomic DNA by electrophoresis, and the fragments isolated from the plasmids by restriction enzyme digestion.

  2. Polymerase chain reaction (PCR). An even more rapid method for copying selected DNA fragments was developed recently. PCR exploits the fact that cells use existing DNA sequences as templates for making new copies. Exponential doubling of copy numbers is thus possible if the copies made in each succeeding round of synthesis are used as templates in the next. Because PCR is carried out entirely in vitro, the transition from one replication cycle to the next can be greatly accelerated compared with rates that are possible in intact organisms, which (1) must copy their entire genome during each replication cycle, and (2) cannot withstand the high temperatures used in PCR to separate the copy from the template DNA after synthesis. DNA sequences of interest can be selected using appropriate priming oligonucleotides (**primers),** short nucleotide sequences that hybridize to the beginning and ending portions of the target DNA sequence. The use of DNA polymerases that specialize in filling in the intervening nucleotides, but that stop copying when the other primer is reached, allows only the target sequence to be copied. Transient heating of the reaction causes the original and copy strand to separate (melt). As the reaction cools, additional primers hybridize to the original and copy strands, using both as templates for the next replication cycle and resulting in another doubling of the copy number. The number of cycles thus determines the number of copies generated.

One key to developing this method was the discovery that bacteria living at high temperatures have polymerases that can withstand heat without denaturing. Another has been the development of **thermal cyclers,** instruments that allow the programming of rapid and repeated temperature shifts so that the time between cycles can be shortened and can occur automatically without handling of the samples. The method can also be used to amplify RNA sequences if viral reverse transcriptase is first used to transcribe the RNA into complementary DNA (cDNA) sequences, which are subsequently subjected to PCR (**RT-PCR**).

MULTIPLE-CHOICE QUESTIONS

Select the single best answer.

1.1. The phase contrast microscope is an optical device used to achieve which of the following objectives?
(A) Convert differences in refractive index into visible differences in light intensity
(B) Illuminate a specimen with a single wavelength of light
(C) Obtain resolution greater than that of the electron microscope
(D) Reveal birefringence of crystalline or fibrillar tissue components
(E) Make quantitative measurements of light intensity
(F) Obtain optical sections of small pieces of tissue
(G) Determine the molecular weight of cellular macromolecules

1.2. Techniques that permit direct observation of living cells include which of the following?
(A) Homogenization and differential centrifugation
(B) Cryofracture and freeze etch
(C) Phase contrast microscopy and tissue culture
(D) Radioautography and transmission electron microscopy
(E) Column chromatography and isoelectric focusing
(F) Blotting and electrotransfer
(G) Polymerase chain reaction

1.3. Which of the following is enhanced by chemical fixation?
(A) Autolysis
(B) Enzyme activity
(C) Structural preservation
(D) Immunohistochemical detection of tissue antigens

(E) Bacterial degradation of histologic specimens
(F) Solubility of tissue proteins
(G) Diffusion of tissue components

1.4. Which of the following techniques were combined to produce the image in Figure 1–2?
(A) Electron microscopy and cryofracture
(B) Cryofracture and radioautography
(C) Light microscopy and Sudan black staining
(D) Electron microscopy and radioautography
(E) Radioautography and light microscopy
(F) Immunohistochemistry and light microscopy
(G) Electron microscopy and enzyme histochemistry

1.5. Which of the following techniques is most commonly used to locate glycogen in cells?
(A) Methylene blue staining
(B) Feulgen's reaction
(C) Periodic acid–Schiff (PAS) reaction
(D) Enzyme histochemistry
(E) Immunohistochemistry
(F) In situ hybridization
(G) Eosin staining

1.6. Frozen sectioning may be required to avoid the removal of which of the following target substances when tissues are prepared for paraffin sectioning?
(A) Basic proteins
(B) Lipids
(C) Enzymes
(D) Carbohydrates
(E) Nucleic acids

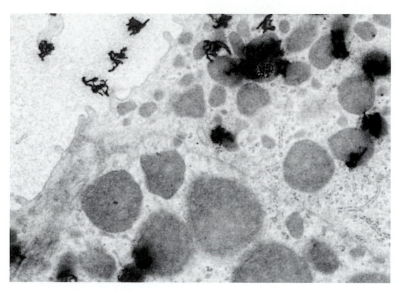

Figure 1–2.

1.7. In situ hybridization is used to demonstrate the pattern of expression of which of the following types of molecule?
(A) Antibody
(B) Carbohydrate
(C) Enzyme
(D) Ion
(E) Lipid
(F) Nucleic acid
(G) Protein

1.8. Radioautography is the method of choice for revealing which of the following in a tissue section?
(A) Sites of synthesis of various molecules
(B) Molecular weights of products of cell synthesis
(C) Diameter of a secretory cell
(D) Number of copies of a particular species of mRNA
(E) Cell structure in unstained tissue sections
(F) Nucleotide sequence of a gene
(G) Amino acid sequence of a protein

1.9. Which of the following compounds is typically used as a marker when coupled with secondary antibodies in electron microscopic immunohistochemical studies?
(A) Fluorescein
(B) Rhodamine
(C) Colloidal gold
(D) Wheat germ agglutinin
(E) Acridine orange
(F) ^{32}P
(G) 3,3′ Diaminobenzidine

1.10. Separation of two hemoglobin isoforms with identical molecular weights but different net charges can be accomplished using which of the following techniques?
(A) Differential centrifugation
(B) Isoelectric focusing
(C) Northern blotting
(D) Polymerase chain reaction
(E) SDS polyacrylamide gel electrophoresis
(F) Southern blotting
(G) Transmission electron microscopy

1.11. Which of the following enzymes is used during DNA cloning procedures to isolate specific genes by cutting DNA at specific nucleotide sequences?
(A) Alkaline phosphatase
(B) DNA ligase
(C) DNA polymerase
(D) Horseradish peroxidase
(E) Restriction nuclease
(F) Reverse transcriptase
(G) RNA polymerase

1.12. Which of the following methods is most appropriate for isolating a large amount of a specific protein from a tissue homogenate using antibodies raised against that protein?
(A) Affinity chromatography
(B) Indirect immunohistochemistry
(C) Isoelectric focusing
(D) Polymerase chain reaction
(E) Two-dimensional gel electrophoresis
(F) Western blotting

1.13. Which of the following terms is used to designate the short oligonucleotide sequences that hybridize to the beginning and ending portions of DNA sequences targeted for rapid copying using the polymerase chain reaction?

(A) Enhancer
(B) Intron
(C) Ligase
(D) Primer
(E) Probe
(F) Promoter
(G) Vector

1.14. Which of the following are small, circular, double-stranded DNA molecules capable of carrying novel nucleotide sequences and entering bacteria and maintaining their integrity while being copied separately from the bacteria's own genome?
(A) DNA polymerases
(B) DNA ligases
(C) Genetic fingerprints
(D) Plasmids
(E) Primers
(F) Probes
(G) Restriction nucleases

1.15. Which of the following procedures entails impregnating a specimen with a solvent of the embedding medium?
(A) Clearing
(B) Dehydration
(C) Fixation
(D) Mounting
(E) Rehydration

1.16. The agents in which of the following procedures are intended to stabilize tissue structure by coagulating proteins and promoting cross-linking?
(A) Clearing
(B) Dehydration
(C) Embedding
(D) Fixation
(E) Mounting
(F) Rehydration
(G) Staining

1.17. Which of the following methods involves immersing fixed tissues in increasing concentrations of ethanol in water?
(A) Clearing
(B) Dehydration
(C) Embedding
(D) Mounting
(E) Rehydration

1.18. Which of the following procedures is designed to localize labeled precursors by causing silver grains to appear in a photographic emulsion overlying the portions of the specimen containing the label?
(A) Clearing
(B) Cryofracture and freeze etch
(C) Enzyme histochemistry
(D) Immunohistochemistry
(E) Polarizing microscopy
(F) Isoelectric focusing
(G) Radioautography

1.19. In which of the following methods is a replica of a dehydrated tissue surface obtained by the deposition of heavy metal followed by a layer of carbon?
(A) Cryofracture and freeze etch
(B) Radioautography
(C) Polarizing microscopy
(D) Density gradient centrifugation
(E) Embedding
(F) Chromatography
(G) Isoelectric focusing

ANSWERS TO MULTIPLE-CHOICE QUESTIONS

1.1. A (III.D.3)
1.2. C (III.D.3 and XI)
1.3. B (Table 1–1; VII and IX)
1.4. D (Table 1–4)
1.5. C (Table 1–3; VII.E.1)
1.6. D (VII.A–F, VIII, and IX)
1.7. F (X)
1.8. A (Table 1–4)
1.9. C (IX.B)
1.10. B (XIV.B; Hemoglobin isoforms are different proteins of essentially the same molecular weight but with slight differences in amino acid sequences and therefore differences in net charge)

1.11. E (XVI.A)
1.12. A (XIII and Table 1–5)
1.13. D (XVI.C.2)
1.14. D (XVI.C.1)
1.15. A (Table 1–1)
1.16. D (Table 1–1)
1.17. B (Table 1–1)
1.18. G (Table 1–4)
1.19. A (Table 1–4)

2

The Plasma Membrane & Cytoplasm

OBJECTIVES

This chapter should help the student to:

- Perceive the inseparability of structure and function in living organisms.
- Know the names and functions of cytoplasmic components.
- Know the subunits of each cytoplasmic component and their roles in its function.
- List the functions of cells and explain the role of each cytoplasmic component in each function.
- Recognize a cell's cytoplasmic components in a micrograph and hence predict the cell's function(s).
- Predict which structures are present in a cell from its function.
- Predict the functional deficit(s) that accompany specific structural aberrations.
- Predict the cytoplasmic component(s) likely to be involved in a functional deficit.
- Explain and give examples of cell differentiation.

MAX-Yield™ STUDY QUESTIONS

1. Compare prokaryotic and eukaryotic cells in terms of the presence of a membrane-limited nucleus, histones, and membranous cytoplasmic organelles, as well as overall size (I.B[1]).
2. List several functions of cell membranes (II.C).
3. List the major biochemical constituents of cell membranes and sketch their organization as described in the fluid mosaic model of membrane structure (II.A and B; Fig. 2–1).
4. Explain why a membrane's phospholipid bilayer appears as a trilaminar structure in transmission electron micrographs (TEMs) (II.A.1).
5. Compare the locations of peripheral and integral membrane proteins in relation to the lipid bilayers of membranes and the methods required to isolate them from these membranes (II.A.2.a and b; Fig. 2–1).
6. Name three membrane receptor types involved in transducing signals across the plasma membrane. Compare them in terms of the number of passes they make through the plasma membrane, the effects of ligand binding on their conformation, and the role of enzymes in signal transduction (II.C.2.a–c; Figs. 2–2, 2–3).
7. How does signal transduction by members of the steroid hormone receptor family differ from that by membrane receptors in terms of the type of ligand, the receptor location, and the ability of the receptor to associate directly with DNA (II.C.2.a–d)?
8. Compare phagocytosis and pinocytosis in terms of the way that vacuoles or vesicles are formed, the types of materials that undergo endocytosis, and the relative size of the vacuoles and vesicles (II.C.3.a and b).
9. List the steps involved in receptor-mediated endocytosis, beginning with the association of a ligand with its cell-surface receptor and ending with the return of the receptor to the cell surface (II.C.3.c; Fig. 2–4).
10. Compare organelles and cytoplasmic inclusions in terms of the presence of limiting membranes, enzyme content, their role (active or passive) in cell function, and their relative constancy in the cytoplasm (III).
11. What is the major function of mitochondria (III.A.2)?

---

[1] See footnote on page 1.

12. Sketch a mitochondrion (III.A; Fig. 2–5) and label or show the location of the following:
    a. Outer mitochondrial membrane   g. Intracristal space
    b. Inner mitochondrial membrane   h. Matrix
    c. Cristae   i. Matrix granules
    d. Inner membrane (F.1) subunits   j. ATP synthase (III.A.1.b)
    e. Intramembrane space (III.A.1.c)   k. Citric acid cycle enzymes
    f. Intercristal space   l. Electron transport system (III.A.1.b)

13. Which substances and structures in the mitochondrial matrix resemble those found in prokaryotic cells and also duplicate eukaryotic cell components found elsewhere in the cell (III.A.1.d)?

14. Compare the mitochondrial cristae of most cells with those of steroid-secreting cells and cells with a high rate of metabolism (III.A.1.b).

15. Briefly describe four major steps in oxidative phosphorylation (III.A.2).

16. What is the primary function of ribosomes (III.B)?

17. List the biochemical and structural components of ribosomes, and name the sites of their synthesis and association (III.B.1).

18. Compare the appearance of ribosomes in light and electron microscopy (III.B.1).

19. How are individual ribosomes held together to form a polyribosome (III.B.2)?

20. What is the difference between the functions of free polyribosomes and those of polyribosomes attached to the rough endoplasmic reticulum (RER) (III.B.2)?

21. Compare the RER and the smooth endoplasmic reticulum (SER) in terms of the presence of ribosomes (III.C.1.a and 2.a), the shape of cisternae (III.C.1.a and 2.a), functions (III.C.1.b and 2.b), and cell types in which each is particularly abundant; include examples (III.C.1.c and 2.c).

22. List the steps in RER-associated protein synthesis and subsequent posttranslational modification, beginning with ribosome attachment to mRNAs for proteins destined for secretion and ending with the budding of transfer vesicles for transport to the Golgi complex (II.D; III.B.2, C.1.a and b).

23. List the functions of the Golgi complex (III.D.2).

24. Sketch a Golgi complex (III.D.1) and label the following:
    a. Cisternae   e. Condensing vacuoles
    b. *Cis* face   f. Secretory granules
    c. *Trans* face   g. Site of selective osmium deposition
    d. Transfer vesicles

25. Compare primary and secondary lysosomes in terms of size, appearance, and contents (III.F.1 and 2).

26. Compare lysosomes (III.F) and peroxisomes (III.G) in terms of size, content, appearance, and function.

27. Compare the contents of autophagosomes with those of heterophagosomes (III.E).

28. Trace the steps in the ingestion and digestion of extracellular materials, beginning with endocytosis (II.C.3) and ending with the formation of residual bodies (III.E and F.1–3).

29. How does an inherited deficiency or complete lack of a particular lysosomal enzyme affect the intracellular concentrations of substrates of that enzyme (III.F)?

30. Compare microfilaments and microtubules in terms of:
    a. Their diameters (III.I.1.a and 2.a)
    b. Their major protein components (III.I.1.a and 2.a)
    c. Their functions (III.I.1.b and 2.b)
    d. The polymerization and depolymerization of their subunits (III.I.1.a,b and 2.a,b)
    e. Their contractile capacity (III.I.1.b and 2.b)
    f. Their location in the cell (III.I.1.c and 2.c)
    g. Their associated motor proteins (III.I.1.b and 2.b)

31. Compare dynein and kinesin in terms of their direction of movement along microtubules and their association with the RER and Golgi complex (III.I.1.b).

32. Explain the main function of intermediate filaments and give their diameter in nanometers (III.I.3.a and b).

33. List five types of intermediate filament proteins and the cell types in which each may be found (III.I.3.a).

34. Compare centrioles, basal bodies, and cilia and flagella in terms of the number and organization of their microtubules (III.J–M).

35. What role do centrioles play in cell function (III.M)?

36. Compare cilia and flagella in terms of their structure, length, motion, and typical number per cell (III.J and K).

SYNOPSIS

I. GENERAL FEATURES OF CELLS

A. Subunits of Life: Cells are the structural and functional units of life (and of disease processes) in all tissues, organs, and organ systems. Each cell's capabilities and limitations are implicit in its structure.

B. Prokaryotes and Eukaryotes: There are two basic cell types. Prokaryotic cells typically are small, single-celled organisms (eg, bacteria) that lack a nuclear envelope, histones, and membranous organelles. Eukaryotic cells exist primarily as components of multicellular organisms. This chapter covers the basic structural and functional features of eukaryotic cells. Specific human cell types are described in later chapters.

C. Cellular Components: Eukaryotic cells have three major components:
1. **Cell membranes** (II) separate a cell from its environment and form distinct functional compartments (nucleus, organelles) in the cell. The outer cell membrane is called the **plasma membrane,** or **plasmalemma.**
2. The **cytoplasm** (III) surrounds the nucleus and is enclosed by the plasma membrane. It contains the structures and substances needed to decode the instructions of DNA and carry on the cell's activities.
3. The membrane-limited **nucleus** (Chapter 3) contains a cell's DNA, which harbors the genetic code for protein synthesis and thus for all of the cell's activities. It also has components that help determine which parts of the genetic code are used and that deliver coded information to the cytoplasm.

D. Cellular Functions: Three activities basic to living organisms are nourishment, growth and development, and reproduction. Functions directed toward these activities are described in this chapter and in Chapter 3. More specialized cell functions receive detailed treatment in subsequent chapters.

II. CELL MEMBRANES

A. Biochemical Components (Fig. 2–1):
1. **Lipids** are present in cell membranes as **phospholipids** (Fig. 2–1, E), **sphingolipids,** and **cholesterol** (Fig. 2–1, A). Phospholipids (eg, lecithin) are the most abundant form. Each phospholipid molecule has a polar (hydrophilic), phosphate-containing head group (Fig. 2–1, G) and a nonpolar (hydrophobic) pair of fatty acid tails (Fig. 2–1, F). Membrane phospholipids are arranged in a bilayer, with their tails directed toward one another at the center of the membrane. In electron micrographs (EMs) of osmium-stained tissue, a single membrane, or **unit membrane,** has two dark outer lines with a lighter layer between them. This **trilaminar** appearance reflects the deposition of reduced osmium on the hydrophilic head groups.
2. **Proteins** may make up more than 50% of membrane weight. Most membrane proteins are globular and belong to one of the following two groups:
   a. Integral membrane proteins (Fig. 2–1, C and D) are tightly lodged in the lipid bilayer; detergents are required to extract them. They are folded, with their hydrophilic amino acids in contact with the membrane phospholipids' phosphate groups and their hydrophobic amino acids in contact with the fatty acid tails. Some protrude from only one membrane surface (Fig. 2–1, D). Others, called **transmembrane proteins** (Fig. 2–1, C), penetrate the entire membrane and protrude from both sides. Some transmembrane proteins, such as **protein-3-tetramer,** are hydrophilic channels for the passage of water and water-soluble materials through hydrophobic regions. Freeze-fracture preparations often split plasma membranes through the hydrophobic region, between the ends of the phospholipid's fatty acid tails (Fig. 2–1). Most integral proteins exposed in the process end up in the side closest to the cytoplasm, termed the **P** (protoplasmic) **face.** The membrane half nearest to the environment, the **E** (ectoplasmic) **face,** usually appears smoother.

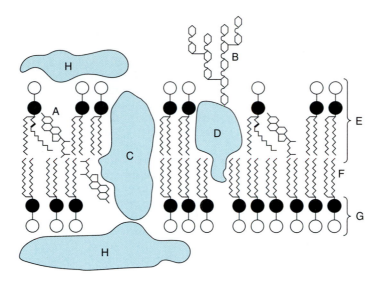

Figure 2–1. Schematic diagram of the biochemical components of plasma membranes (II.A). Labeled components include cholesterol (A), the oligosaccharide moiety (B) of a glycoprotein on the extracellular surface, integral proteins (C and D), phospholipid molecules (E) with their fatty acid tails (F) and polar head groups (G), and peripheral protein (H).

    b. Peripheral membrane proteins (Fig. 2–1, H) are more loosely associated with the inner or outer membrane surface; some are globular, some filamentous. In erythrocytes, examples on the cytoplasmic surface include **spectrin,** which helps maintain membrane integrity, and **ankyrin,** which links spectrin to protein-3-tetramer.

  3. Carbohydrates occur on plasma membranes mainly as oligosaccharide moieties of glycoproteins (Fig. 2–1, B) and glycolipids. Membrane oligosaccharides have a characteristic branching structure and project from the cell's outer surface, forming a surface coat called the **glycocalyx** that participates in cell adhesion and recognition.

B. Membrane Organization: The **fluid mosaic model** describes biologic membranes as "protein icebergs in a lipid sea." Integral proteins exhibit lateral mobility and may rearrange through their association with peripheral proteins, cytoskeletal filaments within the cell (III.I), membrane components of adjacent cells, and extracellular matrix components. Integral proteins may diffuse to and accumulate in one membrane region, a process termed **capping. Membrane asymmetry** refers to differences in chemical composition between the bilayer's inner and outer halves. Oligosaccharides occur only on the plasma membrane's outer surface. Phospholipid asymmetries also occur. The outer half has more phosphatidyl choline and sphingomyelin and the inner half has more phosphatidyl serine and phosphatidyl ethanolamine.

C. Membrane Functions:
  1. Selective permeability. Cell membranes separate the internal and external environments of a cell or organelle, preventing the intrusion of harmful substances, the dispersion of macromolecules, and the dilution of enzymes and substrates. The selective permeability of membranes is essential to maintaining the functional steady state, or **homeostasis,** required for cell survival. Homeostatic mechanisms attributable to the cell membrane maintain optimal intracellular concentrations of ions, water, enzymes, and substrates. Three mechanisms allow selected molecules to cross membranes.
    a. Passive diffusion. Some substances (eg, water) can cross the membrane in either direction, without the need for energy, by following a concentration gradient.
    b. Facilitated diffusion. Some molecules (eg, glucose) are helped across the membrane by a membrane component. This *facilitated* diffusion is often unidirectional, but it follows a concentration gradient and requires no energy.
    c. Active transport. Some nondiffusible molecules can enter or exit a cell even against a gradient. This requires energy, usually in the form of adenosine triphosphate (ATP). One

active transport mechanism is the sodium pump (Na⁺/K⁺-ATPase), which expels sodium ions from a cell even when the sodium concentration is higher outside than inside.

2. **Signal transduction.** Membrane **receptor** proteins with strong binding affinities for signal molecules (eg, neurotransmitters, peptide hormones, and growth factors) are found on cell surfaces. The signal molecule to which a receptor specifically binds is its **ligand.** The receptor transmits the signal (but not the ligand) across the membrane. These receptors are critical to intercellular communication (V.B). Signal transmission depends on the receptor class involved. There are three known receptor classes.

a. **Ion channel–linked receptors** (transmitter-gated ion channels) are long proteins that pass several times through the plasma membrane (Fig. 2–2A). The binding of a neurotransmitter (as ligand) to its receptor at the cell surface induces a conformational change in the receptor that opens (or closes) a transmembrane ion channel. Ligand binding and the resulting opening are brief, permitting regulation. A signal of adequate strength (threshold) allows enough ions to cross the membrane to open additional channels (voltage-gated ion channels) in response to the change in potential. Reaching threshold allows a self-perpetuating membrane depolarization to spread as a wave along the nerve cell surface (9.VII.B.2).

b. **Enzyme-linked receptors** (Fig. 2–2B) constitute a heterogeneous group of transmembrane (typically single-pass) proteins associated with an enzyme (typically a protein kinase) or possessing kinase activity of their own (eg, tyrosine kinase). Protein kinases activate additional proteins by phosphorylating them (III.B.1.b). In this way, the binding of a ligand to its receptor initiates a complicated cascade of enzyme activations. Receptor tyrosine kinases (eg, the insulin receptor) often act by activating an adaptor protein,

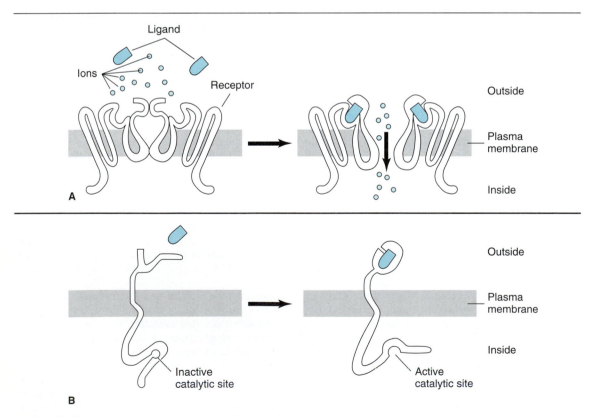

Figure 2–2. Schematic diagrams of signal-transducing membrane receptors. **A.** Ion channel–linked receptor (transmitter-gated ion channel). Ligand binding causes a conformational change in the receptor, allowing ions to pass through the membrane. **B.** Enzyme-linked receptor. Ligand binding causes a conformational change in the receptor that activates an enzyme (eg, kinase) that is part of the receptor (*shown*) or one that is a separate protein associated with the cytoplasmic domain of the receptor (*not shown*).

which subsequently activates the guanosine triphosphate (GTP)-binding protein (G protein) **Ras** by causing it to exchange the guanosine diphosphate (GDP) it binds in its inactive state for GTP. Activated Ras subsequently activates a cascade of cytoplasmic protein kinases, ending with the activation of a gene regulatory protein that changes the pattern of gene expression by the cell. Ras differs from the G proteins described in the next section in that it is monomeric rather than trimeric. It resembles the α subunit in Figure 2–3 and acts without a β or γ subunit.

  c. **G protein–linked receptors** constitute a family of proteins that make seven passes through the membrane (Fig. 2–3). Each receptor binds a different ligand, and the effect of signal binding on the ultimate target protein (either an ion channel [Fig. 2–2A] or an enzyme [Fig. 2–2B]) is indirect and is mediated by a trimeric G protein (Fig. 2–3A). On binding its ligand (Fig. 2–3B), the receptor interacts with and activates the trimeric G protein (Fig. 2–3C). The inactive G protein carries a GDP molecule, which it trades for GTP on activation (Fig. 2–3D and E). The activated G protein complex dissociates from the receptor, allowing the GTP-binding portion to interact with and activate its target (Fig. 2–3F; eg, adenyl cyclase), thus inactivating the G protein (ie, GTP → GDP; Fig. 2–3F).

  d. **Steroid hormone receptor family.** Most peptide hormones and growth factors signal the cell interior through cell-surface receptors. Steroid hormones (eg, hydrocortisone and estrogen), retinoids (vitamin A–related compounds), and vitamin D pass more readily through the membrane and bind to receptors in the nucleus. The thyroid hormone receptor also belongs to this group. This family of nuclear receptors is characterized by a specific ligand-binding site and a DNA-binding site. They typically form dimers and enhance or repress gene expression through their association with the DNA flanking the gene-coding regions.

3. **Endocytosis.** Cells engulf extracellular substances and bring them into the cytoplasm in membrane-limited vesicles by mechanisms described collectively as endocytosis.
  a. In **phagocytosis** (cell eating), the cell engulfs insoluble substances, such as large macromolecules or entire bacteria. The vesicles formed are termed phagosomes.
  b. In **pinocytosis** (cell drinking), the cell engulfs small amounts of fluid, which may contain a variety of solutes. Pinocytotic vesicles are smaller than phagosomes.
  c. In **receptor-mediated endocytosis** (Fig. 2–4), a cell engulfs ligands along with their surface receptors. The binding of ligand to receptor causes the ligand–receptor complexes to collect in a **coated pit,** a shallow membrane depression whose cytoplasmic surface is coated with **clathrin** protein. Invagination and pinching off of the pit creates a **coated vesicle,** which carries the ligand–receptor complexes into the cell. The clathrin coat is released from the vesicle, now termed an **endosome,** and the ligands dissociate from the receptors. The later endosome, or compartment of uncoupling of receptor and ligand **(CURL),** becomes more tubular and divides into two portions, segregating the receptors from the ligands. Receptors may return to the plasma membrane. Ligands are directed to lysosomes.

4. **Exocytosis** ejects substances from the cell. Cells use this process for both secretion and excretion of undigested material. A membrane-limited vesicle or secretory granule fuses with the plasma membrane and releases its contents into the extracellular space, without disrupting the plasma membrane.

5. **Compartmentalization.** Membranes selectively inhibit the passage of most water-soluble substances. The cytoplasm has many membrane-limited compartments (organelles), each with different internal solute concentrations. This compartmentalization prevents the dilution of metabolic intermediates and cofactors in multistep biochemical reactions, and protects sensitive reactions from the intrusion of extraneous substances.

6. **Spatial–temporal organization of metabolic processes.** Some cellular membranes (eg, inner mitochondrial membrane and Golgi complex) contain series of enzymes arranged so that intermediates in multistep metabolic processes are passed from enzyme to enzyme. This arrangement maintains the chronologic order of such processes and sets rate limitations by maintaining local concentrations of intermediates.

7. **Storage, transport, and secretion.** Membrane-limited vesicles isolate certain substances during intracellular processes. Substances in the vesicles may be kept for later use (storage), shuttled from one compartment to another for further processing (transport, II.D), or expelled from the cell (secretion, II.C.4).

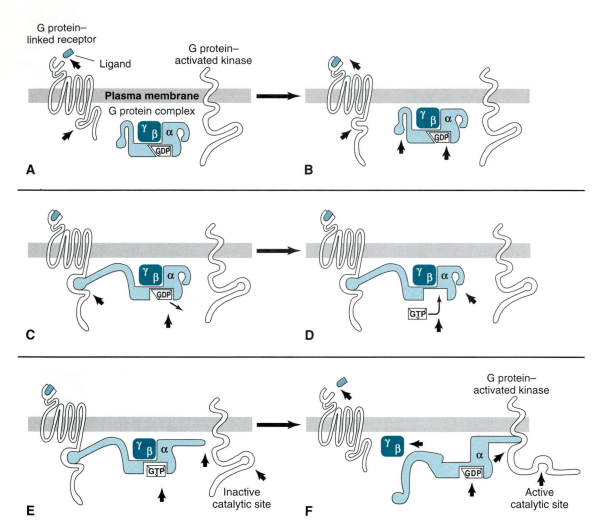

Figure 2–3. Schematic diagram of the activation of a G protein–linked receptor. **A.** Components of the system. **B.** The binding of ligand to receptor causes a conformational change in the receptor that allows the α subunit of the trimeric G protein complex to interact with it. **C.** As the α subunit interacts with the receptor, it undergoes a conformational change, allowing it to release its GDP. **D. GDP is traded for GTP. E.** This causes a further conformational change of the α subunit. **F.** This change allows the subunit to disengage from the receptor, dissociate from the γ and β subunits of the complex, and interact with a G protein–activated kinase. In this interaction, one high-energy bond in the GTP is used to activate the kinase, resulting in the transformation of GTP to GDP.

D. Membrane Flow: The movement of membrane from one organelle to another is called membrane flow and is a general feature of organelle function. Membranes bud as vesicles from an organelle and fuse with other membranes, allowing the amount of membrane in a particular organelle to change without membrane synthesis or breakdown. Vesicle budding is facilitated by protein (eg, β-COP, clathrin) coats that the vesicles acquire on their cytoplasmic surfaces during budding. These coats help shape the bud and ensure its inclusion of molecules that direct vesicle transport toward the proper destination. After a **coated vesicle** is fully formed, the coat is released. The resulting **transfer vesicle** may diffuse to nearby targets (eg, ER to Golgi) or may attach to microtubules for transport over longer distances (eg, neuron cell body to axon terminal). Transmembrane proteins in the budding vesicle membrane, called **SNAREs,** remain after the coat is shed. They appear to be critical in allowing transfer vesicles to recognize, bind, fuse with, and unload their contents at the correct target organelle. Receptors in the membrane

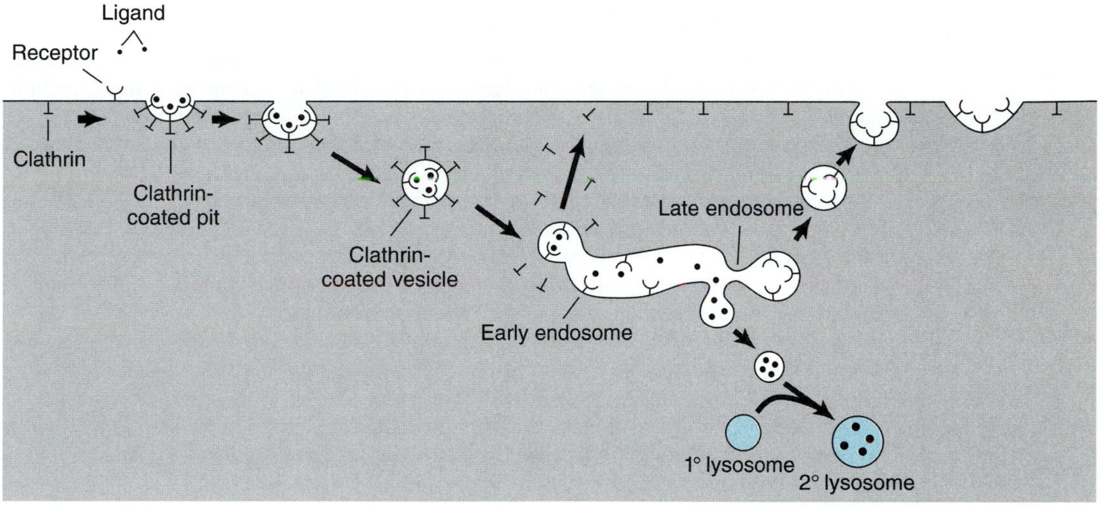

Ligand

Receptor

Clathrin

Clathrin-
coated pit

Clathrin-
coated vesicle

Late endosome

Early endosome

1° lysosome

2° lysosome

Figure 2–4. Schematic diagram of receptor-mediated endocytosis (II.C.4.c).

of the target organelle, called **t-SNAREs,** bind the SNAREs and recruit the additional cytoplasmic proteins required for fusion with the target membrane.

III. CYTOPLASM

Cytoplasmic structures comprise three groups. **Organelles** are membrane-bound, enzyme-containing subcellular compartments (eg, mitochondria). Each organelle has a distinctive structure and unique functions. **Cytoplasmic inclusions** are structures, membrane-limited or not, that are generally more transient than organelles and less actively involved in cell metabolism (eg, lipid droplets). Organelles and inclusions are discussed in sections III.A–H. The **cytoskeleton** (III.I) is a proteinaceous supporting network within the cytoplasm, components of which (microtubules) also form discrete cytoplasmic structures such as centrioles.

A. Mitochondria: The largest organelles, which are known as mitochondria, generate the cell's energy.

  1. **Structure.** Mitochondria are comparable to bacteria in size (typically 2–6 μm in length and 0.2 μm in diameter), and have various shapes (eg, spherical, ovoid, filamentous). Each mitochondrion is bounded by two unit membranes.

     a. **Outer mitochondrial membrane.** (Fig. 2–5, A) This membrane has a smooth contour and forms a continuous but porous covering. It is permeable to small molecules (<5 kDa) owing to large channel-forming proteins called **porins.**

     b. **Inner mitochondrial membrane.** (Fig. 2–5, B) This membrane is less porous (semipermeable) and has many infoldings, or **cristae** (Fig. 2–5, C). The cristae of most mitochon-

Figure 2–5. Schematic diagram of a mitochondrion (III.A). Labeled components include the outer mitochondrial membrane (A), inner mitochondrial membrane (B), cristae (C), mitochondrial matrix in the intercristal space (D), intracristal extension (E) of the intermembrane space (F), matrix granules (G), and inner membrane subunits (F1 subunits) (H).

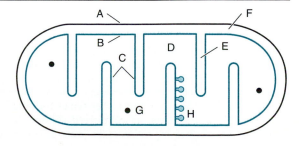

dria are shelflike, but in steroid-secreting cells are tubular. The inner surface is covered by **inner membrane subunits** (Fig. 2–5, H), also called **F1 subunits** (or lollipops, because of their shape); these are sites of mitochondrial **ATP synthase** activity. Intercalated within the inner membrane are components of the **electron transport system,** including enzymes and cofactors that have important roles in mitochondrial function (eg, cytochromes, proton pumps, dehydrogenases, flavoproteins). Mitochondrial ribosomes also associate with the inner surface.

c. **Membrane-limited spaces.** The mitochondrial membranes create two membrane-limited spaces. The **intermembrane space** (Fig. 2–5, F) is located between the inner and outer membranes and is continuous with the **intracristal space** (Fig. 2–5, E), which extends into the cristae. The **intercristal space,** or **matrix space** (Fig. 2–5, D), is enclosed by the inner membrane and contains the mitochondrial matrix.

d. **Mitochondrial matrix.** This matrix contains water, solutes, and large **matrix granules** (Fig. 2–5, G), which play a role in mitochondrial calcium ion concentration. It also contains **circular DNA** and **mitochondrial ribosomes** similar to those of bacteria. The matrix contains numerous soluble enzymes involved in such specialized mitochondrial functions as the **citric acid (Krebs, tricarboxylic acid) cycle, lipid oxidation,** and mitochondrial protein and DNA synthesis.

2. **Function.** Mitochondria provide energy for chemical and mechanical work by storing energy generated from cellular metabolites in the high-energy bonds of ATP. Energy is generated by oxidative phosphorylation, which involves four main steps: (1) the citric acid cycle uses acetyl-CoA generated from pyruvate (from carbohydrates) or fatty acids (from lipids) to generate high-energy electrons, which it donates to NADH and $FADH_2$; (2) these intermediates release the high-energy electrons to the electron transport chain in the inner mitochondrial membrane, which uses their energy to drive pumps that eject protons from the matrix; (3) the proton imbalance establishes an electrochemical gradient across the inner mitochondrial membrane—a form of potential energy; (4) the flow of protons back into the matrix, down the gradient established by the proton pumps, occurs through a channel in ATP synthase, generating energy that ATP synthase captures by converting adenosine diphosphate (ADP) plus inorganic phosphate to ATP. ATP leaves the mitochondrion and releases its stored energy at a variety of intracellular sites. Mitochondria synthesize their own DNA and some proteins. They grow and reproduce by fission or budding and can undergo rapid movement and shape changes.

3. **Location.** Mitochondria occur in nearly all eukaryotic cells, and in most are dispersed throughout the cytoplasm. They accumulate in cell types and intracellular regions with high energy requirements. Cardiac muscle cells are notable for abundant mitochondria. Epithelial cells lining kidney tubules have abundant mitochondria interdigitated between basal plasma membrane infoldings, where active ion and water transport occurs.

B. **Ribosomes:** These protein-synthesizing organelles are of two main types. Mitochondrial (like prokaryotic) ribosomes are smaller (20 nm) than cytoplasmic (eukaryotic) ribosomes (25 nm).

1. **Structure.** Each ribosome type has two unequal **ribosomal subunits,** named for their ultracentrifugal sedimentation (but often called simply "large" and "small"). Mitochondrial ribosomes (70S overall) have a 50S and a 30S subunit; cytoplasmic ribosomes (80S overall) have a 60S and a 40S subunit. Cytoplasmic ribosomes are composed of **ribosomal RNA (rRNA)** synthesized in the nucleolus and many proteins synthesized in the cytoplasm. They are intensely basophilic. Light microscopy reveals cytoplasmic accumulations of ribosomes as basophilic patches, formerly termed **ergastoplasm** in glandular cells and **Nissl bodies** in neurons. In EMs, ribosomes appear as small, electron-dense cytoplasmic granules.

2. **Location and function.** Cytoplasmic ribosomes occur in two forms. **Free ribosomes** are dispersed in cytoplasm. **Polyribosomes,** or **polysomes,** are ribosomes attached to a single strand of **messenger RNA (mRNA)**, permitting synthesis of multiple copies of a protein from the same message. Ribosomes read (translate) the mRNA code and thus play a critical role in assembling amino acids into specific proteins. Polysomes occur free in the cytoplasm (free polysomes) and attached to membranes of the RER; free polysomes synthesize structural proteins and enzymes for intracellular use. Polysomes of RER synthesize proteins to be secreted or sequestered. Various **signal sequences** are encoded in the 5′ end of mRNAs, helping to direct the proteins that contain them to different organelles. The signal sequences

directing secretory proteins to the RER are described in the next section (III.C.1.b). Other signal sequences direct proteins to, and help them enter, the nucleus, mitochondria, and peroxisomes. A common posttranslational modification in the cytoplasm that regulates many cell functions is **protein phosphorylation.** Phosphate groups, transferred from ATP or GTP to proteins by **protein kinases** (eg, tyrosine kinase or serine/threonine kinases), or removed by **protein phosphatases,** can activate or inactivate a wide variety of proteins after they have been synthesized. These reactions are often key steps in signal transduction (II.C.2).

C. **Endoplasmic Reticulum (ER):**   This complex organelle is involved in the synthesizing, packaging, and processing of various cell substances. It is a freely anastomosing network (reticulum) of membranes that form **vesicles,** or **cisternae;** these may be elongated, flattened, rounded, or tubular. Transfer vesicles (II.D) bud from the ER and cross the intervening cytoplasm, delivering their contents to the Golgi complex (III.D) for further processing or packaging. In mature cells, ER occurs in two forms: rough and smooth.

   1. **Rough endoplasmic reticulum.**
      a. **Structure. RER,** also called granular ER, is studded with ribosomes in polysomal clusters. RER cisternae are typically parallel, flattened, and elongated, especially in cells specialized for protein secretion (eg, pancreatic acinar cells, plasma cells), in which the RER is particularly abundant. Ribosomes render the RER basophilic. RER membranes and individual ribosomes are visible only with the electron microscope. Proteins unique to RER membranes include **docking protein** and **ribophorins** (III.C.1.b).
      b. **Function.** The RER synthesizes proteins for sequestration from the cytoplasm, including secretory proteins such as collagen, proteins for insertion into cell membranes (integral proteins; II.A.2.a), and lysosomal enzymes (isolated to prevent **autolysis**). Ribosomes or free polysomes begin reading at the 5′ end of mRNAs and move toward the 3′ end. The 5′ end of mRNAs for secretory and sequestered proteins carries the code for a 20- to 25-amino acid signal sequence. The signal sequence is translated on a free polysome and subsequently interacts with a cytoplasmic **signal recognition particle** (**SRP,** six polypeptides plus a 7S RNA molecule). SRP inhibits further translation until the SRP–polyribosome complex binds to the RER docking protein; subsequently the SRP is released and translation continues. Ribophorins mediate the attachment of the signal sequence and the large ribosomal subunit to the RER membrane and provide a hydrophilic translocation channel for **vectorial discharge** (unidirectional passage) of nascent protein into the RER lumen, where the signal sequence is cleaved by **signal peptidase** and the remainder of the nascent protein is folded and modified. One important posttranslational modification in the ER is **core glycosylation,** in which preassembled oligosaccharides high in mannose are transferred from a lipid carrier (eg, dolichol phosphate) to amino acids, especially asparagine. The oligosaccharides "address" proteins for transport to intracellular destinations.
      c. **Location.** The RER is suspended in the cytoplasm and shows continuity with the nuclear envelope's outer membrane. The RER in protein-secreting epithelial cells often lies in the basal cytoplasm, between the plasma membrane and the nucleus.
   2. **Smooth endoplasmic reticulum (SER).**
      a. **Structure.** The SER lacks ribosomes and thus appears smooth in electron micrographs. SER cisternae are more tubular or vesicular than those of the RER. SER stains poorly, if at all; thus, with the light microscope, it is indistinguishable from the rest of the cytoplasm.
      b. **Function.** Because it lacks ribosomes, SER cannot synthesize proteins. It has many enzymes that are important in lipid metabolism, steroid hormone synthesis, glycogen breakdown (glucose-6-phosphatase), and detoxification. The last occurs by means of enzymatic conjugation, oxidation, and methylation of potentially toxic substances.
      c. **Location.** The SER is suspended in the cytoplasm of many cells and is especially abundant in cells synthesizing steroid hormones (eg, in the adrenal cortex and gonads). It is also abundant in liver cells (hepatocytes), where it is involved in glycogen metabolism and drug detoxification. Specialized SER termed **sarcoplasmic reticulum** is found in striated muscle cells, where it regulates muscle contraction by sequestering and releasing calcium ions.

D. **Golgi Complex:**   The Golgi complex (Golgi apparatus) participates in many activities, particularly those associated with secretion. It is a focal point of membrane flow (II.D) and vesicle traffic among organelles.

1. **Structure.** This membranous organelle comprises three major compartments: (1) a stack of 3 to 10 discrete, slightly curved, flattened cisternae; (2) numerous small vesicles peripheral to the stack; and (3) a few large **condensing vacuoles** at the concave surface of the stack. The *cis* **face** (convex face, forming face) of the stack is usually closest to adjacent dilated ER cisternae and is surrounded by transfer vesicles. Its cisternae stain darkly with osmium. The *trans* **face** (concave face, maturing face) often harbors several condensing vacuoles and generally faces away from the nucleus. It is connected to a system of tubules and vesicles called the *trans* **Golgi network (TGN),** from which secretory and transfer vesicles exit.

2. **Functions.**
   a. **Polysaccharide synthesis.** The Golgi complex contains **glycosyltransferases** that initiate, lengthen, or shorten polysaccharide or oligosaccharide chains one sugar at a time.
   b. **Modification of secretory products.** The *cis* Golgi contains enzymes that glycosylate proteins and lipids and sulfate glycosaminoglycans (GAGs). It is thus important in synthesizing secretory glycoproteins, proteoglycans, glycolipids, and sulfated GAGs. It also participates in marking nascent lysosomal enzymes with mannose-6-phosphate for later segregation by the TGN.
   c. **Sorting of secretory products.** Products synthesized by the RER and modified in the *cis* Golgi are sorted in the TGN. For example, lysosomal enzymes marked with mannose-6-phosphate and secretory proteins destined for constitutive versus regulated exocytosis are segregated into different vesicles.
   d. **Packaging of secretory products.** The TGN packages the segregated products into vesicles. These secretory vesicles, or **secretory granules,** are transported to the plasma membrane for exocytosis (II.C.5).
   e. **Concentration and storage of secretory products.** The Golgi complexes of some cells concentrate and store secretory products prior to secretion. Concentration is a major function of the condensing vacuoles of the TGN, which also often serve as precursors to secretory granules.

3. **Location.** The Golgi complex typically is near the nucleus (juxtanuclear) and is often found near centrioles (which have a role in directing vesicle traffic). Golgi complexes are best developed in neurons and glandular cells, which are specialized for secretion.

4. **Flow of materials through the Golgi complex.** Secretory materials have long been thought to follow a one-way route (*cis* to *trans*) through the Golgi complex. This view currently seems to be an oversimplification. Golgi-associated vesicles differ in their source, destination, function, contents, and surface composition. Certain nonclathrin, vesicle-coating proteins (eg, β-COP) are associated with specific Golgi complex regions, suggesting that various vesicle types fuse with, and bud from, the *cis, trans,* or intermediate Golgi membranes.

E. **Phagosomes:** Phagosomes are membrane-limited vesicles of various sizes containing material destined for lysosomal digestion. **Heterophagosomes** contain the products of heterophagy (ie, extracellular material ingested by phagocytosis). **Autophagosomes** contain the products of autophagy (ie, intracellular material, such as worn or damaged organelles). The digestion of phagosomal contents begins when a phagosome fuses with one or more primary lysosomes to form a secondary lysosome, as described below. (*Note:* Some authors use the term "heterophagosome" to refer to secondary lysosomes [III.F.2].)

F. **Lysosomes:** Lysosomes are spherical, membrane-limited vesicles that may contain more than 50 enzymes each and function as the cellular digestive system. Their enzyme activities distinguish them from other cellular granules. The enzyme most widely used to identify them is **acid phosphatase** because it occurs almost exclusively in lysosomes. Other common enzymes in lysosomes include ribonucleases, deoxyribonucleases, cathepsins, sulfatases, beta-glucuronidase, phospholipases, various proteases, glucosidases, and lipases. An inherited deficiency or lack of a particular lysosomal enzyme can result in life-threatening accumulations of its substrate in the cytoplasm. Lysosomal enzymes usually occur as glycoproteins and are most active at acidic pH. Lysosomes occur in various sizes and electron densities, depending on their activity.

1. **Primary lysosomes** are small (5–8 nm in diameter), with electron-dense contents; they appear as solid black circles in EMs. Enzymes in this storage form of lysosomes (released directly from the TGN) are mostly inactive. Lysosomal enzymes synthesized and core-

glycosylated in the RER are transferred to the Golgi complex for further glycosylation and packaging in vesicles (III.D.2). Primary lysosomes disperse through the cytoplasm. They occur in most cells but are abundant in phagocytic cells (eg, macrophages, neutrophils).

2. **Secondary lysosomes** are larger, less electron-dense, and have a mottled appearance in EMs. They form by the fusion of one or more primary lysosomes with a phagosome. Their primary function is digesting products of heterophagy and autophagy. Lysosomal enzymes that mix with phagosome contents become active. Digestion produces metabolites for cell maintenance and growth (small molecules diffuse into the surrounding cytoplasm) and aids in organelle turnover. Lysosomal enzymes also catabolize some cell synthesis products, thus regulating the quality and quantity of secretory material. Secondary lysosomes occur throughout the cytoplasm in many cells, in numbers that reflect the cell's lysosomal and phagocytic activity.

3. **Residual bodies** are membrane-limited inclusions of various sizes and electron densities associated with the terminal phases of lysosome function. They contain indigestible materials, such as pigments, crystals, and certain lipids. Some cells (eg, macrophages) expel residual bodies as waste, but long-lived cells (eg, nerve, muscle) accumulate them. In the latter, waste-containing residual bodies reflect cellular aging and are termed **"wear-and-tear pigment,"** or **lipofuscin granules.** These granules appear yellow–brown in light microscopy and as electron-dense particles in EMs.

G. **Peroxisomes:** Peroxisomes are membrane-limited, enzyme-containing vesicles slightly larger than primary lysosomes. In rats, they differ from lysosomes because of their electron-dense, granular urate oxidase **nucleoid.** Peroxisomes function in hydrogen peroxide metabolism. They contain urate oxidase, hydroxyacid oxidase, and D-amino acid oxidase, which produce hydrogen peroxide capable of killing bacteria; they also contain catalase, which oxidizes various substrates and uses the hydrogen removed in the process to convert toxic hydrogen peroxide to water. Peroxisomes also participate in gluconeogenesis by assisting in fatty acid oxidation. They occur dispersed in the cytoplasm or in association with SER.

H. **Other Cytoplasmic Inclusions:** Prominent among storage inclusions are spherical **lipid droplets,** which differ in appearance depending on the histologic preparation. **Glycogen granules** are PAS-positive in light microscopy and appear in EMs as rosettes of electron-dense particles. Both lipid droplets and glycogen granules lack a limiting membrane. **Melanin** is a brown pigment widely distributed in vertebrates and often found in electron-dense, membrane-limited granules termed **melanosomes.** It is particularly abundant in epidermal cells and in the retina's pigment layer.

I. **Cytoskeleton:** The cytoskeleton, a mesh of filamentous elements called microtubules, microfilaments, and intermediate filaments, provides structural stability for the maintenance of cell shape. It is also important in cell movement and in the rearrangement of cytoplasmic components.

1. **Microtubules.**

    a. **Structure.** Microtubules (Fig. 2–6) are the thickest (24-nm diameter) cytoskeletal components. In EMs these fine tubular structures vary in length and have dense walls (5-nm thick) and a clear internal space (14-nm across). The walls consist of subunits called **tubulin heterodimers,** each of which comprises one α-**tubulin** and one β-**tubulin** protein molecule. The tubulin heterodimers are arranged in threadlike polymers called **protofilaments,** 13 of which align parallel to one another to form the wall of each microtubule. Each microtubule is polarized, with a plus (+) and minus (−) end. In vivo, they exist in a state of **dynamic instability,** undergoing abrupt changes in length through changes in the balance between polymerization and depolymerization.

    b. **Function.** Microtubules form a remarkable network of roadways in the cell, deploy cytoplasmic organelles (including the ER and Golgi complex), shuttle vesicles from one part of the cell to another, and move chromosomes during mitosis. Their instability is critical to their function. The drug known as **Taxol** interrupts their function by permanently stabilizing them and leads to cell death. Most microtubules anchor by their minus ends in γ-**tubulin rings** that act as nucleation sites in the **cytocenter,** a juxtanuclear region containing the centrioles. Microtubules extend from their nucleation sites by adding GTP-containing tubulin heterodimers to their plus end. The GTP is hydrolyzed by tubulin

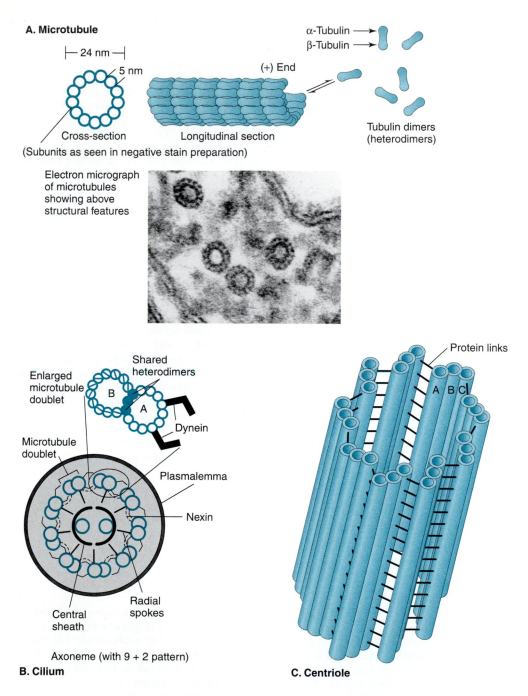

A. Microtubule

B. Cilium

C. Centriole

Figure 2–6. Schematic diagrams of microtubules and their contributions to cilia and centrioles. **A.** Microtubules as seen by the electron microscope. Cross-sections of tubules show a ring of 13 subunits of dimers arranged in a spiral. Changes in microtubule length are caused by the addition or loss of individual tubulin subunits. **B.** A cross-section through a cilium reveals a microtubule complex, or **axoneme**, at its core. An axoneme consists of two central microtubules surrounded by nine microtubule doublets (9 + 2). In the doublets, the A microtubule is complete and consists of 13 subunits, whereas the B microtubule shares two or three heterodimers with the A. When activated by ATP, the dynein arms (which harbor ATPase) link adjacent tubules and provide for the sliding of doublets against each other. **C.** Centrioles consist of nine microtubule triplets linked together in a pinwheel array. In the triplets, the A microtubule is complete and consists of 13 subunits, whereas the B and C microtubules share tubulin subunits. These organelles typically occur in pairs disposed at right angles to each other. (Reproduced, with permission, from Junqueira LC, Carneiro J, Kelley RO: *Basic Histology,* 9th ed. Stamford, CT: Appleton & Lange, 1998.)

to GDP shortly after heterodimer incorporation into a growing microtubule, causing destabilization, depolymerization, and retraction as the GDP-containing heterodimers are released. Rapidly growing tubules develop a **GTP cap** on their plus end, briefly protecting them from retraction, but will depolymerize if they fail to attach to an organelle or a stabilizing microtubule-associated protein **(MAP)** in time. Stabilized microtubules acquire other MAPs; large, ATPase-containing **molecular motors** (eg, **kinesin** and **dynein**) capable of binding cellular structures and "walking" them along the microtubules, provide an intracellular transport system. Kinesin carries its cargoes mainly toward the plus end, and dynein toward the minus end, of stabilized microtubules. Different motors appear to exist for each type of organelle or vesicle. Thus microtubule struts that spread the ER through the cytoplasm with the aid of kinesin, when exposed to **colchicine** (a drug causing net microtubule depolymerization), collapse, allowing the ER to collapse around the nucleus. Conversely, dynein-assisted aggregation of the Golgi complex toward the nucleus fails, causing Golgi dispersion, in the presence of the same drug.

   c. **Location.** Microtubules originate repeatedly from the cytocenter, growing outward through the cytoplasm and retracting if they fail to connect. Those that do connect stabilize and provide a latticework that supports organelle deployment and vesicle traffic throughout the cell. Stabilized microtubule arrays occur in cilia (III.J), flagella (III.K), basal bodies (III.L), centrioles (III.M), and the mitotic spindle apparatus (III.N).

2. **Microfilaments.**

   a. **Structure.** Microfilaments are the thinnest cytoskeletal elements (5- to 7-nm wide) and are more flexible than microtubules. They are filamentous polymers of one of several types of globular **actin** protein monomers. In striated muscle cells, actin filaments form a stable paracrystalline array in association with myosin filaments. Actin filaments in other cells are less stable and can dissociate and reassemble. These changes are regulated in part by calcium ions, cyclic adenosine monophosphate (AMP), and by a host of **actin-binding proteins** in the cytoplasm and attached to the plasma membrane's cytoplasmic surface. In addition to their effects on polymerization and depolymerization, actin-binding proteins arrange microfilaments into the networks and bundles that carry out many of their important functions.

   b. **Function.** Microfilaments are contractile, but to contract they must interact with myosin, the only actin-associated motor protein family. In muscle cells, myosin forms thick filaments. In nonmuscle cells, it exists in soluble form and binds to microfilaments by its globular head, leaving its tail (free end) to attach to the plasma membrane or other cellular components to move them. Each actin monomer harbors a molecule of ATP that promotes binding during polymerization but hydrolyzes to ADP shortly after binding to destabilize the microfilament, unless stabilizing actin–binding proteins are present. Treatment with **cytochalasins** disrupts microfilament organization and interferes with the following functions: endocytosis; exocytosis; formation and contraction of microvilli; cell movement; movement of organelles, vesicles, and granules; cytoplasmic streaming; maintenance of cell shape; and equatorial constriction of dividing cells. **Phalloidin** binds to intact microfilaments, stabilizes them, and interferes with these same functions. Because it fluoresces, phalloidin is often used to localize actin filaments.

   c. **Location.** In nonmuscle cells, microfilaments are distributed as an irregular mesh throughout the cytoplasm. Local accumulations occur as a thin sheath beneath the plasma membrane called the **terminal web;** as parallel strands in cores of microvilli; in the cytoplasm at the leading edge of pseudopods; in association with the plasma membrane, organelles, or other cytoplasmic components; or as a belt ("purse string") around the equator of dividing cells.

3. **Intermediate filaments.**

   a. **Structure.** Intermediate filaments are ropelike and composed of shorter threadlike protein subunits assembled and twisted around one another to form filaments of intermediate thickness (10–12 nm) between microtubules and microfilaments. Their protein subunits are globular at their amino and carboxy terminals, with an elongated, linear central domain. The individual proteins belong to the same family as nuclear lamins (3.II.B) and differ depending on the cell type. *Examples:* **cytokeratins** in epithelial cells, **vimentin** in mesenchymally derived cells (eg, fibroblasts, chondrocytes), **desmin** in muscle cells, **glial fibrillary acidic protein** in glial cells, and **neurofilaments** (intermediate filament bundles) in neurons. The stability and longevity of these proteins, together with their cell-type specificity, make them particularly useful in determining the origin of neoplastic cells.

    b. Function. Intermediate filaments are notable for their tensile strength and durability. Their abundance in cells subjected to mechanical stress (eg, cells of skin, connective tissue, and muscle) indicates that they play a role in stabilizing cell structure and in the many functions that depend on maintaining cell shape.

    c. Location. In most cells, intermediate filaments form a network surrounding the nucleus and extend throughout the cytoplasm. Their ordered arrangement in certain cells (eg, neurons and keratinocytes of the skin) reflects their special role in maintaining cell shape. Cytokeratin-containing tonofilaments of desmosomes (4.IV.B.3) are a good example of such arrangement.

J. Cilia: A ciliated cell typically has hundreds of cilia, which are motile, 5- to 10-μm long, 0.25-μm wide, cell-surface evaginations covered by plasma membrane. Each contains a core, or **axoneme,** composed of nine peripheral microtubule **doublets** surrounding a pair of unjoined microtubules (the **"9 + 2"** arrangement). The peripheral doublets consist of a full A microtubule and a partial B microtubule (Fig. 2–6, B). Attached to each A microtubule are two **dynein arms,** which interact with the B microtubule in the adjacent doublet and drive the sliding of the doublets past one another. Other proteins that crosslink the doublets prevent simple sliding and convert the motion into bending. Ciliary movement occurs in two phases: a forward power stroke, in which the distal part of the cilium remains straight and rigid, and a return or recovery stroke, in which the cilium is more flexible and bent.

K. Flagella: A flagellum is similar to a cilium but it is longer and typically only one or two are present in a cell. In mammals, flagella occur in the tails of spermatozoa, which typically are 50- to 55-μm long and 0.2- to 0.5-μm thick along most of their length. The axoneme of a flagellum is identical to that of a cilium but is separated from the surrounding plasma membrane by large protein masses. Flagellar movement resembles a turning corkscrew more than the whiplike action of the cilia.

L. Basal Bodies: In cells bearing cilia or flagella, centrioles (III.M) migrate to the apical plasma membrane and give rise to basal bodies as in centriole self-duplication. Basal bodies are structurally similar to centrioles (Fig. 2–6, C), with nine microtubule triplets. They are located in the cytoplasm—one at the base of each cilium or flagellum—and serve as the anchoring points and microtubule organizers for these structures.

M. Centrioles:

    1. Structure. A centriole is a cylinder of microtubules, 150 nm in overall diameter and 350- to 500-nm long, containing nine microtubule **triplets** in a pinwheel array (Fig. 2–6, C). Each microtubule in a triplet shares a portion of its neighbor's wall. An interphase (nondividing) cell has a pair of adjacent centrioles with perpendicular long axes, each surrounded by several electron-dense satellites, or **pericentriolar bodies** containing γ-**tubulin.** Cytoplasmic microtubules radiate from the pericentriolar bodies into the cytoplasm.

    2. Function. Centrioles are the cell's structural organizers. Centriole duplication is required for cell division. During mitosis, the centrioles organize the mitotic spindle (III.N). Even in vitro, isolated centrioles control microtubule polymerization; in the cell, centrioles transmit physical organizing forces by means of the microtubules radiating from the pericentriolar bodies. Through their effects on microtubules, centrioles control organelle, vesicle, and granule traffic within the cell. Centrioles also give rise to basal bodies (III.L). However, centrioles are not nucleation sites for cytoplasmic microtubules; the γ-tubulin rings in the pericentriolar bodies serve this function.

    3. Location. Between cell divisions, centrioles lie near the nucleus, often surrounded by Golgi complexes. The centrioles and associated Golgi complexes constitute the **centrosome** (cytocenter), which appears as a clear juxtanuclear zone. During the S phase of interphase (Fig. 3–2), each centriole duplicates, forming a **procentriole** perpendicular to the original. During mitosis, new centriole pairs migrate to opposite cell poles to organize the spindle.

N. Mitotic Spindle Apparatus: In preparation for mitosis, cytoplasmic microtubules depolymerize and repolymerize as the mitotic spindle apparatus. This spindle-shaped microtubule array occurs between two centriole pairs at opposite poles of mitotic cells (3.VI.A.2). Some spindle microtubules (continuous fibers) extend from centriole to centriole. Others (chromosomal fibers) extend from one centriole to the centromere of a chromosome at the equatorial plate. The spindle apparatus is crucial for chromosome separation during mitosis.

IV. CELL FUNCTIONS

Cells are both empowered and constrained by their available resources. The amounts and types of energy and raw materials at their disposal, the information encoded in their genes, and intrinsic and extrinsic factors controlling access to that information are major determinants of cell function. Cells in tissues undergoing growth or repair use more of their resources preparing for, and carrying out, cell division. Fully differentiated cells typically concentrate on more specialized functions, such as secretion and contraction. Maintaining a constant internal environment (**homeostasis**), even in apparently quiescent cells, requires the expenditure of significant amounts of energy and other resources.

A. **Cellular Differentiation:** Refinements in cell structure and function accompany embryonic and fetal development, as well as maturation and aging. This cellular differentiation generally results in a cell's dividing less often and having fewer, but more efficient, capabilities than an embryonic cell. The functions of a differentiated cell can be roughly gauged by the organelles it contains. For example, cells specialized for protein secretion contain abundant RER and a well-developed Golgi complex. Although differentiation can result in dramatic changes, it does not occur suddenly. It occurs in a series of steps, often separated by one or more passes through the cell cycle, and involves interactions among the cell's environment, the metabolic machinery in its cytoplasm, and the information in its DNA. Among the more obvious changes taking place during differentiation are the often dramatic changes in the genes being expressed.

B. **Intercellular Communication:** Tissues, organs, and organ systems are collections of cells and cell products that act in concert to carry out their complex functions. The embryonic cells that ultimately form a tissue develop communication strategies early in embryogenesis. Many types of intercellular communication occur—some direct and some indirect.
   1. **Direct communication.** During cell-to-cell recognition or contact inhibition of cell division, signal transmission may require temporary physical contact. In some tissues, especially in epithelia, cells have more permanent direct contact with their neighbors over large areas of their surface membranes. These areas of contact are often marked by specialized plasma membrane structures called junctional complexes (4.IV.B). Some components of junctional complexes are specialized for attachment (physical communication), and others (gap junctions) provide cytoplasmic channels for the transmission of electric and chemical signals.
   2. **Indirect communication.** Signals also can be transmitted from one cell to another even when the cells are not in contact. In proximal communication, the signal traverses a short distance (eg, hormones, growth factors, or other signal molecules may be produced by one cell type and have effects on another cell type in the same tissue). In distal communication, the signal travels farther (eg, when hormone-producing cells in one tissue elicit responses from targets in different tissue).

C. **Cellular Adhesion:** Many cell functions, especially those involving cell shape and tissue integrity, depend directly or indirectly on cell adhesion. Cell-to-cell adhesion, especially in epithelia, requires linkages between the cytoskeletons of neighboring cells. Intercellular binding is mediated by transmembrane proteins called **cadherins.** The intracellular domains of cadherins are linked to the cytoskeleton by special adaptor proteins, which may form plaques on the cytoplasmic surface of the membrane. Cell–substrate adhesion involves linkages between the cytoskeleton and collagen fibers of the extracellular matrix. The transmembrane proteins in these attachments are called **integrins.** In these junctions, intracellular adaptor proteins attach integrins to the cytoskeleton and extracellular adaptor proteins (eg, laminin and fibronectin) attach integrins to the collagen fibers. Specific examples of cell adhesion are presented in subsequent chapters (4.IV.B,C,E; 5.II.A.1.b and D.2).

MULTIPLE-CHOICE QUESTIONS

Select the single best answer.

2.1. Which of the following best describes the appearance of a unit membrane under a transmission electron microscope?
(A) Junctional complex
(B) Lipid bilayer
(C) Pentalaminar structure
(D) Porous structure
(E) Trilaminar structure

2.2. Which of the following are composed primarily of actin or actin-like proteins?
(A) Basal bodies
(B) Cilia
(C) Cytochalasins
(D) Intermediate filaments
(E) Microfilaments
(F) Microtubules
(G) Molecular motors

2.3. In rat hepatocytes, the crystalline nucleoid (dense core) of peroxisomes is believed to be composed of which of the following substances?
(A) Acid phosphatase
(B) Catalase
(C) D-Amino acid oxidase
(D) Divalent cations
(E) Hydrogen peroxide
(F) Lipofuscin
(G) Urate oxidase

2.4. The synthesis of all proteins appears to be initiated on which of the following cellular components?

(A) Free polyribosomes
(B) Golgi complex
(C) Nucleosomes
(D) Ribophorin
(E) RER
(F) Signal recognition particles
(G) Unit membranes

2.5. Which of the following structures are organelles formed by the fusion of primary lysosomes and phagosomes and also sites of active digestion of phagocytosed materials?
(A) Autophagosomes
(B) Heterophagosomes
(C) Secondary lysosomes
(D) Polysomes
(E) Nucleosomes
(F) Residual bodies
(G) Transfer vesicles

2.6. Which of the following structures is indicated by the arrows in Figure 2–7?
(A) Condensing vacuole
(B) Mitochondrion
(C) Peroxisome
(D) Primary lysosome
(E) Ribosome
(F) Secondary lysosome
(G) Transfer vesicle

2.7. The predominant structure shown in Figure 2–8 is which of the following?
(A) Glycogen granules
(B) Golgi complex
(C) Junctional complex
(D) Microtubules

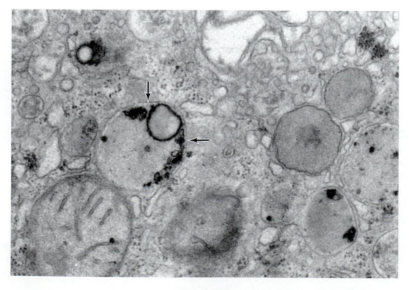

Figure 2–7.

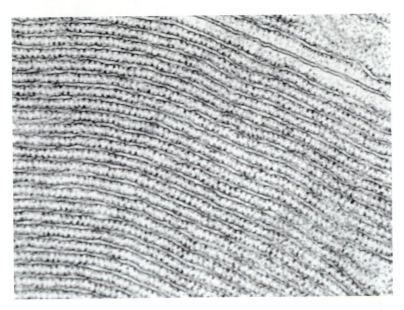

Figure 2–8.

(**E**) Plasma membranes
(**F**) RER
(**G**) SER

2.8. Which of the following structures is indicated by the arrow in Figure 2–9?
(**A**) Autophagosome
(**B**) Condensing vacuole
(**C**) Heterophagosome
(**D**) Nucleosome
(**E**) Peroxisome
(**F**) Primary lysosome
(**G**) Transfer vesicle

2.9. Which of the following pairs of functions is most closely associated with the Golgi complex?
(**A**) Energy metabolism and glycogen synthesis

(**B**) Energy metabolism and lipid metabolism
(**C**) Glycogen synthesis and packaging of secretions
(**D**) Glycosylation and sulfation of secretory products
(**E**) Lipid metabolism and concentration
(**F**) Phagocytosis and receptor recycling
(**G**) Protein synthesis and packaging of secretory products

2.10. Which of the following pairs of cellular structures regulate the distribution of membranous vesicles and organelles through their influence on microtubule polymerization?
(**A**) Centrioles and satellite bodies
(**B**) Chromosomes and mitotic spindles
(**C**) Cilia and flagella
(**D**) Golgi apparatus and endosomes
(**E**) Kinesin and dynein
(**F**) Laminal and intermediate filaments
(**G**) Nuclear envelope and RER

2.11. Which of the following descriptions best characterizes a phagosome?
(**A**) Allows extracellular materials to enter cells without endocytosis
(**B**) Contains densely packed, inactive hydrolytic enzymes
(**C**) Contains aging organelles
(**D**) Forms by budding from a lysosome
(**E**) Is surrounded by membrane derived from the Golgi complex
(**F**) Provides amino acids for protein synthesis by fusing with the RER
(**G**) Transports hormones and growth factors to the nucleus

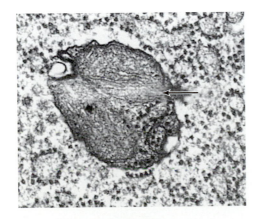

Figure 2–9.

2.12. Which of the following is the most characteristic type of intermediate filament protein found in epithelial cells?
(A) Actin
(B) Cytokeratin
(C) Desmin
(D) Glial fibrillary acidic protein
(E) Integrin
(F) Laminin
(G) Vimentin

2.13. Which of the following is the location of Krebs cycle enzymes and mitochondrial DNA?
(A) Cristae
(B) F1 subunits
(C) Inner mitochondrial membrane
(D) Intracristal space
(E) Matrix granules
(F) Mitochondrial matrix
(G) Outer mitochondrial membrane

2.14. Which of the following is the location of the electron-transport system?
(A) Cristae
(B) F1 subunits
(C) Inner mitochondrial membrane
(D) Intracristal space
(E) Matrix granules
(F) Mitochondrial matrix
(G) Outer mitochondrial membrane

2.15. Which of the following is the site of core glycosylation of secretory proteins?
(A) Golgi complex
(B) Free polyribosomes
(C) Mitochondria
(D) RER
(E) SER

2.16. Which of the following is the site of steroid hormone synthesis?
(A) Golgi complex
(B) Free polyribosomes
(C) Mitochondria
(D) RER
(E) SER

2.17. Which of the following is the site of actin and tubulin synthesis?
(A) Centrioles
(B) Free polyribosomes
(C) Mitochondria
(D) RER
(E) SER

2.18. Which of the following form the purse-string constriction around the equator of mitotic cells?
(A) Centrioles
(B) Intermediate filaments
(C) Microfilaments
(D) Microtubules
(E) Protofilaments

2.19. Which of the following are important components of axonemes?
(A) Centrioles
(B) Basal bodies
(C) Intermediate filaments
(D) Microfilaments
(E) Microtubules

2.20. Which of the following organelles shows the highest concentration of label after subjection to enzyme histochemical methods designed to detect the location of acid phosphatase?
(A) *Cis* Golgi complex
(B) Lysosomes
(C) Mitochondria
(D) Peroxisomes
(E) RER
(F) SER
(G) *Trans* Golgi complex

2.21. Which of the following substances characteristically increases in abundance with increasing age in terminally differentiated cells such as neurons and muscle?
(A) Cyclin
(B) Desmin
(C) Lipofuscin
(D) Osmium
(E) Phalloidin
(F) Ribophorin
(G) Urate oxidase

2.22. A malignant cell of mesenchymal origin contains cytoplasmic filaments composed of vimentin. Which of the following indicates the average diameter of these filaments?
(A) 5–7 nm
(B) 10–12 nm
(C) 24 nm
(D) 50–70 nm
(E) 100–120 nm
(F) 240 nm
(G) 500–700 nm

ANSWERS TO MULTIPLE-CHOICE QUESTIONS

2.1. E (II.A.1)

2.2. E (III.I.2.a)

2.3. G (III.G)

2.4. A (III.B.2. and C.1.b)

2.5. C (III.F.2)

2.6. F (III.F.2; note the partly digested material within)

2.7. F (III.C.1.a)

2.8. A (III.E; note the remnants of the mitochondrion)

2.9. D (III.D.2.a–e)

2.10. A (III.M.1 and 2)

2.11. C (III.E; refers to an *auto*phagosome)

2.12. B (III.I.3.a)

2.13. F (III.A.1.d; Fig. 2–5)

2.14. C (III.A.1.b; Fig. 2–5)

2.15. D (III.C.1.b)

2.16. E (III.C.2.b)

2.17. B (III.B.2)

2.18. C (III.I.2.b and c)

2.19. D (III.J and K; Fig. 2–6)

2.20. B (III.F)

2.21. C (III.F.3)

2.22. B (III.I.3.a)

3

The Nucleus & Cell Cycle

OBJECTIVES

This chapter should help the student to:

- Know the names and functions of the nuclear components.
- Know the subunits of each nuclear component and their roles in its function.
- Explain the process of cell division and its effects on cell morphology.
- Identify the factors and activities controlling the transition from each cell-cycle phase to the next.
- Recognize a cell's nuclear components in a light or electron photomicrograph and hence predict the cell's relative activity.
- Predict a cell's nuclear morphology from its functional characteristics.
- Predict the functional deficit(s) that accompany(ies) specific nuclear or chromosomal aberrations.
- Predict the nuclear component(s) likely to be involved in a functional deficit.
- Explain the role of the nucleus in cell differentiation.

MAX-Yield™ STUDY QUESTIONS

1. List the four major structural components of a nucleus (I.A[1]).
2. Why is the term "nuclear envelope" more appropriate than "nuclear membrane" (II)?
3. List the substances and structures associated with the internal and external surfaces of the nuclear envelope (II.A and B).
4. Name the components of a nuclear pore complex and indicate which component is responsible for anchoring the complex in the envelope and which components form the walls of the complex as it penetrates the envelope (II.C).
5. List several important macromolecules that must traverse the nuclear pores for basic cell functions to be carried out (II.C).
6. Compare euchromatin and heterochromatin (III.B and C) in terms of their appearance in light and electron microscopy, degree of coiling, and involvement in transcriptional activity.
7. List the components of a nucleosome (III.A).
8. List the parts of a nucleolus (IV.A, B.1 and 2).
9. What is the major function of the nucleolus (IV)?
10. In which cell types would you expect to find particularly large or abundant nucleoli (IV)?
11. List in order the phases of mitosis and sketch the appearance and location of the chromosomes during each phase (VI.A.2.a–d).
12. Describe what happens to each of the following during mitosis and indicate the phase(s) during which each change occurs (VI.A.2.a–d):
    - **a.** Nucleolus
    - **b.** Nuclear envelope
    - **c.** Nuclear lamins
    - **d.** Centrioles
    - **e.** Spindle apparatus
    - **f.** Golgi apparatus
    - **g.** Chromatin
    - **h.** Cytoplasmic microtubules
13. Give examples of tissues characterized by a high mitotic rate and of those characterized by a low mitotic rate (VI.B.1).
14. List in order the phases of interphase (VI.B.1–3; Fig. 3–2) and indicate which is associated with:
    - **a.** Most protein and RNA synthesis
    - **b.** Restoration of cell volume

---

[1] See footnote on page 1.

    c. Exit from the cell cycle and entrance into G_0
    d. DNA synthesis and replication
    e. Duplication of the centrioles
    f. Accumulating energy (ATP) for mitosis
15. Diagram the phases of the cell cycle (Fig. 3–2). Include molecules that affect transitions from phase to phase and indicate the points in the cycle at which they act (VI.B.4 and 5.a–c).
16. Beginning with transcription in the nucleus, trace the steps in the synthesis and secretion of a glycoprotein and relate the steps to the organelles involved. (This exercise helps to integrate the information in Chapters 2 and 3.)

SYNOPSIS

I. GENERAL FEATURES OF NUCLEI

    A. Major Components: Each nucleus has a **nuclear envelope, chromatin,** one to several **nucleoli,** and a variable amount of **nucleoplasm.**

    B. Nuclear Function: The nucleus contains in its chromatin a linear code (DNA) for the synthesis of cell components and products, which confers on the cell a range of adaptability to changing environmental conditions and to extrinsic signals such as hormones. It is also indispensable for cellular reproduction, which occurs through a succession of nuclear and cytoplasmic changes comprising the cell-division cycle (or cell cycle). This cycle ends with mitosis (cell division), which produces two identical daughter cells from a single parent. Continuous cycling occurs only with the presence and activity of regulatory molecules that allow progress beyond particular checkpoints.

    C. Nuclear Morphology: Nuclei vary in appearance from tissue to tissue, from cell to cell, and during different phases of the cell cycle. Although some mature cell types (eg, erythrocytes) lack nuclei, at least one nucleus is present during at least one stage in all eukaryotic cells. The microscopic appearance of the nucleus is important in identifying and classifying both normal and diseased cells and tissues. Nuclei display wide variations in:
    1. Nuclear size. The size of the nucleus varies, both in absolute terms and relative to the amount of cytoplasm in the cell (nucleocytoplasmic ratio).
    2. Number per cell. Cells may be enucleate, mononucleate, binucleate, or multinucleate.
    3. Chromatin pattern. The amount and distribution of heterochromatin varies according to cell type and cellular activity.
    4. Nuclear location. The relative position of a nucleus within a cell varies according to cell type and may be basal, central, or eccentric.

II. NUCLEAR ENVELOPE

The nuclear contents are set apart from the cytoplasm by a double membrane called the nuclear envelope and a narrow (40 to 70 nm) intermembrane space called the **perinuclear cisterna,** or perinuclear space.

    A. External Surface: The nuclear envelope's outer surface often is peppered with ribosomes and shows occasional continuities with the RER.

    B. Internal Surface: The inside of the inner membrane is lined with a **fibrous lamina,** which consists of proteins called **lamins** (Fig. 3–1). Cells contain as many as five distinct lamin proteins of two types, A or B, which are structurally related to the cytoplasmic intermediate filament proteins (eg, vimentin and desmin). Lamins form networks that organize the nuclear envelope as well as interactions between the envelope and the chromatin (III); thus lamins may regulate the size, shape, and chromatin pattern of the nucleus and may affect transcription. Associations between lamins and other nuclear components are regulated by enzymatic phospho-

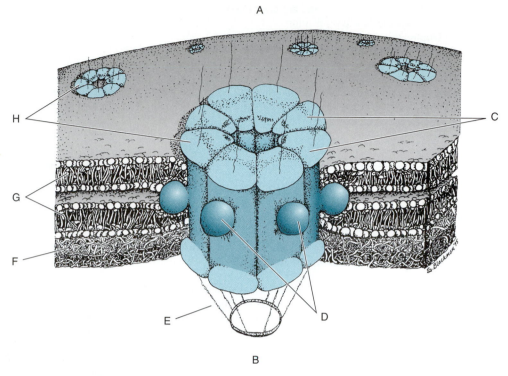

Figure 3–1. Schematic diagram of a nuclear pore complex in the nuclear envelope. Labeled components include the cytoplasm (A), nucleoplasm (B), nucleoporins (C), gp210 (D), nuclear cage (E), nuclear lamina (F), double membranes of the nuclear envelope (G), and nuclear pore complexes (H).

rylation and dephosphorylation of the lamin proteins. For example, the phosphorylation of lamins by the M phase–promoting factor (MPF) causes the nuclear lamina, and thus the nuclear envelope, to disassemble (VI.B.5.b). This disassembly is a prerequisite for mitosis (VI.A.1).

C. **Nuclear Pores:**   The envelope is perforated by many **nuclear pore complexes** (**NPCs;** see Fig. 3–1), each with a diameter of approximately 70 nm. Each NPC is anchored in the nuclear envelope by the glycoprotein **gp210** and is composed of several glycoproteins called **nucleoporins.** Their arrangement gives the NPC its characteristic octagonal shape and allows it to monitor and regulate the passage of substances into (eg, nucleic acids, polymerases, histones, and lamins) and out of (eg, mRNA, tRNA, and ribosomal subunits) the nucleus. Proteins destined to enter the nucleus have a signal sequence called the **nuclear localization signal,** which must bind to a **nuclear import receptor** protein in the cytoplasm to be recognized and allowed to pass through the nuclear pore. This gate-keeping function of the NPC is GTP- and temperature-dependent and can be blocked by the binding of **wheat-germ agglutinin** (a lectin) to the O-linked oligosaccharides of the nucleoporins.

III. CHROMATIN

Nuclear chromatin is intensely basophilic and consists of DNA and associated histone and nonhistone proteins.

A. **Nucleosomes:**   Isolated chromatin appears in electron micrographs as thin strands studded with beadlike particles at regular intervals. Each strand is a double-helical DNA molecule, and the particles are the repeating structural subunits of chromatin, termed **nucleosomes.** Each nucleosome consists of 146 base pairs of the DNA strand coiled around a core of eight **histones** (two copies each of H2A, H2B, H3, and H4). The **linker region** of the strand between two nucleosomes contains an additional 48 base pairs. Another histone (usually H1) binds to the nucleosome surface and to the linker. The H1 histone appears to help the beaded strand coil into a 30-nm thick superhelix with six nucleosomes per turn (a **selenoid**). Further coiling of the selenoid is required to form the condensed form of chromatin (ie, heterochromatin).

B. Chromatin Types: Nuclei containing **heterochromatin** stain darkly with basic dyes. Because the DNA of chromatin must uncoil to be transcribed, dark-staining (heterochromatic) nuclei reflect less DNA transcription activity and the use of less of their total genome. Uncoiled chromatin, termed **euchromatin,** stains poorly and is difficult to distinguish even by electron microscopy (EM). Large, pale-staining (euchromatic) nuclei usually indicate more transcriptional activity and continuous cell division.

C. Chromatin Pattern: The amount and distribution of nuclear chromatin are often used to identify cell types, especially in cells with no characteristic cytoplasmic staining properties. Even in mostly euchromatic nuclei, a rim of heterochromatin is often found on the inner surface of the nuclear envelope associated with the fibrous lamina (II.B). This envelope-associated heterochromatin allows the nuclear boundary to be seen with the light microscope.

D. Chromosomes: Chromosomes are the most condensed form of chromatin and are visible during mitosis. To form chromosomes, selenoids coil further and wind on a central nonhistone protein scaffold. Of the 46 chromosomes present in human cells, 44 (the somatic chromosomes) occur in 22 structurally similar (homologous) pairs. The other pair (sex chromosomes) consists of dissimilar chromosomes (XY) in males and similar chromosomes (XX) in females. In females, only one X chromosome (either of the two) is used by each cell; the inactive X is often visible as a clump of heterochromatin, termed **sex chromatin,** or the **Barr body.** In most cells, the Barr body attaches to the nuclear envelope's inner surface. In a neutrophilic leukocyte, it may appear as a drumstick-shaped appendage of the lobulated nucleus.

E. Karyotyping: A cell's karyotype is its chromosome inventory or an image of its chromosomes arranged by type. Preparing such an image is called karyotyping. Cells in culture are stimulated to enter mitosis with phytohemagglutinin (a plant-derived mitogen). The dividing cells are treated with colchicine to arrest them in metaphase, when the chromosomes are highly coiled and visible. Lysing the cells with a hypotonic solution causes the chromosomes to spread on the slide with little or no overlapping. The chromosome spread is imaged, and chromosome images are selected, paired, and arranged in a specific sequence. Karyotyping allows chromosome cataloging to detect structural abnormalities and deleted or excess chromosomes.

IV. NUCLEOLUS

During interphase (between mitoses), each nucleus typically contains a basophilic body called a nucleolus. Nucleoli synthesize most ribosomal RNA (rRNA). They are usually distinguishable from heterochromatin but may be obscured in very dark nuclei. They are larger and more numerous in embryonic cells, in cells actively synthesizing proteins, and in rapidly growing malignant tumor cells. Some heterochromatin attaches to the nucleolus; the significance of this **nucleolus-associated chromatin** is unknown. The nucleolus disappears in preparation for mitosis and reappears after mitosis is completed. Distinct nucleolar components can be seen with the electron microscope.

A. Pars Amorpha: This pale-staining nucleolar region contains the **nucleolar organizer DNA,** which carries the code for rRNA. In humans, five chromosome pairs have nucleolar organizer regions; thus 10 nucleoli per cell are possible, but fusion of the organizers into fewer, larger nucleoli is more common. Newly synthesized rRNA first appears in this region.

B. Nucleolonema: This light microscopy term refers to a threadlike basophilic substructure of the nucleolus. It contains two rRNA-rich components distinguishable by EM.
  1. The **pars fibrosa** consists of densely packed ribonucleoprotein fibers that are 5 to 10 nm in diameter. These fibers consist of the newly synthesized primary transcripts of the rRNA genes and associated proteins imported from the cytoplasm. Newly synthesized rRNA makes its second appearance in this region.

   2. The **pars granulosa** contains dense granules, 15 to 20 nm in diameter, that represent matur-
      ing ribosomal subunits undergoing assembly for export to the cytoplasm. Newly synthesized
      rRNA makes its third appearance in this region.

V. NUCLEOPLASM

The nucleoplasm is the matrix in which other intranuclear components are embedded. It consists of
enzymatic and nonenzymatic proteins, metabolites, ions, and water. It includes the **nuclear matrix,**
a fibrillar "nucleoskeletal" structure that appears to bind some hormone receptors, and newly synthesized DNA.

VI. NUCLEAR FUNCTIONS

   A. **Cellular Reproduction:**   The reproductive cycle of a cell is termed the **cell cycle** (Fig. 3–2).
      Each complete cycle ends with cell division **(mitosis)** and yields two daughter cells.
      1. **Mitosis and interphase.** Early views of cellular reproduction focused on easily detected
         structural changes that occur during mitosis. The apparently inactive phase between succes-
         sive mitoses seemed a resting period and was dubbed the **interphase.** Yet, even in rapidly
         dividing cells, the duration of mitosis is brief compared with the length of interphase. Cur-

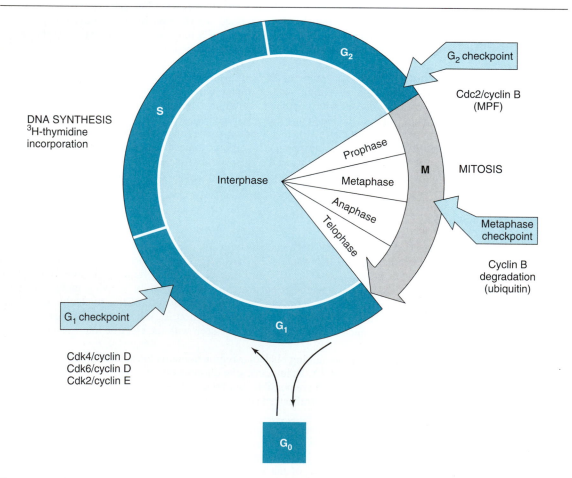

Figure 3–2. Cell cycle of a generalized vertebrate cell. The positions of major checkpoints and the proteins associated
with each of these are indicated.

rently we know that cells carry out important activities during interphase, including those needed to recover from the previous mitosis and to prepare for the next. Both mitosis and interphase currently are viewed as complex and important cell-cycle components and each has been divided into steps to facilitate our understanding.

2. **Steps in cell division (mitosis).** Mitosis is a brief, continuous process. Structural changes observed during this complex process have been used to divide it into four successive phases: prophase, metaphase, anaphase, and telophase.

   a. During **prophase,** chromatin coils to form chromosomes. As the nucleolar organizer DNA coils into its respective chromosomes, the nucleoli disintegrate. The nuclear membrane remains intact. The two centriole pairs migrate to opposite poles of the cell, cytoplasmic microtubules depolymerize, and the mitotic spindle apparatus (2.III.N) begins to assemble between the centriole pairs. Microtubule polymerization, in preparation for the formation of the spindle apparatus, causes the ER and Golgi complex to disintegrate into a multitude of vesicles (2.III.I.1.b).

   b. During **metaphase,** lamin phosphorylation promotes nuclear envelope disintegration. Chromosomes line up at the cell equator between the centriole pairs, and each chromosome splits lengthwise to form a pair of **sister chromatids.** Each chromosome has a **centromere** (late-replicating DNA, or **kinetochore**) to which microtubules of the spindle apparatus attach.

   c. During **anaphase,** replication of kinetochore DNA allows the sister chromatids to separate and move to opposite poles of the now-elliptical cell along the mitotic spindle. The centromere leads, with the chromatin dragging behind, often in a V shape. Chromatid translocation is not fully understood but involves molecular motors (2.III.I.1.b) on the spindle microtubules.

   d. During **telophase,** the chromosomes begin to uncoil. Nucleoli and nuclear envelopes reappear as components of two separate nuclei at opposite ends of the cell. Nuclear envelope reassembly involves the dephosphorylation of nuclear lamins. A **purse-string constriction,** formed by bands of microfilaments beneath the plasma membrane, appears at the equator. Tightening of the constriction eventually divides the cytoplasm and organelles between the daughter cells, a process termed **cytokinesis,** which signals the end of mitosis. After cytokinesis is completed, the spindle apparatus depolymerizes and repolymerizes as the interphase microtubule network. This allows vesicles derived from the Golgi apparatus and ER (VI.A.2.a) to be reassembled as functional organelles in the daughter cells (2.III.I.1.b).

B. **The Cell Cycle:** The cell-cycle model (see Fig. 3–2) accounts for important but less visible changes in the cell between divisions. It retains the four phases of mitosis but focuses on the timing of DNA synthesis and divides interphase into three phases: G_1, S, and G_2.

   1. The G_1 (gap 1) phase of interphase follows the telophase of mitosis. A gap is a period during which no DNA synthesis occurs, as indicated by the fact that no radiolabeled thymidine (^{3}H-thymidine) is incorporated into the cell's DNA. RNA and protein syntheses do occur during the gap phases, and each daughter cell grows to the parent's size. G_1, typically the longest phase of the cycle, is also the most variable in length among different cell types. In rapidly dividing (eg, embryonic and neoplastic) cells, G_1 is short and the transition to subsequent phases is continuous (VI.B.5.a). Cells that are more differentiated may withdraw from the cycle in G_1 and enter a phase called G_0, in which preparations for mitosis are suspended in favor of specialized functions. G_0 cells unable to reenter the cycle (eg, muscle, nerve) are said to be **terminally differentiated.** Other cells in G_0 (eg, hepatocytes, fibroblasts) can reenter the cell cycle in response to growth factors encountered during or after an injury.

   2. During the **S (synthesis)** phase, DNA synthesis and replication occur, as indicated by ^{3}H-thymidine uptake. The centrioles often self-duplicate during this stage.

   3. During G_2 (gap 2), the final preparations for cell division occur; these include repair of damaged DNA, synthesis of tubulin for the spindle apparatus, and ATP accumulation for the energy-expensive mitosis. Very little synthesis occurs during mitosis.

   4. **Proteins controlling cell-cycle progression.** Transitions between cell-cycle phases required for continuous cycling occur only under specific conditions involving the actions of permissive or inhibitory proteins, many of which are highly conserved evolutionarily. Hence, some mammalian cell-cycle–regulating proteins are quite capable of regulating cell-

cycle progression in more primitive cells, such as yeast. **Cyclin-dependent kinases (Cdks)** are enzymes (eg, Cdc2, Cdk2, Cdk4, Cdk6) that phosphorylate serine and threonine residues on other proteins, initiating or blocking activities crucial to cell-cycle progression. Cdks are active only when bound to certain **cyclins** (eg, A, B, C, D, and E) to form **Cdk–cyclin complexes.** During continuous cycling, intracellular Cdk concentrations tend to be constant, whereas cyclins are so named because their concentrations rise and fall as the cell progresses through the cycle. **Growth factors** in a cell's environment can induce cyclin and Cdk synthesis and thus reentry from G_0 into active cycling. **Cyclin-dependent kinase inhibitors** are protein families (eg, Kip/Cip and INK4) whose members are capable of binding to Cdk–cyclin complexes and inhibiting their activity.

5. **Cell-cycle checkpoints.** Certain "checkpoints" in the cycle are susceptible to control by Cdk–cyclin complexes. To progress past these points, enough Cdk–cyclin complexes must be present to overcome inhibition; otherwise, progression may be halted either briefly or for extended periods. Not surprisingly, some aspects of cell-cycle control involve feedback from specific steps in the cycle itself. In this way, progression occurs only when conditions indicate that the previous phase has been successfully completed. If mitosis is entered before DNA replication is completed, the daughter cells are subject to destruction or serious genetic damage. Other aspects of cell-cycle control involve signals from the cell's environment. Excessive growth, even when sufficient nutrients and oxygen are present, may compromise tissue function and be detrimental to the entire organism. Hence, there are two main checkpoints in the mammalian cell cycle: one in G_1, controlling the onset of DNA synthesis (S phase), and one in G_2, controlling the onset of mitosis (M phase). Each checkpoint requires the presence of a different class of cyclin (ie, G_1 cyclins or mitotic cyclins). Presumably, the different complexes phosphorylate different target proteins. In yeast, the same Cdk functions at both checkpoints, requiring only a different cyclin. In mammalian cells, there are multiple Cdk proteins that function at different checkpoints.

a. **G_1 checkpoint (Enter S! or Start!).** For any cell to divide, it must first replicate its DNA. The command to initiate DNA synthesis involves D-type cyclins (D1, D2, and D3) and cyclin E. D-type cyclins typically associate with and activate Cdk4 and Cdk6. The main partner for cyclin E is Cdk2. Progression through the G_1 checkpoint can be inhibited by the association of Kip/Cip proteins with either the D– or E–Cdk complexes. INK4 inhibitors act by selectively binding D–Cdk4 or D–Cdk6 complexes. A well-studied cell-cycle control mechanism involves phosphorylation of the retinoblastoma gene product **(Rb)** by D–Cdk complexes. Active (nonphosphorylated) Rb inhibits the transition from G_1 to S by binding proteins that would otherwise permit progression through this checkpoint. D–Cdk-mediated phosphorylation inactivates Rb, releasing (and activating) the regulatory proteins previously sequestered by Rb and thereby allowing progression to S. Another control mechanism operating at this checkpoint involves the protein **p53,** which accumulates in response to DNA damage. At sufficient concentrations, p53 signals excessive damage to the genetic material and halts the cell cycle in G_1. Interference with inhibition at this checkpoint (through its many regulatory mechanisms) is a common feature of malignant transformation, which allows damaged or mutated cells to reproduce. The complexity of mechanisms controlling this checkpoint may, in part, account for the fact that G_1 is the longest and most variable in duration of cell-cycle phases, even in normal cells. It is also at this checkpoint that cells in G_0 appear to reenter the cycle.

b. **G_2 checkpoint (Enter M!).** Owing to earlier and more extensive studies, G_2 checkpoint regulation is better understood. Early work showed that the command to enter mitosis required a factor **(MPF)** that accumulates in G_2. When added to nondividing cells, MPF allows them to enter mitosis. MPF purification revealed two basic subunits: a Cdk (specifically, **Cdc2**) and a mitotic cyclin. As previously mentioned, cyclins were so named because their concentration rises and falls in cells as they cycle. **Cyclin B** is the main mitotic cyclin. Its levels rise in G_2 and it associates with Cdc2 to form MPF. In addition to the need to accumulate cyclin B, formation of active MPF appears to be delayed by an inhibitory signal from incompletely replicated DNA. This prevents premature entry into mitosis. MPF activation is initiated by phosphorylation and dephosphorylation reactions (eg, by proteins such as **Wee1, MO15,** and **Cdc25**). The apparent ability of activated MPF to activate more MPF accounts for the abrupt entry into mitosis after the gradual accumulation of cyclin B during G_2. Among the key mitotic events carried out by active MPF is the phosphorylation of lamin proteins required for the dissolution of the nuclear envelope (II.B). MPF also phosphorylates the histone H1 (III.A) and thereby may be involved in chromosome supercoiling.

 c. **Metaphase checkpoint (Exit M!).** As important as ensuring that mitosis does not begin until DNA replication is complete, is exiting the mitotic process after it finishes. This involves proteolytic **cyclin degradation.** Cyclins are targeted for degradation by being tagged with **ubiquitin.** Ubiquitinated proteins are selectively degraded by large cytoplasmic enzyme complexes termed **proteasomes.** Cyclin degradation inactivates MPF and also inactivates G_1 Cdk-cyclin complexes.

C. Cellular Differentiation: From the nuclear perspective, differentiation involves shifts or restrictions (or both) of the genes expressed within the cell, in response to both internal and external environmental changes and signals. The fact that cellular differentiation often requires several passes through the cell cycle reflects the molecular complexity involved in repressing the expression of some genes while promoting the expression of others. Activation of repressed genes buried in heterochromatin often must await the uncoiling that accompanies DNA replication during the S phase of the cell cycle. The expression of differentiated functions may be delayed until a cell withdraws from continuous cycling (eg, into G_0), allowing it to devote energy previously needed for the energy-expensive mitosis to specialized functions such as the synthesis of "luxury" (as opposed to maintenance) proteins.

MULTIPLE-CHOICE QUESTIONS

Select the single best answer.

3.1. Because of the frequent presence of ribosomes on its outer surface, the nuclear envelope may be considered a specialized portion of which organelle?
(A) Golgi complex
(B) Nucleolus
(C) Plasma membrane
(D) RER
(E) SER

3.2. Which of the following cellular components retains the greatest structural integrity throughout the cell cycle?
(A) Centriole
(B) Golgi apparatus
(C) Nuclear envelope
(D) Nucleolus
(E) Spindle apparatus

3.3. Which of the following cellular substances is capable of carrying out its cellular functions without ever passing through a nuclear pore?
(A) Histones
(B) Lamin
(C) Messenger RNA (mRNA)
(D) Ribosomal proteins
(E) Ribosomal subunits
(F) RNA polymerase
(G) Tubulin

3.4. Which of the following terms is applied to the smallest chromatin subunit in eukaryotic cells?
(A) Chromosome
(B) Euchromatin
(C) Genes
(D) Heterochromatin
(E) Histone
(F) Nucleosome
(G) Selenoid

3.5. Which of the following proteins is responsible for anchoring the nuclear pore complex in the nuclear envelope?
(A) gp210
(B) Lamin A
(C) Lamin B
(D) Nucleoporins
(E) Wheat-germ agglutinin

3.6. Which of the following histones is typically associated with the linker region of a nucleosome?
(A) H1
(B) H2A
(C) H2B
(D) H3
(E) H4

3.7. Which of the following is the longest and most variable cell-cycle phase in continuously dividing cells?
(A) G_0
(B) G_1
(C) G_2
(D) M
(E) S

3.8. How many somatic chromosomes are in a normal human karyotype?
   (A) 22
   (B) 23
   (C) 24
   (D) 44
   (E) 46

3.9. Which of the following is true of nucleolar organizer DNA?
   (A) Is identical to nucleolus-associated chromatin
   (B) Carries the code for ribosomal RNA
   (C) Is located in the pars granulosa of the nucleolus
   (D) Encodes nuclear lamins
   (E) Is lost during mitosis

3.10. Which of the following is true of the kinetochore?
   (A) Is composed largely of RNA
   (B) Replicates during M phase rather than during S phase
   (C) Is the site at which chromosomes attach to microfilaments
   (D) Is the part of each chromosome farthest from the centriole during anaphase

3.11. Which of the following may be caused by a mutation in the retinoblastoma protein (Rb)?
   (A) Accumulation of p53 in malignant cells
   (B) Repression of Rb's ability to block cell-cycle progression
   (C) Prevention of Rb's ability to phosphorylate MPF
   (D) Accumulation of gp210 in malignant cells
   (E) Inhibition of Cdk production

3.12. Which of the following specifically promotes passage through the G_1 checkpoint of the cell cycle?
   (A) Activated Rb
   (B) Cyclin D
   (C) p53
   (D) MPF
   (E) Ubiquitin

3.13. Which of the following proteins accumulates in cells in response to damage to DNA?
   (A) Cdc2
   (B) Cyclin B
   (C) MPF
   (D) p53
   (E) Rb

3.14. A continuously cycling malignant cell is most likely to be characterized by which of the following?
   (A) Absence of nucleoli in euchromatic nuclei
   (B) Constitutive Cdk4 activity
   (C) Extended G_0 phase
   (D) Heterochromatic nuclei
   (E) Overexpression of p53

3.15. Which of the following predictably increases in concentration during G_2 and is degraded in M?
   (A) Lamin A
   (B) Cyclin B
   (C) Cdc2
   (D) Cyclin D
   (E) Cyclin E

3.16. The cell shown in Figure 3–3 is in which phase of cell division?
   (A) Anaphase
   (B) Interphase
   (C) Metaphase
   (D) Prophase
   (E) Telophase

3.17. Ribosomal RNA (rRNA) is synthesized by which of the following?
   (A) Free polyribosomes
   (B) Nuclear envelope
   (C) Nucleolus
   (D) Phagosomes
   (E) Secondary lysosomes
   (F) RER
   (G) SER

3.18. Ribosomes are not found in mature red blood cells. From this fact and from knowledge of ribosome function and distribution which of the following can be deduced?
   (A) Mature red blood cells are incapable of protein synthesis
   (B) Red blood cells lack nucleoli throughout their life cycle
   (C) Mature red blood cells lack cytoplasmic enzymes
   (D) Mature red blood cells contain inactive RER
   (E) Mature red blood cells contain smooth ER

3.19. Which of the following proteins possesses a nuclear localization signal?
   (A) Adenylate cyclase
   (B) Cadherin
   (C) Integrin
   (D) Kinesin
   (E) Lamin
   (F) t-SNARE
   (G) Ribophorin

Figure 3–3.

3.20. Which of the following is true of the Barr body?
  (**A**) Attaches to the inner surface of the plasma membrane during interphase
  (**B**) Is composed of euchromatin during the mitotic phase of the cell cycle
  (**C**) Its DNA replicates during every other cell cycle in normal males
  (**D**) In normal females, it is a member of a homologous chromosome pair
  (**E**) Occurs when homologous chromosomes fail to separate during anaphase

ANSWERS TO MULTIPLE-CHOICE QUESTIONS

3.1. D (II.A)
3.2. A (VI.A.2.a–d; 2.III.N)
3.3. G (II.C; 2.III.B.2 and I.1)
3.4. F (III.A)
3.5. A (II.C; Fig. 3–1)
3.6. A (III.A)
3.7. B (VI.B.1 and 5.a)
3.8. E (III.D and E)
3.9. B (IV.A and VI.A.2.a)
3.10. B (VI.A.2.b and c)

3.11. B (VI.B.5.a)
3.12. B (VI.B.5.a; Fig. 3–2)
3.13. D (VI.B.5.a)
3.14. B (IV and VI.B.5.a)
3.15. B (VI.B.5.b; Fig. 3–2)
3.16. A (VI.A.2.c)
3.17. C (IV)
3.18. A (IV; 2.III.B.2; 13.VI.A and B.1–5)
3.19. E (II.B and C)
3.20. D (III.D)

Part II: The Four Basic Tissue Types

Epithelial Tissue

4

OBJECTIVES

This chapter should help the student to:

- Know the four basic tissue types.
- Know the structural and functional characteristics that distinguish epithelial tissues from the three other basic tissue types.
- Know the types of epithelial tissues and give examples of sites where each may be found.
- Know the functional capabilities of each epithelial tissue type and relate them to tissue structure.
- Describe the specialized functions of epithelial cell types and give examples of sites where each may be found.
- Recognize epithelia in micrographs and predict their function from their structure and location.
- Know the criteria used to classify glands.
- Know the classes of glands in humans and give examples of sites where each may be found.
- Recognize glands in micrographs or diagrams and identify gland types.

MAX-Yield™ STUDY QUESTIONS

1. List the principal functions of epithelial tissues (II.A[1]).
2. From which embryonic germ layer(s) are epithelial tissues derived? Give examples of epithelia derived from each (II.H; Table 4–1).
3. List the structural and functional characteristics of epithelial tissues that distinguish them from other tissue types. Consider cell polarity (IV); specializations of the apical (IV.A), lateral (IV.B), and basal (IV.C) surfaces; nutrition (II.F); and mitotic rate (II.E).
4. Describe the basal lamina in terms of location, composition, and staining (IV.C.1.a).
5. Which structures and molecules help attach epithelial cells to their basal laminae (IV.C.1.a and 2) and to each other (IV.B.2)?
6. Compare basal laminae and basement membranes (IV.C.1.a).
7. Name four types of junctions found between epithelial cells (IV.B).
8. Which junction(s) named in the answer to question 7 is (are) associated with:
   a. A disklike structure (IV.B.3 and 4)?
   b. A bandlike structure (IV.B.1 and 2)?
   c. Plasma membrane fusion (IV.B.1)?
   d. A sealing effect (IV.B.1)?
   e. Cytokeratins (IV.B.2 and 3)?
   f. Attachment plaques (IV.B.2 and 3)?
   g. Connexons (IV.B.4)?
   h. Integrins (IV.C.1.b and 2)?
   i. Cadherins (IV.B.2 and 3)?
   j. Microfilaments (IV.B.2 and C.1.b)?
   k. Intermediate filaments (IV.B.3)?
   l. Alpha-actinin (IV.B.2)?
9. Which cellular junction is more important in cell-to-cell communication than in cell-to-cell adhesion (IV.B.4)?
10. Compare microvilli, stereocilia, cilia, and flagella (IV.A.1–4) in terms of:
    a. Width and length
    b. The presence of a plasma membrane covering

---

[1] See footnote on page 1.

    c. The presence of microtubules or microfilaments
    d. Motility
    e. The presence of axonemes and basal bodies
    f. Function and location in the body
11. List the types of simple and stratified epithelia and give examples of their locations (III.B.1–8).
12. Compare endocrine and exocrine glands in terms of their embryonic origins and how their products are transported (V.A; Table 4–3).
13. What structural criteria are used to classify exocrine glands (Table 4–5)?
14. Compare the merocrine, holocrine, and apocrine modes of secretion in terms of the part of the cell released, and give an example of each type of gland (Table 4–4).
15. Name the structural modifications and staining properties of epithelial cell types specialized for the following and give examples of each type:
    a. Transport of ions and water (VI.A.1)
    b. Synthesis and secretion of proteins (VI.C.1)
    c. Synthesis and secretion of mucus (VI.C.3)
    d. Synthesis and secretion of steroids (VI.C.5)
16. Compare paracrine (VI.C.2) and endocrine (V.A) cells in terms of the distances their products travel to reach their targets. Give examples of each type of cell.
17. Describe DNES (APUD) cells in terms of what the acronyms DNES and APUD stand for, embryonic origin, structure, secretory product, and distribution (VI.C.2).
18. Give examples of DNES (APUD) cells, and name the substance each produces (VI.C.2).
19. Compare serous and mucous cells in terms of appearance and secretory product (VI.C.3 and 4).

SYNOPSIS

I. THE FOUR BASIC TISSUE TYPES

A tissue is a complex assemblage of cells and cell products that have a common function. The many body tissues are grouped according to their cells and cell products into four basic types: **epithelial, connective, muscular,** and **nervous.**

II. GENERAL FEATURES OF EPITHELIAL TISSUES

Epithelial tissues often are structurally minor but functionally important components of an organ. Glands derive from the invagination and ingrowth of lining epithelia into underlying connective tissue. Composed primarily of epithelial cells, glands are considered a type of epithelial tissue.

  A. **Diversity:** Epithelial tissues range from one to many cell layers and form sheets, solid organs, or glands. Their functions range from protection to secretion and absorption.

  B. **Metaplasia:** When exposed to chronic environmental changes, epithelia undergo metaplasia (ie, they change from one type to another).

  C. **Lining and Covering:** Epithelia cover or line all body surfaces and cavities except articular cartilage in joint cavities. Their function is analogous to that of cell membranes: they (1) separate self from nonself; (2) divide the body into functional compartments; and (3) form barriers that monitor, control, and modify substances that traverse them.

  D. **Basal Lamina:** Epithelia rest on an extracellular basal lamina (or basement membrane) that separates them from an underlying connective tissue layer called the lamina propria.

  E. **Renewal:** Epithelia are continuously renewed and replaced. Cells closest to the basal lamina undergo continuous mitosis, and their progeny replace the surface cells.

  F. **Avascularity:** Blood vessels in the subjacent connective tissue rarely penetrate the basal lamina to invade epithelia.

G. **Cell Packing:**   Epithelial tissues have little intercellular substance. The cells are densely packed, closely apposed, and joined by specialized junctions.

H. **Derivation:**   Ectoderm, mesoderm, and endoderm all give rise to epithelia (Table 4–1).

III. CLASSIFICATION OF EPITHELIAL TISSUES

A. **Classifying Criteria:**   Epithelia are classified by the number of their cell layers and the shape of their surface cells (Table 4–2).

B. **Specific Epithelial Types:**
   1. **Simple squamous epithelium** is a single layer of flat, platelike cells that functions as a semipermeable barrier between compartments. It lines blood vessels (endothelium) and body cavities (mesothelium) and forms the parietal layer of renal corpuscles.
   2. **Simple cuboidal epithelium** is a single layer of blocklike cells that forms the walls of secretory and excretory ducts and regulates ion and water concentration in some of these. It acts as a protective barrier in some locations. Specific examples include kidney tubules and the smaller (intercalary and intralobular) ducts of many glands. It also covers the ovary's free surface and the lens capsule's inner surface.
   3. **Simple columnar epithelium** is a single layer of roughly cylindric cells whose apical (free) surfaces may be covered with cilia or microvilli. It functions in secretion and absorption, and, when ciliated, in the propulsion of mucus. It often acts as a protective barrier. It lines the stomach, intestines, rectum, uterus, and oviducts, as well as the larger ducts of some glands and the papillary ducts of the kidneys.
   4. **Pseudostratified columnar epithelium** is a single layer of cells of variable shape and height, with nuclei at two or more levels. Cells that reach the surface often are ciliated. It forms a protective barrier and, when ciliated, moves surface mucus and trapped debris. **Ciliated pseudostratified columnar epithelium,** or respiratory epithelium, lines the larger diameter respiratory passageways. Pseudostratified columnar epithelium also lines parts of the male reproductive tract, where its apical surfaces often are covered with nonmotile stereocilia (IV.A.4).
   5. **Stratified squamous epithelium** occurs in two forms:
      a. The **keratinized** (cornified) type is a multilayered sheet of cells. The surface cells are squamous, dead, enucleated, and filled with the scleroprotein keratin. Deeper layers have polygonal cells in progressive stages of keratinization. The deepest layer has cuboidal to columnar cells and lies on the basal lamina. Keratinized stratified squamous epithelium is found in the skin (18.I.B.1) and forms a specialized barrier against friction, abrasion, infection, and water loss.

Table 4–1. Epithelial derivatives of embryonic germ layers.

Germ Layer of Origin	Epithelial Derivatives
Ectoderm	1. Skin (keratinized stratified squamous) 2. Sweat glands and ducts (simple and stratified cuboidal) 3. Lining of oral cavity and vaginal and anal canals (nonkeratinized stratified squamous)
Mesoderm	1. Endothelium lining blood vessels (simple squamous) 2. Mesothelium lining body cavities (simple squamous) 3. Linings of genital and urinary ducts and tubules (transitional, pseudostratified columnar, simple cuboidal, simple columnar—depending on location)
Endoderm	1. Lining of esophagus (nonkeratinized stratified squamous) 2. Lining of gastrointestinal tract (simple columnar) 3. Lining of gallbladder (simple columnar) 4. Solid glands such as the liver and pancreas 5. Lining of respiratory system (pseudostratified ciliated columnar → simple ciliated columnar → cuboidal → squamous)

Table 4–2. Terms for classifying covering epithelia.

Criterion	Term	Definitive Features
Number of cell layers	Simple Stratified Pseudostratified	Single cell layer Multiple cell layers Multiple layers of nuclei, but all cells contact basal lamina
Shape of surface cells	Squamous Cuboidal Columnar	Flat, platelike; cells much wider than they are tall Polygonal; cells approximately as wide as they are tall Polygonal; cells taller than they are wide

      b. The **nonkeratinized** (noncornified) type is similar but is thinner. Its surface cells are flat, nucleated, and nonkeratinized. As a protective barrier, nonkeratinized stratified squamous epithelium, also called mucous membrane, is less resistant to water loss than the keratinized type. It lines wet cavities subject to abrasion (eg, mouth, esophagus, vagina, anal canal, and vocal folds).

  6. Stratified cuboidal epithelium typically has two to three layers of cuboidal cells. It is relatively rare and lines the ducts of some glands (eg, salivary, sweat).

  7. Stratified columnar epithelium resembles stratified cuboidal epithelium, but its superficial cells are columnar and may be ciliated. Also rare, it lines the larger ducts of some glands, forms the conjunctiva, and occurs in small patches in some mucous membranes. It sometimes covers the respiratory surface of the epiglottis.

  8. Transitional epithelium is stratified and lines most urinary passages (eg, renal pelvis, ureters, bladder, proximal portions of the urethra). Its surface cells are large and often binucleated. When the bladder is empty, the surface cells appear domelike, giving the epithelium a "cobblestone" appearance; when the bladder is full, the surface cells stretch and flatten.

IV. POLARITY & SPECIALIZATIONS OF EPITHELIAL CELLS

Polarity (structural and functional asymmetry) is characteristic of most epithelial cells. It is most clearly seen in simple epithelia, where each cell has three types of surfaces: an apical (free) surface, lateral surfaces that abut neighboring cells, and a basal surface attached to the basal lamina.

  A. Specializations of the Apical Surface: The cell's apical surface is on the organ's external or internal (lumen) surface. It is specialized to carry out functions that occur at these interfaces, including secretion, absorption, and movement of luminal contents.

    1. Cilia (2.III.J), membrane-covered, cell-surface extensions typically occur in tufts or cover the entire apical surface. They beat in waves, often moving a surface coat of mucus and trapped materials. Ciliated epithelia include ciliated pseudostratified columnar (respiratory) epithelium and the ciliated simple columnar epithelium of the oviducts.

    2. Flagella (2.III.K) are also concerned with movement. Spermatozoa, derived from seminiferous epithelia, are the best examples of flagellated human cells.

    3. Microvilli are plasma membrane–covered cell-surface extensions. Their cores (unlike those of cilia and flagella) contain many parallel actin microfilaments, which are anchored in a dense mat of filaments in the apical cytoplasm called the **terminal web.** By interacting with myosin, the microfilaments contract, shortening the microvilli. An absorptive cell's apical surface is usually covered with microvilli, which greatly increase the surface area when extended. Microvillus–covered, epithelia, which exhibit a striated border, or brush border, include the absorptive simple columnar epithelium lining the small intestines and the absorptive simple cuboidal epithelium lining the kidney's proximal tubules.

    4. Stereocilia are not cilia but are very long microvilli. They are found in the male reproductive tract (epididymis, ductus deferens), where they have an absorptive function, and in the internal ear (hair cells of the maculae and organ of Corti), where they have a sensory function.

  B. Specializations of the Lateral Surfaces: Epithelial cells attach tightly to one another by specialized intercellular junctions. Junctions occur in three major forms: **zonulae** are bandlike and completely encircle the cell; **maculae** are disklike and attach two cells at a single spot; and **gap junctions** are macular in shape but differ in composition and function. A **junctional complex** is a combination of intercellular junctions, typically lying near the cell apex.

1. **Zonula occludens.** The zonula occludens (tight junction or occluding junction) lies near the cell apex and seals off the intercellular space, allowing the epithelium to isolate certain body compartments (eg, it helps keep intestinal bacteria and toxins out of the bloodstream). Its structure, most clearly seen in freeze-fracture preparations, resembles a quilt. The pattern results from the fusion of the two trilaminar plasma membranes of adjacent cells, which forms a pentalaminar structure (as seen in TEM); this fusion requires the attachment of specific integral membrane "tight-junction proteins" to those in adjacent membranes. In some tissues, tight junctions can be disrupted by calcium ion removal or protease treatment.

2. **Zonula adherens.** The zonula adherens (belt desmosome) is typically just basal to the tight junctions and helps maintain the seal. The adhering cells' membranes typically are 20 to 90 nm apart at a zonula adherens, where the gap may be wider than in nonjunctional areas. An electron-dense plaque containing myosin, tropomyosin, alpha-actinin, and vinculin lies on the participating membranes' cytoplasmic surfaces. Actin-containing microfilaments arising from each cell's terminal web insert into the plaques to stabilize the junction. The attachment between the cells is mediated by transmembrane proteins called **cadherins,** which are localized in the membranes between the plaques. The cadherins' intracellular domains insert into and bind to the plaque while their extracellular domains bind with one another in the gap between the cell membranes.

3. **Macula adherens.** A macula adherens, or **desmosome,** is a spot junction consisting of two dense, granular **attachment plaques** composed of several proteins (**desmoplakins**) on the cytoplasmic surfaces of opposing cell membranes. Transverse thin EM sections show dense arrays of **tonofilaments** (cytokeratin intermediate filaments) that insert into the plaques or make hairpin turns and return to the cytoplasm. As in the zonula adherens, cadherins insert into the cytoplasmic plaques and mediate the cell–cell attachment. However, the desmoplakins in the plaques of desmosomes attach to tonofilaments (cytokeratin intermediate filaments) instead of attaching to microfilaments. Thus the intermediate filaments of the adjacent cells are linked by desmosomes through a series of strong protein–protein interactions. The gap between the attached membranes can be greater than 30 nm. Often fibrillar or granular glycoproteins (**desmogleins**) form a dense central line in the intercellular space, stabilizing the interactions between the cadherins. Desmosomes, in patches on most epithelial cells' lateral membranes, form particularly stable attachments but do not prevent the flow of substances between cells.

4. **Gap junction.** A gap junction (nexus) is a disk-shaped structure that is best appreciated by viewing both freeze-fracture and transverse thin EM sections. The intercellular gap is 2-nm wide, and the membrane on each side contains a circular patch of **connexons.** Each connexon is a hexamer of multipass transmembrane proteins. Each has a central 1.5-nm hydrophilic pore. Connexons in one membrane link with those in the other to form continuous pores that bridge the intercellular gap, allowing small molecules (<800 daltons) to pass between cells. As sites of electronic coupling (reduced resistance to ion flow), gap junctions are important in intercellular communication and coordination; they are found in most tissues.

C. **Specializations of the Basal Surface:**   The basal surface contacts the basal lamina. Because it is the surface closest to the underlying blood supply, it often contains receptors for blood-borne factors such as hormones.

1. A **basal lamina** underlies all true epithelial tissues.

a. **Structure.** The basal lamina is a sheetlike structure, typically composed of **type-IV collagen, proteoglycan** (usually heparan sulfate), **laminin** (a glycoprotein that helps bind cells to the basal lamina), and **entactin** (a glycoprotein associated with laminin). The basal lamina exhibits electron-lucent and electron-dense layers termed the **lamina lucida** (lamina rara) and the **lamina densa,** respectively. The lamina densa is a 20- to 100-nm thick fibrillar network; the amount of lamina lucida is variable. Basal lamina components are contributed by the epithelial cells, the underlying connective tissue cells, and (in some locations) muscle, adipose, and Schwann cells. In some sites, a layer of type-III collagen fibers (reticular fibers), produced by the connective tissue cells and termed the **reticular lamina,** underlies the basal lamina. Basal laminae accompanied by reticular laminae are often thick enough to be seen with the light microscope as PAS-positive layers and are sometimes termed **basement membranes.**

    b. **Functions.** The basal lamina forms a sievelike barrier between the epithelium and connective tissue. It aids in tissue organization and helps maintain cell shape through cellular adhesion. Attachment between epithelial cells and their basal lamina is mediated by integral membrane proteins called **integrins.** Specifically, the cytoplasmic domains of integrins are linked to microfilaments by specific actin-binding proteins. Their extracellular domains bind laminin (IV.C.1.a); subsequently, laminin binds tightly to type-IV collagen, establishing a strong physical attachment between the cytoskeleton and the basal lamina.

  2. **Hemidesmosomes** are located on the inner surface of basal plasma membranes and are in contact with the basal lamina in epithelia exposed to extreme stress (eg, stratified squamous). In these strong attachments, integrins mediate a connection between the basal lamina and intermediate filaments of epithelial cells. In these junctions, the adaptor proteins connecting the integrins and the intermediate filaments form a plaque on the cytoplasmic surface of the basal plasma membrane.

  3. **Sodium–potassium ATPase** is a plasma membrane–bound enzyme localized preferentially in the basal and basolateral regions of epithelial cells. It transports sodium out of and potassium into the cell (VI.A.1).

D. Intracellular Polarity: The nucleus and organelles are often found in characteristic regions of epithelial cells, a feature particularly important to glandular cells. For example, in protein-secreting cells, the RER is preferentially located in the basal cytoplasm, the nucleus in the basal-to-middle region just above the RER, and the Golgi complex just above the nucleus. Mature secretory vesicles collect in the apical cytoplasm. The polarized deployment of organelles is regulated by microtubules and attached motor proteins.

E. Cellular Adhesion: If cell attachments (to one another and to the basal lamina) were dependent on the strength of molecular interactions in their plasma membranes, cells and tissues would be torn apart by most of the physical stresses they encounter. Thus interactions between adhesive molecules in the membrane and components of the cytoskeleton are indispensable in maintaining cell and tissue integrity. The common arrangement in adherens junctions, and in attachment to a basal lamina, is the presence of transmembrane adhesion molecules (eg, cadherins, integrins) in the plasma membrane, attaching to stable components of the cytoskeleton through specialized adaptor proteins. The extracellular domains of the cadherins bind to one another in adherens junctions, and those of the integrins bind laminin, and indirectly collagen, in the basal lamina. A similar pattern occurs in the connective tissues, where integrins mediate the connection between the cytoskeleton and collagen of the extracellular matrix by binding to the collagen-binding glycoprotein known as fibronectin (5.II.D.2).

V. GLANDS

Glands are single cells or groups of cells that are specialized for secretion.

A. Exocrine and Endocrine Glands: All glands arise in early development from lining or covering epithelia. Exocrine glands keep their connection with the epithelium in the form of a duct. Endocrine glands (ductless glands) lose their connection with the surface and release their secretions into the bloodstream. Endocrine glands are compared with exocrine glands in Table 4–3; they are described in greater detail in Chapters 20 and 21.

B. Classification of Exocrine Glands: Exocrine glands may be classified according to their structure, secretory product, or mode of secretion (Tables 4–4 and 4–5).

VI. MAJOR TYPES OF EPITHELIAL CELLS

A. Epithelial Cells Specialized for Transport:

  1. **Ion-transporting cells.** Some epithelial cells are specialized for **transcellular transport** (ie, they pump ions across their entire thickness). Sheets of such cells form active barriers to

Table 4–3. Some comparisons of exocrine and endocrine glands.

Category	Exocrine Glands	Endocrine Glands
Transport of secretions	**By ducts.**	**By bloodstream.** No ducts.
Cell number	**Unicellular** (eg, goblet cell) or **multicellular** (eg, salivary glands).	**Unicellular** (eg, DNES cells) or **multicellular** (eg, thyroid gland).
Secretory products	**Proteins** (eg, digestive enzymes), **glycoproteins** (eg, mucus), and some mixtures containing **lipids** (eg, sebum, bile, apocrine sweat, and milk).	**Hormones** of two main types: **peptide hormones** (eg, insulin) and **steroid hormones** (eg, adrenocorticoids). Include **plasma proteins** produced by liver (eg, serum albumin and clotting factors).
Mode of secretion	**Merocrine** (ie, by exocytosis, no loss of cytoplasm); **apocrine** (loss of apical cytoplasm); **holocrine** (entire cell released into duct).	**Merocrine only.**

control ion and water concentrations in body compartments. Tight junctions between the cells restrict backflow. Ion-transporting cells typically have highly infolded basal plasma membranes that interdigitate with numerous mitochondria. Commonly, the ion pump is specific for sodium (ie, it is Na^+/K^+-ATPase), and chloride ions and water follow the sodium passively. Some ion-transporting epithelia exploit this mechanism to concentrate solutes by moving water from one compartment to another. Important ion-transporting epithelia occur in kidney tubules, striated ducts of salivary glands, the gallbladder, the choroid plexus, and the eye's ciliary body.

2. **Cells that transport by pinocytosis.** Epithelial cells specialized for pinocytosis have tight junctions and abundant pinocytotic vesicles. The vesicles transport substances across the cell from luminal to basal surface or vice versa. The best examples are cells lining blood vessels, where transcellular transport is rapid (2–3 minutes).

Table 4–4. Terms for classifying exocrine glands (see Table 4–5 for applications).

Criterion	Term	Definitive Features
Structure	Unicellular	Single secretory cells scattered among other epithelial cells (eg, goblet cells)
	Multicellular	Solid glands; secretions carried by ducts to surface (eg, sweat and salivary glands)
	Simple	Duct is not branched
	Compound	Duct is branched
	Straight	Duct is straight
	Coiled	Duct or secretory component is coiled
	Tubular	Secretory portion is tubular (test tube–shaped; eg, gastric glands)
	Acinar	Secretory portion is rounded or flask-shaped (eg, pancreas)
	Tubuloacinar	Secretory portion has acini branching off a straight tubular portion (eg, sublingual gland)
Type of secretory product	Mucous	Form thick secretion (**mucus**) of highly glycosylated glycoproteins (**mucins;** eg, sublingual gland)
	Serous	Form thin, watery secretion containing proteins and glycoproteins (eg, parotid gland)
	Seromucous	Form mixed secretion of intermediate thickness (eg, submandibular gland)
Mode of secretion	Merocrine	Secretory product exits cell by exocytosis; no loss of cytoplasm (eg, pancreas)
	Apocrine	Secretory product collects in cell apex; entire cell apex is shed (eg, mammary gland)
	Holocrine	Secretory product fills cell; cell lyses, releasing product into duct (eg, sebaceous gland)

Table 4–5. Examples of structural classifications of multicellular exocrine glands.

Duct System	Secretory Portion	Example
Simple	Tubular	Intestinal crypts of Lieberkühn
Simple	Coiled tubular	Eccrine sweat glands of the skin
Simple	Branched tubular	Fundic glands of the stomach
Simple	Branched acinar	Sebaceous glands of the skin
Compound	Tubular	Cardiac glands of the stomach
Compound	Tubuloacinar	Submandibular salivary glands
Compound	Acinar	Exocrine pancreas

B. **Epithelial Cells Specialized for Absorption:**   Specialized absorptive cells lining the digestive tract have many apical microvilli that serve to increase their surface area. Small nutrient molecules diffuse into the microvilli, and microfilament contraction subsequently shortens the microvilli, bringing the nutrients into the cytoplasm. Other nutrients are pinocytosed between microvilli. Absorptive cells with similar specializations occur in the kidney's proximal tubules.

C. **Epithelial Cells Specialized for Secretion:**
   1. **Protein-secreting cells.** Cells synthesizing proteins for segregation and secretion have abundant basophilic RER and a well-developed Golgi complex, and often accumulate secretory granules in the cell apex. Proteins secreted by epithelial cells include the digestive enzymes, produced by pancreatic acinar cells and gastric chief cells; serum albumin, produced by hepatocytes; and protein hormones (eg, parathyroid hormone, which is produced by parathyroid chief cells).
   2. **Polypeptide-secreting cells.** Secreted polypeptides are smaller (fewer amino acids) than secreted proteins. Polypeptide-secreting cells have a small amount of RER, a supranuclear Golgi complex, and an accumulation of 100- to 400-nm secretory granules in their bases. These *a*mine *p*recursor *u*ptake and *d*ecarboxylation **(APUD) cells** concentrate important bioactive amines, such as epinephrine, norepinephrine, and serotonin, in their cytoplasm. They may absorb these amines from the bloodstream or synthesize them from amino acid precursors by means of abundant amino acid decarboxylases. Most APUD cells are unicellular glands scattered among other epithelial cells. They appear to derive mainly from the embryonic neural crest. The number, variety, and wide distribution of cells with these features gave rise to the concept of the **diffuse neuroendocrine system (DNES).** DNES and APUD refer to the same cells. Some APUD polypeptides have paracrine effects on neighboring cells; others enter the bloodstream and have endocrine effects on distant cells. Important APUD polypeptides include glucagon, from pancreatic islet A cells; insulin, from pancreatic islet B cells; gastrin, from the stomach, small intestine, and pancreatic islet G cells; and somatostatin, from the stomach, small intestine, and pancreatic islet D cells. Tumors composed of APUD cells are called apudomas.
   3. **Mucous cells** may be unicellular or may occur in sheets or as solid glands. Histologic characteristics include a light-staining, foamy appearance, owing to the large mucus-containing vesicles concentrated near the cell apex; PAS-positive staining from abundant oligosaccharide residues; predominantly acidophilic staining with H & E; a large supranuclear Golgi complex with distinctive glycosyltransferases; and nuclei and sparse RER in the cell's base.
   4. **Serous cells** are protein-secreting cells that usually are smaller, darker-staining, and more basophilic than mucus-secreting cells. They include pancreatic acinar cells and secretory cells of parotid salivary glands.
   5. **Steroid-secreting cells.** Endocrine cells specialized to secrete steroid hormones are polygonal or rounded, with a central nucleus and pale-staining, acidophilic cytoplasm containing lipid droplets. Their abundant SER contains enzymes for cholesterol synthesis and for converting steroid hormone precursors (eg, pregnenolone) into specific hormones (eg, androgens, estrogens, and progesterone). Their mitochondria have tubular rather than shelflike cristae and contain enzymes that convert cholesterol to pregnenolone. Steroid hormones include testosterone, produced by the testes' interstitial cells; estrogen, from ovarian follicle

cells; progesterone, from granulosa lutein cells of the corpus luteum; and cortisone and aldosterone, from cells of the adrenal cortex.

D. Contractile Epithelial Cells: Contractile epithelial cells, or myoepithelial cells, are stellate or spindle-shaped flattened cells, with fingerlike processes that embrace an acinus or duct. They lie between the epithelial cells and their basal lamina. Their cytoplasm contains abundant microfilaments, myosin, tropomyosin, and cytokeratin intermediate filaments. Several myoepithelial cells may surround a single acinus or duct; their contraction expels exocrine products. Gap junctions between myoepithelial and other cells facilitate synchronous contraction. Myoepithelial cells occur in lacrimal, salivary, mammary, and sweat glands and around the testes' seminiferous tubules.

MULTIPLE-CHOICE QUESTIONS

Select the single best answer.

4.1. Which of the following tissue types is shown in Figure 4–1?
   (**A**) Keratinized, stratified squamous epithelium
   (**B**) Nonkeratinized, stratified squamous epithelium
   (**C**) Pseudostratified ciliated columnar epithelium
   (**D**) Simple ciliated columnar epithelium
   (**E**) Simple ciliated cuboidal epithelium
   (**F**) Simple squamous epithelium
   (**G**) Transitional epithelium

4.2. Which of the following structures is marked with an asterisk (*) in Figure 4–2?
   (**A**) Basal lamina
   (**B**) Desmosome
   (**C**) Hemidesmosome
   (**D**) Gap junction
   (**E**) Squamous epithelium

(**F**) Zonula adherens
(**G**) Zonula occludens

4.3. Which of the following is a characteristic feature of pseudostratified columnar epithelium?
   (**A**) All nuclei lie at the same depth from the surface
   (**B**) All cells border on the lumen
   (**C**) All cells touch the basal lamina
   (**D**) Cells are always of ectodermal origin
   (**E**) All cells are ciliated

4.4. Which of the following is true of epithelial tissues?
   (**A**) Epithelia rest on basal laminae
   (**B**) Epithelia are incapable of metaplasia
   (**C**) Epithelia are highly vascularized
   (**D**) Epithelia derive only from ectoderm
   (**E**) Epithelial cells attach to one another by hemidesmosomes

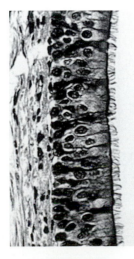

Figure 4–1.

Figure 4–2.

4.5. Which of the following is used in the naming of epithelial types?
- (**A**) Shape of cells in the basal layer
- (**B**) Number of layers of epithelial cells
- (**C**) Presence of a basal lamina
- (**D**) Size of the nuclei
- (**E**) Presence of cytokeratin

4.6. Which of the following is true of holocrine secretion?
- (**A**) Occurs in sebaceous glands
- (**B**) Occurs in endocrine glands
- (**C**) Involves little or no loss of cytoplasm
- (**D**) Involves secretion of chlorine and bromine ions
- (**E**) Is the typical mode of secretion of DNES cells

4.7. Which of the following is true of merocrine glands?
- (**A**) Do not include endocrine glands
- (**B**) Do not include exocrine glands
- (**C**) Secretory cells lack a well-developed Golgi apparatus
- (**D**) Secretory cells lose their apical cytoplasm during secretion
- (**E**) Secretory cells contain secretory granules

4.8. Which of the following is true of the zonula occludens?

- (**A**) Is characterized by the fusion of the outer leaflets of adjacent trilaminar unit membranes into a single pentalaminar unit
- (**B**) Is characterized by the presence of abundant cytokeratin filaments in the vicinity of the junction
- (**C**) Surrounds entire columnar cells in the basal region of their lateral plasma membranes
- (**D**) Is characterized by a dense intracellular plaque
- (**E**) Is characterized by the presence of connexons that link the plasma membranes of adjacent cells

4.9. Which of the following is true of stereocilia?
- (**A**) Are structurally similar to true cilia
- (**B**) Contain an axoneme
- (**C**) Are underlain by a basal body
- (**D**) Contain actin filaments in their core
- (**E**) Contain nine microtubule triplets at their core

4.10. Which of the diagrams in Figure 4–3 represents a simple branched tubular gland?

4.11. Which of the diagrams in Figure 4–3 represents a compound tubuloacinar gland?

4.12. Which of the diagrams in Figure 4–3 represents a compound acinar gland?

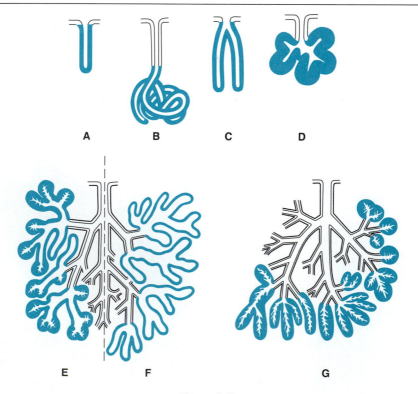

Figure 4–3.

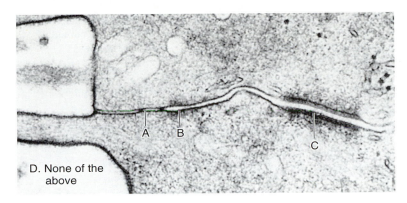

D. None of the
   above

Figure 4–4.

4.13. Which of the following cell types is best characterized as having abundant RER?
(A) Ion-transporting cells
(B) Mucus-secreting cells
(C) Peptide-secreting cells
(D) Protein-secreting cells
(E) Steroid-secreting cells

4.14. Which of the following cell types is best characterized as having abundant SER?
(A) Ion-transporting cells
(B) Mucus-secreting cells
(C) Peptide-secreting cells
(D) Protein-secreting cells
(E) Steroid-secreting cells

4.15. Which of the following often contain mitochondria with tubular cristae?
(A) Ion-transporting cells
(B) Mucus-secreting cells
(C) Peptide-secreting cells
(D) Protein-secreting cells
(E) Steroid-secreting cells

4.16. Which of the following contain abundant basal mitochondria between infoldings of the basal plasma membrane?
(A) Ion-transporting cells
(B) Mucus-secreting cells
(C) Peptide-secreting cells
(D) Protein-secreting cells
(E) Steroid-secreting cells

4.17. Which of the letters in Figure 4–4 represents a desmosome?

4.18. Which of the letters in Figure 4–4 represents a gap junction?

4.19. Which of the letters in Figure 4–4 represents a zonula adherens?

4.20. Which of the letters in Figure 4–4 represents a zonula occludens?

4.21. Which of the following is true of cilia?
(A) Are covered by plasma membrane

(B) Possess microfilaments in their cores
(C) Are anchored in the terminal web
(D) Are characteristically associated with absorptive epithelia
(E) Only one typically occurs per cell

4.22. Which of the following is a characteristic component of the basal lamina that binds directly to integrins in the epithelial cell's basal plasma membrane?
(A) Desmoglein
(B) Desmoplakin
(C) Heparan-sulfate proteoglycan
(D) Laminin
(E) Type-IV collagen

4.23. Which tissue type lines the ducts in Figure 4–5?
(A) Pseudostratified columnar
(B) Nonkeratinized stratified squamous
(C) Simple ciliated columnar
(D) Stratified cuboidal
(E) Transitional

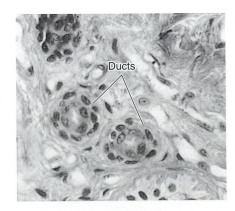

Figure 4–5.

4.24. Which tissue type lines most of the urinary passages and bladder?
 (A) Nonkeratinized stratified squamous
 (B) Pseudostratified columnar
 (C) Simple ciliated columnar
 (D) Stratified cuboidal
 (E) Transitional

ANSWERS TO MULTIPLE-CHOICE QUESTIONS

4.1. C (III.B.4; note the apical cilia and multiple layers of nuclei)
4.2. G (IV.B.1)
4.3. C (III.B.4)
4.4. A (II.A–H; Table 4–1)
4.5. B (III.A; Table 4–2)
4.6. A (Table 4–4)
4.7. E (Table 4–4)
4.8. A (IV.B.1, 2, and 4)
4.9. D (IV.A.3 and 4)
4.10. C (Table 4–4)
4.11. E (Table 4–4)
4.12. G (Table 4–4)

4.13. D (VI.C.1)
4.14. E (VI.C.5)
4.15. E (VI.C.5)
4.16. A (VI.A.1)
4.17. C (IV.B.3)
4.18. D (IV.B.4)
4.19. B (IV.B.2)
4.20. A (IV.B.1)
4.21. A (IV.A.1–3; VI.A.1)
4.22. D (IV.C.1.a and E)
4.23. D (III.B.6)
4.24. E (III.B.8; the lining comprises two layers [is stratified]; the top layer is cuboidal)

Connective Tissue

5

OBJECTIVES

This chapter should help the student to:

- List the structural and functional features of connective tissue that distinguish it from other basic tissues.
- Know the biochemical composition and sites of synthesis of the extracellular matrix components and how they associate with one another.
- Know the structure and function of the cell types found in connective tissue.
- Compare connective tissue types in terms of the types, amounts, and arrangement of their components.
- Relate the composition of each connective tissue type to its specific functions.
- Name body sites where each connective tissue type occurs, and relate the location of each type to its function.
- Recognize connective tissue cells and connective tissue types in micrographs of tissues or organs, and predict their functions.
- Predict the functional consequences of a given structural defect in a connective tissue.

MAX-Yield™ STUDY QUESTIONS

1. List the three major classes of connective tissue components (I.C[1]).
2. List the general functions of connective tissues (I.A; IV.A).
3. Name the germ layer(s) from which connective tissue cells are derived and the embryonic tissues composed of undifferentiated connective tissue cells (I.E; II.E.1.a; III.D).
4. List the two major classes of macromolecules that constitute ground substance (II.D).
5. List the glycosaminoglycans commonly found in ground substance (II.D.1).
6. Name the important structural glycoproteins found in connective tissue ground substance and describe their functions (II.D.2).
7. Give some common causes of edema and describe the effect of each on the pressures that act on water in capillaries (IV.B).
8. Name the three main connective tissue fiber types (II.A–C) and compare their:
   a. Protein composition (II.A.1.a, B, and C)
   b. Protein subunit arrangement (II.A.1.a, B, and C)
   c. Distinctive amino acids (II.A.1.a, B, C.1.a and b)
   d. Degree of glycosylation (II.B)
   e. Arrangement and appearance (II.A, B, C.1.b and 2)
   f. Diameter (II.A–C)
   g. Physical properties and function (II.A.2 and 4, B and C.3)
   h. Staining properties (II.A.3, B and C.2)
   i. Location (II.A.5, B, and C.4)
9. Name the major types of collagen (II.A.2) and compare them in terms of:
   a. Protein composition
   b. Tendency to form fibers or fibrils
   c. Tissue distribution
   d. The cell responsible for their synthesis

---

[1] See footnote on page 1.

10. Describe collagen synthesis and assembly, indicating the intracellular or extracellular location for each step (II.A.1.a and b). *Suggestion:* Diagram a cell with numbered steps.
11. Compare procollagen and tropocollagen in terms of structure and location (II.A.1.b).
12. Describe the roles of these enzymes in fiber synthesis, assembly, and turnover:
    a. Collagenase (IV.E)
    b. Elastase (II.C.1.b)
    c. Signal peptidase (2.III.C.1.b)
    d. Lysyl oxidase (II.A.1.b and C.1.b)
    e. Procollagen peptidase (II.A.1.b)
    f. Proline hydroxylase (II.A.1.a)
13. List the cell types found in connective tissues and indicate which type is most common (II.E).
14. Compare fibroblasts and fibrocytes (II.E.1.b) in terms of shape, nuclear morphology, and activity (mitotic and synthetic).
15. From which type of circulating blood cells are macrophages derived (II.E.2.b)?
16. Name the organelles that are abundant in macrophages and indicate the major function they provide (II.E.2.b).
17. Describe mast cells (II.E.2.a) in terms of:
    a. Shape and size
    b. Staining properties
    c. Granule contents
    d. Cause of degranulation
    e. Effects of degranulation
    f. Role in allergic reactions
18. Describe plasma cells (II.E.2.c) in terms of:
    a. Shape
    b. Staining properties
    c. Nuclear morphology
    d. Major cytoplasmic organelles
    e. Major secretory product
    f. Role in immunity
    g. Blood cell precursor
19. List the leukocyte types commonly found in connective tissue (II.E.2.d).
20. Name three types of connective tissue proper. Compare them in terms of function and location (III.A.1, 2.a and b).
21. Compare loose (areolar) and dense connective tissue (III.A.1 and 2) in terms of:
    a. Abundance
    b. Matrix composition
    c. Number of cells
    d. Flexibility and resistance to stress
    e. Collagen turnover rate (IV.E)
22. Compare dense regular and dense irregular connective tissues in terms of collagen bundle arrangement and location (III.A.2.a and b).
23. Describe reticular connective tissue in terms of its characteristic appearance, the organs in which it commonly occurs, its predominant cell type, and its function (III.B).
24. Describe elastic connective tissue in terms of its composition, primary cell type, and location (III.C).
25. Describe mucous connective tissue in terms of its primary matrix component, primary cell type, consistency, and location (III.D).
26. Discuss the active and passive roles of connective tissue in defending against pathogen invasion (IV.A.2.a and b).
27. Name two bacterial enzymes that digest specific matrix components (IV.A.2.a).
28. Which connective tissue cells contribute to wound repair by removing and replacing damaged tissue (IV.A.3)?
29. List the effects of the following on connective tissue structure and function (III.C and D):
    a. Hydrocortisone
    b. ACTH
    c. Hypothyroidism
    d. Ascorbic acid

SYNOPSIS

I. GENERAL FEATURES OF CONNECTIVE TISSUES

A. Functions: Connective tissue functions, determined chiefly by their mechanical properties, include binding together, compartmentalization, support, physical and immunologic protection, and storage (see also IV.A).

B. Types: The tissues described in this chapter (III) include loose and dense collagenous connective tissue (**connective tissue proper**), reticular connective tissue, elastic connective tissue, and mucous connective tissue. Adipose tissue, cartilage, and bone are specialized connective tissues discussed in Chapters 6, 7, and 8, respectively. Integrative multiple-choice questions pertaining to all connective tissues are included after Chapter 8. Blood, another specialized connective tissue, is discussed in Chapters 12 and 13.

C. **Three Fundamental Components:** Connective tissue types differ in appearance, but all consist of **cells, fibers,** and **ground substance.** Connective tissue types and subtypes are classified according to the amounts, types, and arrangement of these components.

D. **Extracellular Matrix:** Fibers and ground substance comprise the extracellular matrix. The abundant matrix of connective tissues largely determines their mechanical properties. The two fiber types are collagen and elastic. The ground substance, in which the fibers and cells are embedded, is composed mainly of glycosaminoglycans (GAGs) dissolved in tissue fluid. Matrix viscosity and rigidity are determined by the amount and types of cross links among the matrix components. Fiber and ground substance components are synthesized and secreted by connective tissue cells (mainly fibroblasts), and the fibers are assembled in the extracellular space.

E. **Embryonic Origin:** All connective tissue cell types derive from embryonic mesenchyme. Mesenchyme derives mainly from embryonic mesoderm. Craniofacial mesenchyme derives from the neural crest (mesectoderm).

II. COMPONENTS OF CONNECTIVE TISSUE

A. **Collagen Fibers:** Collagen is the body's most abundant protein. There are many types, some of which form fibers. Collagen fibers often form bundles, ranging from 0.5 to 15 μm in diameter.
1. **Synthesis and assembly.**
   a. **Intracellular steps.** Free polysomes reading collagen mRNA attach to the RER (2.III.B.2 and C.1.b), and **protocollagen** polypeptides are deposited in the cisternae. Each protocollagen, or **alpha chain,** has a mass of approximately 28 kDa and contains approximately 250 amino acids; every third amino acid is **glycine.** Proline and lysine residues in the chains are hydroxylated by proline and lysine hydroxylases (possibly in the SER) to form **hydroxyproline** and **hydroxylysine,** rare amino acids present in large amounts in collagen. Core sugars (galactose and glucose) attach to the hydroxylysine residues in the ER. With the aid of **registration peptides** at the ends of the alpha chains, three chains coil around one another to form a triple-helical molecule called **procollagen.** Further glycosylation may occur in the Golgi complex, where procollagen is packaged. Golgi vesicles release procollagen into the extracellular space by exocytosis.
   b. **Extracellular steps.** In the extracellular space, the enzyme **procollagen peptidase** cleaves the registration peptides from procollagen, converting it to **tropocollagen.** Nearby cells align tropocollagen molecules in a staggered fashion to form collagen fibrils, and also arrange fibrils into fibers. Cell attachments to the fibers are mediated by plasma membrane integrins that bind to the matrix glycoprotein fibronectin, which in turn binds to the collagen. The extracellular enzyme **lysyl oxidase** stabilizes the nascent fibers by cross-linking lysine and hydroxylysine residues in adjacent tropocollagens.
2. **Collagen types.** Of the 20 known collagen types, seven are of particular importance. Their structures differ in the amino acid sequences of their alpha chains. **Type I,** the most abundant and widespread, forms large fibers and fiber bundles. It occurs in tendons, ligaments, bone, dermis, organ capsules, and loose connective tissue. **Type II** occurs in adults only in the cartilage matrix (some occurs in embryonic notochord) and forms only thin fibrils. **Type III** resembles type I, but is more heavily glycosylated and stains with silver. Often occurring with type I, type III forms networks of thin fibrils (reticular fibers) that surround and support soft flexible tissues (adipocytes, smooth muscle cells, nerve fibers). It is the major fiber of hematopoietic tissues (eg, bone marrow, spleen) and of reticular laminae of epithelial basement membranes. **Type IV** is the major collagen type in basal laminae. It does not form fibers or fibrils. **Type V** occurs in placental basement membranes and blood vessels. **Type X** is found in the matrix surrounding hypertrophic chondrocytes of growth plate cartilage in sites of future bone formation [8.III.C.1.b.(1)(d)]. **Types IX and XI** occur in association with type II in cartilage.

3. **Histologic appearance.**
    a. **Light microscopy.** Collagen occurring in large or small bundles of fibrils or as individual fibrils stains pink in H & E–stained sections. In sections stained with Masson's trichrome, collagen fibers stain green. Thin fibers (eg, type III) stain darkly with silver stains, but thicker bundles do not. Collagen molecules that do not form fibers or fibrils (eg, type IV) are distinguishable from ground substance only by immunohistochemistry.
    b. **Electron microscopy.** All collagen fibrils and fibers have stripes at 64-nm intervals along their length. This periodicity reflects the staggering of tropocollagen molecules.
4. **Mechanical properties.** The most important mechanical property of collagen fibers is tensile strength, which is, weight for weight, greater than that of steel.
5. **Location.** Collagen fibers occur in all connective tissues and in reticular laminae of basement membranes. In bone, their lacunar regions (spaces between overlapping tropocollagen) act as nucleation sites for hydroxyapatite crystals.

B. **Reticular Fibers:** Reticular fibers resemble collagen fibers but are thinner (0.1–1.5 μm), are more highly glycosylated, and form delicate silver-staining networks instead of bundles. The supportive fiber networks allow motile cells to move about in loosely arranged (eg, hematopoietic) tissues. Reticular fibers consist mainly of type-III collagen and glycoproteins.

C. **Elastic Fibers:** Elastic fibers range in diameter from 0.1 to 10 μm. They consist of the amorphous protein **elastin,** in which are embedded many protein **microfibrils.**
1. **Synthesis and assembly.**
    a. **Intracellular steps.** Microfibrillar proteins and proelastin are synthesized on RER and secreted separately. Proelastin contains large amounts of three hydrophobic amino acids—glycine, proline, and valine—which accounts for elastin's insolubility. Microfibrillar protein contains mostly hydrophilic amino acids.
    b. **Extracellular steps.** Proelastin molecules polymerize extracellularly to form elastin chains. Lysyl oxidases subsequently catalyze the conversion of some elastin lysine residues to aldehydes, three of which condense with a fourth, unaltered lysine to form **desmosine** and **isodesmosine.** These amino acids, rare except in elastin, cross-link individual elastin chains. Elastin subsequently associates with many microfibrils to form a branching and anastomosing network of elastic fibers. Owing to elastin's unusual composition, its turnover requires the specialized enzyme **elastase.**
2. **Histologic appearance.** Because it has few charged amino acids, elastin stains poorly with standard ionic dyes. Special stains, such as Verhoeff's stain or Weigert's resorcin-fuchsin stain, are used in light microscopic preparations. In EM preparations, both the elastin and microfibrils can be visualized.
3. **Mechanical properties.** Elastic fibers can stretch to 150% of their length without breaking and return to their original length.
4. **Location.** Elastic fibers occur where their mechanical properties are needed to allow tissues to stretch or expand and return to their original shape (eg, in arterial walls, the lungs' interalveolar septa, bronchi and bronchioles, vocal ligaments and cartilages, and ligamenta flava of the vertebral column).

D. **Ground Substance:** Ground substance primarily consists of two glycoconjugate classes: proteoglycans and glycoproteins. Tissue fluids and salts also are present.
1. **Proteoglycans** consist of a **core protein** to which **glycosaminoglycans (GAGs)** are attached. GAGs of proteoglycans are straight-chain polymers of repeating amino sugar heterodimers made up of a hexosamine (glucosamine or galactosamine) and a uronic acid (glucuronic or iduronic acid). Five major classes of GAGs, differing in their sugars, exist in connective tissues: **hyaluronic acid** (which does not form proteoglycans), **chondroitin sulfate, dermatan sulfate, keratan sulfate,** and **heparan sulfate** (see also 7.II.A.1.b).
2. **Glycoproteins** are proteins to which shorter, branched oligosaccharide chains are covalently bound. Glycoproteins of ground substance are much smaller than proteoglycans. *Examples:* **fibronectin,** which mediates cell adhesion to the extracellular matrix, and **laminin,** a basal lamina component that mediates epithelial cell adhesion. For both of these adaptor proteins, linkages between the matrix and the cytoskeleton are mediated by integrins.

E. **Cells:** Connective tissue cells can be grouped into two classes: fixed and wandering.

1. **Fixed cells** are native to the tissue in which they are found.
   a. **Mesenchymal cells** are the precursors of connective tissue cells. Embryonic mesenchyme comprises a loose network of stellate cells and abundant intercellular fluid. Some mesenchymal cells remain undifferentiated in adult connective tissue and constitute a reserve population of stem cells called **adventitial cells,** which are difficult to distinguish from fibroblasts.
   b. **Fibroblasts** are the predominant cells in connective tissue proper. They synthesize, secrete, and maintain all major extracellular matrix components. Structurally, fibroblasts are of two types, one of which resembles mesenchymal cells. This type is stellate, with long cytoplasmic processes and a large, ovoid, pale-staining nucleus. The cells are mitotically active, with abundant RER and Golgi complexes. This cell type is important in producing collagen and other matrix components. Cells of the second type are less active and are termed **fibrocytes** because they are more mature. Fibrocytes are smaller and spindle-shaped, with a dark, elongated nucleus and fewer organelles. They may revert to the fibroblast state and participate in tissue repair.
   c. **Reticular cells** make up a functionally diverse yet morphologically similar group. They produce reticular fibers (II.B) that form the netlike stroma of hematopoietic, lymphoid, and adipose tissues (III.B). Some actively phagocytose antigenic material and cellular debris. Some are antigen-presenting cells, which process antigens and display them on their surfaces to help activate immunocompetent cells. Reticular cells are typically stellate with long, thin cytoplasmic processes. Each has a central, pale, irregularly rounded nucleus and a prominent nucleolus. In the cytoplasm, mitochondrial number and Golgi complex and RER development are variable. Some reticular cells, particularly those with less-developed organelles, may be stem cells of various blood types.
   d. **Adipose cells** or **adipocytes** are mesenchymal derivatives specialized for lipid storage (see Chapter 6).
2. **Wandering cells** are immigrant cells, usually from blood or bone marrow (Chapters 12–14). Some retain their original characteristics and may eventually leave the connective tissue; most differentiate and become permanent residents.
   a. **Mast cells.** These large (20- to 30-μm) cells derive from bone marrow precursors and have abundant basophilic cytoplasmic granules that are electron-dense in EMs. Mast cells also have many small plasma membrane folds and a well-developed Golgi complex. The granules, which often obscure the small central nucleus, contain **heparin, histamine,** and eosinophilic chemotactic factor of anaphylaxis (ECF-A). Mast cells have surface receptors for the IgE antibodies that trigger **degranulation,** the exocytosis of the granule contents that initiates the local inflammation of allergic reactions.
   b. **Macrophages** are large, stellate cells derived from monocytes that infiltrate connective tissue and develop into phagocytes. Resident macrophages can proliferate and form additional macrophages. Dye particles injected into the body are engulfed by these cells and accumulate in cytoplasmic granules. Otherwise, these cells are difficult to detect in H & E–stained sections. Macrophages contain many lysosomes, which aid in digesting phagocytosed materials, and a well-developed Golgi complex. They help maintain connective tissue integrity by removing foreign substances and cell debris, and they participate in the immune response by presenting phagocytosed antigens to lymphocytes. To remove large foreign objects, such as splinters, macrophages may fuse to form **multinuclear giant cells.** Monocyte-derived phagocytes, which together constitute the **mononuclear phagocyte system,** include the macrophages (lymphoid organs, lungs, serous cavities, and connective tissue), as well as Kupffer cells (liver), osteoclasts (bone), and microglial cells (central nervous system).
   c. **Plasma cells** differentiate from antigen-stimulated B lymphocytes. As the primary producers of circulating antibodies, they are the main effectors of the humoral immune response. They are sparsely distributed throughout the body but are abundant in areas susceptible to penetration by bacteria. Plasma cells are large and ovoid, with an eccentric nucleus and abundant RER. The characteristic "clock face" nucleus results from a large, central nucleolus and large heterochromatin clumps regularly spaced around the nuclear envelope's inner surface envelope. These cells often exhibit a clear juxtanuclear area (cytocenter) containing a well-developed Golgi complex and centrioles.
   d. **Other blood-derived connective tissue cells.** Many wandering cell types originate in the bone marrow and are carried to connective tissue by the blood and lymph. Blood-

derived cells found in connective tissues include the leukocytes (ie, white blood cells, including **lymphocytes, monocytes, neutrophils, eosinophils,** and **basophils**), which have roles in the immune response and are described in detail in Chapters 12 through 14.

III. CONNECTIVE TISSUE TYPES

A. **Connective Tissue Proper:** Connective tissue proper, found in most organs, is characterized by a predominance of fibers (mainly type-I collagen). Its varied functions chiefly relate to binding cells and tissues into organs and organ systems. Its subclasses are based on the type, density, and orientation of its fibers. (Cell and fiber types and ground substance composition are summarized in Table 5–1.)

1. **Loose connective tissue (areolar tissue)** appears disorganized. It consists of a loose network of different fiber types, on which many fixed and wandering cells are suspended. The abundant ground substance is only moderately viscous. This flexible yet delicate tissue surrounds and suspends vessels and nerves, underlies and supports most epithelia, and fills spaces between other tissues. It also supports the serous membranes (mesothelia) of the pleura, pericardium, and peritoneum. Always well-vascularized, areolar tissue conveys oxygen and nutrients to avascular epithelia. Its cells function in immune surveillance for foreign substances entering the body through the blood or epithelia.

2. **Dense connective tissue.** Collagen fibers (nearly all type I) predominate in dense connective tissue. The cells are mainly mature fibroblasts (fibrocytes). The ground substance is similar to that of areolar tissue but is less abundant (see Table 5–1). There are two types: **regular,** with a ropelike arrangement of fiber bundles, and **irregular,** with a fabriclike arrangement.

   a. **Dense regular connective tissue.** The fibers in this tissue are tightly packed into parallel bundles, between which are a few attenuated fibroblasts. The condensed, flattened fibroblast nuclei occur in rows between the fibers; the cytoplasm is virtually indistinguishable with the light microscope. There is little room for ground substance, which nevertheless permeates the tissue. The collagen fibers' tensile strength makes them ideal for transmitting mechanical force over long distances with a minimal use of material and space. This tissue transmits the force of muscle contraction, attaches bones, and protects other tissues and organs. It is found in tendons, ligaments, periosteum, perichondrium, deep fascia, and organ capsules.

   b. **Dense irregular connective tissue.** The components of this tissue are identical to those in dense regular connective tissue (see Table 5–1). Dense irregular connective tissue seems poorly organized, but the complex woven pattern of its collagen bundles resists tensile stress from any direction. It covers fragile tissues and organs, protecting them from multidirectional mechanical stresses. It occurs in the reticular layer of the dermis and in most organ capsules.

B. **Reticular Connective Tissue:** Reticular fibers (type-III collagen) form a delicate network on which cells (the predominant element) are suspended. Reticular cells attach to and cover the fibers with their long, thin processes. Other cells (eg, lymphocytes) are suspended in the network's spaces (see Table 5–1). There is little ground substance. Reticular connective tissue supports motile cells and filters body fluids. It occurs mainly in hematopoietic tissues, such as bone marrow, spleen, and lymph nodes.

C. **Elastic Connective Tissue:** In H & E–stained sections, elastic tissue resembles dense regular (collagenous) connective tissue. Fibers predominate and most are elastic. Elastic fibers are collected in thick, wavy, parallel bundles. The bundles are separated by loose collagenous tissue and fibroblasts with attenuated cytoplasm and condensed, oblong nuclei. Other connective tissue cells may be present in small numbers (see Table 5–1). The ground substance is sparse and similar to that of other dense connective tissues. Elastic connective tissue provides flexible support and predominates in the vertebral column's ligamenta flava and the suspensory ligament of the penis.

D. **Mucous Connective Tissue:** This tissue has few cells and fibers distributed randomly in the abundant ground substance, which has a syrupy to jellylike consistency and is composed

Table 5–1. Characteristics of connective tissue (CT) types.

Connective Tissue	Tissue Type	Cells	Fibers	Ground Substance	Organization	Functions	Locations
Embryonic CT	Mesenchyme	Mesenchymal cells.	Relatively few. Type-I and type-III collagen.	Watery. Mainly tissue fluid. Some glycoprotein and glycosaminoglycan.	Loose array of stellate cells. Large intercellular spaces filled with ground substance.	Embryonic connective tissue. Forms all connective tissue listed below.	Throughout vertebrate embryos. Much in head and neck derives from neural crest, rest from mesoderm.
	Mucous	Mesenchymal cells and fibroblasts.	Small number. Mainly collagen. Few elastic and reticular.	Syrupy to jellylike. Mainly hyaluronic acid and glycoproteins.	Cells and fibers distributed randomly in abundant ground substance.	Forms elastic cushion to protect nearby structures from pressure.	Wharton's jelly of umbilical cord. Pulp of developing teeth. Nucleus pulposis of intervertebral disks.
CT Proper	Loose (Areolar)	Fibroblasts, fibrocytes, mesenchymal cells, mast cells, macrophages, adipocytes, plasma cells, and leukocytes, (basophils, eosinophils, lymphocytes, and neutrophils).	Collagen (type I), elastic, and reticular (type-III collagen).	Moderately viscous. Hyaluronan, glycoproteins, sulfated glycosaminoglycans, and proteoglycans.	More cellular than dense CT. Cells suspended in disorganized network of loosely interwoven fibers. Interstices filled with ground substance.	Suspends, supports, and protects vessels, nerves, and epithelia.	Around vessels and nerves. Under covering and lining epithelia (ie, lamina propria). Supports serous membranes. Fills potential spaces between other tissues.
	Dense Regular	Predominantly mature fibroblasts (fibrocytes). Other types listed for loose CT may be present, but not in abundance.	Almost all type-I collagen. Elastic and reticular fibers often present.	Similar to loose CT, but in smaller quantities.	Fibers predominate. Large collagen fibers packed into parallel bundles. Cells (fibrocytes) scattered between fiber bundles with long axes parallel to the fibers.	Transmits mechanical force of muscles. Binds bones to one another. Forms protective cover for some organs.	Tendons, ligaments, periosteum, perichondrium, joint capsules, deep fascia, and some organ capsules.
	Dense Irregular				Fiber bundles of differing sizes woven into collagenous mat or fabric.	Resists tensile stress from all directions. Protects fragile organs.	Reticular layer of dermis. Most organ capsules.
Specialized CT	Reticular	Reticular cells. Presence of other types depends on location.	Reticular fibers (type-III collagen).	Very little. Composition unclear.	Delicate network. Reticular cells attach to and cover fibers with long, thin cell processes. Other cells suspended in spaces.	Suppports motile cells. Important in filtration of blood and lymph nodes.	Mainly hematopoietic tissues (eg, bone marrow, spleen, and lymph nodes).
Specialized CT	Elastic	Fibroblasts predominate. Other types listed for loose CT present, but not in abundance.	Elastic fibers predominate. Some collagen.	Sparse. Composition similar to that of dense CT.	Fibers predominate in parallel wavy bundles separated by collagen and fibroblasts. Resembles dense regular CT, but fibers are wider and more refractile.	Provides flexible support.	Ligamenta flava of vertebral column, vocal ligament, and suspensory ligament of the penis.

chiefly of hyaluronan (see Table 5–1). Mucous tissue yields readily to pressure but returns to its original shape; thus, it is useful for protecting underlying structures from excess pressure. It is the predominant component (Wharton's jelly) of the umbilical cord, of the intervertebral disks' nucleus pulposis, and of the pulp of young teeth.

IV. HISTOPHYSIOLOGY OF CONNECTIVE TISSUE

A. Functions:

1. **Support.** Structural support is the major function of connective tissue, which forms the framework on which all other body tissues are assembled. Its physical properties allow it to bind, to fill spaces, and to separate functional units of other tissues and organs. It thus maintains functional units in their proper three-dimensional relationships, allowing the maintenance and coordination of all body functions.

2. **Defense.**
   a. **Physical.** The viscosity of the extracellular matrix, which is due largely to the presence of hyaluronan, slows the spread of many bacteria and foreign particles. Sheets of tightly packed, often interwoven, collagen fibers, as in organ capsules, help to confine local infections. However, some bacteria secrete enzymes that hydrolyze matrix components (eg, staphylococci, clostridia, streptococci, and pneumococci secrete **hyaluronidase,** and *Clostridium perfringens* secretes **collagenase**).
   b. **Immunologic.** Foreign bodies that penetrate epithelia are intercepted by immunoresponsive cells inhabiting the underlying connective tissue (II.E.2). These cells activate local immune responses (inflammation) and mobilize the immune system to supply additional cells by means of the bloodstream. Recruited cells migrate through capillary and venule walls into the connective tissue, a process called **diapedesis.**

3. **Repair.** Rapidly closing breaches in the body's protective barriers is an important connective tissue function. Injury stimulates invasion of the site by immunocompetent cells and also stimulates the proliferation of fibroblasts. Macrophages remove clotted blood, damaged tissue, and foreign material while fibroblasts secrete matrix materials to fill the breach. Rapidly formed collagenous matrices that close wounds are often less well-organized than original tissues and form scars. Small scars may be completely remodeled; larger scars are only partially remodeled.

4. **Storage.** Reserves of water and electrolytes, especially sodium, are stored in extracellular matrix, owing to the high polyanionic charge density of GAGs. Energy reserves in the form of lipids are stored in adipocytes.

5. **Transport.** Except in the central nervous system, most blood and lymphatic vessels are surrounded by loose connective tissue, which is thus a crossroads for transporting substances to and from other tissues.

B. Edema: The water in tissue fluid is forced to exit the blood from the arterial ends of capillary beds by means of hydrostatic pressure (arterial pressure). Fluid loss to the tissues increases the blood solute concentration at the venous ends of capillaries; this increased colloid osmotic pressure, along with the venous ends' lower hydrostatic pressure, draws most of the lost fluid back into the blood. Excess fluid remaining in the tissue is normally drained away by lymphatic capillaries; thus no net change occurs in the amount of blood or tissue fluid. Edema (ie, an accumulation of excess tissue fluid) accompanies pathologic conditions that cause the following: (1) increased hydrostatic pressure in capillaries caused by obstructed venous blood flow (eg, congestive heart failure); (2) decreased colloid osmotic pressure in the blood caused by a lack of blood proteins (eg, starvation); (3) increased hydrostatic pressure in the tissue caused by a blockage of lymphatic drainage (by parasites or tumor cells); or (4) increased colloid osmotic pressure in the tissue caused by an excessive accumulation of GAGs in the matrix (edema caused by this condition is called **myxedema**).

C. Hormonal Effects: Cortisol (hydrocortisone), produced by the adrenal glands under the influence of pituitary adrenocorticotropic hormone (ACTH), inhibits connective tissue fiber synthesis and retards local inflammatory and immune responses. Cortisol or synthetic cortisone therefore reduces local heat, redness, and tenderness but delays and impairs wound healing. Insufficient thyroid hormone (hypothyroidism) causes the connective tissue matrix to accumulate excess GAGs, which leads to myxedema.

D. Nutritional Factors: As a cofactor of proline hydroxylase (II.A.1.a), **vitamin C** (ascorbic acid) is required for collagen synthesis. Vitamin C deficiency leads to **scurvy,** which is characterized by the weakening of all connective tissue. Proline hydroxylase activity also requires iron, molecular oxygen, and α-ketoglutarate (see also 8.III.D.3).

E. Collagen Renewal: Collagen is a very stable protein with a slow turnover (rate of removal and replacement); its turnover is slowest in tendons and other dense connective tissues, and most rapid in loose connective tissue. Macrophages and neutrophils release **collagenase,** which breaks down old collagen; new collagen is synthesized by fibroblasts. With age, extracellular collagen becomes increasingly cross-linked, and its turnover slows in all connective tissues.

MULTIPLE-CHOICE QUESTIONS

Select the single best answer.

5.1. Which of the following connective tissue components is a part of the extracellular matrix but not of the ground substance?
(A) Capillaries
(B) Collagen fibers
(C) Fibroblasts
(D) Fibronectin
(E) GAGs
(F) Hyaluronan
(G) Proteoglycans

5.2. Which of the following cell types produces and maintains all extracellular matrix components of connective tissue?
(A) Fibroblast
(B) Lymphocyte
(C) Macrophage
(D) Mesenchymal cell
(E) Mast cell
(F) Neutrophil
(G) Plasma cell

5.3. Which of the following is true of dense regular connective tissue?
(A) Is composed primarily of fibroblasts
(B) Is composed primarily of ground substance
(C) Is the predominant tissue type in most organ capsules
(D) Contains more mast cells than any other type of connective tissue
(E) May be found in tendons

5.4. Which of the following are the three basic components of all types of connective tissue?
(A) Arteries, veins, and capillaries
(B) Cells, fibers, and ground substance
(C) Fibroblasts, reticular fibers, and proteoglycan aggregates
(D) Mast cells, lymphocytes, and adipocytes
(E) Type-II collagen, hyaluronan, and fibronectin

5.5. Which of the following fiber types is most commonly associated with the formation of a fine supportive network in the soft tissues of the body?
(A) Elastic fibers
(B) Purkinje fibers
(C) Reticular fibers
(D) Type-I collagen
(E) Type-II collagen

5.6. Which of the following tissue types is shown in Figure 5–1?
(A) Dense irregular connective tissue
(B) Dense regular connective tissue
(C) Elastic connective tissue
(D) Loose (areolar) connective tissue
(E) Mesenchyme
(F) Mucous connective tissue
(G) Reticular connective tissue

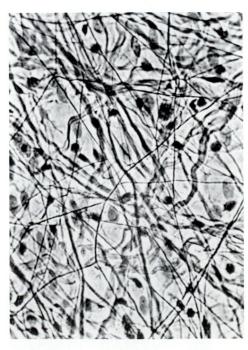

Figure 5–1.

5.7. Which of the following tissue types is shown in Figure 5–2?
(**A**) Dense irregular connective tissue
(**B**) Dense regular connective tissue
(**C**) Elastic connective tissue
(**D**) Loose (areolar) connective tissue
(**E**) Mesenchyme
(**F**) Mucous connective tissue
(**G**) Reticular connective tissue

5.8. Which of the following is the site of assembly for collagen bundles?
(**A**) Extracellular space
(**B**) Golgi apparatus
(**C**) Nucleus
(**D**) RER
(**E**) Transfer vesicles

5.9. Which of the following is true of mucous connective tissue?
(**A**) Has a thin watery consistency
(**B**) Is sometimes referred to as Wharton's jelly
(**C**) Cushions and protects blood vessels in the elderly
(**D**) Is composed primarily of type-III collagen
(**E**) Lacks connective tissue cells

5.10. Which of the following is a fibroblast secretion that is synthesized mainly in the Golgi complex?
(**A**) Fibronectin
(**B**) Glycosaminoglycan
(**C**) Procollagen

Figure 5–2.

(**D**) Proelastin
(**E**) Tropocollagen

5.11. Which of the following is true of elastic fibers?
(**A**) Can be stretched to 150% of their length and subsequently will spring back to their original size
(**B**) Are ensheathed by type-IV collagen fibers
(**C**) Are extremely soluble in aqueous solutions
(**D**) Are notably absent in the cartilages of the nose and ear
(**E**) Can be distinguished from bundles of type-I collagen fibers with standard hematoxylin and eosin staining

5.12. Which of the following comprises every third amino acid in collagen?
(**A**) Desmosine
(**B**) Glycine
(**C**) Hydroxylysine
(**D**) Hydroxyproline
(**E**) Leucine
(**F**) Lysine
(**G**) Proline

5.13. Which of the following amino acids is preferentially glycosylated during collagen synthesis?
(**A**) Desmosine
(**B**) Glycine
(**C**) Hydroxylysine
(**D**) Hydroxyproline
(**E**) Leucine
(**F**) Lysine
(**G**) Proline

5.14. Which of the following is a cross-linked amino acid that is characteristic of the protein elastin?
(**A**) Desmosine
(**B**) Glycine
(**C**) Hydroxylysine
(**D**) Hydroxyproline
(**E**) Leucine
(**F**) Lysine
(**G**) Proline

5.15. Which of the following cells characteristically synthesize and secrete immunoglobulins?
(**A**) Fibroblasts
(**B**) Lymphocytes
(**C**) Macrophages
(**D**) Mast cells
(**E**) Mesenchymal cells
(**F**) Plasma cells
(**G**) Reticular cells

5.16. Which of the following is the predominant cell type in connective tissue proper?
(**A**) Fibroblast
(**B**) Lymphocyte

(**C**) Macrophage
(**D**) Mast cell
(**E**) Mesenchymal cell
(**F**) Plasma cell
(**G**) Reticular cell

5.17. Which of the following cells characteristically contain secretory granules filled with heparin and histamine?
(**A**) Fibroblasts
(**B**) Lymphocytes
(**C**) Macrophages
(**D**) Mast cells
(**E**) Mesenchymal cells
(**F**) Plasma cells
(**G**) Reticular cells

5.18. Which of the following cells characteristically contain the most lysosomes?
(**A**) Fibroblasts
(**B**) Lymphocytes
(**C**) Macrophages
(**D**) Mast cells
(**E**) Mesenchymal cells
(**F**) Plasma cells
(**G**) Reticular cells

5.19. Which of the following cells characteristically contain a nucleus with a clock-face pattern of heterochromatin?
(**A**) Fibroblasts
(**B**) Lymphocytes

(**C**) Macrophages
(**D**) Mast cells
(**E**) Mesenchymal cells
(**F**) Plasma cells
(**G**) Reticular cells

5.20. Which of the following is true of tropocollagen?
(**A**) Includes a registration peptide
(**B**) Includes a signal sequence
(**C**) Is secreted by fibroblasts
(**D**) Lacks hydroxyproline
(**E**) Is found in the extracellular matrix

5.21. Which of the following is true about the initiation of collagen synthesis?
(**A**) Begins on a free polyribosome
(**B**) Begins with reading of the code for the registration peptide
(**C**) Begins on the RER
(**D**) Begins in the extracellular matrix
(**E**) Begins with the codon for hydroxyproline

5.22. Which of the following conditions affecting connective tissue structure and function cannot be remedied or prevented by attention to diet?
(**A**) Myxedema
(**B**) Osteomalacia
(**C**) Rickets
(**D**) Scurvy

ANSWERS TO MULTIPLE-CHOICE QUESTIONS

5.1. B (I.D; II.D)
5.2. A (II.E.1.b)
5.3. E (III.A.2 and 2.a)
5.4. B (I.C)
5.5. C (II.B)
5.6. D (III.A.1)
5.7. A (III.A.2.b)
5.8. A (II.A.1.b)
5.9. B (III.D)
5.10. B (II.D.1 and E.1.b; 2.III.D.2.a and b; 7.I.C)
5.11. A (II.C.1–4)

5.12. B (II.A.1.a)
5.13. C (II.A.1.a)
5.14. A (II.C.1.b)
5.15. F (II.E.2.c)
5.16. A (II.E.1.b)
5.17. D (II.E.2.a)
5.18. C (II.E.2.b)
5.19. F (II.E.2.c)
5.20. E (II.A.1.b)
5.21. A (II.A.1.a)
5.22. A (IV.C and D)

Adipose Tissue

6

OBJECTIVES

This chapter should help the student to:

- Relate the functions of adipose tissue to its structural characteristics.
- Describe adipose tissue as a connective tissue in terms of its cells, fibers, and ground substance.
- Know the differences and similarities between the two types of adipose tissue.
- Recognize the type of adipose tissue present in a micrograph of a tissue or organ.

MAX-Yield™ STUDY QUESTIONS

1. Explain why the body must store fuel (I.A[1]).
2. List the functions of unilocular (white, yellow) adipose tissue (II.A–C).
3. Describe collagen and reticular fiber distribution in adipose tissue (I.B).
4. Compare unilocular and multilocular adipose tissue in terms of:
   a. Cell size (II.A; III.A)
   b. Cytoplasmic lipid distribution (II.A; III.A)
   c. Nuclear shape and location (II.A; III.A)
   d. Number of mitochondria (II.A; III.A)
   e. Organelle distribution (II.A; III.A)
   f. Precursor cell (II.D; III.D)
   g. Function (II.C; III.C)
   h. Vascular supply (II.A)
   i. Autonomic nerve distribution (II.C.2.c; III.C)
   j. Abundance (II.A; III.A)
   k. Location (II.B; III.B)
5. Name the chief biochemical constituent of the lipid droplets in adipocytes (II.C).
6. List the factors that lead to:
   a. Increased lipid storage and synthesis by adipocytes (lipogenic factors; II.C.1)
   b. Increased lipid mobilization by adipocytes (lipolytic factors; II.C.2)
7. During starvation, which lipid deposits begin mobilization first and which are mobilized last (II.C.2)?
8. What accounts for the color of brown adipose tissue (III.A) and of yellow adipose tissue (II.A)?

SYNOPSIS

I. GENERAL FEATURES OF ADIPOSE TISSUE

A. A Tissue and an Organ: Adipose tissue, or fat, is a connective tissue that is specialized to store fuel. If we were unable to store fuel, all of our time would be spent obtaining food. The cytoplasm of fat cells, or **adipocytes,** contains large triglyceride deposits in the form of one or more lipid droplets with no limiting membranes. Together, adipocyte clusters scattered throughout the body constitute an important metabolic organ that varies widely in size and distribution, depending on such factors as age, sex, and nutritional status.

B. General Organization: Clusters of adipocytes are divided into lobes and lobules by collagenous connective tissue septa of variable density. Individual cells are surrounded by a reticular fiber network. Ground substance is sparse.

---

[1] See footnote on page 1.

C. Two Types: There are two basic types of adipose tissue: white adipose tissue, or white fat, and brown adipose tissue, or brown fat. A white adipocyte contains a single large lipid droplet; a brown adipocyte contains many small droplets.

II. WHITE ADIPOSE TISSUE

A. Distinguishing Features: White adipose tissue, the more abundant of the two types, is also termed **unilocular adipose tissue**—a reference to the single fat droplet in each cell. In mature adipocytes, the droplet is so large that it displaces the nucleus and remaining cytoplasm to the cell periphery. Cell diameter varies from 50 to 150 μm. Adipocytes in histologic sections have a **signet-ring** appearance because the lipid is washed away during preparation, leaving only a flattened nucleus and a thin rim of cytoplasm. The cytoplasm near the nucleus contains a Golgi complex, mitochondria, a small amount of RER, and free ribosomes. The cytoplasm in the thin rim contains SER and pinocytotic vesicles. This tissue is sometimes termed yellow adipose tissue or **yellow fat;** dietary carotenoids accumulate in the lipid droplets, causing the tissue to appear yellow. White fat is richly vascularized, but not as richly as brown fat.

B. Distribution:
1. **Subcutaneous fat** (hypodermis) is the layer of white adipose tissue that is present under the skin, except in the eyelids, penis, scrotum, and most of the external ear (there is some fat in the earlobe). In infants, it forms a thermal insulating layer of uniform thickness that covers the entire body and is known as the **panniculus adiposus.** In adults, it thickens or thins in selected areas, depending on age, sex, and dietary habits. Where it thins, it resembles areolar tissue. In males, the fat layer thickens over the nape of the neck, deltoids (shoulders), triceps brachii (back of the upper arm), lumbosacral region (lower back), and buttocks. In females, additional fat is deposited in the breasts, buttocks, and hips and over the anterior aspect of the thighs.
2. **Intraabdominal fat.** Fat deposits of variable size surround blood and lymphatic vessels in the abdomen's omentum and mesenteries. Other accumulations occur in retroperitoneal areas (eg, around the kidneys).
3. **Other locations.** Other prominent fat accumulations occur within the eye orbits, around major joints (eg, knees), and in pads in the palms and soles.

C. Functional Characteristics: Adipocytes store fatty acids in **triglycerides** (esters of glycerol and three fatty acids). The triglycerides stored in both white and brown fat undergo continuous turnover. Released fatty acids serve as a source of chemical energy (the predominant source in resting muscle) and as raw materials used in the synthesis of phospholipids. Turnover is regulated by several histophysiologic factors, which shift the equilibrium toward fat uptake or mobilization, depending on the level of, and need for, circulating fatty acids.
1. **Factors that enhance lipid uptake (lipogenic influences).**
   a. **Dietary abundance.** Dietary fats are absorbed by intestinal epithelial cells and carried, in particles called **chylomicrons,** by lymphatic vessels to the blood (15.VII.B.3.c). Chylomicron triglycerides are hydrolyzed by lipoprotein lipases in adipose tissue capillaries; the released fatty acids are absorbed by adipocytes and resynthesized into triglycerides for storage. Dietary glucose can be converted in the liver to fatty acids, which the blood subsequently carries to adipocytes in triglycerides of **very low-density lipoproteins (VLDLs).** Glucose is also directly absorbed from blood and converted into triglycerides or glycerol by adipocytes.
   b. **Hormones.** Insulin increases glucose uptake by adipocytes and enhances triglyceride synthesis from carbohydrates.
2. **Factors that enhance lipid mobilization (lipolytic influences).** When blood levels of fatty acids and glucose fall below homeostatic levels (eg, during starvation or prolonged exercise), adipocytes break down triglycerides and release stored fatty acids and glycerol into the blood. Lipid mobilization occurs first from subcutaneous, mesenteric, and retroperitoneal adipose tissues and last from deposits in the hands, feet, and retroorbital fat pads.
   a. **Hormone-sensitive lipases.** Peptide hormones and norepinephrine increase adipocyte cyclic adenosine monophosphate (cAMP) levels (see Chapter 20). Hormone-sensitive lipases in adipocyte cytoplasm are activated by cAMP and cleave fatty acids from stored triglycerides.

    b. Hormones. Adrenocorticotropic hormone (ACTH), released by the anterior pituitary gland, stimulates adipocytes to release free fatty acids. Other hormones with variable lipolytic ability are glucagon, growth hormone, and thyroid hormone. Sex-dependent regional sensitivity of adipose tissue to circulating androgens and estrogens strongly influences sex-dependent differences in fatty acid uptake and mobilization by adipocytes.

    c. Innervation. Interruption of the autonomic nerve supply to adipose tissues decreases fat loss from affected areas, which suggests that the autonomic nervous system (ANS) is important in fatty acid mobilization. Autonomic nerve fibers that supply white fat terminate only on blood vessel walls, but in brown fat, they also directly contact the adipocytes. Exogenous norepinephrine can double the blood levels of free fatty acids by stimulating lipolysis in adipose tissue. Synthetic adrenergic agonists (eg, in asthma inhalers) also can reduce fat deposits by stimulating fat mobilization.

D. Histogenesis: Unilocular adipocytes derive from mesenchymal precursors resembling fibroblasts. The appearance of numerous small cytoplasmic lipid droplets signals their transformation into lipoblasts. As lipid accumulation continues, small droplets fuse until a single lipid droplet forms.

III. BROWN ADIPOSE TISSUE

A. Distinguishing Features: Brown fat is called **multilocular adipose tissue** because of the multiple small lipid droplets in its cells. Brown adipocytes are smaller than white adipocytes and have a spherical, central nucleus. They contain many mitochondria; mitochondrial cytochromes are chiefly responsible for the tan to red–brown tissue color. Loose connective tissue septa give brown adipose tissue a lobular appearance resembling a gland in histologic section. The vascular supply (which contributes to the color) is very rich, as is the autonomic nerve supply. Many unmyelinated nerve fibers contact the adipocytes.

B. Distribution: Brown fat is less abundant than white fat at all ages. Young and middle-aged adults have little or none, but fetuses, newborns, and the elderly have accumulations in the axilla, near the carotid artery and the thyroid gland, and around the renal hilus.

C. Functional Characteristics: Brown fat has many of the same functional capabilities as white fat, but its metabolic activity is more intense and can generate heat. In excessively cold conditions, autonomic stimulation can uncouple oxidative phosphorylation from adenosine triphosphate (ATP) synthesis in the many mitochondria; in such circumstances, the released energy dissipates as heat. The numerous vessels supplying this tissue carry the heat to the body. Brown fat is important in hibernating animals and in human infants before other thermoregulatory mechanisms develop.

D. Histogenesis: The multilocular adipocytes of brown fat derive from mesenchymal precursors that assume an epithelial shape and arrangement. The multiple small fat droplets that appear during development do not coalesce during maturation.

MULTIPLE-CHOICE QUESTIONS

Select the single best answer.

6.1. The fat in which of the following locations is mobilized *last* during prolonged starvation?
(A) Eyelids
(B) Fat pads of the hands and feet
(C) Mesenteric fat deposits
(D) Retroperitoneal fat deposits
(E) Subcutaneous fat deposits

6.2. Which of the following hormones produced by the adrenal medulla, and by certain autonomic neurons, has a dramatic lipolytic effect in adipose tissue?
(A) Androgens and estrogens
(B) Adrenocorticotropic hormone (ACTH)
(C) Growth hormone
(D) Insulin
(E) Norepinephrine
(F) Thyroid hormone

6.3. Which of the following is the preferred substrate for hormone-sensitive lipase?
(**A**) Chylomicron
(**B**) Free fatty acid
(**C**) Glucose
(**D**) Glycerol
(**E**) Triglyceride

6.4. Which of the following is a large, fat-containing particle formed by the intestinal epithelium?
(**A**) Chylomicron
(**B**) Free fatty acid
(**C**) Glucose
(**D**) Glycerol
(**E**) Triglyceride

6.5. Which of the following is the primary form of stored lipid in adipocytes?

(**A**) Chylomicron
(**B**) Free fatty acid
(**C**) Glucose
(**D**) Glycerol
(**E**) Triglyceride

6.6. Which of the following describes brown adipose tissue that is not shared by white or yellow adipose tissue?
(**A**) Is widely distributed in adults
(**B**) Is highly vascular
(**C**) Contains large adipocytes with a "signet-ring" appearance in histologic section
(**D**) Can generate heat
(**E**) Is derived from embryonic mesenchyme
(**F**) Contains a single large fat droplet in the cytoplasm of each adipocyte
(**G**) Reticular cells and fibers make up the tissue's stroma

ANSWERS TO MULTIPLE-CHOICE QUESTIONS

6.1. B (II.C.2)
6.2. E (II.C.2.a–c; 21.II.B.2.a and 3)
6.3. E (II.C.2.a)

6.4. A (II.C.1.a)
6.5. E (II.C)
6.6. D (II.A–D)

Cartilage 7

OBJECTIVES

This chapter should help the student to:

- Know the similarities and differences among the three cartilage types.
- Relate the functions of the three cartilage types to their structure and location.
- Know the steps in the histogenesis and growth of cartilage.
- Relate chondrocyte ultrastructure to the synthesis and maintenance of extracellular matrix.
- Recognize the cartilage type present in a micrograph and identify its components.

MAX-Yield™ STUDY QUESTIONS

1. Compare the three types of cartilage in terms of:
   a. Type, amount, and arrangement of cells, fibers, and ground substance (Table 7–1)
   b. Location in the body (Table 7–1)
   c. Histogenesis (II.A.3, B.2, and C.2[1])
   d. Function (Table 7–1)
2. Sketch a typical proteoglycan aggregate of cartilage ground substance (Fig. 7–1). Label the following components and indicate which of them are GAGs:
   a. Hyaluronan
   b. Link protein
   c. Core protein
   d. Chondroitin sulfate
   e. A proteoglycan molecule
3. Compare capsular (territorial) matrix and intercapsular matrix in terms of location, composition, and staining properties (II.A.2).
4. Describe the structure and function of the perichondrium (I.B).
5. List the functions of chondrocytes and name the organelles involved in each function (I.C).
6. List the factors known to increase or decrease the synthesis and secretion of sulfated GAGs (eg, chondroitin sulfate) by chondrocytes (II.A.4).
7. Name the types of growth that occur in hyaline cartilage and compare them in terms of the location of their dividing cells and their importance in expansion of girth, replacement of worn articular cartilage, and bone lengthening at the epiphyseal plate (II.A.4).
8. What is the major structural difference between articular and other hyaline cartilage (II.A.6)?
9. How are the chondrocytes of articular cartilage supplied with nutrients and oxygen (II.A.2)?
10. What process, other than cell division, increases cartilage mass during growth (II.A.4)?
11. Describe the origin of chondrocytes that fill and repair a cartilage fracture (II.A.5).
12. Describe an intervertebral disk (III.A and B) in terms of:
    a. Location
    b. Function
    c. Tissue in the annulus fibrosus
    d. Tissue in the nucleus pulposus
    e. Embryonic origin of the nucleus pulposus

---

[1] See footnote on page 1.

SYNOPSIS

I. GENERAL FEATURES OF CARTILAGE

Cartilage is a **skeletal connective tissue** characterized by firmness and resiliency. It forms the fetal skeleton and persists where its mechanical properties are needed. Most fetal cartilage is replaced by bone.

A. Composition: Like all connective tissues, cartilage consists of cells, fibers, and ground substance. The extracellular matrix predominates and determines its mechanical properties. Type-II collagen is a characteristic cartilage matrix component. The abundant ground substance is firm and gel-like. Cartilage cells are termed chondrocytes.

B. Blood Supply: Most cartilage is enveloped by a dense connective tissue layer, the perichondrium, which contains the vascular supply and fibroblastlike stem cells from which additional chondrocytes arise. Few vessels (or nerves) occur in cartilage; thus, ground substance composition controls the percolation of nutrients and oxygen to chondrocytes from perichondrial vessels.

C. Cells: Under the light microscope, chondrocytes are round, with an eccentric nucleus, prominent nucleolus, and basophilic cytoplasm. In EMs, chondrocyte surfaces exhibit characteristic projections and infoldings. The RER and Golgi complex are well-developed; as the cell grows, the Golgi complex enlarges and its cisternae fill with secretory material. Lipid droplets occur in the cytoplasm. Chondrocytes synthesize and secrete extracellular matrix fibers and ground substance: collagen is synthesized on the RER, and GAGs are assembled and sulfated in the Golgi complex. Owing to their meager oxygen supply, chondrocytes produce much of their energy by anaerobic glycolysis.

II. THE THREE TYPES OF CARTILAGE

Hyaline cartilage, elastic cartilage, and **fibrocartilage** differ in appearance and mechanical properties, owing to differences in extracellular matrix composition (Table 7–1). No distinction is made among the cells in different cartilage types.

A. Hyaline Cartilage: Hyaline cartilage, the most common type in both fetus and adult, is white and translucent when fresh, with a firm, gel-like consistency.
   1. Composition.
      a. Fibers. Hyaline cartilage matrix contains thin type-II collagen fibrils. Their small size and a refractive index close to that of the ground substance make them difficult to distinguish with the light microscope. Type-II collagen contains more hydroxylysine than does type I.
      b. Ground substance, the predominant tissue component, comprises the following:
         (1) GAGs, which are mostly chondroitin sulfates and hyaluronan but also include small amounts of keratan sulfate and heparan sulfate.
         (2) Proteoglycans, which are core proteins with covalently bound GAG side chains.
         (3) Proteoglycan aggregates (Fig. 7–1), which are proteoglycans noncovalently linked to long chains of hyaluronan by **link protein.**
         (4) Glycoproteins, including link protein, fibronectin, and chondronectin, which attach various matrix components to one another and cells to the matrix.
         (5) Tissue fluid, an ultrafiltrate of blood plasma.
   2. Organization. The firmness of hyaline cartilage results from extensive cross-linking among its components. **Link protein** attaches the core proteins of proteoglycans to long hyaluronan chains to form **proteoglycan aggregates** (see Fig. 7–1). The GAG side chains of the proteoglycans associate with type-II collagen fibrils. The chondrocytes are embedded in the matrix either singly or in **isogenous groups** of two to eight cells derived from one parent cell. The potential space occupied by each chondrocyte, called a **lacuna,** is visible only after the cell's death or after shrinkage during tissue processing. The chondrocytes at the core

Table 7–1. Characteristics of cartilage types.

Cartilage	Cell Type	Fiber Type	Ground Substance	Organization	Functions	Locations
Hyaline cartilage	Chondrocytes	Type-II collagen	Predominant tissue component. Includes GAGs (mainly chondroitin sulfate, smaller amounts of keratan and heparan sulfates), proteoglycan aggregates, and glycoprotein (fibronectin, chondronectin, and link protein).	Cells and fibers embedded in abundant ground substance. Notable lack of capillaries. Cells may occur in isogenous groups. Fibers difficult to distinguish from ground substance. Extensive cross-linking among ground substance components (proteoglycan aggregates) and between fibers and ground substance.	Useful as fetal skeleton, owing to ability to provide support and grow rapidly. Maintains airways in respiratory passages. Cushions and provides low-friction surface in joints.	Fetal skeleton; articular and costal cartilages; laryngeal, tracheal, and bronchial cartilages.
Elastic cartilage	Chondrocytes	Elastic fibers, type-II collagen	Same as hyaline cartilage.	Organization identical to that of hyaline cartilage except for presence of a dense network of elastic fibers.	Provides flexible support. Semi-rigid; returns to original shape after being deformed.	External ear, auditory tubes, epiglottis, and corniculate and cuneiform cartilages of larynx.
Fibrocartilage	Chondrocytes	Type-I collagen, type-II collagen	Similar to hyaline cartilage but with equal amounts of chondroitin and dermatan sulfates.	Cross between cartilage and dense regular connective tissues. Chondrocytes usually in rowlike isogenous groups surrounded by hyaline matrix. Nests of chondrocytes lie between dense bundles of large type-I collagen fibers.	Attaches bone to bone and provides restricted mobility under great mechanical stress.	Annulus fibrosus of intervertebral disks, pubic symphysis, and bone–ligament junctions. Always associated with dense connective tissue.

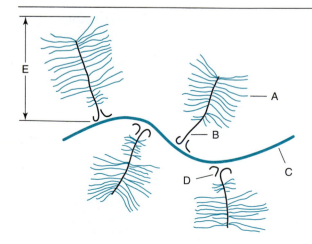

Figure 7–1. Schematic diagram of a proteoglycan aggregate. Labeled components include sulfated glycosaminoglycan (A), core protein (B), hyaluronan (C), link protein (D), and a complete proteoglycan (E).

of a tissue mass are spherical; those at the periphery are flattened or elliptic. The matrix immediately surrounding the chondrocytes, called the **capsular (territorial) matrix,** is more basophilic and PAS-positive than the **intercapsular (interterritorial) matrix,** owing to the higher sulfated GAG concentration and lower collagen concentration. Except for articular (joint) cartilage, hyaline cartilage is surrounded and nourished by perichondrium. Articular cartilage is nourished by the synovial fluid in the joint cavity (8.IV.B).

3. **Histogenesis.** All cartilage derives from embryonic mesenchyme. During hyaline cartilage development, mesenchymal cells retract their cytoplasmic extensions and assume a rounded shape, becoming more tightly packed and forming a **mesenchymal condensation,** or precartilage condensation. The increased cell-to-cell contact stimulates commitment to cartilage differentiation, which progresses from the center outward. Cells at the condensation's core are the first to become **chondroblasts** and secrete cartilage matrix. After it is surrounded by cartilage matrix, a chondroblast is termed a **chondrocyte.** Peripheral mesenchyme condenses around the developing cartilage mass to form the fibroblast-containing, dense regular connective tissue of the perichondrium.

4. **Growth.** Cartilage grows by two processes; both involve mitosis and the deposition of additional matrix. Matrix synthesis is enhanced by growth hormone, thyroxine, and testosterone and is inhibited by estradiol and excess cortisone. **Interstitial growth** involves the division of existing chondrocytes and gives rise to the **isogenous groups;** it is important in forming the fetal skeleton and continues in the epiphyseal plates and articular cartilages (II.A.6). **Appositional growth** involves the differentiation into chondrocytes by chondroblasts and stem cells on the perichondrium's inner surface. It is responsible for increases in the girth of the cartilage masses.

5. **Repair.** Repair of cartilage fractures involves an invasion of the breach by mesenchymal stem cells from the perichondrium, which subsequently differentiate into chondrocytes. If the gap is large, a dense connective tissue scar may form.

6. **Function and location.** Its ability to grow rapidly while maintaining its rigidity makes hyaline cartilage an ideal fetal skeletal tissue. As fetal cartilage is replaced by bone, hyaline cartilage remains in the **epiphyseal plates** at the ends of long bones, allowing these bones to lengthen from birth to adulthood. At all ages, hyaline cartilage without a perichondrium covers the articular surfaces of bone, where its resistance to compression and its smooth texture provide cushion and a low-friction surface. Hyaline cartilage is the most abundant and widely distributed cartilage type in the body. The costal (rib) cartilages, most of the laryngeal cartilages, the cartilaginous rings supporting the trachea, and the irregular cartilage plates in the walls of the bronchi are hyaline cartilage.

B. **Elastic Cartilage:**   Elastic cartilage is yellow when fresh and is more flexible than hyaline.

1. **Composition and organization.** Elastic cartilage resembles hyaline but contains, in addition to type-II collagen fibers, a dense network of branching elastic fibers. This network is densest at the core of the cartilage mass and, when stained with an elastic stain (eg, Verhoeff or Weigert), may obscure tissue organization. The chondrocytes characteristically occur in isogenous groups. A perichondrium is present.

2. **Histogenesis and growth.** Elastic cartilage develops from a primitive connective tissue containing wavy fibril bundles that differ from both elastin and collagen in protein composition. Fibroblasts subsequently secrete elastin, and the fiber bundles are transformed into branching elastic fibers by an unknown mechanism. Chondrocyte development, the production of other matrix materials, and further growth resemble that of hyaline cartilage.

3. **Function and location.** Elastic cartilage provides flexible support. It occurs alone and with hyaline cartilage; the two may gradually blend into each other at the border between them to form one cartilage mass. In humans, elastic cartilage occurs in the external ear, the external auditory canals and auditory tubes, the epiglottis, and the corniculate and cuneiform cartilages of the larynx.

C. **Fibrocartilage:**   Fibrocartilage is intermediate in character between hyaline cartilage and dense connective tissue.

1. **Composition and organization.** Fibrocartilage is characterized by abundant type-I collagen fibers; at low magnification, it resembles dense connective tissue. The ground substance contains equal amounts of dermatan sulfate and chondroitin sulfate (see Table 7–1). The capsular matrix resembles that of hyaline cartilage and contains some type-II collagen. The

chondrocytes are distributed in columnar isogenous groups between the densely packed type-I collagen bundles. No perichondrium can be distinguished.

2. **Histogenesis and growth.** Where strong mechanical stresses occur, fibrocartilage develops from dense regular connective tissue through the transformation of fibroblasts or fibroblast-like precursors into chondrocytes. Fibrocartilage growth has not been closely examined.

3. **Function and location.** Fibrocartilage is always associated with dense connective tissue, and the border between the two is indistinct. Its combination of cartilaginous ground substance and dense collagen bundles allows fibrocartilage to resist deformation under great stress, a quality that is important in attaching bone to bone and providing restricted mobility. Sites in humans include the annulus fibrosus of the intervertebral disks, the symphysis pubis, and some bone–ligament junctions.

III. INTERVERTEBRAL DISKS

Intervertebral disks act as cushions between the vertebrae, allowing limited movement of the vertebral column. They are bound to the vertebrae by ligaments. Each disk has two parts.

A. Annulus Fibrosus: This outer ring consists of fibrocartilage and is covered on its outer surface by the dense connective tissue of associated ligaments. The fibrocartilage is arranged in concentric layers, with the collagen bundles in each layer oriented at right angles to those in the next. This organization may appear as a "herringbone" pattern when seen through a light microscope at low power.

B. Nucleus Pulposus: This structure forms the center of the disk and derives from embryonic notochord. It consists of mucous connective tissue, with a few fibers and rounded cells embedded in syrupy, hyaluronan-rich ground substance. The nucleus pulposus is smaller in adults than in children because it is partially replaced by fibrocartilage.

MULTIPLE-CHOICE QUESTIONS

Select the single best answer.

7.1. What distinguishes cartilage from most other connective tissues?
(A) Its extracellular matrix contains collagen
(B) Its predominant cell type is a mesenchymal derivative
(C) Its predominant cell type secretes both fibers and ground substance
(D) It lacks blood vessels
(E) It functions in mechanical support

7.2. Which of the following areas in cartilage is collagen-poor and sulfated glycosaminoglycan-rich?
(A) Annulus fibrosus
(B) Capsular matrix
(C) Epiphyseal plate
(D) Intercapsular matrix
(E) Perichondrium

7.3. In which of the following ways does articular cartilage differ from most other hyaline cartilage?

(A) It contains isogenous groups of chondrocytes
(B) It lacks blood vessels
(C) It lacks a perichondrium
(D) It contains type-II collagen
(E) It is derived from embryonic mesenchyme
(F) It contains proteoglycan aggregates in its ground substance
(G) It undergoes mainly appositional growth

7.4. Which of the following tissues is shown in Figure 7–2?
(A) Elastic cartilage
(B) Fibrocartilage
(C) Hyaline cartilage
(D) Mesenchyme

7.5. Which of the following occurs first in cartilage histogenesis?
(A) Appositional growth
(B) Differentiation of chondroblasts into chondrocytes
(C) Formation of mesenchymal condensations
(D) Interstitial growth
(E) Secretion of the matrix

Figure 7–2.

7.6. Which of the following is true of fibrocartilage?
(A) Contains large numbers of elastic fibers
(B) Seldom contains isogenous groups of chondrocytes
(C) Is found in the epiphyses of the long bones in young children
(D) Is structurally intermediate between dense connective tissue and cartilage
(E) The collagen in its matrix is mainly type II
(F) Is the cartilage of the epiphyseal plates
(G) Is the predominant cartilage of the external ear (auricle)

7.7. Which of the following is true of elastic cartilage?

(A) Is the primary skeletal tissue in the fetus
(B) Has no identifiable perichondrium
(C) Is found in the annulus fibrosus of intervertebral disks
(D) Is the most widely distributed cartilage type in the body
(E) The collagen in its matrix is mainly type II

7.8. Which of the following is true about the appositional growth of cartilage?
(A) Involves only cell division
(B) Involves the differentiation of perichondrial cells into chondroblasts
(C) Is the predominant type of growth in epiphyseal plates
(D) Is the form of cartilage growth most likely to result in the formation of isogenous groups
(E) Involves mainly the division of preexisting chondrocytes

7.9. The link protein of cartilage proteoglycan aggregates serves to attach which two matrix components to one another?
(A) Chondroitin sulfate and collagen
(B) Chondroitin sulfate and core protein
(C) Chondroitin sulfate and hyaluronan
(D) Collagen and hyaluronan
(E) Core protein and collagen
(F) Core protein and hyaluronan

7.10. The intense basophilia of the capsular matrix in hyaline cartilage is caused mainly by the synthetic activity of which of the following chondrocyte organelles?
(A) Centriole
(B) Golgi complex
(C) RER
(D) SER
(E) Plasma membrane

ANSWERS TO MULTIPLE-CHOICE QUESTIONS

7.1. D (I.B)
7.2. B (II.A.2)
7.3. C (II.A.2)
7.4. C (II.A.1.a, B.1 and C.1; note the apparent lack of fibers)
7.5. C (II.A.3)

7.6. D (II.C)
7.7. E (II.B.1)
7.8. B (II.A.4)
7.9. F (II.A.1.b.[3]; Fig. 7–1)
7.10. B (I.C; II.A.2)

Bone

8

OBJECTIVES

This chapter should help the student to:

- Describe bone as a connective tissue in terms of its cells, fibers, and ground substance.
- Compare bone cell types in terms of their origin, structure, and primary functions.
- Relate the physical properties of bone tissue to specific tissue components.
- List the bone tissue types and name the sites where each may be found.
- Compare the two processes of bone histogenesis in terms of embryonic tissue of origin, intermediate steps, structure of the mature tissue, and location in the body.
- Compare the steps of bone histogenesis with those of fracture repair.
- Know the alterations in tissue structure that occur during bone growth and remodeling.
- Explain the effects of nutrients and hormones on bone tissue structure and function.
- Identify bone types, cell types, and named structures in micrographs of bone tissue.
- List the types of joints and compare them in terms of their structure, mobility, and location.

MAX-Yield™ STUDY QUESTIONS

1. List the functions of bone (I.B[1]).
2. Describe two methods of preparing bone for microscopy that are necessitated by its hardness (III.A). Which method resembles bone resorption (III.D.1.a)?
3. List the functions of osteoblasts and the organelle(s) associated with each function (III.A.1.b).
4. Describe osteoblast cytoplasmic staining and name the cell components stained (III.A.1.b).
5. Describe the relationships among osteoprogenitor cells, osteoblasts, and osteocytes (III.A.1.a–c).
6. Compare osteocytes (III.A.1.c) with osteoblasts (III.A.1.b) in terms of:
   a. Shape
   b. Length of filopodia
   c. Amount of RER
   d. Location
   e. Rate of matrix synthesis
7. How are osteocytes located far from capillaries able to survive when nutrients, oxygen, and wastes cannot diffuse through calcified bone matrix (III.A.1.c)?
8. Describe osteoclasts (III.A.1.d) in terms of:
   a. Size
   b. Number of nuclei
   c. Precursor cells
   d. Staining properties
   e. Organelles present
   f. Major function
   g. Substances secreted
   h. Location and function of ruffled border
   i. Reaction to parathyroid hormone
   j. Reaction to calcitonin
9. List the inorganic components of bone matrix. Which two are most abundant (III.A.2.b)?
10. Describe the composition of the organic matter (osteoid) of bone matrix (III.A.2.a.[1] and [2]).
11. Compare endosteum and periosteum in terms of location, thickness, number of layers, and cell types present (II.B).
12. Compare compact and spongy bone in terms of the presence of cavities and trabeculae, histologic structure under high-power magnification, and location (III.B.1 and 2).

---

[1] See footnote on page 1.

13. Compare primary and secondary bone (III.C.1 and 2) in terms of:
    a. Relative permanence
    b. Type prevalent in adults
    c. Orientation of collagen fibers
    d. Cellularity (cell-to-matrix ratio)
    e. Presence of lamellae
    f. Relative mineral content
14. Sketch an osteon (haversian system) in cross-section (III.C.2.b; Fig. 8–1) and label the following:
    a. Haversian canal
    b. Endosteum
    c. Blood vessel
    d. Nerve
    e. Lymphatic vessel
    f. Lamellae
    g. Lacunae
    h. Osteocytes
    i. Filopodia
    j. Canaliculi
15. Compare haversian and Volkmann's canals in terms of contents, orientation, and encirclement by bony lamellae (III.C.2.b).
16. Beginning with mesenchyme, list the steps in intramembranous bone formation (III.C.1.a).
17. Beginning with mesenchyme, list the steps in endochondral bone formation (III.C.1.b).
18. Compare your answers to questions 16 and 17. At what point do the two processes diverge? At what point do they reconverge? What is the major difference between the two types?
19. Beginning with the zone of resting cartilage and ending with the zone of ossification, name in order the zones of endochondral bone formation seen in longitudinal section of an epiphyseal plate (III.C.1.b.[1].[c]–[f]). In which zone are there large isogenous groups of chondrocytes?
20. Compare calcified cartilage matrix and bone matrix in terms of staining properties (III.C.1.c).
21. How do long bones grow in length and width (III.C.3)?
22. Describe the steps in long bone fracture repair (III.C.4). Compare this with your answer to question 17.
23. Using what you have learned about the mechanisms of bone growth and remodeling, describe the cellular events that must occur in the bony alveolus (socket) of a tooth to allow a permanent reorientation of that tooth through the application of braces (III.C.1.a and 2.a).
24. Describe the effects of the following on bone tissue:
    a. Increased circulating parathyroid hormone (III.D.1.a)
    b. Increased circulating calcitonin (III.D.1.b and 3.b)
    c. Low levels of calcium in the blood (III.D.1.a)
    d. High levels of calcium in the blood (III.D.1.b)
    e. Dietary deficiency of protein and vitamin C (III.D.3.a and f)
    f. Vitamin D deficiency (III.D.3.c)
    g. Vitamin A deficiency (III.D.3.d)
    h. Vitamin A excess (III.D.3.e)
    i. Insufficient growth hormone production in children (III.D.4.b)
    j. Excess growth hormone production in children (III.D.4.b)
    k. Excess growth hormone production in adults (III.4.b)
25. Compare osteomalacia (III.D.3.b) and osteoporosis (III.D.2) in terms of etiology and effect on the mineral-to-matrix ratio in bone.
26. Compare synarthroses and diarthroses in terms of the movement they permit (IV.A and B).
27. Compare synostoses, synchondroses, and syndesmoses in terms of the tissue between the bones. Give examples of body sites where each occurs (IV.A).
28. Draw a schematic diagram of a diarthrosis (Fig. 8–2) and label the following:
    a. Articular cartilage
    b. Bone
    c. Periosteum
    d. Joint (articular) capsule
    e. Fibrous layer of the capsule
    f. Synovial membrane
    g. Joint cavity
29. Describe synovial fluid in terms of its location, the tissue that produces it, the cell type that produces its hyaluronan, its composition, and its function (IV.B).

SYNOPSIS

I. GENERAL FEATURES OF BONE

Bone is the main constituent of the adult skeletal system. Like cartilage, it is a skeletal connective tissue that is specialized for support and protection.

A. Composition: All mature bone tissue contains cells (osteocytes, osteoblasts, and osteoclasts), fibers (type-I collagen), and ground substance. It differs from other connective tissues primarily in having large inorganic salt deposits in its matrix, which account for its hardness.

B. Functions: Bone is second to cartilage in its ability to withstand compression and second to enamel in hardness. It supports and protects fragile tissues and organs, harbors hematopoietic tissue (bone marrow; see Chapter 13), and forms levers and pulleys that multiply and focus the contractile forces of muscle. The constant turnover of bone tissue results from a balance between the activities of the bone-forming osteoblasts and the bone-resorbing osteoclasts and allows bone matrix to function as an important storage site for calcium and other essential minerals (see III.D).

C. Types of Bone Tissue: Bone tissue is classified by its architecture as spongy or compact and by its fine structure as primary (woven) or secondary (lamellar). All bone tissue begins as primary bone, but nearly all is eventually replaced by secondary bone. The distinction between intramembranous and endochondral bone is based on histogenesis but is not microscopically detectable in mature bone.

D. Terminology: The term "bone" refers to both a tissue and an organ. Individually named elements of the adult skeleton, bones are organs composed largely of bone tissue that also contain other connective tissues, bone marrow, blood vessels, and nerves (II).

II. BONES

The adult skeleton comprises more than 200 bones, which, together with cartilage and ligaments, form the body's supportive framework.

A. Shape: Bones are classified by their shape (eg, long bones, flat bones) and the process by which they form (endochondral bones, membrane bones). Most exhibit protuberances that serve as attachment sites for muscles, tendons, and ligaments.

B. Surfaces: The outer surfaces of bones are covered by a double-layered connective tissue coat known as the **periosteum.** The periosteum's outer, or **fibrous, layer** is dense connective tissue; its inner, or **osteogenic, layer** is looser tissue containing bone cell precursors. **Sharpey's fibers** are periosteal collagen fibers that penetrate bone matrix to anchor periosteum to bone. The internal surfaces of bones are covered by a thinner, condensed reticular connective tissue (5.III.B) called **endosteum,** which contains bone and blood cell precursors. The endosteum lines marrow cavities and extends into haversian canals (III.C.2.b).

C. Parts of Long Bones: Most bones (eg, femur) are categorized as long bones; knowledge of their parts is important in the study of regional bone histology. The **diaphysis** is the long bone's shaft, and the **epiphyses** are its bulbous ends. The diaphysis is cylindrical, with walls of compact bone (III.B.2) and a central marrow cavity lined with endosteum. Each epiphysis contains mostly spongy bone. Where bones contact other bones to form movable joints (IV.B), their surfaces are covered by articular cartilage.

III. BONE TISSUE

A. Composition: Bone is a connective tissue composed of cells, fibers, and ground substance. **Bone matrix,** which contains abundant mineral salts, is the chief tissue component. Bone's hardness makes it difficult to section. Obtaining thin sections involves grinding bone slices until translucent, or demineralizing fixed bone by immersion in dilute acid or calcium-chelating agents (eg, EDTA). Demineralized bone can be sectioned and stained by standard methods.
  1. **Bone cells.**
    a. **Osteoprogenitor cells** are stem cells found in endosteum and periosteum. These spindle-shaped cells have ovoid to elongate nuclei and unremarkable cytoplasm. Two types are distinguishable in EMs: one forms osteoblasts and the other forms osteoclasts. Os-

teoblast precursors derive from mesenchyme and have sparse RER and Golgi complexes. Osteoclast precursors derive from blood monocytes and have abundant free ribosomes and mitochondria.

b. **Osteoblasts,** the major bone-forming cells, are cuboidal; each possesses a large, round nucleus and a basophilic cytoplasm. These cells form one-cell-thick sheets resembling simple cuboidal epithelium on surfaces where new bone is deposited. Osteoblasts exhibit high alkaline-phosphatase activity and have the well-developed RER and Golgi complex typical of protein-secreting cells. They synthesize and secrete all the organic components of bone matrix (III.A.2.a) and participate in bone mineralization. After they are surrounded by matrix, osteoblasts are mature and are called osteocytes.

c. **Osteocytes** are terminally differentiated bone cells found in cavities in the bone matrix called **lacunae.** Their long, thin cytoplasmic processes, called **filopodia,** radiate from the cell body in fine extensions of the lacunar cavity called **canaliculi.** Osteocytes are isolated from one another by the impermeable bone matrix and contact one another at the tips of their filopodia through gap junctions. This arrangement provides limited cytoplasmic continuity between the cells and explains how osteocytes obtain nutrients and oxygen and dispose of wastes at considerable distances from blood vessels. Although they are incapable of mitosis, osteocytes retain synthetic and resorptive capacity, by means of which they turn over and maintain nearby bone matrix. Osteocyte death results in bone breakdown, or resorption (III.A.1.d). Osteocytes recently derived from osteoblasts lie near bone surfaces in rounded lacunae; older cells lie farther from the surface in flattened lacunae.

d. **Osteoclasts** are bone-resorbing cells lying on bony surfaces in shallow depressions termed **Howship's lacunae.** They are large and multinucleated (2–50 per cell), with an acidophilic cytoplasm that contains many lysosomes and mitochondria and a well-developed Golgi complex. The cell surface facing the depression exhibits a **ruffled border** of plasma–membrane infoldings, forming many compartments between the cell and the bone surface. The cells release acid, collagenase, and other lytic enzymes into the compartments; these break down bone matrix and release minerals, a process called **bone resorption.** Osteoclasts respond to **parathyroid hormone (PTH)** (III.D.1.a) by enlarging their ruffled borders and increasing their activity; together, these responses result in increased blood calcium levels. PTH's effect is mediated by a signal from the osteoblasts. Calcitonin (III.D.1.b), which decreases blood calcium, reduces surface ruffling and osteoclast activity. Although their immediate precursors lie in the endosteum and periosteum, osteoclasts ultimately form by fusion of blood monocyte derivatives and are considered components of the mononuclear phagocyte system.

2. **Bone matrix.** Bone matrix contains organic components, or **osteoid,** and inorganic components, or **bone mineral.**

a. **Organic components. Osteoid** (fibers and unmineralized ground substance) constitutes approximately 50% of bone volume and 25% of bone weight.

   (1) **Fibers.** Osteoid is 90 to 95% type-I collagen. The overlapping pattern of staggered tropocollagen (5.II.A.1.b and 5) results in periodic gaps (lacunar regions), which contain as much as 50% of the hydroxyapatite crystals (mineral).

   (2) **Ground substance.** Hydroxyapatite crystals and collagen fibers are embedded in the acidic ground substance of proteins, carbohydrates, and some proteoglycans and lipids. The proteins include glycoproteins, phosphoproteins, sialoproteins (eg, osteopontin), and those containing γ-carboxyglutamic acid. The carbohydrates (GAGs) include chondroitin and keratan sulfates. Some ground substance components are hydroxyapatite crystal nucleation sites.

b. **Inorganic components.** Bone mineral accounts for approximately 50% of bone volume and 75% of bone weight. It consists chiefly of calcium and phosphate, with some bicarbonate, citrate, magnesium, and potassium and trace amounts of other metals. Calcium and phosphate form needlelike crystals of **hydroxyapatite** ($Ca_{10}[PO_4]_6[OH]_2$). Hydrated ions at the crystal surface form an enveloping **hydration shell,** through which ions are exchanged between the crystal and surrounding body fluids (III.D.1.a).

B. **Organization:** The two basic organizational classes of adult bone, spongy and compact, are similar in composition and microscopic appearance but differ in overall architecture.

1. **Spongy bone,** also called **cancellous bone,** forms a fine three-dimensional lattice with

many open spaces. The branching and anastomosing slips of bone between the spaces, termed **trabeculae,** or **spicules,** align along the lines of stress to which the bones are subjected, maximizing their weight-bearing capacity. Spongy bone fills the epiphyses of mature long bones and short bones (eg, phalanges) and lies between the thick plates, or tables, of the skull's flat bones, where it is called **diploë.** It may be either primary or secondary bone (III.C.1 and 2).

2. **Compact bone,** also called **dense bone,** or **cortical bone,** lacks the large spaces and trabeculae of spongy bone. It forms the thick diaphyseal cylinder of long bones, a thin covering around the epiphyses, and the tables of the skull's flat bones. Compact bone is always secondary bone (III.C.2).

C. **Histogenesis, Remodeling, Growth, and Repair:**
1. **Primary bone.** The first bone tissue to appear during new bone formation, or during fracture repair, is termed primary bone, or **woven bone.** This immature bone, always spongy, is subsequently replaced by secondary bone, except near the skull sutures and in alveolar bone of the mandible and maxilla. Its collagen fibers do not form concentric rings (III.C.2) but exhibit an irregular "woven" appearance. It is less mineralized than secondary bone, making it more radiolucent (penetrable by x-rays), and it has a higher osteocyte-to-matrix ratio. Primary bone can develop by means of intramembranous or endochondral bone formation.
    a. **Intramembranous** bone formation occurs within membranelike **mesenchymal condensations.** The cells in such connective tissue membranes differentiate into osteoblasts and begin to synthesize and secrete osteoid, which later becomes mineralized. This initial site of bone formation is the **primary ossification center.** Osteoblasts surround themselves with bone matrix, forming spicules that eventually fuse into a spongy lattice of primary bone. Mesenchyme between the spicules participates in bone marrow development. Only a few human bones form entirely in this way; most are flat and called **membrane bones.** Membrane bones of the skull are the frontal and parietal bones, the mandible, and the maxilla. The term "membrane bone" also refers to the tissue type formed by this mechanism. Membrane bone forms parts of some bones, such as the temporal and occipital bones of the skull and the periosteal bone collar of endochondral bones.
    b. **Endochondral bone formation** involves replacing cartilage with bone and occurs in all except membrane bones; thus, it is easiest to remember which bones are membrane bones and that the remainder are **endochondral bones** (or "cartilage bones").
    (1) **Basic steps in the formation of an endochondral bone.**
        (a) **Cartilage model.** In the embryo, a hyaline cartilage model resembling the bone to be formed develops first.
        (b) **The periosteal bone collar.** Capillaries penetrate the perichondrium, and mesenchymal cells on its inner surface become osteoprogenitor cells. Some differentiate into osteoblasts and secrete bone matrix, creating primary bone spicules just inside the perichondrium (now the periosteum). The spicules eventually fuse to form a thin periosteal bone collar of membrane bone around the cartilage model. Thus, ironically, the first bone tissue in an endochondral bone forms by intramembranous ossification.
        (c) **Proliferation.** While the periosteal bone collar forms, structural and functional changes begin in the cartilage model. Chondrocytes near the collar proliferate rapidly, forming stacks (isogenous groups) of flattened cells parallel to the bone's long axis.
        (d) **Hypertrophy.** The chondrocytes hypertrophy rapidly into large, rounded cells that are not separated by matrix. The result is tubelike superlacunae filled with columns of hypertrophic chondrocytes, which secrete **type-X collagen.**
        (e) **Calcification.** As hypertrophy progresses, the strips of cartilage matrix between the tubular superlacunae begin to calcify. Thus oxygen, nutrients, and cellular wastes can no longer diffuse through the matrix, and the hypertrophic chondrocytes die.
        (f) **Formation of the primary marrow cavity.** Dead cells and part of the calcified cartilage matrix are removed by chondroclasts (large, multinucleated cells resembling osteoclasts). Tunnels at the center of the developing bone, created by chondrocyte proliferation and hypertrophy and enlarged by chondroclasts, become the bone's primary marrow cavity.

(g) **The periosteal bud** is a small cluster of blood vessels and perivascular tissue from the periosteum that penetrates the primary marrow cavity. This bud and its branches invade the tunnels left by the dead chondrocytes. Osteoprogenitor cells and bone marrow stem cells, delivered by the invading blood vessels, are deposited on the calcified cartilage matrix surface.

(h) **Ossification.** This term requires attention to context. In its broadest sense, ossification is synonymous with bone formation. Here, in a more restricted connotation, it refers to the final steps (ie, osteoid deposition followed by mineralization). Osteoprogenitor cells divide and differentiate into osteoblasts, which deposit primary bone on the calcified cartilage matrix strips. The primary bone and the residual calcified cartilage are subsequently resorbed and replaced by secondary bone (III.C.2).

(2) **Ossification centers.** The previous steps may occur more than once during the formation of a bone. In long bones, the process occurs first in mid-diaphysis, forming the **primary ossification center. Secondary ossification centers** form later, by the same process, in the epiphyses. The region between primary and secondary ossification centers is the **metaphysis.** Ossification centers enlarge until all that separates them is a thin plate with resting cartilage at its center—the epiphyseal plate. The primary and secondary ossification centers of a bone should not be confused with primary and secondary bone. In some bones, **tertiary ossification centers** subsequently form the bony tubercles and ridges to which large muscle groups or ligaments attach. In humans, the first bone to ossify is the clavicle.

c. **Histologic appearance of developing endochondral bone.** The microscopic structure of the metaphyses of developing endochondral bones has five overlapping zones. The **zone of resting cartilage** is typical hyaline cartilage and is farthest from the primary marrow cavity. The **zone of proliferation** contains stacks (isogenous groups) of flattened chondrocytes. In the **zone of hypertrophy,** chondrocytes in the stacks are enlarged and rounded. The **zone of calcification,** in H & E–stained sections, is characterized by a more basophilic matrix. Overlap often results in a single **zone of hypertrophy and calcification.** The **zone of ossification** borders on the primary marrow cavity. It is characterized by intensely acidophilic osteoid, osteocytes within the bone matrix, and a monolayer of basophilic osteoblasts on the newly formed primary bone's surface.

2. **Secondary bone.** In adults, both dense and spongy bone are **secondary bone, or lamellar bone.**

a. **Secondary bone formation (remodeling).** Osteoclasts erode the primary bone matrix; blood vessels, nerves, and lymphatics invade the cavity formed by the erosion; and osteogenic cells in the perivascular connective tissue are deposited on the cavity's walls. Osteoblasts descended from these cells, along with osteocytes released from their lacunae during resorption, deposit secondary bone in concentric layers, or **lamellae,** the oldest of which are farthest from the vessels. Owing to its greater organization, secondary bone is more efficient than the primary bone it replaces. Remodeling helps reshape growing bones to adapt to changing stresses and loads; it occurs continuously, even in adults, as secondary bone is eroded and replaced by new secondary bone.

b. **Microscopic appearance of secondary bone** (Fig. 8–1). Secondary bone is a collection of densely packed bony cylinders, each with a central endosteum-lined **haversian canal** containing lymphatic and blood vessels, nerves, and loose connective tissue. The cylinder surrounding each canal consists of a series of concentric lamellae. Collagen fibers in each lamella are positioned parallel to one another and nearly perpendicular to those in adjacent lamellae, an arrangement that strengthens the tissue. Osteocytes lie between the lamellae in rows of lacunae; their filopodia lie in canaliculi extending radially from each lacuna. A haversian canal, its contents, and the surrounding system of osteocytes and lamellae are termed a **haversian system, or osteon.** Vascular connections between osteons are established by **Volkmann's canals,** which run perpendicular to haversian canals and cut across the lamellae. Osteons may bifurcate but lie roughly parallel and are held together by **cementing substance,** which fills the spaces between the cylinders. Often an old osteon is only partly eroded before a new one begins to form, so that wedge-shaped portions of old lamellae, called **interstitial lamellae**, appear between recently formed osteons.

3. **Bone growth.** Bones grow from birth to early adulthood. During growth, bone tissue is continuously remodeled. Growth occurs in two directions. Growth in the **length** of long

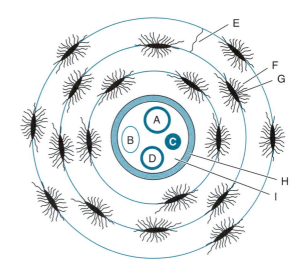

Figure 8–1. Schematic diagram of an osteon (haversian system). Labeled components include the vein (A), lymphatic vessel (B), nerve (C), artery (D), lamella (E), canaliculus (F), lacuna (G), endosteum (H), and haversian canal (I).

bones occurs primarily by means of chondrocyte division in the epiphyseal zone of proliferation, under the influence of growth hormone. Childhood growth hormone levels cause cartilage to be produced in the epiphyseal plates as fast as it can be replaced by endochondral bone formation. At puberty, growth hormone levels decline and endochondral bone gradually overtakes and replaces the remaining cartilage, a process termed **epiphyseal closure.** Growth in **girth** occurs by proliferation and differentiation of osteoprogenitor cells in the periosteum's inner layer and bone deposition on the bone's outer surface.

4. **Bone repair.** Bone fractures tear vessels in the periosteum, endosteum, and haversian and Volkmann's canals, causing local hemorrhage and clot formation between the bone's broken ends. The periosteum and endosteum provide macrophages and fibroblasts; the macrophages remove the clot, and the fibroblasts fill the breach with fibrous connective tissue. Some connective tissue cells differentiate into chondrocytes; this tissue eventually becomes a **callus,** containing islands of fibrocartilage and hyaline cartilage that serve as a model for bone formation. The presence of cartilage in the callus is typical of endochondral bones (eg, long bones), whereas flat membrane bones (eg, the mandible) often heal without cartilage formation. Beginning in the subperiosteal region (as soon as 2 days after an injury in young people), the callus is replaced by primary bone, which is subsequently remodeled and replaced by secondary bone. The time required for complete healing depends on the site and extent of the injury and is longer in older people.

D. **Histophysiology of Bone:**

1. **Calcium reserve.** The skeleton contains 99% of the body's calcium, which serves as a cofactor for many enzyme systems and is important in muscle contraction, transmission of nerve impulses, blood clotting, and cell adhesion. Blood and tissue calcium concentrations must be maintained within narrow limits, and bone serves as the calcium reservoir, storing excess calcium and releasing it when needed.

a. **Calcium mobilization.** Calcium release (mobilization) occurs by two mechanisms. **Rapid mobilization** involves ion transfer between hydroxyapatite crystals and interstitial fluid along a concentration gradient. This occurs readily where bone has a high surface-to-volume ratio (ie, around spicules of primary bone and in spongy secondary bone). The second mechanism involves **PTH** and is rapid, although slower than the first. Parathyroid chief cells (21.V.A.2) sense a decrease in blood calcium and release PTH, which increases the number of osteoclasts and activates those that exist. Equilibrium shifts toward bone breakdown **(resorption)** and the release of calcium to the blood. PTH also inhibits bone deposition by osteoblasts and reduces renal calcium excretion (see 19.II.B.4). Excessive PTH production (hyperparathyroidism) causes bone calcium depletion, elevation of blood calcium, and abnormal calcium deposition in soft tissues, especially in the kidneys and arterial walls.

b. **Calcium deposition** (storage) is promoted by **calcitonin,** a hormone secreted by the thyroid gland's parafollicular C cells (21.IV.C). Calcitonin opposes PTH effects: it enhances osteoid synthesis by the osteoblasts and the deposition of calcium. The rapid ion exchange that characterizes calcium mobilization is also involved in calcium deposition.

2. **Osteoporosis** reflects decreased bone formation or increased bone resorption. Common in chronically immobilized patients and postmenopausal women, it is characterized by decreased bone mass and a normal mineral-to-matrix ratio. Osteoporosis should not be confused with osteomalacia (see next section), in which the mineral-to-matrix ratio is subnormal.

3. **Nutritional factors.**
   a. **Protein deficiency** reduces collagen synthesis, inhibiting bone growth and maintenance.
   b. **Calcium deficiency** leads to incomplete bone–matrix calcification and, if prolonged, to bone resorption. In growing children, this causes **rickets** (ie, bone deformities, including bowed legs). In adults, it causes **osteomalacia** (ie, insufficient calcification of newly deposited bone), weakening but not deforming bones. Such bones are more susceptible to fracture and are slower to repair than healthy bones. Osteomalacia may be exacerbated by pregnancy, because of the fetal demand for calcium. This disease is characterized by a mineral-to-matrix ratio that is subnormal.
   c. **Vitamin D deficiency** reduces blood calcium. Vitamin D aids intestinal absorption of dietary calcium and reduces renal calcium excretion. Deficiency of this vitamin has effects similar to those of dietary calcium deficiency.
   d. **Vitamin A deficiency** slows bone growth and affects bone cell distribution. Poor coordination between skull and brain growth rates may cause abnormally high pressure on the brain, thereby damaging the central nervous system.
   e. **Vitamin A excess** slows cartilage growth and accelerates ossification. An excess before birth, especially during cartilage model formation, causes skeletal deformities and deletions. An excess in childhood or adolescence causes premature epiphyseal closure and small stature.
   f. **Vitamin C deficiency** inhibits bone growth and slows fracture repair because ascorbic acid is required for normal collagen synthesis.

4. **Hormonal factors.**
   a. **PTH and calcitonin.** See sections III.D.1.a and b.
   b. **Growth hormone,** produced by the anterior pituitary gland (20.III.A.2.a), stimulates overall growth, especially that of epiphyseal cartilage. Childhood growth hormone deficiency causes **pituitary dwarfism,** and an excess causes **gigantism.** Excess growth hormone in adults causes **acromegaly,** which is characterized by excessive bone thickening.
   c. **Sex steroids** (androgens and estrogens) have complex, but generally stimulatory, effects on bone formation. They affect the timing of ossification center formation and epiphyseal closure. Precocious sexual maturity, owing to increased sex hormone synthesis by tumors, may cause early epiphyseal closure and short stature. Sex hormone deficiency may delay puberty and epiphyseal closure, resulting in tall stature.

IV. JOINTS

Joints, or **arthroses,** are complex connective tissue structures that join bones to form the skeletal system. There are two main types.

A. **Synarthroses:** These joints, which permit little or no movement, are divided into three subclasses. **Synostoses** fuse and immobilize bones. *Example:* skull sutures of the elderly. **Synchondroses** join bones by cartilage and permit slight movement. *Example:* the joints between the ribs and sternum and those that comprise the pubic symphysis. **Syndesmoses** join bones by dense connective tissue, permitting slight movement. *Example:* skull sutures of younger people and inferior tibiofibular articulation.

B. **Diarthroses:** These are movable joints, such as those between long bones (Fig. 8–2). The articulating surfaces are covered by **articular cartilage** (hyaline cartilage without a perichondrium), which provides a smooth surface. The bones are joined by a two-layered connective tissue **joint capsule** that seals the **articular cavity** from surrounding tissues. The outer, or **fibrous,**

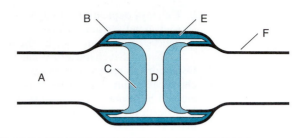

Figure 8–2. Schematic diagram of a diarthrosis (movable joint). Labeled components include the bone (A), fibrous layer of the joint capsule (B), articular cartilage (C), joint cavity (D), synovial layer of the joint capsule (E), and periosteum (F).

layer of dense connective tissue is continuous with the periosteum and supports the joint. The inner layer is the **synovial membrane** and contains two main cell types. The phagocytic **A cells** contain abundant lysosomes and help clear debris formed during friction between the articular cartilages. **B cells** contain abundant RER and help produce the **synovial fluid** that fills the articular cavity. This fluid is viscous, owing to hyaluronan, and lubricates the articular cartilage, further reducing friction. Some diarthroses (eg, the knee) are reinforced by ligaments inside or outside the articular cavity (eg, cruciate and collateral ligaments, respectively), and most are stabilized by surrounding muscles and tendons.

MULTIPLE-CHOICE QUESTIONS

Select the single best answer.

8.1. Which of the following cell types in bone is most likely derived from blood monocytes?
(A) Endothelial
(B) Fibroblast
(C) Osteoblast
(D) Osteoclast
(E) Osteocyte

8.2. Which of the following impedes the distribution of nutrients and oxygen to osteocytes?
(A) Bone matrix
(B) Canaliculi
(C) Filopodia
(D) Gap junctions
(E) Haversian canals
(F) Periosteum

8.3. Which of the following bones begins to ossify first in humans?
(A) Clavicle
(B) Femur
(C) Hard palate
(D) Parietal diploë
(E) Seventh cervical vertebra

8.4. Which of the following is true of haversian canals but false regarding Volkmann's canals?
(A) Are surrounded by concentric bony lamellae
(B) Carry nerve fibers
(C) Carry blood vessels
(D) Are found in compact bone
(E) May be found in the diaphyses of adult long bones

(F) Are lined by endosteum
(G) Are lined by periosteum

8.5. Which of the following is true of endosteum?
(A) Is composed of two layers: osteogenic and fibrous
(B) Is continuous with the joint capsule
(C) Is attached to the surface of bone by Sharpey's fibers
(D) Lines the marrow cavity
(E) Contains mature osteocytes

8.6. Which of the following is true of compact (dense) bone?
(A) Is the predominant bone tissue in the epiphyses of adult long bones
(B) Forms the diploë
(C) Can be either primary or secondary bone
(D) Lines the marrow cavity
(E) Is also called cancellous bone
(F) Is characterized by the presence of osteons
(G) Is typically more radiolucent than other types of bone

8.7. Which of the following is true of primary bone?
(A) Is the first bone tissue to appear during both intramembranous and endochondral bone formation
(B) Is also termed lamellar bone
(C) Has a higher mineral content than secondary bone
(D) Contains collagen fibers regularly arranged in layers of parallel bundles, which spiral in opposite directions in adjacent layers

(E) Is the product of bone remodeling
(F) May appear as either spongy or compact bone
(G) Is characterized by the presence of Volkmann's canals

8.8. Which of the following is true of intramembranous bone formation?
(A) Involves zones of proliferation, hypertrophy, and calcification
(B) Is the process by which the periosteal bone collar is formed
(C) Yields the epiphyseal plates between primary and secondary ossification centers
(D) Forms most of the bony skeleton in humans
(E) Involves the formation of osteoid by osteoclasts
(F) Involves the direct formation of secondary bone without a primary bone model
(G) Forms small long bones such as those in the fingers

8.9. Which type of joint contains a joint cavity filled with synovial fluid?
(A) Diarthrosis
(B) Synchondrosis
(C) Syndesmosis
(D) Synostosis

8.10. Which of the following hormones is most responsible for increasing osteoclast number and activity?
(A) ADH
(B) Calcitonin
(C) Estrogen
(D) Parathyroid hormone
(E) Testosterone

For questions 8.11 through 8.26, select the letter in Figure 8–3 that matches each of the following.

8.11. Osteoblast
8.12. Osteocyte
8.13. Mesenchyme
8.14. Osteoid
8.15. Osteoclast
8.16. Calcified bone matrix
8.17. Uncalcified bone matrix
8.18. Most basophilic cell type
8.19. Most acidophilic cell type
8.20. Direct precursor of osteocyte
8.21. Site of bone marrow development
8.22. Most active in synthesis of osteoid
8.23. Contains abundant lysosomes
8.24. Cell most responsible for bone resorption
8.25. Cell with most abundant RER
8.26. Characteristically exhibits a ruffled border

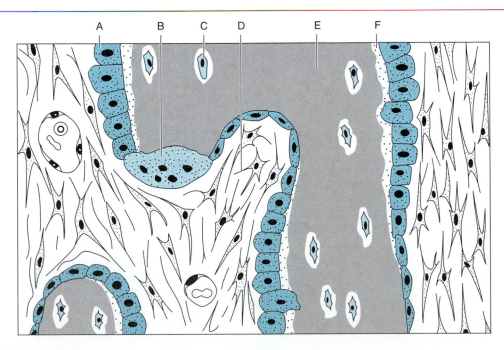

Figure 8–3.

ANSWERS TO MULTIPLE-CHOICE QUESTIONS

8.1. D (III.A.1.d)
8.2. A (III.A.1.c)
8.3. A (III.C.1.b.[2])
8.4. A (III.C.2.b)
8.5. D (II.B)
8.6. F (III.B.1 and 2)
8.7. A (III.C.1, C.2.a and b)
8.8. B (III.C.1.a and b)
8.9. A (IV.A and B; Fig. 8–2)
8.10. D (III.D.1.a and b and 4.a–c)
8.11. A (III.A.1.b)
8.12. C (III.A.1.c)
8.13. D (III.A.1.a, and C.1.a)

8.14. F (III.A.1.b, 2.a, and C.1.b.[1][h])
8.15. B (III.A.1.d)
8.16. E (III.C.1.b.[1][h])
8.17. F (III.A.1.b, 2.a, and C.1.b.[1][h])
8.18. A (III.A.1.b)
8.19. B (III.A.1.d)
8.20. A (III.A.1.b)
8.21. D (III.C.1.a)
8.22. A (III.A.1.b)
8.23. B (III.A.1.d)
8.24. B (III.A.1.d)
8.25. A (III.A.1.b)
8.26. B (III.A.1.d)

INTEGRATIVE MULTIPLE-CHOICE QUESTIONS: CONNECTIVE TISSUES

Select the single best answer.

CT.1. Which of the following cell types is capable of uncoupling oxidative phosphorylation and producing heat?
(**A**) Chondrocytes
(**B**) Fibroblasts
(**C**) Mast cells
(**D**) Mesenchymal cells
(**E**) Multilocular adipocytes
(**F**) Osteoclasts
(**G**) Unilocular adipocytes

CT.2. Which of the following is the silver-stained connective tissue type shown in Figure CT–1?
(**A**) Dense irregular connective tissue
(**B**) Elastic connective tissue
(**C**) Fibrocartilage
(**D**) Loose (areolar) connective tissue
(**E**) Mesenchyme
(**F**) Fibrocartilage
(**G**) Spongy bone

CT.3. Which of the following tissue types is shown in Figure CT–2?
(**A**) Dense regular connective tissue
(**B**) Dense (compact) bone
(**C**) Elastic connective tissue
(**D**) Elastic cartilage
(**E**) Fibrocartilage
(**F**) Hyaline cartilage
(**G**) Reticular connective tissue

CT.4. Which of the following cell types secretes glycosaminoglycan but not collagen?
(**A**) Fibroblasts
(**B**) Chondroblasts
(**C**) Mast cells
(**D**) Reticular cells
(**E**) Osteoblasts

CT.5. Which of the following cell types typically releases the contents of its lysosomes into the extracellular space?
(**A**) Fibroblasts
(**B**) Macrophages
(**C**) Mast cells
(**D**) Mesenchyme cells
(**E**) Neutrophils
(**F**) Osteoclasts
(**G**) Reticular cells

CT.6. Which of the following is the predominant protein in the fibers of bone matrix?
(**A**) Type-I collagen
(**B**) Type-II collagen
(**C**) Type-III collagen
(**D**) Type-IV collagen
(**E**) Type-V collagen
(**F**) Type-X collagen

CT.7. Which of the following is the predominant protein in the fibers of hematopoietic tissues?
(**A**) Type-I collagen
(**B**) Type-II collagen
(**C**) Type-III collagen

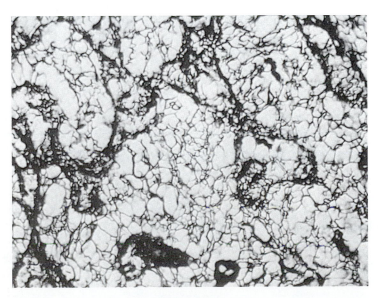

Figure CT–1.

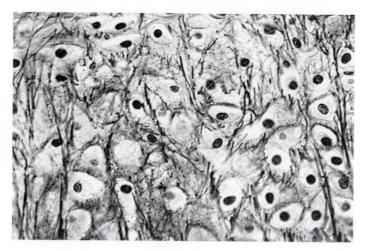

Figure CT–2.

(D) Type-IV collagen
(E) Type-V collagen
(F) Type-X collagen
CT.8. Which of the following is the predominant collagen type found in basal laminae?
(A) Type-I collagen
(B) Type-II collagen
(C) Type-III collagen
(D) Type-IV collagen
(E) Type-V collagen
(F) Type-X collagen
CT.9. Which of the following proteins is characteristically found only in hypertrophic cartilage?
(A) Type-I collagen
(B) Type-II collagen
(C) Type-III collagen
(D) Type-IV collagen
(E) Type-V collagen
(F) Type-X collagen
CT.10. Which of the following is the predominant fiber type in fibrocartilage?
(A) Type-I collagen
(B) Type-II collagen

(C) Type-III collagen
(D) Type-IV collagen
(E) Type-V collagen
(F) Type-X collagen
CT.11. Which of the following tissues is characterized by the fewest embedded capillaries?
(A) Bone
(B) Dense regular connective tissue
(C) Hyaline cartilage
(D) Loose connective tissue
(E) Reticular connective tissue
(F) Unilocular adipose tissue
CT.12. Which of the following is the initial tissue-forming process shared by cartilage, endochondral bone, and membrane bone?
(A) Appositional growth
(B) Callus formation
(C) Type-II collagen secretion
(D) Interstitial growth
(E) Mesenchymal condensation

ANSWERS TO INTEGRATIVE MULTIPLE-CHOICE QUESTIONS

CT.1. E (6.III.A and C)
CT.2. F (5.II.A.2 and II.B)
CT.3. D (7.II.B.1)
CT.4. C (5.II.E.1.b and c and 2.a; 7.I.C; 8.III.A.1.b and 2.a)
CT.5. F (8.III.A.1.d)
CT.6. A (8.III.A.2.a.[1])

CT.7. C (5.II.A.2 and B)
CT.8. D (5.II.A.2)
CT.9. F (5.II.A.2; 8.III.C.1.b.[1][d])
CT.10. A (7.II.C.1)
CT.11. C (8.I.B)
CT.12. E (7.II.A.3; 8.III.C.1.a and b)

Nerve Tissue

9

OBJECTIVES

This chapter should help the student to:

- List the structural and functional features distinguishing nerve tissue from other basic tissue types.
- List the cell types that make up nerve tissue and describe the structure, function, location, and embryonic origin of each.
- Describe in detail how neurons receive, propagate, and transmit signals.
- Describe a neuron's organelles in terms of their location and roles in neuronal impulse transmission and neuronal repair.
- Describe synapses in terms of their structural components, function, and classification.
- Describe nervous system organization in terms of the structure, functions, distribution, and distinguishing features of its subsystems.
- Describe the structure and function of the meninges.
- Describe the response of nerve tissue to injury.
- Recognize the type of nerve tissue displayed in a micrograph and identify its cells and cell processes.

MAX-Yield™ STUDY QUESTIONS

1. List the basic functions of nerve tissue (I.A.1; VII.B[1]).
2. Compare the central nervous system (CNS) and peripheral nervous system (PNS) (I.D.1; Table 9–1) in terms of:
   a. Major components (organs)
   b. Names given to a collection of nerve cell bodies
   c. Names given to a collection of nerve cell fibers
   d. Supporting cells (neuroglia) present
   e. Cells responsible for myelination
   f. Cells that invest unmyelinated fibers
   g. Embryonic origin
3. Compare gray matter and white matter (Table 9–1) in terms of:
   a. Predominant neuronal components (cell bodies, axons, dendrites)
   b. Amount of myelin present
   c. Predominant astrocyte type (III.A.1)
   d. Abundance of synapses
4. List two basic subdivisions of the autonomic nervous system (ANS) (I.D.2; Table 9–2).
5. Compare the sympathetic and parasympathetic nervous systems (I.D.2; Fig. 9–1; Table 9–2) in terms of:
   a. Locations of cell bodies of preganglionic neurons
   b. Locations of cell bodies of postganglionic neurons
   c. The neurotransmitter released by axons of postganglionic neurons
   d. Primary function (sensory or motor)
6. Beginning with the formation of the neural plate, list the basic steps in nervous system development (I.E).

---

[1] See footnote on page 1.

7. List the cell types derived from the embryonic neural crest (I.E).
8. Compare the dura mater, arachnoid, and pia mater (I.G) in terms of:
   a. Location
   b. Attachments (eg, periosteum, brain, spinal cord)
   c. Tissue type
   d. Presence of blood vessels
9. Describe the blood–brain barrier in terms of its structural correlates and its function (I.H; III.A.1).
10. Compare multipolar, bipolar, and pseudounipolar neurons (II.C and D; Table 9–3) in terms of their:
    a. Number of axons
    b. Number of dendrites
    c. Typical function
    d. Location in the body (include examples)
11. Compare axons (II.C) and dendrites (II.B) in terms of:
    a. Number per neuron
    b. Relative length
    c. Presence of surface projections
    d. Primary function
    e. Presence of Nissl bodies (RER and ribosomes)
    f. Degree of branching
    g. Variation in diameter as a function of distance from the perikaryon
    h. Content of synaptic vesicles
12. Draw a terminal bouton and its associated synapse (see Fig. 9–2) and label the synaptic vesicles, mitochondria, presynaptic membrane, synaptic cleft, and postsynaptic membrane.
13. Compare protoplasmic astrocytes and fibrous astrocytes in terms of their location and the length and diameter of their cell processes (III.A.1).
14. Compare astrocytes, oligodendrocytes, and microglia (III.A.1–3) in terms of:
    a. Nuclear shape, size, and staining intensity
    b. Relative number of cell processes
    c. Ability to form myelin
    d. Relationship to the mononuclear phagocyte system
15. Describe ependymal cells in terms of their embryonic origin and location (III.A.4).
16. Compare neurons (II) and neuroglia (III) in terms of:
    a. Cytoplasmic staining properties and visibility (II.A; III.A)
    b. Nuclear size (III.A)
    c. Relative number in the CNS (III.A)
    d. Capacity for proliferation in adults (VIII.A and B.1)
    e. Embryonic origin (I.E)
    f. General function (I.A.1 and 2)
17. Which part of a Schwann cell forms the myelin sheath, and what is the predominant biochemical constituent of myelin (III.B.1)?
18. Compare myelinated and unmyelinated axons of the peripheral nervous system in terms of:
    a. Impulse conduction velocity (I.B)
    b. Diameter (Table 9–5)
    c. Number of axons ensheathed by a single Schwann cell (III.B.1)
    d. Presence of nodes of Ranvier (III.B.1)
    e. Action potential (diffusion versus saltatory conduction) (VII.B.2 and 5)
19. Compare Schwann cells (III.B.1) and oligodendrocytes (III.A.2) in terms of:
    a. Location
    b. Number of axons each can myelinate
    c. Number of cells per internode
    d. Whether they ensheathe unmyelinated axons
20. Compare craniospinal and autonomic ganglia (V; Table 9–4) in terms of:
    a. Location
    b. Primary function (motor or sensory)
    c. Class of neurons present
    d. Distribution of ganglion cell bodies
    e. Nuclear shape and position in ganglion cell bodies
    f. Completeness of satellite cell layer

21. How do the ganglion cells of the spiral (acoustic) ganglion differ from those in craniospinal ganglia (Table 9–4)?
22. How do intramural ganglia differ from other autonomic ganglia (Table 9–4)?
23. Draw a peripheral nerve cross-section (VI; Fig. 9–3) and label the epineurium, perineurium, endoneurium, myelin sheaths, and axons.
24. Compare the inside and outside of a resting state neuron in terms of K^+ and Na^+ ion concentrations, and approximate resting membrane potential in millivolts (VII.B.1).
25. How does a neuron maintain its resting membrane potential (VII.B.1)?
26. Beginning with an excitatory synaptic stimulus, list the events leading to the generation of an action potential (VII.B.2).
27. After depolarization, how does a neuron reestablish its resting membrane potential (VII.B.2 and 3)? Does this process require ATP? How long does it take?
28. Beginning with the spread of an action potential into a terminal bouton, list the sequence of events leading to depolarization of a postsynaptic membrane (IV.A–C).
29. What happens to acetylcholine after it binds to receptors in the postsynaptic membrane (IV.C)?
30. List several neurotransmitters (Table 9–3). Which one is inhibitory?
31. Compare nerve fibers (types A, B, and C) in terms of myelination, fiber diameter, and internode length (Table 9–5).
32. What happens to the following after a nerve is cut (VIII.A, B.1 and 2)?
    a. Distal stumps of injured axons
    b. Myelin distal to the cut
    c. Schwann cells distal to the cut
    d. Nissl bodies of perikarya
    e. Volume of perikarya
    f. Position of nuclei in perikarya
    g. Proximal stumps of injured axons
33. If a large gap exists between the proximal and distal cut ends of a nerve (VIII.B.2), what may happen to the neurites and effector organ?

SYNOPSIS

I. GENERAL FEATURES OF NERVE TISSUE & THE NERVOUS SYSTEM

A. Two Classes of Cells: Nerve tissue consists of **neurons** that transmit electrochemical impulses and the supporting cells that surround them. It contains little extracellular material.

1. **Neurons** (II). These cells are specialized to receive, integrate, and transmit electrochemical messages. Each has a cell body, also called the **soma** ("body") or **perikaryon** ("around the nucleus"), comprising the nucleus and the surrounding cytoplasm and plasma membrane. Each neuron has a variable number of **dendrites** (cytoplasmic processes that collect incoming messages and carry them toward the soma) and a single **axon** (a cytoplasmic process that transmits messages to the target cell). Axons of most neurons have a **myelin sheath** formed by supporting cells and interrupted by gaps called **nodes of Ranvier.** Myelinated axon segments between the gaps are called **internodes.**

2. **Supporting cells** (III) are called **neuroglia** ("nerve glue") or **glial cells.** Their functions include structural and nutritional support of neurons, electrical insulation, and enhancement of impulse conduction velocity (VII.B.2 and 5).

B. Impulse Conduction: Signals (impulses) are propagated as a wave of depolarization along the plasma membrane of the dendrites, soma, and axon. Depolarization involves channels (ionophores) in the membrane, which allow ions (eg, Na^+, K^+) to enter or exit the cell. In unmyelinated axons, depolarization occurs in waves over the entire surface. In myelinated axons, depolarization occurs only at nodes of Ranvier, jumping from node to node (**saltatory conduction**). Impulse conduction is therefore faster in myelinated axons.

C. Synapses: Signals pass from neuron to **target cell** by specialized junctions called synapses. A target may be another neuron or a cell in the end-organ (eg, gland or muscle). At **chemical synapses** (IV), signals are transmitted by the exocytosis of **neurotransmitters**—chemicals such as acetylcholine that cross a narrow gap (**synaptic cleft**) between cells to initiate target cell depolarization. At the less common **electrical synapses,** signals are transmitted by ions flowing through a gap–junction-like complex.

D. **Subsystems of the Nervous System:** The nervous system comprises two overlapping pairs of subsystems.
1. The **central** and **peripheral nervous systems** are defined mainly by location. The **central nervous system (CNS)** includes the brain and spinal cord. The **peripheral nervous system (PNS)** includes all other nerve tissue. See Table 9–1 for terminology associated with the CNS and PNS and structural comparisons.
2. The **autonomic** and **somatic nervous systems** are defined according to function but also have distinctive anatomic features. Each has CNS and PNS components. The **autonomic nervous system (ANS;** Fig. 9–1) controls involuntary visceral functions (eg, glandular secretion, smooth muscle contraction) and has both motor and sensory pathways, although visceral sensory pathways typically are not considered part of the ANS. Each **motor pathway** consists of two neurons that synapse in a peripheral autonomic ganglion (V; Fig. 9–1). The cell body of the first **(preganglionic)** neuron is in the CNS; the cell body of the second **(postganglionic)** neuron is in the autonomic ganglion. The cell bodies of the **sensory neurons** are in craniospinal ganglia (V) and have processes that extend peripherally. The ANS is subdivided into the **sympathetic** and **parasympathetic** nervous systems, whose structure and functions are compared in Table 9–2. When they innervate the same end-organ, sympathetic and parasympathetic nerves usually have opposing effects. The **somatic nervous system** includes all nerve tissue except that of the ANS. It controls somatosensory perception (eg, touch, heat, cold) and somatomotor (voluntary) functions (eg, skeletal muscle contraction). Acetylcholine is the most common somatic neurotransmitter.

E. **Embryonic Development of Nerve Tissue:** All neurons and supporting cells derive from ectoderm. Cells of the early embryo's midline dorsal ectoderm are induced by the underlying notochord to form a thickened **neural plate.** The plate's lateral border thickens and the center invaginates, forming a troughlike **neural groove.** As the groove deepens, the lateral borders contact each other to close the groove and form the **neural tube.** Cells lining the tube elongate to form a mitotically active pseudostratified columnar epithelium (neuroepithelium), and they eventually form the layers that generate the entire CNS. As the neural groove closes, cells at its lateral borders proliferate to form two columnar masses that eventually lie dorsal to the neural tube and form the **neural crest.** Neural crest cells migrate away from the neural tube and form the PNS, including the sensory neurons of the craniospinal ganglia (V), the postganglionic neurons of the ANS, the Schwann cells of peripheral nerves, and the satellite cells of ganglia. Neural crest cells also form the meninges (I.G) and the craniofacial mesenchyme. Neural crest derivatives described in other chapters include the odontoblasts of developing teeth (15.III.C.5.b and E), the skin's melanocytes (18.II.B.2), and the adrenal medulla's chromaffin and ganglion cells (21.II.B.2).

Table 9–1. Comparisons of central and peripheral nervous systems and associated terminology.

Comparison	Central Nervous System	Peripheral Nervous System
Components	Brain and spinal cord	Peripheral nerves, ganglia, and nerve plexuses
Term(s) for collections of nerve cell bodies	**Gray matter** (groups of cell bodies in gray matter are called **nuclei**)	**Ganglia** (V) (eg, spinal or dorsal root ganglia and sympathetic chain ganglia)
Term for collections of myelinated axons	**White matter**	**Peripheral nerves**
Types of supporting cells present	Astrocytes, oligodendrocytes, microglia, and ependymal cells	Schwann cells and satellite cells
Cell type that forms myelin	Oligodendrocytes	Schwann cells
Supporting cell type that invests unmyelinated fibers	None	Schwann cells
Embryonic orgin	Neural ectoderm	Neural crest

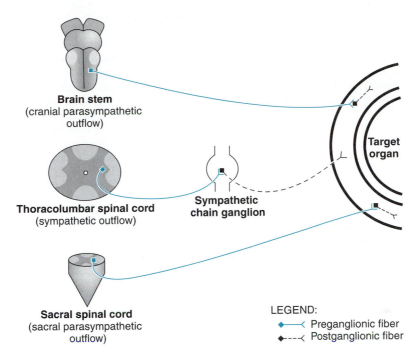

Figure 9–1. Schematic diagram of autonomic motor pathways. Note that the sympathetic (thoracolumbar) outflow involves short preganglionic and long postganglionic fibers, whereas the parasympathetic (craniosacral) outflow involves long preganglionic and short postganglionic fibers.

F. **Aging and Repair:** Mature neurons are incapable of mitosis and are often used as examples of terminally differentiated cells. Aging neurons may contain abundant lipofuscin pigment. The inability of neurons to divide makes repair of nerve tissue more difficult than repair of most other tissues. Neuron cell bodies lost through injury or surgery cannot be replaced, but if an axon is severed or crushed and the cell body remains intact, axonal regeneration is possible (VIII). Supporting cells, unlike neurons, can divide if stimulated by injury.

Table 9–2. Comparisons of motor pathways of the sympathetic and parasympathetic divisions of the autonomic nervous system.

Comparison	Sympathetic Motor Pathways	Parasympathetic Motor Pathways
Location of cell bodies of preganglionic motor neurons	**Thoracolumbar outflow.** Intermediolateral cell column of spinal cord gray matter in spinal segments T1 to L3.	**Craniosacral outflow.** *Cranial division:* Nuclei of the medulla and midbrain. Axons leave CNS through cranial nerves III, VII, IX, and X. *Sacral division:* Intermediate gray matter of sacral spinal cord segments.
Relative length of preganglionic axons	Short (some preganglionic fibers extend to adrenal medulla and are thus longer)	Long
Typical preganglionic neurotransmitter	Acetylcholine	Acetylcholine
Location of cell bodies of postganglionic motor neurons	Sympathetic chain ganglia (close to spinal column)	Intramural ganglia and other ganglia close to target organs
Relative length of postganglionic axons	Long	Short
Typical postganglionic neurotransmitter	Epinephrine or norepinephrine	Acetylcholine

G. Meninges: The brain and spinal cord are separated from the bony compartments housing them (skull and vertebral canal) by three connective tissue layers termed the **meninges.** The outer layer, or **dura mater,** is dense connective tissue bound tightly to the periosteum of the surrounding bone. The middle layer, or **arachnoid,** has two components: (1) a layer of loose connective tissue in contact with the dura mater, and (2) many connective tissue trabeculae (strands) attaching the arachnoid to the underlying pia mater. The spaces between the arachnoid trabeculae contain **cerebrospinal fluid.** Projections of the arachnoid into sinuses in the dura are called arachnoid villi. The innermost layer, or **pia mater,** is a thin, richly vascularized layer of loose connective tissue that is firmly attached to the surface of the brain or spinal cord but is separated from the neurons by neuroglial cell processes. Ramified, cuboidal, epithelium-covered projections of the pia mater into the brain's ventricles are collectively termed the **choroid plexus;** they produce the cerebrospinal fluid by selective ultrafiltration of blood plasma.

H. Blood–Brain Barrier: CNS tissue receives oxygen and nutrients from capillaries in the pia mater. These capillaries are relatively impermeable because (1) their endothelial cells lack fenestrations and are joined at their borders by tight junctions, and (2) they are partly surrounded by the cytoplasmic processes of neuroglia called astrocytes (III.A.1). These features contribute to a structural and functional barrier that protects CNS neurons from many extraneous influences and prevents certain antibiotics and chemotherapeutic agents from reaching the CNS.

II. NEURONS

A. Cell Body: The cell body (soma, perikaryon) is the neuron's synthetic and trophic center. It can receive signals from axons of other neurons through synaptic contacts on its plasma membrane and subsequently relay them to its axon. The nucleus typically is large, central, and euchromatic. It has a prominent nucleolus and heterochromatin around the nuclear envelope's inner surface. The cytoplasm of the soma contains many organelles, including mitochondria, lysosomes, and centrioles. The abundant free and RER-associated polyribosomes appear as clumps of basophilic material collectively called Nissl bodies. The Golgi complex, which is well developed, packages (and glycosylates) neurotransmitters in neurosecretory, or synaptic, vesicles. Once packaged, the vesicles are transported by molecular motor proteins down the axon to the terminal bouton (II.C). Neurotubules (microtubules) and bundles of neurofilaments (intermediate filaments) are found throughout the perikaryon and extend into the axon and dendrites.

B. Dendrites: These extensions of the soma increase the surface available for incoming signals. The farther they are from the soma, the thinner they are, owing to successive branching. Much of their surface often is covered with synaptic contacts, and some have sharp projections, termed **dendritic spines,** or **gemmules,** that act as synaptic sites. Dendrites lack Golgi complexes but contain small amounts of other organelles found in perikarya.

C. Axon: Each neuron has one axon, a complex cell process that carries impulses away from the soma. An axon is divisible into several regions. The **axon hillock,** the part of the soma leading into the axon, differs from the rest of the perikaryon in that it lacks Nissl bodies. An entire axon is usually not visible in sectioned material, but its origin is distinguishable from that of dendrites by the absence of Nissl-related basophilia. The **initial segment** is the part of a myelinated axon between the axon hillock's apex and the beginning of the myelin sheath. It is characterized by a thin layer of electron-dense material, or **dense undercoating,** beneath the plasma membrane and it contains neurotubule and neurofilament bundles originating in the axon hillock. The **axon proper** is the axon's main trunk. Unlike those of dendrites, axons' diameters tend to be constant along their entire length. The larger an axon's diameter, the more likely it is to be myelinated and the higher its rate of impulse conduction. Some axons have branches, termed **collaterals,** which may contact other neurons or even return to the soma of origin to modulate their own subsequent depolarization. The **axoplasm** (cytoplasm) contains few organelles other than some mitochondria and parallel bundles of neurotubules and neurofilaments. It has limited metabolic activity, but it conveys metabolic products to and from the axon terminals (VII.A). Signal transmission (VII.B) relies heavily on the asymmetric ion distribution (potential differences) on either side of the **axolemma,** the axonal plasma membrane. Many axons undergo branching (arborization) near their terminations. The degree of **terminal arborization** depends on axon size and the function of the axon. Each terminal branch ends in an enlargement called a **terminal end-bulb** or **terminal bouton.** Swellings in an axon's wall before its termination are termed **boutons en passage.** Each bouton contains many mitochondria and neurosecretory vesicles. A specialized region of its plasma membrane, the **presynaptic membrane,** forms part of a synapse (IV).

D. **Classification of Neurons:**   Overlapping classifications describe the wide variety of neurons in terms of their structure and function (Table 9–3).

Table 9–3. Classification of neuron types.

Criterion	Types and Description	Examples
Configuration of cell processes	**Multipolar:** Most abundant. Two or more dendrites. Dendrites radiate in many directions. Those of Purkinje cells extend in flat, fanlike array from soma.	**Motor neurons** of ventral horn of spinal cord gray matter, **pyramidal cells** of cerebral cortex, and **Purkinje cells** of cerebellar cortex.
	Bipolar: Single dendrite arising from pole of soma opposite the axon. **Special sensory** function.	Retina, olfactory mucosa, and cochlear (spiral) and vestibular ganglia of inner ear.
	Pseudounipolar: Single T-shaped process. Both branches of T resemble axons in structure and function. Impulses are carried only by processes, bypassing the soma at branch-point of T. **General sensory** function.	**Sensory neurons** of dorsal root and in most cranial ganglia. Begin in embryo as bipolar neurons; axon and dendrite subsequently fuse.
	Unipolar: Single short axon and no dendrites.	**Photoreceptor cells** (rods and cones; 24.V.E.1).
Cell size	**Golgi type I:** Large soma and long axon.	**Motor neurons** of spinal cord and **pyramidal cells** of cerebal cortex.
	Golgi type II: Small soma and short axon that undergoes extensive terminal arborization close to soma.	**Interneurons** of the spinal cord.
Function	**Motor neurons:** Carry impulses to end organs. Induce or inhibit muscle contraction and glandular secretion. Both the somatic and autonomic nervous systems have motor components.	**Multipolar neurons** of spinal cord gray matter and autonomic ganglia. **Pyramidal cells** of cerebral cortex. **Purkinje cells** of cerebellar cortex.
	Sensory neurons: Receive impulses generated by stimulation of peripheral sensory cells and organs and carry them toward central nervous system.	**Bipolar neurons** of special sensory ganglia, **pseudounipolar neurons** of craniospinal ganglia, and **unipolar neurons** (rods and cones) of retina.
	Interneurons: Carry signals between (1) motor neurons, (2) sensory neurons, and (3) motor and sensory neurons.	**Golgi type II neurons** in brain and spinal cord that coordinate neural activity and mediate reflexes.
Neurotransmitter released	**Cholinergic neurons** release acetylcholine.	Most somatic motor neurons (at neuromuscular synapses) and parasympathetic motor neurons.
	Adrenergic and noradrenergic neurons release adrenaline (epinephrine) and noradrenaline (norepinephrine) respectively.	Most postganglionic sympathetic motor neurons.
	GABAergic neurons release GABA, an inhibitory neurotransmitter.	Some neurons of cerebellum, cerebral cortex, and hippocampus.
	Dopaminergic neurons release dopamine.	Some neurons of hypothalamus (20.III.E.2).
	Serotonergic neurons release serotonin (5-hydroxytryptamine).	Parasympathetic neurons in gut and pineal gland.
	Glycinergic neurons release glycine.	Some neurons of spinal cord.

III. SUPPORTING CELLS

By providing neurons with structural and functional support, these cells play a passive role in neural activity. Positioned between the blood and the neurons, they define compartments and monitor materials passing between them. It is difficult to maintain neurons in tissue culture without adding supporting cells. As indicated in Table 9–1, different supporting cell types are found in the CNS and PNS.

A. Supporting Cells of the CNS: There are approximately 10 neuroglial cells per neuron in the CNS. Glial cells are generally smaller than neurons. Their processes, although abundant and extensive, are indistinguishable without special stains. Identification is based on nuclear morphology. The supporting cells in the CNS are the **macroglia** (astrocytes and oligodendrocytes), the microglia, and ependymal cells.

1. **Astrocytes** are the largest glial cells. Their nuclei, also the largest, are irregular, spherical, and pale-staining, with a prominent nucleolus. Their branching cytoplasmic processes are tipped by **vascular end-feet;** these surround capillaries of the pia mater and are important components of the blood–brain barrier (I.H). **Protoplasmic astrocytes (mossy cells)** are more common in gray matter. They have ample granular cytoplasm and short, thick, highly branched processes. **Fibrous astrocytes** are more common in white matter. Silver stains reveal fibrous material in their cytoplasm. Their long, thin processes are less branched than those of protoplasmic astrocytes.

2. **Oligodendroglia,** or **oligodendrocytes,** the most numerous glial cells, occur in both gray and white matter. Their spherical nuclei range between those of astrocytes and those of microglia in size and staining intensity. Like the Schwann cells of the PNS, oligodendrocytes form myelin and occur in rows to myelinate entire axons. Unlike a Schwann cell, each may provide myelin for segments of several axons. Unmyelinated axons of the CNS are not sheathed (compare with III.B.1).

3. **Microglia,** the smallest and rarest of the glia, occur in both gray and white matter. Their nuclei are small and often bean-shaped, and their chromatin is so condensed that they often appear black in H & E–stained sections. Their processes are shorter than those of astrocytes and are covered with thorny branches. Microglial cells may derive from mesenchyme, or they may be **glioblasts** (immature oligodendrocytes) of neuroepithelial origin. Some microglia are components of the mononuclear phagocyte system and have phagocytic capabilities. When neural injury is unaccompanied by vascular injury, phagocytic cells in the lesioned area appear to derive from macroglia.

4. **Ependymal cells** derive from ciliated neuroepithelial cells lining the neural tube (I.E). In adults, they retain their epithelial nature and some cilia, and they line the neural tube derivatives (the brain's ventricles and aqueducts and the spinal cord's central canal). The lining resembles a simple columnar epithelium, but ependymal cells have basal cell processes extending deep into the gray matter. The ependymal lining is continuous with the cuboidal epithelium of the choroid plexus (I.G).

B. Supporting Cells of the PNS:

1. **Schwann cells** are the supporting cells of peripheral nerves. One Schwann cell may envelop segments of several **unmyelinated** axons or provide a segment of a single **myelinated** axon with its myelin sheath. Each myelinated axon segment is surrounded by multiple layers of a Schwann cell process with most of its cytoplasm squeezed out; the remaining multilayered Schwann cell plasma membrane, called **myelin,** consists mainly of phospholipid. Gaps between the myelin sheath segments are the **nodes of Ranvier.** Ovoid or flattened Schwann cell nuclei lie peripheral to the axon they support. They are more euchromatic than the fibrocyte nuclei scattered among the axons.

2. **Satellite cells** are specialized Schwann cells in craniospinal and autonomic ganglia (V), where they form a one-cell-thick covering over the cell bodies of the neurons (ganglion cells). Their nuclei are spherical, with mottled chromatin. In sections, the nuclei typically appear as a "string of pearls" surrounding the much larger ganglion cell bodies.

IV. SYNAPSES (CHEMICAL)

Synapses are specialized junctions by means of which stimuli are transmitted from a neuron to its target cell. Artificially stimulated axons can propagate a wave of depolarization in either direction, but the signal can travel in only one direction across a synapse, which functions as a unidirectional signal valve. Synapses are named according to the structures they connect (eg, axodendritic, axosomatic, axoaxonic, and dendrodendritic). The three major structural components of each synapse are the presynaptic and postsynaptic membranes and the synaptic cleft between them (Fig. 9–2).

A. Presynaptic Membrane: This is the part of the terminal bouton membrane closest to the target cell. It includes an electron-dense thickening into which many short intermediate filaments insert, as in a hemidesmosome. In response to stimulation, neurosecretory vesicles in the bouton fuse with the presynaptic membrane and exocytose their neurotransmitters into the synaptic cleft. Neurosecretory vesicles occur only in the presynaptic component of the junction. Vesicle membrane added to the presynaptic membrane is recycled by endocytosis of the membrane lateral to the synaptic cleft. Intact vesicles do not cross the cleft.

B. Synaptic Cleft (Synaptic Gap): This is a fluid-filled space, generally 20-nm wide, between the presynaptic and postsynaptic membranes. It is shielded from the rest of the extracellular space by supporting cell processes and basal lamina material that binds the presynaptic and postsynaptic membranes together. Some clefts are traversed by dense filaments that link the membranes and perhaps guide neurotransmitters across the gap.

C. Postsynaptic Membrane: This thickening of the plasma membrane of the target cell (eg, neuron or muscle) resembles the presynaptic membrane but also contains receptors for neurotransmitters. When enough receptors are occupied, hydrophilic channels open, depolarizing the postsynaptic membrane (VII.B.2). Neurotransmitter (eg, acetylcholine) remaining in the cleft after stimulating the postsynaptic neuron (or other target cell) is degraded by enzyme (eg, acetylcholinesterase) in the cleft. Degradation products undergo endocytosis by coated pits (2.II.C.3.c) in the bouton membrane, lateral to the presynaptic thickening. Removal of excess transmitter allows the postsynaptic membrane to reestablish its resting potential and prevents continuous activation of the target cell in response to a single stimulus.

V. GANGLIA

Peripheral clusters of neuron cell bodies, called ganglia, are of two major types: **craniospinal ganglia** and **autonomic ganglia.** Each ganglion contains large **ganglion** (neuron) **cell** bodies surrounded by **satellite cells.** Cell processes are supported by Schwann cells with smaller, elongated, pale-staining nuclei. Condensed fibroblast nuclei occur in the capsule and are scattered throughout the ganglion itself. Table 9–4 compares the key structural and functional features of the two main ganglion types.

VI. PERIPHERAL NERVES

Peripheral nerves contain myelinated and unmyelinated axons, Schwann cells, and fibroblasts, but lack neuron cell bodies. Nuclei seen in peripheral nerve cross-sections belong to Schwann cells (large, pale-staining) or to fibrocytes (mature fibroblasts; small, dark-staining). Each peripheral

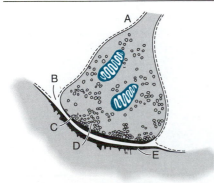

Figure 9–2. Schematic diagram of a terminal bouton and synapse. Labeled components include the bouton (A) surrounded by a thin basal lamina (dashed line) that extends into the narrow synaptic cleft (B), the presynaptic membrane (C), neurosecretory vesicles (D), and the postsynaptic membrane (E).

Table 9–4. Comparisons of the two main types of ganglia.

Comparison	Craniospinal Ganglia	Autonomic Ganglia
Examples and location	*Two types:* **ganglia of cranial nerves** in head, and **dorsal root** (or **spinal**) **ganglia** associated with spinal cord.	*Two types:* **sympathetic ganglia** lie closer to the CNS in sympathetic chain, or in separate clusters called prevertebral ganglia. **Para-sympathetic ganglia** occur farther from CNS, often as part of organs they supply (see Table 9–2), and lack connective tissue capsules typical of other ganglia.
General function	Sensory	Motor.
Neuron type (ganglion cells)	Mainly pseudounipolar. Because impulses bypass cell body (see Table 9–3), no synapses occur in these ganglia. Spiral (auditory) ganglia have bipolar neurons (see Table 9–3).	Large, multipolar neurons.
Ganglion cell distribution	Concentrated peripherally in ganglion around a core of myelinated and unmyelinated cell processes.	Randomly distributed throughout ganglion.
Ganglion cell nuclei	Large, spheric, and mostly central.	Large, ovoid, and mostly eccentric.
Relative number of satellite cells	More numerous and spheric; completely surround soma of each ganglion cell.	Less numerous. Discontinuous layer (interrupted by cell processes of multipolar neurons).

nerve (Fig. 9–3) is surrounded by a dense connective tissue sheath, or **epineurium,** branches of which penetrate the nerve, dividing the nerve fibers into bundles, or **fascicles.** The sheath surrounding each fascicle is called the **perineurium.** Fine slips of reticular connective tissue from the perineurium penetrate the fascicles to surround each nerve fiber, forming the **endoneurium.** Branches of blood vessels in the epineurium penetrate the nerve along with the connective tissue. The three main fiber types in peripheral nerves (A, B, and C) are compared in Table 9–5.

VII. HISTOPHYSIOLOGY OF NERVE TISSUE

A. Axoplasmic (Axonal) Transport: Movement of metabolic products through the axoplasm, which can be **fast** (eg, 400 mm/day) or **slow** (eg, 1 mm/day), involves neurotubules and neurofilaments. Anterograde or orthograde axoplasmic transport moves newly synthesized products and synaptic vesicles toward the axon's terminal arborization and can be fast or slow. Retrograde axoplasmic transport, the return of worn materials to the perikaryon for degradation or reutilization, is usually fast.

B. Signal Generation and Transmission: The basic function of nerve tissue is to generate and transmit signals in the form of nerve impulses, or **action potentials,** from one part of the body to another. The arrangement of neurons in chains and circuits allows the integration of

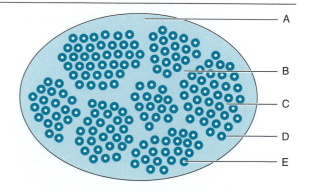

Figure 9–3. Schematic diagram of a peripheral nerve in cross-section. Labeled coverings include the epineurium (A), the perineurium (B), the endoneurium (C), and the myelin sheath (D) surrounding each axon (E).

Comparison	Type A Fibers	Type B Fibers	Type C Fibers
Relative diameter	Large	Medium	Small
Myelination	Myelinated	Myelinated	Unmyelinated
Internode length	Long	Shorter	None
Impulse conduction velocity	Fast	Medium	Slow

Table 9–5. Fiber types in peripheral nerves.

simple on–off signals into complex information. The microscopic structure of nerve tissue (eg, axon diameter, presence or absence of myelin) exploits physicochemical phenomena to regulate the rate and sequence of signal transmission.

1. **Resting membrane potential.** The K^+ concentration is 20-fold higher inside neurons than outside, whereas the Na^+ concentration is 10-fold higher outside than inside. Because the plasma membrane is more permeable to K^+ than to other ions, K^+ ions tend to leak out until the accumulated positive charge outside the cell inhibits further K^+ movement. In this state of equilibrium, the inside of the cell is negatively charged (−40 to −100 mV) relative to the outside; this potential difference (voltage) across the membrane is the **resting membrane potential.** Energy-requiring pumps in the plasma membrane help maintain the resting potential, keeping the neuron ready to receive and transmit signals. The best known pump is Na^+/K^+-ATPase, which exchanges internal Na^+ for escaped K^+ when ATP is available.

2. **Firing and propagating action potentials.** The binding of excitatory neurotransmitters (eg, acetylcholine) to receptors in the postsynaptic membrane allows positive ions to enter the cell, reducing the potential difference across the membrane. When this **membrane depolarization** reaches a critical level, or **threshold,** integral membrane proteins acting as voltage-sensitive Na^+ channels (voltage-gated channels) open, allowing Na^+ ions to rush in and reverse the membrane potential in one region of the membrane. This is the **firing** of the action potential. Incoming Na^+ ions diffuse to nearby sites, causing threshold depolarization and opening the Na^+ channels in these areas; thus, a wave of depolarization spreads along the neuron surface. Spread of the wave of depolarization is termed **propagation** of the action potential. The firing of an action potential is an "all-or-none" event and does not occur unless the threshold is reached.

3. **Refractory period.** Reversal of the membrane potential at threshold opens voltage-gated K^+ channels and allows K^+ ions to exit the cell, returning the membrane to its resting potential **(repolarization).** An even greater potential difference **(hyperpolarization)** may be achieved before stabilizing at normal resting levels. The refractory period is the 1- to 2-ms interval between the firing of the action potential and the restoration of the resting potential, during which another impulse cannot be generated. Na^+/K^+-ATPase helps restore the normal balance of ions across the membrane during this period.

4. **Direction of signal transmission.** For action potentials fired by neurotransmitters crossing a synapse, the sequence of depolarization is usually dendrites → soma → axon → synapse → next neuron (or end-organ). This is termed **orthodromic spread.** Two factors normally prevent **antidromic spread** along the axon toward the soma: (1) the region directly behind the newly depolarized region the axon is refractory, and (2) the signal cannot be propagated in a reverse direction across a synapse. In response to action potentials fired artificially by electric stimulation of an axon, both orthodromic and antidromic spread occur, but the antidromic spread has no effect because it cannot cross a synapse.

5. **Saltatory conduction.** Depolarization of myelinated axons occurs only at nodes of Ranvier, where insulation is reduced and Na^+ and K^+ channels are concentrated. The action potential must therefore jump from node to node along the axons, a phenomenon called saltatory conduction. The result is faster impulse conduction, less change in ion concentration, and thus a lower energy requirement for recovery of resting potential.

6. **Blocking signal transmission.** Cold, heat, and pressure on a nerve can block impulse conduction. Local anesthetics allow more complete and reversible impulse blocking by disturbing the resting potential. Some poisons block ion channels and prevent propagation of the action potential.

VIII. RESPONSE OF NERVE TISSUE TO INJURY

A. Damage to the Cell Body: Because mature neurons cannot divide, dead neurons cannot be replaced. Neurons not connected with other functioning neurons or end-organs are useless, and mechanisms have evolved to dispose of them. Thus, if a neuron makes synaptic contact with only one other neuron and the latter is destroyed, the former undergoes autolysis, a process termed **transneuronal degeneration.** Most neurons, however, have multiple connections.

B. Damage to the Axon: Regeneration can occur in axons injured or severed far enough from the soma to spare the cell. Partial degeneration, and subsequent regeneration, follow.

1. **Degeneration.** A crushed or severed axon degenerates both distal and proximal to the injury. **Distal** to the injury, both the axon and myelin sheath degenerate completely because the connection with the soma has been lost. During this **wallerian, descending,** or **secondary degeneration,** which takes approximately 2 to 3 days, nearby Schwann cells proliferate, phagocytose degenerated tissue, and invade the remaining endoneurial channel. **Proximal** to the injury, degeneration of the axon and myelin sheath is similar but incomplete. This **retrograde, ascending,** or **primary degeneration** proceeds for approximately two internodes before the injured axon is sealed. The cell body also changes in response to injury. The perikaryon enlarges; **chromatolysis,** or dispersion of Nissl substance, occurs; and the nucleus moves to an eccentric position. Proximal degeneration and cell body changes take approximately 2 weeks.

2. **Regeneration** begins during the third week after injury. As the perikaryon gears up for increased protein synthesis, the Nissl bodies reappear. The axon's proximal stump gives off a profusion of smaller processes called **neurites;** one of these grows into the endoneurial channel while the others degenerate. In the channel, the neurite grows 3–4 mm/day, guided and subsequently myelinated by the Schwann cells. Growth is maintained by orthograde axoplasmic transport of material synthesized in the soma. When the neurite tip reaches its termination, it connects with its end-organ or another neuron in the chain. If a severed nerve's cut ends are matched by fascicle size and arrangement and sutured together by their epineurial sheaths within 3 to 4 weeks, innervation often can be restored. If the gap between the cut ends is too wide, the neurites may fail to find endoneurial sheaths and may grow out in a potentially painful disorganized swelling called a **neuroma.** Target organs deprived of innervation often atrophy.

MULTIPLE-CHOICE QUESTIONS

Select the single best answer.

9.1. Which of the following is true of the thorny spines (gemmules) that project from dendrites?
(A) Often have a larger diameter than the dendrites themselves
(B) Typically contain numerous synaptic vesicles
(C) Represent the sites of synaptic contact with bouton terminaux
(D) Are found mainly at the nodes of Ranvier
(E) Are also called neurites

9.2. Which part of the neuron serves as its trophic center?
(A) Axon
(B) Axon hillock
(C) Cell body (soma)
(D) Dendrite
(E) Terminal arborization

9.3. Which of the following structures is shown in Figure 9–4?
(A) Dura mater
(B) Gray matter
(C) Peripheral nerve
(D) Spinal ganglion
(E) White matter

9.4. Which of the following is true of synaptic vesicles?
(A) Are found in terminal boutons
(B) Are found in the synaptic cleft
(C) Contain lysosomal enzymes
(D) Bud directly from the RER
(E) Are found in dendrites near the postsynaptic membrane

9.5. The nuclei of which of the following cells are labeled "n" in Figure 9–5?
(A) Fibrous astrocytes
(B) Microglia
(C) Multipolar neurons

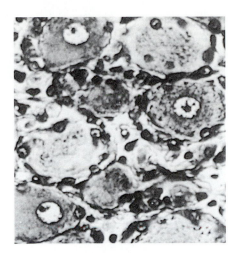

Figure 9–4.

(D) Oligodendrocytes
(E) Protoplasmic astrocytes
(F) Pseudounipolar neurons
(G) Schwann cells

9.6. Which of the following is true of Nissl bodies?
(A) Are components of the RER that lack ribosomes
(B) Undergo chromatolysis after injury to the axon
(C) Are found in the axon hillock
(D) Synthesize glycosaminoglycans
(E) Are PAS positive

9.7. Which of the following is true of resting-state axons?
(A) The K^+ concentration inside the axon is greater than that outside the axon
(B) The Na^+/K^+ pump in the plasma membrane is inactive

(C) The Na^+ concentration outside the axon is lower than that inside the axon
(D) The interior of the axon is positively charged compared with the exterior

9.8. Which of the following is true of synapses?
(A) Permit transmission of a nerve impulse in either direction
(B) When gamma aminobutyric acid (GABA) serves as the neurotransmitter, they generally have an excitatory effect
(C) Can occur between a dendrite and a neuron cell body
(D) Contain some basal lamina in the synaptic cleft

9.9. Which of the following cell types myelinate axons in the central nervous system?
(A) Ependymal cells
(B) Fibrous astrocytes
(C) Microglia
(D) Oligodendrocyte
(E) Protoplasmic astrocytes
(F) Satellite cells
(G) Schwann cells

9.10. Which of the following cell types is found predominantly in white matter?
(A) Ependymal cells
(B) Fibrous astrocytes
(C) Microglia
(D) Oligodendrocyte
(E) Protoplasmic astrocytes
(F) Satellite cells
(G) Schwann cells

9.11. Which of the following cell types may provide myelin for several axons simultaneously?
(A) Ependymal cells
(B) Fibrous astrocytes
(C) Microglia
(D) Oligodendrocytes
(E) Protoplasmic astrocytes
(F) Satellite cells
(G) Schwann cells

9.12. Which of the following cell types surrounds neuronal perikaryons located in ganglia?
(A) Ependymal cells
(B) Fibrous astrocytes
(C) Microglia
(D) Oligodendrocytes
(E) Protoplasmic astrocytes
(F) Satellite cells
(G) Schwann cells

9.13. Which of the following cell types may derive from blood monocytes?
(A) Ependymal cells
(B) Fibrous astrocytes
(C) Microglia
(D) Oligodendrocytes
(E) Protoplasmic astrocytes
(F) Satellite cells
(G) Schwann cells

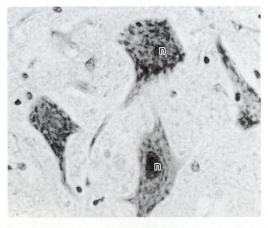

Figure 9–5.

9.14. Which of the following is the most abundant glial cell type in the central nervous system?
(A) Ependymal cells
(B) Fibrous astrocytes
(C) Microglia
(D) Oligodendrocytes
(E) Protoplasmic astrocytes
(F) Satellite cells
(G) Schwann cells

9.15. Which of the following is most tightly attached to the periosteum of the skull and the spinal column?
(A) Arachnoid mater
(B) Dura mater
(C) Gray matter
(D) Pia mater
(E) White matter

9.16. Which of the following forms trabeculae through which cerebrospinal fluid flows?
(A) Arachnoid mater
(B) Dura mater
(C) Gray matter
(D) Pia mater
(E) White matter

9.17. Which of the following consists mainly of axons and neuroglial cells?
(A) Arachnoid mater
(B) Dura mater
(C) Gray matter
(D) Pia mater
(E) White matter

9.18. Which of the labels in Figure 9–6 corresponds to the postsynaptic membrane?
(A) Arachnoid mater
(B) Dura mater
(C) Gray matter
(D) Pia mater
(E) White matter

9.19. Which of the following is true of dendrites?
(A) Each has a constant diameter along its entire length
(B) Each neuron has only one
(C) Each ends with a typical terminal bouton
(D) Are found only in white matter
(E) May contain Nissl bodies
(F) Contain neurosecretory vesicles
(G) Are typically myelinated

9.20. Which of the following is true of unmyelinated axons of the central nervous system?

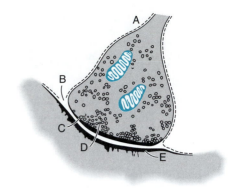

Figure 9–6.

(A) May be covered by Schwann cells
(B) Several may be covered by a single oligodendrocyte
(C) May have nodes of Ranvier
(D) Occur in both gray matter and white matter
(E) Have a faster conduction velocity than myelinated axons

9.21. The use of colchicine in cancer chemotherapy has what effect on the neurons of adults?
(A) None
(B) Slows the rate of neuronal cell division
(C) Halts all axoplasmic flow
(D) Prevents axonal degeneration
(E) Inhibits axonal regeneration

9.22. Which of the following terms is applied to collections of neuron cell bodies (somata) in the central nervous system?
(A) Ganglia
(B) Neuroglia
(C) Nodes
(D) Nuclei
(E) White matter

9.23. After an injury involving transection of an axon, which of the following changes occurs distal to the site of transection?
(A) Chromatolysis
(B) Complete axonal degeneration
(C) Degeneration of the myelin sheath back to the initial segment
(D) Movement of the nucleus to an eccentric position in the soma
(E) Neurite outgrowth from the distal stump

ANSWERS TO MULTIPLE-CHOICE QUESTIONS

9.1. C (II.B and C)
9.2. C (II.A)
9.3. D (V; Table 9–4; note the large rounded cells with central nuclei)
9.4. A (IV.A–C)
9.5. C (II.D; Table 9–3; note the multiple processes on each cell)
9.6. B (II.A and C)
9.7. A (VII.B.1)
9.8. D (IV.A–C; Table 9–3)
9.9. D (Table 9–1; III.A.2)
9.10. B (Table 9–1; III.A.1)
9.11. D (III.A.2)

9.12. F (III.B.2)
9.13. C (III.A.3)
9.14. D (III.A.2)
9.15. B (I.G; Table 9–1)
9.16. A (I.G; Table 9–1)
9.17. E (I.G; Table 9–1)
9.18. E (IV.C; Fig. 9–2)
9.19. E (II.B and C)
9.20. D (III.A.2 and B.1; Table 9–1)
9.21. E (I.G, VII.A, VIII.B.2; 2.III.I.1.b; 3.III.E)
9.22. D (Table 9–1)
9.23. B (VII.B.1 and 2)

10

Muscle Tissue

OBJECTIVES

This chapter should help the student to:

- Know the three major muscle tissue types and compare their structure, function, and location.
- Know the function(s) of muscle and the measures required to sustain life without it.
- Know the relationships among muscle fascicles, muscle fibers, myofibrils, and myofilaments.
- Explain the roles of T tubules and the sarcoplasmic reticulum in striated muscle function.
- Describe muscle stimulation, contraction, and relaxation at molecular, cellular, and tissue levels.
- Recognize the type of muscle tissue present in a micrograph and describe its function(s).

MAX-Yield™ STUDY QUESTIONS

1. Compare skeletal, cardiac, and smooth muscle in terms of:
   a. Cell size and shape (Table 10–1)
   b. Overlap of adjacent cells (IV.C[1])
   c. Presence of striations (I.F; Table 10–1)
   d. Ratio and arrangement of thick and thin filaments (II.B.1.c.[1]; IV.B.1.c; Table 10–1)
   e. Presence of distinct myofibrils (II.G; III.B.1; IV.B.1.c; Table 10–1)
   f. Intracellular membrane systems (eg, triads, dyads, caveolae) (II.B.2; III.B.1; IV.B.2; Table 10–1)
   g. T-tubule placement (II.B.2; III.B.1; Table 10–1)
   h. Nuclear number and position (Table 10–1)
   i. Motor control (voluntary or involuntary; Table 10–1)
   j. Motor end-plates (myoneural junctions; Table 10–1)
   k. Intercalated disks (III.B.2; Table 10–1)
   l. Capillary abundance (III.C)
2. Compare muscles, fascicles, fibers, and myofibrils of skeletal muscle (II.G; Fig. 10–2) in terms of their largest structural subunits and the structure that ensheathes each.
3. Sketch a longitudinal section of two resting sarcomeres attached end to end and label the thin filaments, thick filaments, A band, I band, Z line, H band, and M line (Fig. 10–2).
4. Which of the bands or lines in question 3 contain the following?
   a. Thin filaments only (II.B.1.d)   d. α-Actinin (II.B.1.c.[2])
   b. Thick filaments only (II.B.1.d)   e. No actin (II.B.1.d)
   c. Both thick and thin filaments (II.B.1.d) f. No myosin (II.B.1.d)
5. Sketch the arrangement of myofilaments (Fig. 10–2) in a cross-section of a sarcomere cut through (1) the H band lateral to the M line, (2) the A band lateral to the H band, and (3) the I band.
6. Sketch a longitudinal section through two adjacent sarcomeres during contraction (II.D; Fig. 10–4). Which bands or lines shrink (compared with your drawing for question 3)?
7. How are thin filaments attached to the Z lines in skeletal muscle (II.B.1.c.[2]) and to dense bodies in smooth muscle (IV.B.1.a)?
8. Compare thick filaments and thin filaments (II.B.1.a and b) in terms of their proteins and the names and arrangement of their subunits or components.
9. Compare troponin and tropomyosin (II.B.1.a.[2] and [3]) in terms of their structure, association with thin filaments, and function during contraction.

---

[1] See footnote on page 1.

10. Sketch a myoneural junction (motor end-plate) and label the terminal bouton, synaptic (acetylcholine) vesicles, presynaptic membrane, postsynaptic membrane, junctional folds, primary synaptic cleft, secondary synaptic clefts, and the basal lamina (Fig. 10–3).

11. How does the localized membrane depolarization caused by acetylcholine binding to the postsynaptic membrane of the myoneural junction spread throughout the muscle fiber (II.D; Fig. 10–4)?

12. Beginning with an impulse traveling down the axon of a motor neuron, list the events of skeletal muscle fiber stimulation, contraction, and relaxation (II.D and E; Fig. 10–4).

13. Compare red and white skeletal muscle fibers (II.B.3) in terms of:
   a. Myoglobin content
   b. Cytochrome content
   c. Contraction rate
   d. Main energy source
   e. Capacity for sustained activity
   f. Location

14. List the intercalated disk's components. Describe the structure and function of each (III.B.2).

15. Compare atrial and ventricular cardiac muscle cells (III.B.3.a and b) in terms of T-tubule number, cell size, and presence of small cytoplasmic granules.

16. Compare vascular and nonvascular smooth muscle in terms of the type(s) of intermediate filaments they contain (IV.B.3.a and b).

SYNOPSIS

I. GENERAL FEATURES OF MUSCLE TISSUE

A. Terminology: Special terms applied to muscle include the prefixes **sarco** and **myo.**

B. Specialization for Contraction: Muscle cells are structurally and functionally specialized for contraction, which requires two types of special protein filaments called myofilaments; these include thin filaments containing **actin** and thick filaments containing **myosin.**

C. Mesodermal Origin: Nearly all muscle arises from mesoderm. Mesenchymal cells differentiate into muscle cells through a process involving an accumulation of myofilaments in the cytoplasm and the development of special membranous channels and compartments. *Exception:* Smooth muscles of the iris arise from ectoderm.

D. Cell Shape: The length of muscle cells, which sometimes reaches 4 cm, is greater than their width. Muscle cells are therefore often called **muscle fibers,** or **myofibers.**

E. Organization: Muscle tissues are groups of muscle cells organized by connective tissue. This arrangement allows the groups to act together or separately, generating mechanical forces of varying strength. The muscles of the body (eg, biceps brachii) are organs made up of highly organized muscle tissue (II.G).

F. Types of Muscle Tissue: The main muscle tissue types are **striated muscle,** which includes both skeletal and cardiac muscle, and **smooth muscle.** Smooth muscle (IV) occurs mainly in the walls of hollow organs (eg, intestines and blood vessels); its contraction is slow (often occurring in waves) and involuntary. In histologic section, it lacks the banding pattern, or striations, seen in the other two types. **Skeletal muscle** (II) occurs mainly in association with bones, which act as pulleys and levers to multiply the force of its quick, strong, voluntary contractions. **Cardiac muscle** (III) occurs exclusively in the heart; its contractions are quick, strong, rhythmic, and involuntary. Characteristics of the different muscle types are summarized in Table 10–1.

II. SKELETAL MUSCLE

A. Histogenesis: Skeletal muscle arises from mesenchyme of mesodermal origin. The mesenchymal cells retract their cytoplasmic processes and assume a shortened spindle shape to become **myoblasts;** these fuse to form multinucleated **myotubes.** Myotubes elongate by incorpo-

Table 10–1. Distinguishing characteristics of muscle types.

Features	Skeletal Muscle (Striated)	Cardiac Muscle (Striated)	Smooth Muscle (Nonstriated)
Cells	Thick, long, unbranched, cylindric	Branched, cylindric	Small, spindle-shaped
Nuclei per cell	Many; peripheral	One or two; central	One; central
Filament ratio	Six thin/one thick	Six thin/one thick	Twelve thin/one thick
Sarcoplasmic reticulum and myofibrils	Highly organized sarcoplasmic reticulum surrounds myofibrils	Less organized sarcoplasmic reticulum; no distinct myofibrils	Poorly organized sarcoplasmic reticulum; no distinct myofibrils
T tubules	At A–I band junctions; form triads	At Z lines; form dyads	None
Motor end-plates	Present	Absent	Absent
Motor control	Voluntary	Involuntary	Involuntary
Other	Prominent fascicles	Intercalated disks at cell-to-cell junctions	Abundant caveolae
	Thick perimysium and epimysium		Cells overlap

rating additional myoblasts while myofilaments accumulate in their cytoplasm. Eventually, the accumulated myofilaments organize into myofibrils (II.B.1.c) and displace the nuclei and other cytoplasmic components peripherally.

B. Skeletal Muscle Cells: Mature skeletal muscle fibers are elongated, unbranched, cylindrical, multinucleated cells. The flattened, peripheral nuclei lie just under the **sarcolemma** (muscle cell plasma membrane); most of the organelles and **sarcoplasm** (muscle cell cytoplasm) are near the poles of the nuclei. The sarcoplasm contains many mitochondria, glycogen granules, and an oxygen-binding protein called **myoglobin,** and it accumulates lipofuscin pigment with age. Mature skeletal muscle fibers cannot divide.

1. **Myofilaments** in skeletal muscle fibers are of two major types.
    a. **Thin (actin) filaments** (Fig. 10–1) have several components.
        (1) **Filamentous actin (F-actin)** is a polymeric chain of **globular actin (G-actin)** monomers. Each thin filament contains two F-actin strands wound in a double helix.
        (2) **Tropomyosin** is a long, thin, double-helical polypeptide that wraps around the actin double helix, lies in the grooves on its surface, and spans seven G-actin monomers.
        (3) **Troponin** is a complex of three globular proteins. **TnT (troponin T)** attaches each complex to a specific site on each tropomyosin molecule, **TnC (troponin C)** binds calcium ions, and **TnI (troponin I)** inhibits interaction between thin and thick filaments.

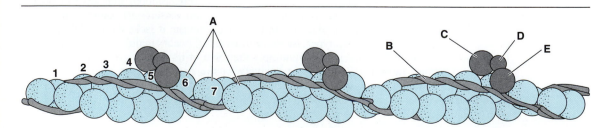

Figure 10–1. Schematic diagram of an assembled thin filament. Labeled components include globular actin (G-actin) monomers (A) assembled into a chain or polymer of filamentous actin (F-actin), a double-helical strand of tropomyosin (B) lying in the grooves of the F-actin, and the three components of the troponin complex, TnT (C), TnC (D), and TnI (E). Note that each tropomyosin molecule spans seven G-actin monomers. (Reproduced, with permission, from Junqueira LC, Carneiro J, Kelly RO: *Basic Histology,* 9th ed. Stamford, CT: Appleton & Lange, 1998.)

b. **Thick filaments.** A myosin molecule is a long, golf club–shaped polypeptide. A thick (myosin) filament is a bundle of myosin molecules whose shafts point toward and overlap in the bundle's middle and whose heads project from the bundle's ends (Fig. 10–2). This arrangement leaves a headless region in the center of each filament corresponding to the H band (II.B.1.d). Papain (a proteolytic enzyme) cleaves myosin into two pieces, at a point near the head. The piece containing most of the thin shaft is termed **light meromyosin;** the head and associated section of the shaft make up **heavy meromyosin.** The head portion of heavy meromyosin has an ATP-binding site and an actin-binding site, both of which are necessary for contraction. Heavy and light meromyosins, which are enzyme-generated fragments, should not be confused with **heavy and light myosins,** which are distinct gene products (proteins) that combine to form a thick myosin filament (see Fig. 10–2, *13*).

c. **Myofilament organization.** Skeletal muscle banding (II.B.1.d) reflects the grouping of its thick and thin myofilaments into parallel bundles called **myofibrils.** Each muscle fiber may contain several myofibrils, depending on its size.

   (1) **Myofibrils in cross-section.** EM images of myofibrils in cross-section reveal large and small dots corresponding to thick and thin filaments, respectively. Sections containing both filament types have six thin filaments in hexagonal array around each thick filament. Each thick filament shares two of its surrounding thin filaments with each adjacent thick filament to form a repeating crystalline pattern (see Fig. 10–2, *9*).

   (2) **Myofibrils in longitudinal section.** At both light and electron microscopic levels, each myofibril exhibits repeating, linearly arranged, functional subunits called **sarcomeres,** whose bands (striations) run perpendicular to the myofibril's long axis. The sarcomeres of each myofibril lie in register with those in adjacent myofibrils so that their bands appear continuous. The sarcomere is separated from its neighbors at each end by a dense Z line, or Z disk. A major Z-disk protein, α-**actinin,** anchors one end of the thin filaments and helps maintain spatial distribution. The thin filaments extend toward the middle of the sarcomere. The center of each sarcomere is marked by the M line, which holds the thick filaments in place. Desmin-containing intermediate filaments are found in both M lines and Z disks. The thick filament bundles lie at the center of each sarcomere, are bisected by the M line, and overlap the thin filaments' free ends. The overlap between thick and thin filaments produces the banding pattern and differs depending on the myofibrils' state of contraction (see Fig. 10–2).

d. **Bands.** Under the light microscope, skeletal muscle fibers exhibit alternating light- and dark-staining bands that run perpendicular to the cells' long axes. The light-staining bands contain only thin filaments and are known as I **bands** (isotropic) because they do not rotate polarized light. Each I band is bisected by a Z line. Thus, each sarcomere has two half I bands, one at each end (see Fig. 10–2, *4* and *5*). One dark-staining band lies in the middle of each sarcomere and shows the position of the thick filament bundles; this is known as an **A band** (anisotropic) because it is birefringent (rotates polarized light). At the EM level, each A band has a lighter central region, or **H band,** which is bisected by an **M line.** The H band lies between the thin filaments' free ends and contains only the shafts of myosin molecules. The darker peripheral parts of the A bands are regions of overlap between the thick and thin filaments and contain the myosin heads. Interaction between the myosin heads of the thick filaments and the thin filaments' free ends causes muscle contraction (see Fig. 10–2).

2. **Sarcoplasmic reticulum** is the SER of striated muscle cells and is specialized to sequester calcium ions. In skeletal muscle, this anastomosing complex of membrane-limited tubules and cisternae ensheathes each myofibril. At each A–I **band junction,** a tubular invagination of the sarcolemma, termed a **transverse tubule** (or **T tubule**), penetrates the muscle fiber and overlies the surface of the myofibrils. On each side of the T tubule lies an expansion of the sarcoplasmic reticulum termed a **terminal cisterna.** Two terminal cisternae and an intervening T tubule comprise a **triad.** Triads are important in initiating muscle contraction (II.D).

3. **Types of skeletal muscle fibers.** The three basic skeletal muscle fiber types differ in myoglobin content, number of mitochondria, and speed of contraction. In humans, most skeletal muscles are mixtures of these fiber types. Initially, muscle fiber types were distinguished by

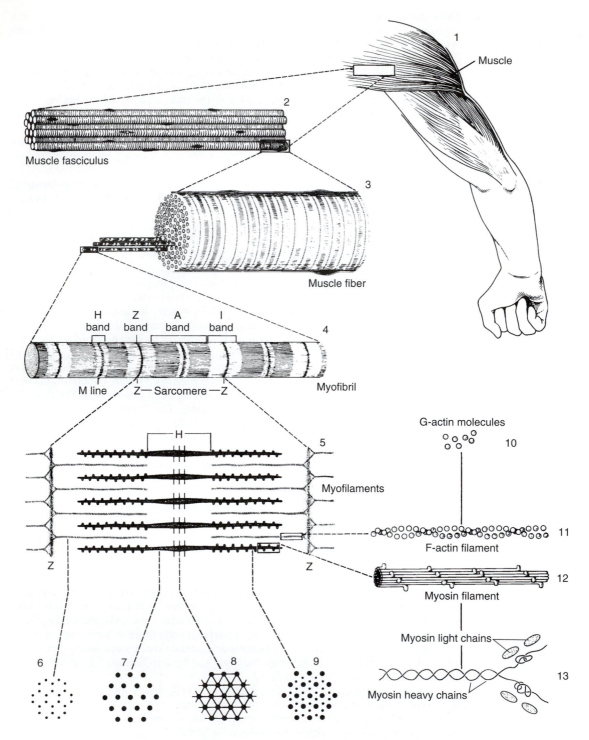

Figure 10–2. Schematic diagram showing the various levels of organization of skeletal muscle. Numbers 6 through 9 show the arrangement of the myofilaments in cross-sections through different regions of a sarcomere. (Drawing by Sylvia Colard Keene. Reproduced, with permission, from Bloom W, Fawcett DW: *A Textbook of Histology,* 9th ed. Philadelphia: WB Saunders, 1968.)

enzyme histochemistry targeting the fiber-type–specific myosin ATPase activity. Currently, immunohistochemistry targeting fiber-type–specific expression of four main **myosin heavy chain (MHC) isotypes** (I, IIA, IIB, and IIX) provides more reliable information.

    a. **Red fibers** contain more myoglobin and mitochondria. Their contraction in response to nervous stimulation is slow and steady, which has resulted in their designation as **slow fibers.** They predominate in postural muscles and occur in large numbers in certain limb muscles. Slow fibers are characterized by a predominance of MHC type I.

    b. **White fibers** contain less myoglobin and fewer mitochondria. They react quickly, with brief, forceful contractions but cannot sustain contraction for long periods; thus, they are termed **fast fibers.** These fibers predominate in the extraocular muscles. Fast fibers are characterized by a predominance of one or more MHC type-II isoforms.

    c. **Intermediate fibers** have structural and functional characteristics between those of red and white fibers but are a subclass of the latter. They are dispersed among the red and white fibers in muscles where either type predominates. As a subclass of fast fibers, intermediate fibers contain mostly MHC type-II isoforms.

C. **Motor End-Plates:** A motor end-plate, or **myoneural junction,** is a collection of specialized synapses of a motor neuron's terminal boutons with a skeletal muscle fiber's sarcolemma (Fig. 10–3). It transmits nerve impulses to muscle cells, initiating contraction. Each myoneural junction has three major components:

    1. The **presynaptic (neural) component** is the terminal bouton. Although extensions of Schwann cell cytoplasm cover the bouton, the myelin sheath ends before reaching it. The bouton contains mitochondria and acetylcholine-filled synaptic vesicles. The part of the bouton's plasma membrane directly facing the muscle fiber is the **presynaptic membrane.**

    2. The **synaptic cleft** lies between the presynaptic membrane and the opposing postsynaptic membrane and contains a continuation of the muscle fiber's **basal lamina.** It also contains acetylcholinesterase, which degrades the neurotransmitter so that when neural stimulation ends, contraction ends. The **primary synaptic cleft** lies directly beneath the presynaptic membrane and communicates directly with a series of **secondary synaptic clefts** created by infoldings of the postsynaptic membrane.

    3. The **postsynaptic (muscular) component** includes the sarcolemma (postsynaptic membrane) and the sarcoplasm directly under the synapse. The postsynaptic membrane contains acetylcholine receptors and is thrown into numerous **junctional folds.** The sarcoplasm beneath the folds contains nuclei, mitochondria, ribosomes, and glycogen, but lacks synaptic vesicles.

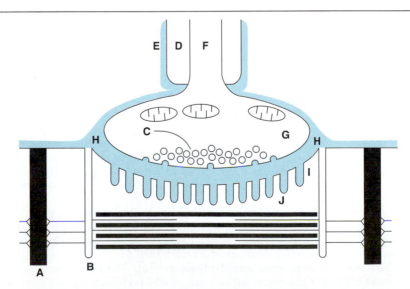

Figure 10–3. Schematic diagram of a synapse at a myoneural junction. Labeled components include the Z disk (A), transverse tubule (or T tubule) (B), synaptic vesicles (C), myelin sheath (D), basal lamina (E), axon (F), terminal bouton (G), primary synaptic cleft (H), secondary synaptic cleft (I), and junctional folds (J).

D. Mechanism of Contraction: According to the **sliding-filament hypothesis,** skeletal muscle contraction involves a multistep cascade whereby the completion of each step initiates the succeeding step (Fig. 10–4). The many steps occur nearly instantaneously. Because each step depends on the one that precedes it, a disease process that interferes with even a single step can interrupt the entire cascade and result in paralysis.

E. Relaxation: When neural stimulation ends, all of the membranes repolarize, allowing the sarcoplasmic reticulum to sequester Ca^{2+} from the sarcoplasm by active transport. This removes Ca^{2+} from the TnC and returns the TnI to a position in which it inhibits binding of the myosin head to the actin filament.

F. Energy Production: Muscles use glucose (from stored glycogen and from the blood) and fatty acids (from the blood) to form the ATP and phosphocreatine that provide chemical energy for contraction. When ATP is not available, actin–myosin binding becomes stabilized, accounting for **rigor mortis,** the muscular rigidity that occurs shortly after death.

G. Organization of the Skeletal Muscles: Each muscle (eg, biceps brachii) is a bundle of muscle **fascicles** surrounded by a sheath of dense connective tissue termed the **epimysium.** Each fascicle is a bundle of muscle fibers surrounded by a dense connective tissue sheath called the **perimysium,** comprising septumlike inward extensions of epimysium. Each muscle fiber is a bundle of myofibrils surrounded by the sarcolemma, which is in turn surrounded by a delicate connective tissue sheath termed the **endomysium,** which consists of a basal lamina and a loose mesh of reticular fibers. Each myofibril is a bundle of myofilaments surrounded by an investment of sarcoplasmic reticulum, with a triad at both A–I junctions of each sarcomere. The connective tissue investments are continuous with one another.

H. Muscle–Tendon Junctions: The attachment of muscle to tendon must be secure to prevent the muscle from tearing away during contraction. The tendon's collagen fibers blend with the epimysium and penetrate the muscle along with the perimysium. Near the junction with the tendon, the ends of the muscle cells taper and exhibit many infoldings of their sarcolemmas. Collagen and reticular fibers enter the infoldings, penetrate the basal lamina, and attach directly to the outer surface of the sarcolemma. The attachment of actin filaments to the inner surface of the sarcolemma helps stabilize the association between the collagen fibers and the muscle cell.

I. Pattern of Innervation: Each motor neuron has a single axon that may terminate on a single muscle fiber or undergo terminal branching (arborization) and terminate on multiple muscle fibers. A motor neuron and all the muscle fibers it innervates (1 to >100) comprise a **motor unit.** Muscles responsible for delicate movements (eg, extraocular muscles) are composed of many small motor units; those responsible for coarser movements (eg, gluteus maximus) are composed of fewer large motor units.

III. CARDIAC MUSCLE

A. Histogenesis: Cardiac muscle arises as parallel chains of elongated splanchnic mesenchymal cells in the walls of the embryonic heart tube. Cells in each chain develop specialized junctions between them and often branch and bind to cells in nearby chains. As development continues, the cells accumulate myofilaments in their sarcoplasm. The branched network of myoblasts forms interwoven bundles of muscle fibers, but cardiac myoblasts do not fuse.

B. Cardiac Muscle Cells: Cardiac muscle fibers are long, branched cells with one or two ovoid central nuclei. The sarcoplasm near the nuclear poles contains many mitochondria and glycogen granules and some lipofuscin pigment. Mitochondria lie in rows between the myofilaments, whose arrangement yields striations as in skeletal muscle.

1. **Sarcoplasmic reticulum and T-tubule system.** The sarcoplasmic reticulum of cardiac muscle fibers is less organized than that of skeletal muscle and does not subdivide myofilaments into discrete myofibrillar bundles. Cardiac T tubules occur at Z lines instead of A–I junctions. In most cells, cardiac T tubules associate with a single expanded cisterna of the sarcoplasmic reticulum; thus, cardiac muscle contains **dyads** instead of triads.

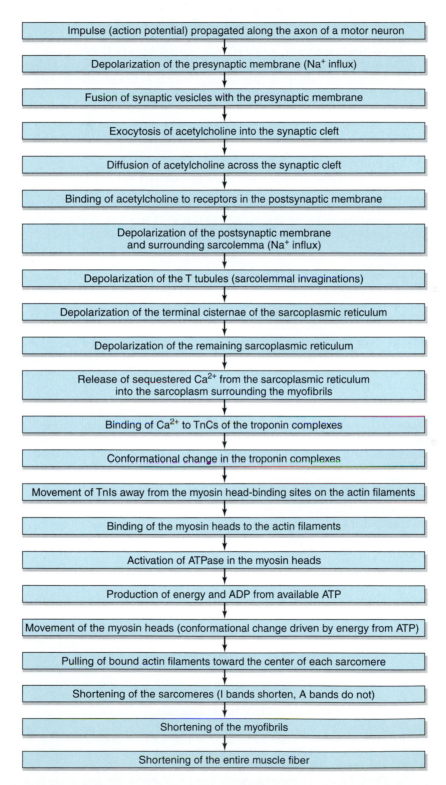

Figure 10–4. Flow diagram showing the cascade of events leading to the neural induction of skeletal muscle contraction. Each step is dependent on the successful completion of those that precede it.

2. **Intercalated disks.** These unique histologic features of cardiac muscle appear as dark transverse lines between the muscle fibers and represent specialized junctional complexes. In EMs, intercalated disks exhibit three major components arranged in a stepwise fashion.
   a. The **fascia adherens,** similar to a zonula adherens (4.IV.B.2), is found in the vertical (transverse) portion of the step. Its α-actinin anchors the thin filaments of the terminal sarcomeres.
   b. The **macula adherens (desmosome;** 4.IV.B.3) is the second component of the junction's transverse portion. It prevents detachment of the cardiac muscle fibers from one another during contraction.
   c. The **gap junctions** (4.IV.B.4) of intercalated disks form the horizontal (lateral) portion of the step. They provide electrotonic coupling between adjacent cardiac muscle fibers and pass the stimulus for contraction from cell to cell.
3. **Types of cardiac muscle fibers.**
   a. **Atrial cardiac muscle fibers** are small and have fewer T tubules than ventricular cells. They comprise many small membrane-limited granules that contain a precursor of **atrial natriuretic factor,** a hormone secreted in response to increased blood volume that opposes the action of aldosterone (19.II.B.4; 21.II.A.3.a). It acts on the kidneys to cause sodium and water loss, reducing blood volume and blood pressure.
   b. **Ventricular cardiac muscle fibers** are larger cells with more T tubules and no granules.

C. **Organization of Cardiac Muscle:** Owing to the abundant capillaries in the endomysium, cardiac muscle fibers appear more loosely arranged than those of skeletal muscle. The whorled arrangement of cardiac muscle fibers in the heart wall accounts for the myocardium's ability to "wring out" blood in the heart chambers (11.III.B.2.b).

D. **Mechanism of Contraction:** Although the arrangement of the sarcoplasmic reticulum and T-tubule complex of cardiac muscle fibers differs from that of skeletal muscle, the composition and arrangement of myofilaments are almost identical. Thus, at the cellular level, skeletal and cardiac muscle contractions are essentially the same.

E. **Initiation of Cardiac Muscle Contraction:** Unlike skeletal muscle fibers, which rarely contract without direct motor innervation, cardiac muscle fibers contract spontaneously with an intrinsic rhythm. The heart receives autonomic innervation through axons that terminate near, but never form synapses with, cardiac muscle cells. The autonomic stimulus cannot initiate contraction but can speed up or slow down the intrinsic beat. The initiating stimulus for contraction is normally provided by a collection of specialized cardiac muscle cells called the **sinoatrial node;** it is delivered by specialized cells called **Purkinje fibers** to the other cardiac muscle cells. The stimulus is passed between adjacent cells through the gap junctions of the intercalated disks. The gap junctions establish ionic continuity among cardiac muscle fibers, allowing them to work together as a functional syncytium (11.III.E).

IV. SMOOTH MUSCLE

A. **Histogenesis:** Most smooth muscle cells differentiate from mesoderm in the walls of developing hollow organs of cardiovascular, digestive, urinary, and reproductive systems. During differentiation, the cells elongate and accumulate myofilaments. Smooth muscles of the iris arise from ectoderm.

B. **Smooth Muscle Cells:** Mature smooth muscle fibers are spindle-shaped cells with a single central ovoid nucleus. The sarcoplasm at the nuclear poles contains many mitochondria, some RER, and a large Golgi complex. Each fiber produces its own basal lamina, which consists of proteoglycan-rich material and type-III collagen fibers.
1. **Myofilaments.**
   a. **Thin filaments.** Smooth muscle actin filaments are similar to those of skeletal and cardiac muscle. They are stable and are anchored by α-actinin to **dense bodies** associated with the plasma membrane.
   b. **Thick filaments.** Smooth muscle myosin filaments are less stable than those in striated muscle; they form in response to contractile stimuli (IV.D). Unlike the thick filaments in striated muscle cells (II.B.1.b), those in smooth muscle have heads along most of their length and bare areas at the ends of the filaments.

    c. **Organization of the myofilaments.** The thick and thin filaments run mostly parallel to the cell's long axis, but they overlap much more than those of striated muscle, accounting for the absence of cross-striations. The greater overlap results from the unique organization of the thick filaments and permits greater contraction. The ratio of thin to thick filaments in smooth muscle is approximately 12:1, and the arrangement of the filaments is less regular and crystalline than in striated muscle (II.B.1.c).

2. **Sarcoplasmic reticulum.** Smooth muscle cells contain a poorly organized sarcoplasmic reticulum that participates in Ca^{2+} sequestration and release but does not divide the myofilaments into myofibrillar bundles. Abundant surface-associated membrane-limited vesicles, **caveolae,** aid in Ca^{2+} uptake and release. The small size and slow contraction of these fibers make an elaborate stimulus-conducting system unnecessary; these fibers have no T tubules, dyads, or triads.

3. **Types of smooth muscle fibers.** Although they are similar in morphology, these cells can be classified according to developmental, biochemical, and functional differences.

    a. **Visceral smooth muscle** derives from splanchnopleural mesenchyme and occurs in the walls of respiratory, digestive, urinary, and reproductive organs. In addition to thick myosin and thin actin filaments, its sarcolemma-associated dense bodies are linked by **desmin**-containing intermediate filaments. Owing to their poor nerve supply, the cells transmit contractile stimuli to one another through abundant gap junctions, acting as a functional syncytium. Contraction is slow and in waves. Visceral smooth muscle is classified as **unitary smooth muscle.**

    b. **Vascular smooth muscle** differentiates in situ from mesenchyme around developing blood vessels. Its intermediate filaments contain **vimentin,** as well as desmin. It functions like visceral smooth muscle and is also classified as unitary, although its waves of contraction are localized and not sustained.

    c. **Smooth muscle of the iris.** The sphincter and dilator pupillae muscles are unique. Their cells derive from ectoderm and have a rich nerve supply. They are classified as **multiunit smooth muscle** because the cells can contract individually; they are capable of precise and graded contractions.

C. Organization of Smooth Muscle: Unlike striated muscle fibers, which abut end to end, smooth muscle fibers overlap and attach by fusing their endomysial sheaths. The sheaths are interrupted by gap junctions, which transmit the ionic currents that initiate contraction. Smooth muscle fibers form fascicles smaller than those in striated muscle. The fascicles, each surrounded by a meager perimysium, are often organized in layers separated by the thicker epimysial connective tissue. Fibers in adjacent layers may lie perpendicular to one another.

D. Mechanism of Contraction: Smooth muscle contraction involves a modified sliding-filament mechanism. First, the myosin filaments appear and the actin filaments are pulled toward and between them. Continued contraction involves the formation of more myosin filaments and further sliding of the actin filaments. The sliding actin filaments pull the attached dense bodies closer together, shortening the cell. Unlike striated muscle fibers, individual smooth muscle fibers may undergo partial peristaltic, or wavelike, contractions. During relaxation, the myosin filaments disintegrate.

E. Initiation of Smooth Muscle Contraction: Like cardiac muscle fibers, smooth muscle fibers are capable of spontaneous contraction that may be modified by autonomic innervation. Motor end-plates are absent. Neurotransmitters diffuse from terminal expansions of the nerve endings between smooth muscle cells to the sarcolemma. Sympathetic (adrenergic) and parasympathetic (cholinergic) endings are present and exert antagonistic (reciprocal) effects. In some organs, contractile activity is enhanced by cholinergic nerves and decreased by adrenergic nerves; in others, the opposite occurs.

V. RESPONSE OF MUSCLE TO INJURY

The response of muscle to injury depends on the muscle type. The wound closure mechanism always involves the proliferation of perimysial and epimysial fibroblasts and the synthesis of connective tissue matrix.

A. **Skeletal Muscle:** Small, mononucleate **satellite cells** are scattered in adult skeletal muscles within the basal lamina (endomysium) of mature fibers. Mature skeletal muscle fibers are incapable of mitosis, but the normally quiescent satellite cells can divide after muscle injury, differentiate into myoblasts, and fuse to form new skeletal muscle fibers.

B. **Cardiac Muscle:** Cardiac muscle has little regenerative ability after early childhood. Lesions of the adult heart are repaired by replacement with connective tissue scars.

C. **Smooth Muscle:** Smooth muscle contains a population of relatively undifferentiated mononucleate smooth muscle precursors that proliferate and differentiate into new smooth muscle fibers in response to injury. The same mechanism appears to be involved in adding new muscle to the myometrium as the uterus enlarges during pregnancy to accommodate the growing fetus.

MULTIPLE-CHOICE QUESTIONS

For questions 10.1 through 10.8, select the single best answer.

10.1. Which of the following is true of intercalated disks?
(A) Are found only in smooth muscle
(B) Are autonomic myoneural junctions
(C) Consist of desmosomes, fascia adherens, and gap junctions
(D) Are located at the M line
(E) Are the middle components of the triads

10.2. Which globular protein in the troponin complex inhibits the binding of myosin to actin?
(A) TnA
(B) TnC
(C) TnI
(D) TnM
(E) TnT

10.3. Which filamentous protein winds around F-actin and lies in grooves in its surface?
(A) Desmin
(B) Myoglobin
(C) Myosin
(D) Tropomyosin
(E) Troponin

10.4. Which of the following is true of motor units?
(A) Are found only in cardiac muscle
(B) Are largest in muscles responsible for delicate movements
(C) Consist of a muscle fiber and all the nerves that supply it
(D) Consist of a motor neuron and all the muscle fibers it supplies
(E) Are the same as myoneural junctions

10.5. Figure 10–5 represents a cross-section through which part of a sarcomere?
(A) A band
(B) I band
(C) H band

Figure 10–5.

(D) M line
(E) Z line

10.6. Which of the following is true of sarcoplasmic reticulum of skeletal muscle?
(A) Is a specialized form of SER
(B) Expands to form terminal cisternae near the Z line
(C) Is an invagination of the sarcolemma
(D) Carries the acetylcholine receptors of the postsynaptic membrane
(E) Forms a sheath around each myofilament

10.7. Which of the following is true of skeletal muscle myosin molecules?
(A) Can be cleaved into light and heavy myoglobin by the proteolytic enzyme papain
(B) Are attached to the Z line by α-actinin
(C) Each has a head component that includes an actin-binding site
(D) Can shorten to half their resting length during contraction
(E) Have tail components that overlap in the I band

10.8. Which of the following is true of T tubules?
(A) Are evaginations of the sarcoplasmic reticulum
(B) Sequester calcium ions during muscle relaxation
(C) Carry depolarization to the muscle fiber interior

(D) Are found overlying the A–I band junction in cardiac muscle cells

(E) Contain a rich supply of acetylcholine receptors

10.9. Which of the letters in Figure 10–6 represents the H band?

10.10. Which of the letters in Figure 10–6 represents the myosin-containing filament?

10.11. Which of the letters in Figure 10–6 represents the subregion of the A band that shortens during contraction?

10.12. The basal lamina of a muscle fiber is a component of which of the following?
(A) Endomysium
(B) Epimysium
(C) Fascia
(D) Perimysium
(E) Sarcoplasmic reticulum

10.13. Which of the following structures surrounds a muscle fascicle?
(A) Endomysium
(B) Epimysium
(C) Fascia
(D) Perimysium
(E) Sarcoplasmic reticulum

10.14. Which of the following characteristics is unique to skeletal muscle?
(A) Contains abundant caveolae
(B) Contains intercalated disks
(C) Contains sarcomeres
(D) Contains sarcoplasmic reticulum
(E) Contains triads
(F) Contains tropomyosin
(G) Contains T tubules

10.15. Which of the following characteristics is unique to cardiac muscle?
(A) Cells contain centrally located nuclei
(B) Cells are often branched
(C) Cells are striated
(D) Cells are multinucleated

(E) Cells lack T tubules
(F) Cells contain distinct myofibrils
(G) Cells contain sarcoplasmic reticulum

10.16. Which of the following characteristics is unique to smooth muscle?
(A) T tubules lie across Z lines
(B) Each thick filament is surrounded by six thin filaments
(C) Thin filaments attach to dense bodies
(D) Cells are multinucleated
(E) Cells lack T tubules
(F) Cells have distinct myofibrils
(G) Cells contain centrally located nuclei

10.17. Which of the following is true of thick filaments?
(A) Contain actin
(B) Contain tropomyosin
(C) Are found in the I band
(D) Contain ATPase
(E) Are anchored in the Z lines

10.18. Which of the following is true of thin filaments?
(A) Globular heads project from each
(B) Contain myosin
(C) Are found in the A band
(D) Are anchored in the M lines
(E) Are found in the H band

10.19. Membrane depolarization occurs in skeletal muscle fibers in which order after acetylcholine crosses the synaptic cleft?
(A) T tubule, terminal cisternae, sarcoplasmic reticulum, sarcolemma
(B) Sarcoplasmic reticulum, T tubule, presynaptic membrane, sarcolemma
(C) Terminal cisternae, sarcolemma, sarcoplasmic reticulum, caveolae
(D) Sarcolemma, T tubule, terminal cisternae, sarcoplasmic reticulum
(E) Sarcolemma, sarcoplasmic reticulum, T tubule, terminal cisternae

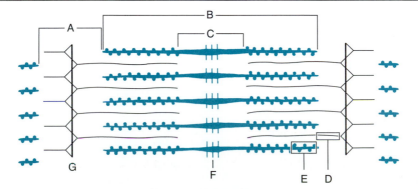

Figure 10–6.

10.20. Which of the following tissue types most often replaces cardiac muscle damaged by a myocardial infarction?
(**A**) Cardiac muscle
(**B**) Dense connective tissue
(**C**) Elastic tissue
(**D**) Skeletal muscle
(**E**) Smooth muscle

10.21. Which of the following is true of synaptic clefts of myoneural junctions?

(**A**) Are devoid of basal lamina
(**B**) Each includes one primary and many secondary clefts
(**C**) Are structurally identical to large gap junctions
(**D**) Are abundant in cardiac muscle
(**E**) Contain many mitochondria

ANSWERS TO MULTIPLE-CHOICE QUESTIONS

10.1. C (III.B.2)
10.2. C (II.B.1.a.[3])
10.3. D (II.B.1.a.[2]; Fig. 10–1)
10.4. D (III)
10.5. A (Fig. 10–2)
10.6. A (II.B.2)
10.7. C (II.B.1.b and d; Figs. 10–2 and 10–4)
10.8. C (II.B.2.D and E; Figs. 10–3 and 10–4)
10.9. C (II.B.1.d; Fig. 10–2)
10.10. E (II.B.1.d; Fig. 10–2)
10.11. C (II.B.1.c.[2] and d; Fig. 10–2)

10.12. A (II.G)
10.13. D (II.G)
10.14. E (II.B.2; III.B.1; Table 10–1)
10.15. B (III.A; Table 10–1)
10.16. C (IV.B.1.a; Table 10–1)
10.17. D (II.B.1.b; Fig. 10–4)
10.18. C (II.B.1.a and d; Fig. 10–2)
10.19. D (II.B.2 and C.1–3; Figs. 10–3 and 10–4)
10.20. B (V.B)
10.21. B (II.C.2)

INTEGRATIVE MULTIPLE-CHOICE QUESTIONS: BASIC TISSUE TYPES

Select the single best answer.

BT.1. Which of the following tissues is shown in Figure BT–1?
(A) Cardiac muscle
(B) Dense regular connective tissue
(C) Nerve tissue
(D) Smooth muscle
(E) Skelatal muscle

BT.2. Which of the following epithelial tissues is shown in Figure BT–2?
(A) Pseudostratified columnar
(B) Simple cuboidal
(C) Stratified cuboidal
(D) Stratified squamous
(E) Transitional

BT.3. Which of the following are the two predominant tissue types in Figure BT–3?
(A) Cardiac muscle and dense irregular connective tissue
(B) Dense regular connective tissue and nerve tissue
(C) Smooth muscle and dense irregular connective tissue
(D) Stratified squamous epithelium and dense irregular connective tissue
(E) Transitional epithelium and dense irregular connective tissue

BT.4. Which of the following tissues includes examples that derive from all three embryonic germ layers?
(A) Connective tissues
(B) Epithelial tissues
(C) Muscle tissues
(D) Nerve tissues

BT.5. Which of the following tissue types does not include secretion as a major function of its constituent cells?
(A) Connective tissue
(B) Epithelial tissue
(C) Muscle tissue
(D) Nerve tissue

BT.6. In which of the following tissue types are extracellular matrix components more abundant than cellular components?
(A) Connective tissues
(B) Epithelial tissues
(C) Muscle tissues
(D) Nerve tissues

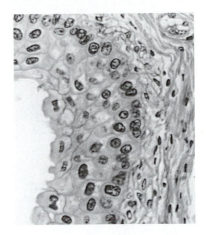

Figure BT–2.

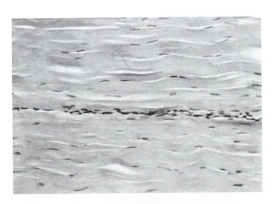

Figure BT–1.

Figure BT–3.

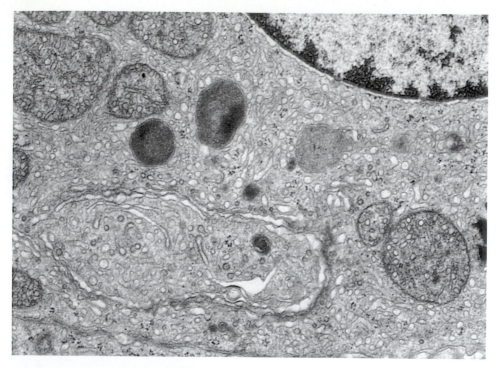

Figure BT–4.

BT.7. Which of the following tissue types is characterized by cells with specializations on their apical surfaces, including cilia, stereocilia, and microvilli?
(A) Connective tissue
(B) Epithelial tissue
(C) Muscle tissue
(D) Nerve tissue

BT.8. Which of the following products is most likely synthesized by the cell shown in Figure BT–4?

(A) Collagen
(B) Glycosaminoglycans
(C) Neurotransmitters
(D) Peptide hormones
(E) Steroid hormones

BT.9. Which of the following tissues is shown in Figure BT–5?
(A) Bone
(B) Mesenchyme
(C) Mucous connective tissue
(D) Nerve tissue
(E) Stratified squamous epithelium

BT.10. What are the two predominant tissue types in Figure BT–6?
(A) Cardiac muscle and nerve
(B) Cardiac muscle and smooth muscle
(C) Dense regular connective tissue and transitional epithelium
(D) Skeletal muscle and glandular epithelium
(E) Smooth muscle and dense regular connective tissue

BT.11. Which of the following tissues is characterized by abundant capillaries?
(A) Articular cartilage
(B) Cardiac muscle
(C) Dense regular connective tissue
(D) Elastic cartilage
(E) Stratified squamous epithelium

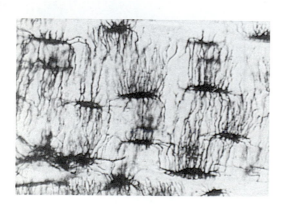

Figure BT–5.

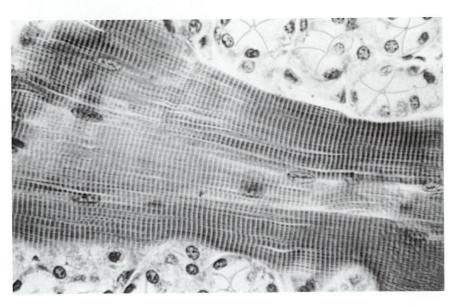

Figure BT–6.

ANSWERS TO INTEGRATIVE MULTIPLE-CHOICE QUESTIONS

BT.1. B (5.III.A.2.a)
BT.2. E (4.III.B.8)
BT.3. D (4.III.B.5; 5.III.A.2.b)
BT.4. B (4.II.H; Table 4–1)
BT.5. C (4.II.A; 5.II.E.1.b; 9.II.A; 10.I.B)
BT.6. A (4.II.G; 5.I.D; Table 5–1; 9.I.A; 10.II.G, III.C, and IV.C)

BT.7. B (4.IV.A)
BT.8. E (4.VI.C.5)
BT.9. A (8.III.C.2.b; Fig. 8–1)
BT.10. D (4.VI.C.3; 10.II.B.1.d)
BT.11. B (4.II.F; 5.III.A.2.a; 7.I.B; 10.III.C)

Part III: Organs & Organ Systems

Circulatory System

11

OBJECTIVES

This chapter should help the student to:

- Name the types, subtypes, and major functions of each circulatory system component.
- Name the three tunics that make up the walls of all circulatory system components, and know the tissue type in each tunic.
- Compare circulatory system components in terms of size and wall structure.
- Relate the wall structure of each circulatory system component to its major functions.
- Describe the heart's impulse-generating and -conducting system in terms of structure, function, location, and how the impulse is conveyed to the cardiac muscle fibers.
- Recognize the types of vessels present in a micrograph and identify all of their structural components.
- Distinguish between cardiac muscle and Purkinje fibers, and identify the endocardium, myocardium, epicardium, and valves in micrographs of the heart.
- Predict the functional consequences of a structural defect in any circulatory system component.

MAX-Yield™ STUDY QUESTIONS

1. List the general functions of the circulatory system (I.A[1]).
2. Name the two vascular systems that make up the circulatory system (I.B.1 and 2).
3. Name the four types of components that make up the blood vascular system (I.B.1).
4. Name the three types of components that make up the lymphatic vascular system (I.B.2).
5. Name two points of functional contact between the blood and lymphatic vascular systems. How does fluid from the blood vascular system enter the lymphatic vascular system? How does lymph enter the blood vascular system (IV; V.C)?
6. Name the three layers (tunics) that comprise blood vessel walls and the tissues characteristic of each (I.C.1–3).
7. Which of the layers named in answer to question 6 are absent in capillaries (I.C.1–3; II.A)?
8. Name the four major types of blood capillaries (II.A.3.a–c) and compare them in terms of diameter and the presence of fenestrae, a continuous basal lamina, and phagocytic cells in and around the capillary wall.
9. Describe three ways in which the exchange of substances (eg, proteins, fluid, salts) may occur across capillary walls between blood and tissue fluid (II.A.4).
10. Describe the action of capillary endothelial cells on angiotensin I, bradykinin, serotonin, prostaglandins, norepinephrine, thrombin, thrombus formation, and lipoproteins (II.A.2.a).
11. List the three main classes of arteries according to diameter (Table 11–1) and compare them in terms of their relative abundance (II.A.1); the composition of their tunica intima, media, and adventitia; and their function.

---

[1] See footnote on page 1.

12. Describe the carotid and aortic bodies (II.E) in terms of location, receptor class, and function.
13. List some physiologic phenomena in which arteriovenous anastomoses participate (II.G).
14. List the three main classes of veins according to their diameter (Table 11–2) and compare them in terms of their relative abundance (II.A.1); the composition of their tunica intima, media, and adventitia; and their function.
15. Compare arteries (II.B; Table 11–1) and veins (II.C; Table 11–2) in terms of:
    a. Valves
    b. Internal elastic lamina
    c. Elastin
    d. Relative tunica media thickness
    e. Smooth muscle in tunica adventitia
    f. Relative tunica adventitia thickness
    g. Relative total wall thickness
    h. Vasa vasorum (II.H)
16. Describe the innervation of blood vessels (II.H).
17. Name the heart's three tunics, the blood vessel tunic to which each corresponds, and the basic tissue type in each (III.B.1–3). Which is the thickest tunic?
18. Name the components of the heart's fibrous skeleton and describe their functions (III.C).
19. Compare the structure of valves in the heart (III.D) with that of valves in veins (Table 11–2).
20. List, in order, the components of the heart's impulse-conducting system through which an electric stimulus must pass to cause contraction of the ventricular myocardium (III.E). **Hint:** Include the cardiac muscle cells themselves.
21. Compare Purkinje fibers (III.E.5) and typical cardiac muscle cells (III.B.2.a,b and E.6) in terms of:
    a. Diameter
    b. Conduction velocity
    c. Myofilament amount and location
    d. Intercalated disks
22. In a cross-section of the heart, would you find Purkinje fibers closer to the epicardium or to the endocardium (III.B.1)?
23. How does neural control of cardiac muscle differ from that of skeletal muscle (III.H)?
24. Compare the effects of sympathetic and parasympathetic innervation on heart rate (III.H).
25. Compare lymphatic capillaries with blood capillaries (V.B) in terms of:
    a. Epithelial lining
    b. Fenestrations
    c. Zonulae occludens
    d. Basal lamina
26. In what aspects do large lymphatic vessels resemble veins (I.C.1–3; V.A)?
27. Name the two largest lymphatic vessels in the body (V.C).

SYNOPSIS

I. GENERAL FEATURES OF THE CIRCULATORY SYSTEM

A. General Function: The circulatory system is responsible for the transport and homeostatic distribution of oxygen, nutrients, wastes, body fluids and solutes, body heat, and immune system components.

B. The Two Subsystems:
1. The **cardiovascular system** is a closed system of tubes, through which the blood circulates with the aid of an in-line pump. It has four types of components: the **heart,** a muscular pump; the **arteries,** which carry blood from the heart to the tissues; the **veins,** which return blood from the tissues to the heart; and the **capillaries,** which intervene between the arteries and veins, allowing an exchange of nutrients, oxygen, and waste products between the blood and other tissues.
2. The **lymphatic vascular system** comprises another set of vessels, in which **lymph** (excess tissue fluid, cellular debris, and lymphocytes) moves in only one direction (toward the junction of the lymph vessels with the large veins in the neck). This system lacks a separate pump and includes three vessel types. **Lymphatic capillaries** are blind-ended, endothelial tubes that collect lymph from the intercellular spaces. **Lymphatic vessels** collect lymph from lymphatic capillaries. **Lymphatic ducts** collect lymph from smaller lymphatic vessels and empty into the jugular and subclavian veins.

C. **Walls of Blood and Lymphatic Vessels:** Circulatory system components are hollow, with an open channel, or **lumen,** at their center. They are described in terms of their wall structure (II). Vessel walls typically have three concentric layers, or tunics. In lymphatic vessels, tunic borders are less distinct than those in blood vessels. Local weakening of vessel walls as a result of embryonic defects, disease, or lesions may cause a thin-walled outpocketing, or **aneurysm,** that may rupture, causing a hemorrhage.

1. The **tunica intima** is the innermost layer and borders the lumen. The intima of arteries and veins and that of the heart (the endocardium) are virtually identical. It consists of **endothelium** (a simple squamous epithelium bordering the lumen, underlaid by a thin basal lamina) and **subendothelial connective tissue.** Capillaries consist solely of endothelium. In arteries, the intima is separated from the tunica media by a fenestrated layer of elastin, the **internal elastic lamina.**

2. The **tunica media,** or middle layer, consists mainly of circumferential **vascular smooth muscle** fibers. Arteries generally have a thicker media (more muscle and elastic fibers) than do veins or lymphatic vessels. Large arteries often exhibit an **external elastic lamina** between the media and the tunica adventitia. The heart's media (myocardium) is much thicker than that of the largest artery (aorta) and consists of cardiac muscle.

3. The **tunica adventitia,** the outermost layer, consists chiefly of type-I collagen and elastic fibers that anchor the vessel in the surrounding tissues. In veins, the adventitia is the thickest layer; in large veins, it may contain longitudinal smooth muscle. In all large vessels, the adventitia contains small blood vessels (**vasa vasorum**) that supply oxygen and nutrients to cells in the vessel wall too far from the lumen to be nourished by diffusion. The heart's outer layer (epicardium) is not an adventitia but rather a **serosa** (connective tissue covered on its outer surface by a **mesothelium**). The smooth surface reduces friction between the beating heart and surrounding structures.

II. BLOOD VESSELS

Blood vessels are classified according to type and size. Comparisons are based on structure (Fig. 11–1) and function and often focus on the tunics' thickness and composition (Tables 11–1 and 11–2).

A. **Blood Capillaries:** These are the smallest vascular channels, with an average diameter of 7 to 9 µm. Their walls consist of a simple squamous epithelial (endothelial) cell sheet rolled into a tube and surrounded by a thin basal lamina. The cells attach to one another at their borders by junctional complexes, including tight (occluding) junctions and gap junctions. Some blood capillaries have fenestrations (pores) in their endothelial linings.

1. **Capillary beds.** The arterial tree's basic plan is such that a few large-diameter vessels branch to feed an increasing number of smaller-diameter vessels. Capillaries are the smallest vessels and hence the most numerous. They commonly occur as components of a profusion of anastomosing (interconnecting) channels termed a **capillary bed** (see Fig. 11–1).

2. **Cells of capillaries.**

   a. **Endothelial cells,** the chief structural components of capillaries, are simple squamous epithelial cells of mesenchymal origin joined by intercellular junctions (including zonulae occludens) to form a tube. The nucleus causes each cell to bulge into the capillary lumen, but the cell thins toward its periphery to as little as 0.2 µm. Abundant pinocytotic vesicles occur throughout the cytoplasm; organelles and filaments collect near the nucleus. Key functions carried out by endothelial cells of capillaries and larger vessels include: (1) converting angiotensin I to angiotensin II (**angiotensin** regulates blood pressure by causing arterial smooth muscle contraction); (2) inactivating bioactive compounds (eg, bradykinin, serotonin, prostaglandins, norepinephrine, and thrombin) and thus regulating their effects; (3) breaking down lipoproteins (lipolysis) to yield triglycerides and cholesterol (for energy metabolism, hormone synthesis, and cell membrane assembly); (4) preventing **thrombus** (clot) formation (endothelial cells release prostacyclin, an inhibitor of platelet aggregation; damage to these cells may induce local clotting by decreasing prostacyclin release and uncovering the basal lamina, whose collagen stimulates thrombogenesis); and (5) participating in capillary transport (II.A.4).

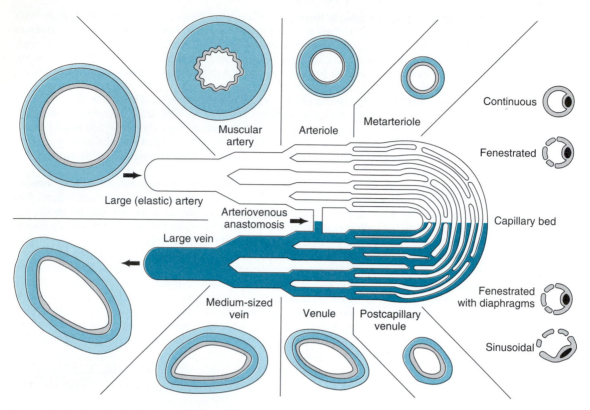

Figure 11–1. Simplified schematic diagram of the vessels of the blood vascular system. Schematic cross-sections of the various types of vessels are also shown. Compare the relative thickness of the three tunics in the cross-sections: intima (*light gray*), media (*medium color*), and adventitia (*light color*).

    b. Pericytes, or **adventitial cells,** are small mesenchymal cells scattered along capillaries. Each is surrounded by its own basal lamina and clings by long cytoplasmic processes to capillary surfaces. These mesenchymal stem cells may be contractile and may differentiate into a variety of cell types.

3. **Types of capillaries.** Capillaries, like all vessels, are classified by wall structure.
    a. Continuous capillaries have a smooth, nonporous, endothelial lining. The cells attach tightly by junctional complexes. Continuous capillaries occur in muscles, brain, and peripheral nerves.
    b. Fenestrated capillaries have endothelial cells perforated by pores **(fenestrae).** The pores may be open or covered by thin **diaphragms** that limit the size of macromolecules able to pass. Fenestrated capillaries occur in tissues where a rapid exchange between the tissues and blood is required. Fenestrated capillaries occur in the kidneys, intestines, and endocrine glands.
    c. Sinusoidal capillaries (1) have unusually wide lumens (30 to 40 μm); (2) follow a tortuous path; (3) have gaps between their endothelial cells, often allowing cells to pass; (4) have many fenestrations; (5) often have phagocytes interspersed among their endothelial cells; and (6) have discontinuous basal laminae.

4. **Transport across capillary walls.** Capillaries are **exchange vessels** because capillary beds serve as sites for the exchange of oxygen, nutrients, and many other substances between blood and tissues. Transcapillary transport mechanisms are not well understood, but morphologic bases exist for at least four types. **Fenestrae** penetrate the endothelium, allowing passive diffusion; some are covered by diaphragms. **Intercellular clefts** are spaces between neighboring endothelial cells, especially in sinusoidal capillaries, through which particles and even cells may pass. **Pinocytosis** is the process by which small amounts of plasma or tissue fluid are endocytosed by endothelial cells. This mechanism is followed by the trans-

Table 11–1. Comparison and classification of arteries.

Type, Functions, and Examples	Tunica Intima	Tunica Media	Tunica Adventitia
Elastic, large, or conducting arteries: Largest-diameter arteries in the body (eg, **aorta**). Conduct blood away from heart. Important in maintaining constant pressure in arterial system. When heart's left ventricle contracts **(systole),** blood is expelled into aorta under high hydrostatic pressure, causing elastic arteries to stretch. As ventricle relaxes **(diastole),** ventricular pressure drops and elastic walls of large arteries recoil (contract), converting force that expanded them back to hydrostatic pressure.	Is thicker than in muscular arteries. Endothelial cells of all vessels larger than capillaries contain rodlike **Weibel–Palade granules** that store components of factor VIII. Endothelium rests on thin basal lamina, underlain by thick sub-endothelial connective tissue. A porous internal elastic lamina may be present between intima and media, but is hard to distinguish owing to abundant elastin in media.	Contains abundant elastin, as concentric, fenestrated membranes that increase in number with age. Several circular layers of smooth muscle fibers lie between elastic membranes. Muscle cells are interwoven with reticular (type-III collagen) fibers and immersed in sparse chondroitin sulfate ground substance.	Is thin relative to vessel diameter. Contains elastic and type-I collagen fibers and external elastic lamina that may be hard to distinguish.
Muscular, medium-sized, or distributing arteries: Vary in diameter. Occur in many tissues and organs as major distributing branches of elastic arteries. Examples include arteries of limbs (eg, brachial) and abdominal cavity (eg, superior mesenteric).	Contains typical endothelium and sub-endothelial connective tissue. Prominent internal elastic lamina appears as wavy, refractile line between intima and media.	Is thick; contains as many as 40 layers of smooth muscle. Collagen, elastic fiber, and proteoglycan amounts vary (the larger the artery, the more elastin in media).	Is relatively thin and contains mostly collagen fibers.
Arterioles: Small arterial vessels, 0.5 mm or less in diameter.	Contains typical endothelium. Often lacks sub-endothelial connective tissue and internal elastic lamina.	Comprises one to five layers of smooth muscle encircling vascular lumen.	Is very thin and composed of collagen fibers.
Metarterioles: Small branches of arterioles. Constriction can regulate blood flow in capillaries. **Precapillary sphincters,** rings of smooth muscle around metarterioles at capillary origins, can halt capillary flow.	Contains typical endothelium. No subendothelial connective tissue or internal elastic lamina.	Is an incomplete single layer of smooth muscle.	Is indistinguishable.

Table 11–2. Comparison and classification of veins.

Type, Functions, and Examples	Tunica Intima	Tunica Media	Tunica Adventitia
Large veins: Largest-diameter veins in body (eg, **superior** and **inferior venae cava**). Conduct blood toward heart.	Is well developed and includes thick layer of sub-endothelial connective tissue. Extensions of intima protrude into lumens of large veins as valves.	Contains several layers of smooth muscle cells and abundant reticular and collagen fibers. Elastin is sparse.	Is the best-developed layer in large veins. Contains abundant collagen and longitudinal bundles of smooth muscle that strengthen vessel wall to prevent excessive distention.
Small and medium-sized veins: Narrower than large veins, these have thinner walls. Examples include saphenous (leg) and hepatic portal (abdominal cavity) veins.	Contains typical endothelium. Less subendothelial tissue and fewer valves than in large veins. No internal elastic lamina.	Is thin relative to vessel diameter. Has few elastic fibers.	Relatively thick, but unlike large veins contains little if any muscle. Contains mostly collagen.
Venules: Smaller version of general vein morphology. **Postcapillary venules** receive blood leaving capillaries.	Contains typical endothelium, but lacks valves.	Is very thin.	Is very thin. Contains mostly collagen.

port of membrane-bound **pinocytotic vesicles** across the endothelial cytoplasm in either direction. **Diapedesis** is the process by which some leukocytes pass from blood into tissues. It may involve the opening of junctions between endothelial cells by means of locally released substances (eg, histamine, which is involved in inflammation and increases vascular permeability).

B. Arteries: Arteries have a thicker tunica media than do veins. The media is best exemplified in medium-sized (muscular) arteries. Large (elastic) arteries contain more elastin in their media and adventitia than any other vessels. Arteries are also distinguished by refractile, eosinophilic internal and external elastic laminae. In most sites, veins accompany arteries. In cross-sections through paired vessels, arteries appear rounder than veins, with thicker walls and smaller lumens. For more details, see Table 11–1.

C. Veins: In cross-sections, veins often appear collapsed. They have thinner walls than arteries and may contain more erythrocytes in sectioned tissue. They have a thicker adventitia, which in larger veins may contain longitudinal smooth muscle. Veins contain valves that help maintain unidirectional blood flow. These extensions of the intima into the lumen consist of a fibroelastic connective tissue core covered on both sides by endothelium. Blood pressure is low in veins. Valves help ensure return of blood to the heart and help prevent blood pooling. Pooling (stasis) can lead to clot formation and obstruct blood flow. For more details, see Table 11–2.

D. Portal Vessels: Portal vessels carry blood from one capillary (or sinusoidal) bed to another without first returning it to the heart. Examples include the hepatic portal vein between the intestines and the liver, the hypophyseal portal veins in the pituitary, and the efferent arterioles of the renal cortex.

E. Carotid and Aortic Bodies: These unencapsulated chemoreceptors comprise clumps and cords of epithelioid cells permeated by fenestrated and sinusoidal capillaries. Carotid bodies lie at the bifurcation of the common carotid artery. The left aortic body is in the wall of the aorta, near the origin of the subclavian artery. The right aortic body is in the angle between the common carotid and subclavian arteries. Changes in blood oxygen, CO_2, or pH levels generate nerve impulses in their rich supply of unmyelinated nerve endings. The glossopharyngeal nerve transmits these signals to the brain, where they elicit responses that maintain homeostasis.

F. Carotid Sinus: This unencapsulated mechanoreceptor at the bifurcation of the common carotid consists of a dilation of the arterial lumen (sinus) and a thinned media, whose outer portion contains many large nerve endings. The sinus acts as a baroreceptor, responding to increased blood pressure by generating impulses that are carried by the glossopharyngeal nerve to the brain, where they elicit peripheral vasodilation and reflexive slowing of the heart.

G. Arteriovenous Anastomoses: These direct connections between arteries and veins regulate blood flow by smooth muscle contraction. When they are open, more blood passes directly from arteries to veins, bypassing the capillary bed. Complex anastomoses between arterioles and venules, called glomera, occur mainly in the finger pads, nail beds, and ears. The arterioles of glomera lack an internal elastic lamina and have more smooth muscle in their media, which can contract to completely or partially close the vessels. Arteriovenous anastomoses permit efficient management of blood distribution during stress, heavy exertion, and temperature changes. They also help regulate blood pressure and other physiologic processes, such as erection and menstruation.

H. Blood and Nerve Supply to Blood Vessels: Oxygen, nutrients, and wastes are not able to reach all cells in the walls of large arteries and veins by simple diffusion from the lumen. The vasa vasorum ("vessels of the vessels") form a capillary network to distribute blood to cells in the vessel walls. The walls of all blood vessels except capillaries and some venules contain a rich nerve supply. Unmyelinated vasomotor fibers (sympathetic fibers) arise in the sympathetic ganglia, ramify in the adventitia, and terminate in small knoblike endings in the media. Arteries usually contain more of these fibers, which stimulate smooth muscle contraction. Small intraadventitial ganglia occur in the aorta and in some other large arteries. Myelinated fibers occur in bundles in the adventitia. Their unmyelinated (free) nerve endings appear to be sensory. Many terminate in the adventitia; some extend to the intima.

III. HEART

A. **Chambers:** The heart has four chambers: two **atria,** thinner-walled chambers located at the base (top) of the heart, which collect returning blood, and two **ventricles,** thicker-walled chambers located in the body and apex of the heart, which redistribute the collected blood. See the following section (IV) for a description of the route of blood through these chambers.

B. **Tunics:** The walls of the heart have three layers, or tunics.
  1. The **endocardium** (inner layer) is homologous to the intima of vessels and has three major components. The innermost layer is the **endothelium,** which is underlain by a thin, continuous basal lamina. Surrounding this is a layer of **subendothelial connective tissue** with elastic fibers and some smooth muscle cells. The **subendocardium** is a layer of areolar tissue with small blood vessels, nerves, and, in the ventricles, branches of the impulse-conducting system (bundle branches and Purkinje fibers; III.E.1–5).
  2. The **myocardium** is the middle layer. This layer consists mainly of cardiac muscle fibers and carries out the forceful contractions that allow the heart to serve as a pump. It is homologous to the much thinner media of vessels. It contains the impulse-conducting system and parts of the cardiac skeleton (III.C). Each cardiac muscle fiber is surrounded by an endomysium, and each fascicle of fibers is surrounded by perimysium. The muscles in the atria and ventricles differ in some important respects.
      a. **Atrial cardiac muscle** is arranged in overlapping networks (musculi pectinati), giving the inner surface of the atria a woven appearance. Muscle cells in the outer myocardium form a complex helical pattern around the chamber, resembling the arrangement in the ventricles. Collagen and elastic fibers are interspersed among the muscle cells. Compared with ventricular cardiac muscle, atrial cells (1) are somewhat smaller, (2) have many granules containing **atrial natriuretic factor,** (3) have a less extensive T-tubule system, (4) have more gap junctions, (5) conduct impulses at a higher rate, and (6) contract more rhythmically.
      b. **Ventricular cardiac muscle** comprises complex layers of cells wound helically around the ventricular cavity. This arrangement aids in "wringing out" the heart during contraction, which maximizes the percentage of blood in the cavity that is expelled during a contraction (ie, the ejection fraction). The superficial muscle layers surround both ventricles, whereas the deeper muscle layers surround each ventricle individually and contribute to the interventricular septum. The cells of these inner and outer layers also may differ in their metabolic activity. Elastic connective tissue is less abundant in ventricular than in atrial myocardium.
  3. The **epicardium,** or visceral pericardium, is the outermost tunic. Although it occupies the same relative position as the tunica adventitia, it is a **serosa** rather than an adventitia. It consists of a single layer of squamous mesothelial cells, a thin basal lamina, and a layer of subepicardial connective (areolar) tissue that binds the epicardium to the myocardium. The smooth mesothelial surface reduces the friction between the heart and the surrounding structures that is generated during contraction.

C. **Cardiac Skeleton:** The dense fibrous connective tissue scaffolding into which the cardiac muscle fibers insert and from which the cardiac valves extend is the cardiac skeleton, or fibrous skeleton of the heart. It has three major groups of components. The **annuli fibrosae** are dense connective tissue rings that surround and reinforce the valve openings in the atrioventricular canals and at the origins of the aorta and pulmonary artery. The **trigona fibrosae** are two triangular dense connective tissue masses, occasionally containing some cartilage, that lie between the two groups of annuli fibrosae. The **septum membranaceum** is a dense fibrous plate forming the top of the otherwise muscular interventricular septum. Together with the arrangement of the muscle fibers, the fibrous skeleton directs the force of myocardial contraction so that the heart wrings out the blood in its chambers. Parts of the skeleton may calcify during disease and aging.

D. **Cardiac Valves:** These control the direction of blood flow through the heart. Each is a fold of endocardium enclosing a platelike core of dense connective tissue that is anchored in, and continuous with, the annuli fibrosae. The **tricuspid valve,** between the right atrium and ventricle, has three cusps (flaps). The free edge of each cusp is anchored to **papillary muscles** in the floor of each ventricle by fibrous cords called **chordae tendineae.** The **bicuspid,** or **mitral valve,** which is between the left atrium and ventricle, has two cusps, each anchored by chordae tendineae to papillary muscles in the ventricle floor. The **semilunar valves,** each composed of three semilunar cusps, are not attached by chordae tendineae. Each has a characteristic thickening (nodule) at the center of its free edge. The two semilunar valves are the aortic valve, between the left ventricle and the aorta, and the pulmonary valve, between the right ventricle and the pulmonary artery.

E. Impulse-Generating and -Conducting System: This system comprises unusual cardiac muscle cells specialized to initiate and conduct electrochemical impulses. The distribution of these cells allows the impulses they carry to coordinate myocardial contraction.

1. The **sinoatrial (SA) node,** or **pacemaker node,** is a small cell mass in the right atrium's median wall, near the opening of the superior vena cava. All cardiac muscle cells contract spontaneously; those with the fastest intrinsic rhythm lead neighboring cells to contract faster. Because the SA node's cells have the fastest intrinsic rhythm, they set the pace for the rest of the heart. Autonomic nerve fibers and ganglia located near the SA node do not directly dictate heart rhythm but can modulate heart rate. Impulses generated in the SA node travel slowly through ordinary atrial cardiac muscle to the atrioventricular node. This slow conduction allows the atria to complete their contraction before the ventricles begin theirs.

2. The **atrioventricular (AV) node** is a cell cluster on the right side of the interatrial septum. As an impulse leaves the AV node, it passes rapidly along the atrioventricular bundle.

3. The **AV bundle (of His)** is a bundle of specialized cardiac muscle fibers, 15-mm long and 2- to 3-mm-wide, passing from the interatrial septum into the interventricular septum. It gives off a smaller bundle (bundle branch) to each ventricle.

4. The **right** and **left bundle branches** travel a short distance before branching to form Purkinje fibers.

5. **Purkinje fibers** are cardiac muscle cells specialized to conduct electrochemical impulses. They are wider than typical cardiac muscle cells, with sparse myofilaments concentrated at the cell periphery. They are generally wider than bundle branch cells and, like typical cardiac muscle cells, are connected by intercalated disks and may have one or two central nuclei. Impulses are transmitted through gap junctions between the Purkinje fibers and the cardiac muscle cells they contact.

6. **Ventricular cardiac muscle cells** are the last link in the impulse conduction chain. They not only contract in response to the impulse, but also propagate (albeit more slowly) the impulses they receive from Purkinje fibers and pass them on to their neighbors. Thus, the cardiac musculature functions effectively as a syncytium, its cells contracting as one in a synchronous, coordinated manner.

F. Blood Supply to the Heart: The coronary arteries arise near the aorta's origin and supply oxygen-rich blood to the myocardium. Blockage of a coronary vessel or its branches by a thrombus or atherosclerotic plaques (fatty deposits in the media and intima) may rob the tissue supplied by the vessel of oxygen and nutrients. This **ischemia** can lead to localized tissue necrosis, called an **infarction.** Tissues with high energy and oxygen demands, such as the brain and myocardium, are particularly susceptible to infarction. The capillary density in cardiac muscle is greater than in skeletal muscle and is a diagnostic feature of this tissue in histologic section. Most of the venous blood returns through the coronary sinus to the superior vena cava as it enters the heart.

G. Lymphatics of the Heart: The myocardium contains abundant lymphatic capillaries. These begin as blind-ended tubes in the myocardium (near the endocardium) and drain into larger lymphatic vessels in the epicardial connective tissue.

H. Innervation of the Heart: Many myelinated and unmyelinated autonomic motor fibers (sympathetic and parasympathetic) enter the heart's base (top) and ramify, forming plexuses and innervating several ganglia. There are no myoneural junctions in the heart. The ANS adjusts the heart rate to meet changing demands by various organs and tissues. Generally, sympathetic stimulation increases and parasympathetic stimulation decreases the heart rate.

IV. ROUTE OF THE BLOOD

The route the blood takes through the cardiovascular system may be summarized as follows. Venous blood returns to the heart through the **superior** and **inferior venae cava.** It enters the right atrium, which contracts and forces blood through the tricuspid valve into the right ventricle. Contraction of the right ventricle forces blood through the pulmonary (semilunar) valve into the pulmonary artery, through which it reaches the capillaries surrounding the lungs' alveoli. Here, the blood picks up oxygen and releases carbon dioxide and other volatile wastes. Newly oxygenated blood is collected in the pulmonary veins and carried to the left atrium, which contracts to force it through the bicuspid (mitral) valve and into the left ventricle. The left ventricle subsequently contracts, forcing blood through the aortic (semilunar) valve and into the **aorta** for distribution to the body. The aorta gives off numerous branches (distributing arteries) through which blood passes to arteries of successively smaller diameters (muscular arteries, arterioles) until it reaches the capillary beds, where it releases its oxygen and nutrients to the tissues and picks up carbon dioxide and other metabolic by-products. Some fluid also escapes from the capillaries into intercellular tissue spaces; most of this excess tissue fluid returns to the capillary lumen before the blood leaves the tissue. The blood in the capillary bed enters the venules and subsequently enters veins of increasing diameters (medium-sized veins, large veins), finally returning to the heart through the largest veins, the superior and inferior venae cava.

V. LYMPHATIC VESSELS

A. **Lymphatic Vessels and Ducts:** The walls of these vessels and ducts resemble those of veins. The beaded appearance of lymphatic ducts and vessels reflects the presence of valves that control the direction of lymph flow. The adventitia is thin and lacks smooth muscle. The media contains both longitudinal and circular smooth muscle, but longitudinal fibers predominate.

B. **Lymphatic Capillaries:** Like blood capillaries, these are simple squamous endothelial tubes. Unlike blood capillaries, they have a greater diameter (as wide as 100 μm) and a thinner discontinuous basal lamina. They lack fenestrations and have fewer tight junctions than blood capillaries.

C. **Route of the Lymph:** The route the lymph takes is unidirectional. Excess tissue fluid not returned to the blood capillaries (IV) is collected by blind-ended lymphatic capillaries in the region of the blood capillary beds and carried through lymphatic vessels to lymphatic ducts. There is one major lymphatic duct for each side of the body: the thoracic duct on the left and the right lymphatic duct on the right. The lymphatic ducts return lymph to the blood by emptying into the venous system at the junction of the jugular and subclavian veins in the neck. The lymphatic system is discussed further in Chapter 14.

MULTIPLE-CHOICE QUESTIONS

Select the single best answer.

11.1. Which of the following is true of continuous capillaries?
(A) Possess unusually wide lumens
(B) Have abundant fenestrations
(C) Typically follow a tortuous (twisting) course
(D) Are common in brain and muscle tissue
(E) Often have phagocytic cells inserted between the endothelial cells of their lining

11.2. Which of the following is the thickest layer in the walls of veins?
(A) Internal elastic lamina
(B) Subendothelial connective tissue
(C) Tunica adventitia
(D) Tunica intima
(E) Tunica media

Figure 11–2.

11.3. Which of the structures in Figure 11–2 are the Purkinje cells (fibers)?

11.4. Which of the following is true of pericytes?
(A) Are specialized cardiac muscle cells
(B) Are attached to the outside of capillaries
(C) Are specialized smooth muscle cells
(D) Are terminally differentiated
(E) Are multinucleated

11.5. Which of the following structures are indicated by the single arrows in Figure 11–3?
(A) Coated pits
(B) Fenestrations closed by diaphragms
(C) Junctional complexes
(D) Micropinocytotic vesicles
(E) Nuclear pores

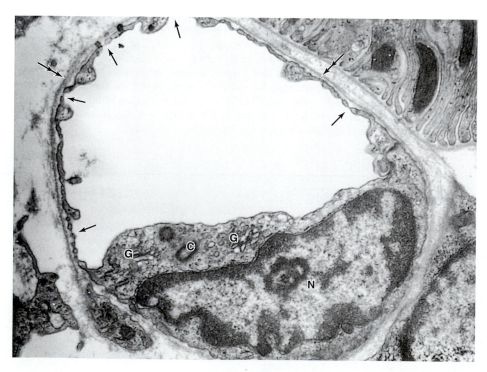

Figure 11–3.

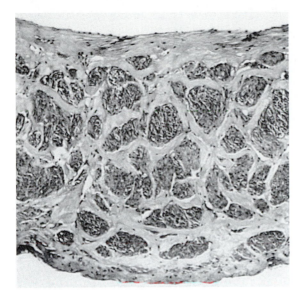

Figure 11–4.

11.6. Figure 11–4 shows a cross-section of a portion of the wall of which type of vessel?
(A) Aorta
(B) Arteriole
(C) Medium-sized vein
(D) Muscular artery
(E) Vena cava

11.7. Which of the following is true of the tunica intima?
(A) Includes a layer of dense connective tissue
(B) Contains some small capillaries
(C) Is separated from the tunica media of arteries by the external elastic lamina
(D) Includes a layer of simple squamous epithelium
(E) Is a fenestrated sheet of elastin
(F) Is homologous to the pericardium of the heart
(G) Is absent in large elastic arteries

11.8. Which of the following is the predominant basic tissue type in the tunica media?
(A) Connective tissue
(B) Epithelium
(C) Muscle
(D) Nerve

11.9. Passage of blood cells across capillary walls occurs through which of the following structures?
(A) Fenestrae
(B) Gap junctions
(C) Intercellular clefts
(D) Pinocytotic vesicles

11.10. Which of the following is true of lymphatic vessels?
(A) Resemble arteries more than veins
(B) Typically lack valves
(C) Have sharp (distinct) borders between their tunics
(D) Have longitudinal smooth muscle in their tunica media

11.11. Which of the labels in Figure 11–5 corresponds to the sinoatrial node?

11.12. Which of the labels in Figure 11–5 corresponds to the Purkinje fiber system?

11.13. Which of the following structures is most likely to contain longitudinal smooth muscle in its tunica adventitia?
(A) Aorta
(B) Arteriole
(C) Blood capillary
(D) Medium-sized artery
(E) Thoracic duct
(F) Vena cava
(G) Venule

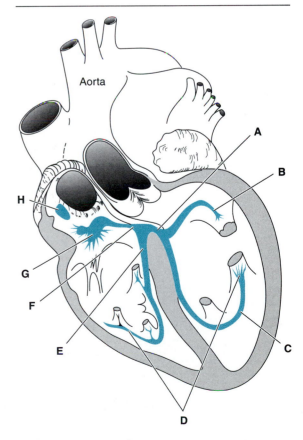

Figure 11–5.

11.14. Which of the following structures is likely to contain the most distinctive internal elastic lamina?
(A) Aorta
(B) Arteriole
(C) Blood capillary
(D) Medium-sized artery
(E) Thoracic duct
(F) Vena cava
(G) Venule

11.15. Which of the following structures is likely to contain the most vasa vasorum in its tunica media?
(A) Aorta
(B) Arteriole
(C) Blood capillary
(D) Medium-sized artery
(E) Thoracic duct
(F) Vena cava
(G) Venule

11.16. Which of the following structures contains the most elastin in its tunica media?
(A) Aorta
(B) Arteriole
(C) Blood capillary
(D) Medium-sized artery
(E) Thoracic duct
(F) Vena cava
(G) Venule

11.17. Which of the following is true of the ventricles?
(A) Are located at the base of the heart
(B) Their myocardial cells contain abundant granules

(C) Receive blood directly from the venae cava and the pulmonary veins
(D) Their walls contain the right and left bundle branches
(E) Contain more abundant elastic fibers than do the atria

11.18. Which of the following is true of blood capillaries?
(A) Carry lymphocytes
(B) Are blind-ended tubes
(C) Contain valves
(D) Characteristically lack occluding junctions between their lining cells
(E) Are lined by cells derived from endoderm

11.19. The sensory nerve fibers of arteries enter through the adventitia. Those that penetrate the farthest into the wall may extend into which of the following layers?
(A) External elastic lamina
(B) Lumen
(C) Tunica adventitia
(D) Tunica intima
(E) Tunica media

11.20. How are electrical impulses passed from Purkinje fibers to ventricular cardiac muscle fibers?
(A) Chemical synapses
(B) Diffusion through the endomysium
(C) Gap junctions
(D) Saltatory conduction
(E) Tight junctions (membrane fusion)

ANSWERS TO MULTIPLE-CHOICE QUESTIONS

11.1. D (II.A.3.a–c)
11.2. B (II.C; Table 11–2)
11.3. B (III.E.5)
11.4. B (II.A.2.b)
11.5. B (II.A.3.b; Fig. 11–1)
11.6. E (II.C; Fig. 11–1; Table 11–2; note the smooth muscle bundles in the thick adventitia)
11.7. D (I.C.1; Table 11–1)
11.8. C (I.C.1–3)
11.9. C (II.B.4)
11.10. D (I.B.2; V.A)
11.11. H (III.E.1)

11.12. D (III.E.5)
11.13. F (II.C; Tables 11–1 and 11–2)
11.14. D (II.B; Table 11–1)
11.15. F (II.H; larger vessels have more vasa vasorum. These penetrate deeper into the walls of veins owing to the lower oxygen tension in the blood they carry.)
11.16. A (II.B; Table 11–1)
11.17. D (III.A.2.a and b; III.E; IV)
11.18. A (III.A.4; V.B; 14.III.B.1.a)
11.19. D (II.H)
11.20. C (III.E.5)

Peripheral Blood

12

OBJECTIVES

This chapter should help the student to:

- Know the name, structure, and function of each of the formed elements in blood.
- Know the percentage contributed by each cell type to peripheral blood cell numbers (as determined by a differential cell count) and to blood volume (as determined by hematocrit).
- Know the percentage of normal blood volume contributed by plasma.
- Know the composition of plasma and distinguish between plasma and serum.
- Describe the sequence of events of clot formation, including the roles of the platelets and various plasma proteins.
- Identify the formed elements in a micrograph of a blood smear.

MAX-Yield™ STUDY QUESTIONS

1. What is the approximate total blood volume of adult humans (I.A[1])?
2. Name the two major components of blood (I.A).
3. Name the three classes of formed elements in blood (III.A–C).
4. Compare serum and plasma in terms of the procedures for isolating them from whole blood (I.A and D; IV.A) and their fibrinogen and serotonin content (IV.A).
5. Define "hematocrit" and give the normal range of values for adult humans (I.D).
6. List several types of plasma proteins (II.B.1; IV.B).
7. To which class of plasma proteins do the circulating antibodies (immunoglobulins) secreted by plasma cells belong (II.B.1)?
8. What is the normal diameter of a human erythrocyte (III.A.1)?
9. What is the functional significance of the biconcave shape of normal erythrocytes (III.A.1)?
10. Compare erythrocytes of sickle cell anemia with normal erythrocytes (III.A.3) in terms of:
    a. Hemoglobin (Hb) type (including amino acid composition)
    b. Effect of low oxygen tension on hemoglobin solubility
    c. Effect of low oxygen tension on cell shape and flexibility
11. What components of erythrocyte plasma membranes determine blood group (eg, MN, ABO) (III.A.4)?
12. Describe hemoglobin (III.A.3) in terms of:
    a. Primary function
    b. Number of subunits per molecule
    c. Number of hemes per molecule
    d. Type of metal ion associated with the heme
    e. Types present in blood after birth
    f. Predominant type in adults
    g. Predominant type in the fetus
13. Describe mature erythrocytes (III.A.1 and 2) in terms of:
    a. Organelles present
    b. Capacity for protein synthesis
    c. Energy metabolism
    d. Site of production
    e. Duration in circulation
    f. Site(s) of removal from circulation
14. List the five types of leukocytes in peripheral blood (III.B.1.a,b and 2.a–c). Indicate which are agranulocytes and which are granulocytes (III.B.1 and 2).

---

[1] See footnote on page 1.

15. Compare granulocytes and agranulocytes (III.B.1 and 2) in terms of the presence and relative amount of specific and azurophilic granules and the shape of their nuclei.
16. In addition to blood, leukocytes are normal components of what tissue type (III.B.1 and 2.a–c)? How do leukocytes in this tissue differ from those in the blood?
17. What is the normal number of leukocytes per microliter of blood (provide a range), and the predominant leukocyte type, in the peripheral blood of adult humans (III.B.1 and 2.a–c)?
18. What percentage of the leukocytes in normal adult blood are neutrophils (III.B.2.a)? Lymphocytes (III.B.1.a)? Monocytes (III.B.1.b)? Eosinophils (III.B.2.b)? Basophils (III.B.2.c)?
19. Compare the three types of granulocytes in terms of staining properties, the size and contents of their specific granules, the average number of nuclear lobes in a mature cell, the diameter of a mature cell, and their function (Table 12–1).
20. Describe the agranulocytes most commonly found in the blood in terms of their cytoplasm (amount and staining properties), nuclei (shape and staining properties), cell diameter, basic function, and their ability to leave and reenter the circulation (III.B.1.a and b; Table 12–1).
21. Name the two granule types in neutrophils. Compare their size and contents (Table 12–1).
22. Sketch an eosinophil's specific granule as seen in a transmission EM (Table 12–1). Label the unit membrane, internum, and matrix.
23. List the effects of parasitic infection, allergic reaction, and increased corticosteroid production on circulating eosinophil numbers (III.B.2.b).
24. Name the connective tissue cell type that resembles the basophil in terms of staining properties, granule contents, and function (III.B.2.c).
25. Of the three sizes of circulating lymphocytes, which is the most common (Table 12–1)?
26. Name the two functional classes of lymphocytes found in the blood. Compare them in terms of:
    a. Type of immunity (humoral or cell-mediated) primarily associated with each (III.B.1.a.[1][a])
    b. Primary or central lymphoid organs in which each type develops in humans (III.B.1.a.[3])
    c. Predominant type found circulating in the blood (III.B.1.a)
27. Name three functionally distinct T-lymphocyte (T cell) types found in blood (III.B.1.a.[1][b]).
28. During rejection of a transplanted heart, which T cells kill the cardiac muscle cells in the transplant (III.B.1.a.[4].[b])? Are they memory or effector cells (III.B.1.a.[1])?
29. Name the B-lymphocyte effector cell that forms after antigenic stimulation (III.B.1.a.[1][a]).
30. List the advantages of forming memory cells (III.B.1.a.[1]).
31. Name the diffuse system of cells derived from blood monocytes. What functional capacity do these cells have in common (III.B.1.b)?
32. Draw a platelet and label the hyalomere, granulomere, granules, and marginal bundle of microtubules (III.C; Fig. 12–2).
34. To which component of the endothelial basal lamina and subendothelial connective tissue do platelets attach during primary aggregation (IV.C.1)?
35. Platelets attaching to a damaged vessel wall release the contents of their alpha and delta granules (III.C). What is the effect of these substances on the unattached platelets in the blood (IV.C.3) and on the cascade of interactions involved in forming fibrin (IV.B and C.3)?
36. Compare blood clot (thrombus) components (IV.C.3) with those of a platelet plug (IV.C.1).
37. Which process helps prevent clots from occluding vessels until damage is repaired (IV.D)?
38. The inactive proenzyme plasminogen is synthesized and added to the blood plasma by the liver (IV.E). State:
    a. How it is converted to its active form
    b. The name and function of its active form
    c. The cells in which the substances that activate plasminogen are synthesized

SYNOPSIS

I. GENERAL FEATURES OF THE BLOOD

A. Two Components: Humans have a total blood volume of approximately 5 L (depending on body size). Blood is divisible into two parts: the **formed elements** (III), which include blood cells and platelets, and the **plasma,** or liquid phase (II), in which the formed elements are suspended and in which a variety of important proteins, hormones, and other substances are dissolved.

B. **Basic Cell Types:**   There are two basic blood cell types: the erythrocytes, or red blood cells (III.A), and the leukocytes, or white blood cells (III.B).

C. **Clotting:**   Outside the blood vessels, blood undergoes clot formation or coagulation (IV), which plays an important role in repairing damaged vessels and preventing blood loss.

D. **Hematocrit:**   When anticoagulants (heparin, citrate) are added, blood samples can be separated in a centrifuge into three major fractions. Erythrocytes are the densest of these and end up at the bottom. The **hematocrit** is the percentage of packed erythrocytes per unit volume of blood. In adults, normal hematocrit values vary from 35–50 and are sex-dependent. Leukocytes are less dense, less numerous (approximately 1% of blood volume), and form a thin white or grayish layer over the erythrocytes, called the **buffy coat.** Over the buffy coat is a thin layer of platelets. The least dense fraction is the clear layer of plasma, which constitutes 42 to 47% of blood volume and overlies the buffy coat.

E. **Differential Cell Count:**   Blood is also studied by spreading a drop on a slide to produce a single layer of cells (blood smear). The cells are stained, differentiated by type, and counted to reveal disease-related changes in their relative numbers. The smears are usually stained with Romanowsky-type dye mixtures containing eosin and methylene blue.

F. **Staining Properties:**   The descriptions of blood cell–staining properties in this chapter refer to the cell's appearance after staining with Romanowsky-type mixtures (eg, Wright or Giemsa). Blood cells and their components exhibit four major staining properties that allow the cell types to be distinguished. **Basophilia** is an affinity for methylene blue; basophilic structures stain purple to black. **Azurophilia** is an affinity for the oxidation products of methylene blue called azures; azurophilic structures stain reddish purple. **Eosinophilia,** or **acidophilia,** is an affinity for eosin; eosinophilic structures stain salmon pink to orange. **Neutrophilia** is an affinity for a complex of dyes (originally thought to be neutral) in a stain mixture; these structures stain salmon pink to lilac.

II. COMPOSITION OF PLASMA

A. **Water:**   Plasma contains 90% water by volume.

B. **Solutes:**   Plasma contains 10% solutes by volume. These solutes include plasma proteins, other organic compounds, and inorganic salts.
    1. **Plasma proteins.** Plasma contains many soluble proteins (7% by volume). **Albumin** is the most abundant plasma protein (3.5–5.0 g/dL of blood) and is mainly responsible for maintaining blood's osmotic pressure. Water-insoluble substances (eg, lipids) are carried in plasma associated with albumin. Alpha, beta, and gamma **globulins** are globular proteins dissolved in the plasma. Gamma globulins include antibodies, or **immunoglobulins. Fibrinogen** is converted by blood-borne enzymes into **fibrin** during clot formation. Fibrinogen is synthesized and secreted by the liver.
    2. **Other organic compounds.** Other organic molecules in plasma (2.1% by volume) include nutrients such as amino acids and glucose, vitamins, and a variety of regulatory peptides, steroid hormones, and lipids.
    3. **Inorganic salts.** Inorganic salts in plasma (0.9% by volume) include blood electrolytes such as sodium, potassium, and calcium salts.

III. FORMED ELEMENTS

A. **Erythrocytes:**   Erythrocytes (Fig. 12–1), also called **red blood cells,** or **RBCs,** are the most abundant formed elements in blood (4–$6 \times 10^6/\mu L$). Their presence in most tissues and organs makes them useful in estimating the size of other structures (through estimates of multiples or fractions of RBC diameter).

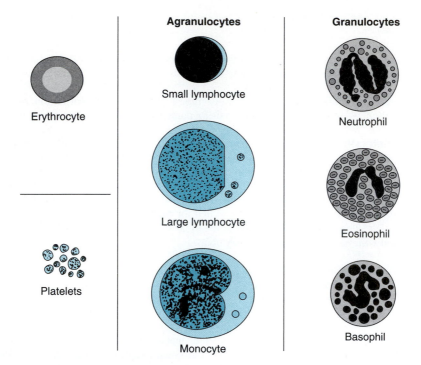

Figure 12–1. Formed elements of peripheral blood. Note the relative sizes of the cells and the shape, pattern, and darkness of the nuclei. Note the relative sizes of granules in the leukocytes. The diagrams are drawn roughly to scale. Refer to text for detailed descriptions of the individual cell types.

1. **Normal structure and function.** RBCs are structurally and functionally specialized to transport oxygen from the lungs to other tissues. Their cytoplasm contains the oxygen-binding protein hemoglobin. Their small diameter (7–8 μm) and biconcave shape (in humans) help to maximize their surface-to-volume ratio, facilitating oxygen exchange. Because of the opposing concavities at the center of a normal RBC, the staining intensity appears to be reduced in this region, creating a **central pallor.** An increase in the size of this pale-staining region indicates an abnormally low amount of hemoglobin in the cytoplasm. Mature RBCs lack nuclei and organelles, which they lose during differentiation. Their lack of mitochondria necessitates the use of anaerobic glycolysis for the energy needed to maintain hemoglobin function. Without ribosomes, glycolytic enzymes and other proteins cannot be renewed. Mature erythrocytes therefore have a limited life span (120 days) in the circulation before their removal by spleen and bone marrow macrophages.

2. **Abnormalities. Anisocytosis** refers to a high percentage of RBCs with size variations. **Macrocytes** are larger than 9 μm in diameter. **Microcytes** are smaller than 6 μm. In some disease states, nuclear fragments, or **Howell-Jolly bodies** (**Cabot rings** when they form circles), remain in otherwise mature RBCs. Some RBCs recently released from the bone marrow contain a small amount of residual RER and ribosomes that can be precipitated into blue, netlike structures with the vital dye brilliant cresyl blue. When these **reticulocytes** constitute more than approximately 1% of the circulating RBCs, they indicate an increased demand for oxygen-carrying capacity (eg, from a loss of RBCs caused by hemorrhage, anemia, or recent ascent to a high altitude).

3. **Hemoglobin.** Each hemoglobin molecule consists of four polypeptide subunits, each of which includes an iron-containing **heme** group. Hemoglobin can bind reversibly to oxygen, forming **oxyhemoglobin,** and to carbon dioxide, forming **carbaminohemoglobin.** Hemoglobin binds irreversibly to carbon monoxide, however, forming **carboxyhemoglobin,** which reduces the blood's oxygen-carrying capacity. Hemoglobin (Hb) exists in different forms, distinguishable by their amino acid sequence. In humans, only three forms are nor-

mal during postnatal life: **HbA₁** constitutes 97%, **HbA₂** 2%, and **HbF** 1% of the hemoglobin of healthy adults. HbF makes up 80% of a newborn's hemoglobin; however, this proportion gradually decreases until normal adult levels are reached at approximately 8 months of age. **HbS** is an abnormal form of HbA that is found in patients with **sickle cell anemia;** it differs by a single amino acid substitution in the beta chain (valine in HbS, glutamine in HbA). Unlike HbA, HbS becomes insoluble at low oxygen tensions and crystallizes into inflexible rods that deform the RBCs, giving them the characteristic sickle shape. When the rigid sickled cells pass through narrow capillaries, they cannot bend as normal RBCs do. They may become trapped, obstructing blood flow, or rupture, decreasing the number of RBCs available for oxygen transport (anemia).

4. **Plasmalemma and stroma.** When placed in a hypotonic solution, RBCs swell and release their hemoglobin into the surrounding solution, a process termed **hemolysis;** they leave behind an empty shell, or **red cell ghost,** composed of the plasmalemma and the stroma. The stroma consists of proteins such as **spectrin** that are associated with the plasmalemma's inner surface; it maintains the RBC's biconcave shape. The plasmalemma's outer surface is covered by a carbohydrate-rich glycocalyx, which contains genetically determined antigens that allow **blood types** (including A, B, O, and M and N groups) to be distinguished. The major RBC integral membrane glycoprotein is **glycophorin.**

B. **Leukocytes:** Leukocytes, or white blood cells, are nucleated and are larger and less numerous (6000–10,000/µL) than erythrocytes. Leukocytes can be divided into two main groups, granulocytes and agranulocytes, according to their granule content. Each group can be further divided based on size, nuclear morphology, ratio of nuclear to cytoplasmic volume, and staining properties. Two classes of cytoplasmic granules occur in leukocytes: specific and azurophilic granules. **Specific granules** occur only in granulocytes; their staining properties (neutrophilic, eosinophilic, or basophilic) distinguish the three granulocyte types. **Azurophilic granules** occur in both agranulocytes (III.B.1) and granulocytes (III.B.2); their lytic enzymes suggest that they function as lysosomes. Unlike the RBCs, all leukocytes can exit the capillaries by squeezing between endothelial cells (a process termed **diapedesis**) and enter the surrounding connective tissue in response to infection or inflammation. Extravascular leukocyte activity is cell-type specific.

1. **Agranulocytes** (see Fig. 12–1) have unsegmented nuclei. These **mononuclear leukocytes** lack specific granules but contain azurophilic granules (0.05–0.25 µm in diameter).

a. **Lymphocytes** constitute a diverse class of cells; they have similar morphologic characteristics but a variety of highly specific functions. They normally account for 20 to 25% of adult white blood cells but are characterized by a broad range of normal variation (20 to 45%). Lymphocytes are also found outside blood vessels, grouped in lymphatic organs (see Chapter 14) or dispersed in connective tissues. They respond to invasion of the body by foreign substances and organisms and assist in their inactivation. Unlike other leukocytes, lymphocytes never become phagocytic and may recirculate after leaving the bloodstream. The two major functional classes of lymphocytes are T cells and B cells. Lymphocytes in the blood are predominantly (approximately 80%) T cells. (See Table 12–1 for structural comparisons with other leukocytes.)

(1) **Memory cells and effector cells.** When stimulated by an antigen, lymphocytes undergo **blast transformation,** a process of enlargement, and **clonal expansion,** a series of mitotic divisions. Some of the daughter cells, called memory cells, return to an inactive state but can respond more quickly to the next encounter with the same antigen. Other daughter cells, called effector cells, become activated to carry out an immune response to the antigen. Effector cells may derive from either B cells or T cells. Although circulating B and T cells are morphologically indistinguishable, they carry different cell-surface components (antigens recognized by other species) and can be identified by special procedures.

(a) **B lymphocytes** differentiate into **plasma cells** (5.II.E.2.c), which secrete antigen-binding molecules (antibodies or immunoglobulins) that circulate in the blood and lymph and serve as a major component of humoral immunity.

(b) **T-lymphocyte** derivatives serve as the major cells of the **cellular immune response.** They produce a variety of **cytokines** (eg, interferon) that influence the activities of macrophages and of other leukocytes involved in an immune response. There are several types. **Cytotoxic (killer) cells** secrete substances that kill other cells and in some cases kill by direct contact; they play the principal role in graft rejection. **Helper T** cells enhance the activity of some B cells and other T cells. **Suppressor T** cells inhibit the activity of some B cells and other T cells.

(2) **Null cells** are circulating cells that morphologically resemble lymphocytes but exhibit neither B-cell nor T-cell surface antigens. They may represent circulating stem cells of lymphocytes or other blood cell types (13.I.A and VII.C).

(3 The **primary (central) lymphoid organs** include the **thymus,** where lymphocyte precursors are programmed to become T cells, and in birds, the **bursa of Fabricius,** where lymphocyte precursors are programmed to become B cells. Humans have no bursa; our B cells are programmed in the **bone marrow.**

b. **Monocytes** are often confused with large lymphocytes. They are large and constitute only 3 to 8% of the white blood cells in healthy adults. Monocytes occur only in the blood, but remain in circulation for less than a week before migrating through capillary walls to enter other tissues or to become incorporated in the lining of sinuses. Outside the bloodstream, they become phagocytic and do not recirculate. The **mononuclear phagocyte system** consists of monocyte-derived phagocytic cells throughout the body. Examples include the liver's Kupffer cells and some connective tissue macrophages. (See Table 12–1 for monocyte structure and comparisons with other leukocytes.)

2. **Granulocytes** (see Fig. 12–1) have segmented nuclei and are described as **polymorphonuclear leukocytes (PMNLs).** Depending on the cell type, the mature nucleus may have two to seven lobes connected by thin strands of nucleoplasm. Granulocyte types are distin-

Table 12–1. Distinguishing features of leukocyte structure.

Cell Type	Diameter	Nucleus	Cytoplasm
Lymphocyte	Small: 6–8 μm (chief circulating form).	Is spheric, often flattened on one side, densely heterochromatic, purplish blue to black.	Thin rim around nucleus, pale basophilia, many ribosomes, sparse ER, few mitochondria, small Golgi, few azurophilic granules, no specific granules.
	Medium and large: 8–18 μm (often antigen-activated cells).	Is large, less heterochromatic, reddish purple.	More abundant, pale basophilia, many ribosomes, sparse ER, few mitochondria, small Golgi, few azurophilic granules, no specific granules.
Monocyte	12–15 μm in circulation; as large as 20 μm in tissues.	Usually is kidney- or horseshoe-shaped, eccentric. Chromatin is less-dense, with smudgy reddish-purple appearance and 2–3 nucleoli.	Abundant, faint blue–gray, many small azurophilic granules, no specific granules, many small mitochonria, well-developed Golgi, sparse RER and polyribosomes.
Neutrophil	Approximately 12 μm in circulation; as large as 20 μm in tissues.	Contains condensed chromatin and is multilobed (usually three lobes, more than five **[hypersegmented]** in aging cells); small heterochromatic **drum stick** may extend from one lobe. Represents female **Barr body** (inactive X chromosome).	Abundant small (0.3–0.8 μm), salmon pink, specific (neutrophilic) granules; fewer reddish-purple azurophilic granules. Specific granules contain alkaline phosphatase and bactericidal cationic proteins called **phagocytins.** Abundant glycogen also is present.
Eosinophil	9 μm in circulation; as large as 14 μm in tissues.	Contains condensed chromatin and typically two lobes, often partly obscured by abundant specific granules.	Abundant large (0.5–1.5 μm), brightly eosinophilic, specific granules that are specialized lysosomes carrying peroxidase, acid phosphatase, cathepsin, ribonuclease, **major basic protein (MBP,** eosinophilic antiparasitic agent) and fewer, reddish-purple azurophilic granules. In EMs, specific granules are ovoid with a dense **internum** surrounded by an electron-lucent **externum.**
Basophil	10–12 μm; smaller than neutrophils.	Contains condensed chromatin and typically three lobes, often in an S shape, partially or completely obscured by abundant, dark specific granules.	Less abundant, variable-sized (0.3–1.5 μm), reddish-violet to black specific (basophilic) granules and fewer, reddish-purple azurophilic granules. Granules contain heparin and histamine to be released in response to allergic stimuli.

ER = endoplasmic reticulum; RER = rough endoplasmic reticulum; EMs = electron micrographs.

guished by their size and staining properties and by the EM appearance of the abundant specific granules in their cytoplasm. These granules are membrane-limited and bud off of their small Golgi complex. Each granulocyte also contains a few mitochondria, free ribosomes, and sparse RER. (See Table 12–1 for granulocyte structure and comparisons with other leukocytes.)

a. **Neutrophils** are the most abundant circulating leukocytes. They constitute 60 to 70% of the white blood cells, and are characterized by a limited range of normal variation (50 to 75%). They are also found outside the bloodstream, especially in loose connective tissue. Neutrophils are the first line of cellular defense against bacterial invasion. After they leave the bloodstream, they spread out, develop ameboid motility, and become active phagocytes. Unlike lymphocytes, neutrophils are terminally differentiated and hence incapable of mitosis.

b. **Eosinophils** constitute only 1 to 4% of the circulating leukocytes in healthy adults. They may exit the bloodstream by diapedesis, spread out, and move about in the connective tissues. They are capable of limited phagocytosis, with a preference for antigen–antibody complexes. Circulating eosinophil numbers increase during allergic reactions and parasitic infections and rapidly decrease during corticosteroid treatment.

c. **Basophils** are the least numerous circulating leukocytes, constituting 0 to 1% in healthy adults. Basophils may exit the circulation but are capable of only limited ameboid movement and phagocytosis. Extravascular basophils are seen at sites of inflammation and are important cells at sites of **cutaneous basophil hypersensitivity.** Despite similarities to mast cells (5.II.E.2.a), basophils differ ultrastructurally.

C. **Platelets:** Platelets (Fig. 12–2), or **thrombocytes,** the smallest formed elements, are disklike cell fragments that vary in diameter from 2 to 5 μm. In humans, they lack nuclei and originate by budding from large cells in the bone marrow called megakaryocytes (13.VI). They range in number from 150,000 to 300,000/μL of blood and have a life span of approximately 8 days. In blood smears they appear in clumps. Each platelet has a peripheral **hyalomere** that stains a faint blue and a dense central **granulomere** containing a few mitochondria, glycogen granules, and various purple granules. **Delta granules** are 250 to 300 nm in diameter and contain calcium ions, pyrophosphate, ADP, and ATP; they take up and store serotonin. **Alpha granules** are 300 to 500 nm in diameter and contain fibrinogen, platelet-derived growth factor, and other platelet-specific proteins. **Lambda granules** (lysosomes) are 175 to 200 nm in diameter and contain only lysosomal enzymes. The hyalomere contains a **marginal bundle** of microtubules that maintains the platelet's discoid shape. The glycocalyx is unusually rich in glycosaminoglycans and is associated with adhesion, the major function of platelets. Platelets plug wounds and contribute to the cascade of molecular interactions among the various clotting factors dissolved in the plasma (see next section).

IV. CLOT FORMATION

A. **The Clot and Serum:** Clotted blood consists of two parts: (1) the clot, or **thrombus,** which includes the formed elements and some proteins formerly dissolved in the plasma, and (2) **serum,** a clear yellow liquid similar to plasma but lacking fibrinogen and containing more serotonin.

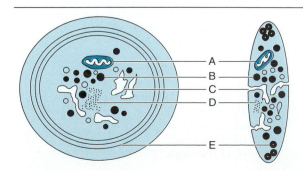

Figure 12–2. Frontal (*left*) and cross- (*right*) sections of a platelet. Labeled components include a mitochondrion (A), various granules (B), an open canalicular system (C), clusters of glycogen granules (D), and the marginal bundle of microtubules (E).

B. Clotting Factors: Clotting involves a cascade of interactions among several plasma proteins and ions (clotting factors I–XIII). The cascade can be initiated by either of two converging pathways, each of which results in the conversion of fibrinogen to fibrin by the enzyme **thrombin.** In the **intrinsic pathway,** cascade initiation occurs when factor XII is activated in response to contact with collagen under the endothelium (which indicates damage to the vessel's endothelial lining). In the **extrinsic pathway,** cells in a damaged vessel wall or the surrounding tissue release the clot-promoting substance thromboplastin (factor III), which combines with blood calcium and factor VII to activate factor X, a plasma protein. Factor X is a point of convergence of the two pathways and, in its activated form, promotes the conversion of prothrombin (factor II) to thrombin. Both factor X and prothrombin are synthesized by the liver and require vitamin K as a cofactor in their synthesis. Thrombin enzymatically converts plasma fibrinogen (factor I, released by platelets and the liver) into fibrin; this explains why the concentration of fibrinogen is lower in serum than in plasma. Other factors act as clot promoters and accelerators or help stabilize the fibrin. An inherited abnormality in factor VIII results in the clotting disorder known as **hemophilia.**

C. The Role of Platelets:
1. **Primary aggregation.** Platelets in the damaged region attach to collagen revealed by the discontinuity in the vessel wall, forming a **platelet plug.**
2. **Secondary aggregation.** Platelets in the plug release the contents of their alpha and delta granules. This release of **serotonin** explains why the concentration of serotonin is higher in serum than in plasma. Serotonin, a vasoconstrictor, restricts blood flow to the damaged area by causing contraction of vascular smooth muscle.
3. **Blood coagulation.** Platelets release fibrinogen in addition to that normally found in the plasma. The fibrinogen is converted into fibrin, which forms a dense fibrous mat to which more platelets and other blood cells attach. The resulting **clot** plugs the hole in the vessel wall.

D. Clot Retraction: The clot (thrombus) initially bulges into the vessel lumen but subsequently contracts and condenses through the interactions of **thrombosthenin** (a contractile protein released by the platelets) and platelet actin, myosin, and ATP.

E. Clot Removal: As the vessel wall heals and the protection afforded by the clot is no longer needed, the clot is removed by the enzyme plasmin. Plasmin is formed by the action of **plasminogen activators** (released by endothelial cells) on the plasma proenzyme plasminogen (synthesized by the liver). Enzymes from the platelets' lambda granules (lysosomes) also aid in clot digestion.

MULTIPLE-CHOICE QUESTIONS

Select the single best answer.

12.1. Which of the following is the approximate life span of an erythrocyte in the circulation?
(A) 8 days
(B) 20 days
(C) 5 weeks
(D) 4 months
(E) 1 year

12.2. Which of the following is true of the darker-staining central region of human platelets?
(A) Contains the marginal bundle of microtubules
(B) Contains the platelet nucleus
(C) Contains the platelet granules

(D) Is termed the hyalomere
(E) Is termed the central pallor

12.3. Which of the following is the biochemical constituent of the erythrocyte cell surface that is primarily responsible for determining blood type (eg, A,B,O; M,N)?
(A) Carbohydrate
(B) Lipid
(C) Nucleic acid
(D) Protein

12.4. Which of the following cell types is shown in Figure 12–3?
(A) Basophil
(B) Eosinophil
(C) Erythrocyte

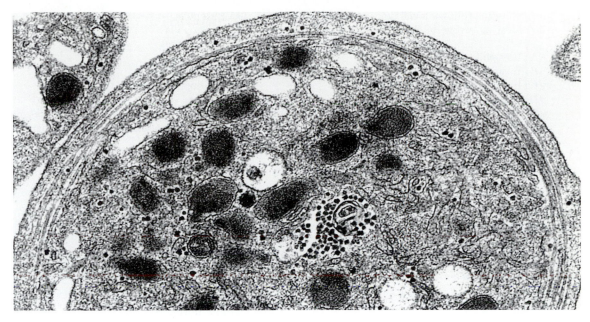

Figure 12–3.

(**D**) Neutrophil

(**E**) Platelet

12.5. Which of the following is true of specific granules?

(**A**) May be found in lymphocytes

(**B**) May be found in monocytes

(**C**) May be found in platelets

(**D**) May function as lysosomes

(**E**) May contain phagocytins or major basic protein

12.6. Which of the following is true of null cells?

(**A**) Comprise 80% of circulating lymphocytes

(**B**) Are terminally differentiated B lymphocytes

(**C**) Are terminally differentiated T lymphocytes

(**D**) Are neither T nor B lymphocytes

(**E**) Are inactive helper cells

(**F**) Are phagocytic

(**G**) Have segmented nuclei

12.7. A differential cell count of a blood smear from a patient with a parasitic infection is likely to reveal an increase in the circulating numbers of which of the following cell types?

(**A**) Basophils

(**B**) Eosinophils

(**C**) Erythrocytes

(**D**) Lymphocytes

(**E**) Monocytes

(**F**) Neutrophils

(**G**) Platelets

12.8. Which of the following cell types is most likely to be absent in a differential cell count of a blood smear from a normal patient?

(**A**) Basophils

(**B**) Eosinophils

(**C**) Erythrocytes

(**D**) Lymphocytes

(**E**) Monocytes

(**F**) Neutrophils

(**G**) Platelets

12.9. Large numbers of which of the following cell types in a heart muscle biopsy from a heart transplant recipient is a sign of graft rejection?

(**A**) Basophils

(**B**) Eosinophils

(**C**) Erythrocytes

(**D**) Lymphocytes

(**E**) Monocytes

(**F**) Neutrophils

(**G**) Platelets

12.10. Which of the following cell types is capable of returning to the circulation after leaving the blood to enter the connective tissue?

(**A**) Basophils

(**B**) Eosinophils

(**C**) Erythrocytes

(**D**) Lymphocytes

(**E**) Monocytes

(**F**) Neutrophils

(**G**) Platelets

12.11. Which of the following cell types normally is the most numerous of the circulating leukocytes?
(A) Basophils
(B) Eosinophils
(C) Erythrocytes
(D) Lymphocytes
(E) Monocytes
(F) Neutrophils
(G) Platelets

12.12. Which of the following cells is an agranulocyte that becomes phagocytic after it enters the connective tissues?
(A) Basophil
(B) Eosinophil
(C) Erythrocyte
(D) Lymphocyte
(E) Monocyte
(F) Neutrophil
(G) Platelet

12.13. Which of the following cell types is shown in Figure 12–4?
(A) Basophils
(B) Eosinophils
(C) Erythrocytes
(D) Lymphocytes
(E) Monocytes
(F) Neutrophils
(G) Platelets

12.14. Which of the following cell types has cytoplasmic granules that contain heparin and histamine?

(A) Basophils
(B) Eosinophils
(C) Erythrocytes
(D) Lymphocytes
(E) Monocytes
(F) Neutrophils
(G) Platelets

12.15. Which of the following substances is present in higher concentrations in plasma than in serum?
(A) Albumin
(B) Fibrinogen
(C) Glucose
(D) Immunoglobulin
(E) Major basic protein
(F) Serotonin
(G) Thrombosthenin

12.16. Which of the following substances is primarily responsible for maintaining the osmotic pressure of blood?
(A) Albumin
(B) Fibrinogen
(C) Glucose
(D) Immunoglobulin
(E) Major basic protein
(F) Serotonin
(G) Thrombosthenin

12.17. Which of the following terms refers to the percentage of packed erythrocytes per unit volume of blood?
(A) Differential count
(B) Central pallor

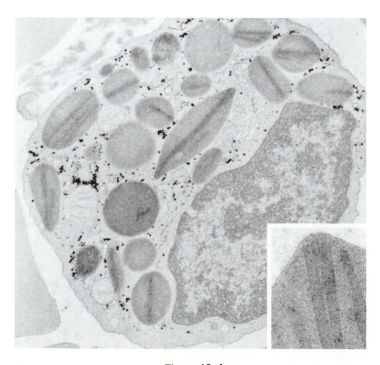

Figure 12–4.

(C) Hemoglobin
(D) Hematocrit
(E) Hematopoiesis
(F) Hematoma
(G) Hemolysis

12.18. Which of the following is the predominant form of hemoglobin present in human erythrocytes at birth?
(A) Carboxyhemoglobin
(B) HbA_1
(C) HbA_2
(D) HbB
(E) HbF
(F) HbS

12.19. When the oxygen tension in the blood of patients with sickle cell anemia decreases as it does at high altitudes, which of the following changes occurs?
(A) The solubility of the hemoglobin in their erythrocytes increases
(B) Sickled erythrocytes return to a normal biconcave shape
(C) Increased erythrocyte flexibility causes excessive packing in capillary beds
(D) The oxygen-carrying capacity of the sickled cells increases
(E) Hematocrit values decrease owing to hemolysis

12.20. Circulating leukocyte counts often increase in patients suffering from serious bacterial infections. Which of the following cell types is likely to account for most of this increase?
(A) Basophils
(B) B lymphocytes
(C) Eosinophils
(D) Monocytes
(E) Neutrophils
(F) Null cells
(G) T lymphocytes

12.21. An inherited abnormality affecting which clotting factor is the most common cause of the clotting disorder known as hemophilia?
(A) Factor VII
(B) Factor VIII
(C) Factor X
(D) Fibrinogen (Factor I)
(E) Prothrombin (Factor II)
(F) Thromboplastin (Factor III)
(G) Vitamin K

ANSWERS TO MULTIPLE-CHOICE QUESTIONS

12.1. D (III.A.1)
12.2. C (III.C)
12.3. A (III.A.4)
12.4. E (III.C; Fig. 12–2)
12.5. E (III.B.2.a,b, and C; Table 12–1)
12.6. D (III.B.1.a.[2])
12.7. B (III.B.2.b)
12.8. A (III.B.1.a–b and 2.a–c)
12.9. D (III.B.1.a.[1][b])
12.10. D (III.B.1.a)
12.11. F (III.B.2.c)

12.12. E (III.B.2.a and b)
12.13. B (III.B.2.b)
12.14. A (III.B.2.c)
12.15. B (II.B; III.B.2 and 3, C)
12.16. A (II.B.1)
12.17. D (I.D)
12.18. E (III.A.3)
12.19. E (III.A.3)
12.20. E (III.B.2.a)
12.21. B (IV.B)

13

Hematopoiesis

OBJECTIVES

This chapter should help the student to:

- Describe the structural and functional characteristics of a stem cell.
- Compare mature circulating blood cells and hematopoietic stem cells.
- Distinguish between the monophyletic and polyphyletic theories of hematopoiesis.
- Know the general structural characteristics of hematopoietic tissues and describe the changes that occur in bone marrow composition with age.
- Describe the life cycle of each formed element of blood, from stem cell to death.
- Describe the hormonal control of erythropoiesis.
- Name the phases of intrauterine hematopoiesis, the sites where each occurs, and differences in the erythrocytes produced during each phase.
- Recognize the erythrocyte and granulocyte precursors in micrographs of bone marrow. Name the stage immediately preceding and immediately following each cell.

MAX-Yield™ STUDY QUESTIONS

1. Describe pluripotent hematopoietic stem cells (I.A[1]) in terms of:
   a. Two names for these cells in scientific nomenclature
   b. Embryonic germ layer of origin
   c. Degree of differentiation (maturity)
   d. Ability to produce a variety of cell types
   e. Frequency of cell divisions
2. Name the two types of bone marrow (I.C) and compare them in terms of hematopoietic activity, relative number of adipocytes, the most abundant form in infants and in adults, and sites in the body where they occur in adults (III.A).
3. List the structural components of active bone marrow (other than developing blood cells) in terms of the cell types present (III.A.1), the type of capillaries present (III.A.2), and the type of connective tissue present, including the predominant collagen type (III.A.1).
4. List the functions of active bone marrow other than hematopoiesis (III.A.3).
5. List three organs containing macrophages that actively destroy old red blood cells (III.A.3).
6. Name three by-products of the breakdown of hemoglobin and describe the fate of each (III.A.3).
7. What are the effects of hypoxia and hemorrhage on yellow bone marrow (I.D)?
8. During the differentiation and maturation of erythrocytic cells, which general changes (increase, decrease, or no change) are observed in the following:
   a. Cell volume and diameter (IV.A)
   b. Nuclear volume and diameter (IV.A)
   c. Amount of heterochromatin in the nucleus (IV.A)
   d. Size and visibility of the nucleoli (IV.B.1 and 2)
   e. Number of polyribosomes in the cytoplasm (IV.A)
   f. Cytoplasmic basophilia (IV.A)
   g. Amount of hemoglobin in the cytoplasm (IV.A)
   h. Cytoplasmic acidophilia (IV.A)
   i. Number of mitochondria in the cytoplasm (IV.B.5)

---

[1] See footnote on page 1.

9. Beginning with the first recognizable cell type in the erythroid series, list, in order, the six stages of erythrocyte differentiation (IV.B).

10. Return to your list of stages in question 9 and indicate at which stage(s) or between which stages the following events occur:
   a. Cells divide (IV.B.1–4)
   b. Intense RNA synthesis takes place (IV.B.1 and 2)
   c. Cytoplasmic basophilia reaches its peak (IV.B.2 and 3)
   d. Hemoglobin synthesis accelerates (IV.B.2)
   e. Patches of cytoplasmic acidophilia appear; cytoplasm acquires a grayish tinge (IV.B.3)
   f. Hemoglobin synthesis peaks and begins to decline (IV.B.4)
   g. Capacity for mitosis is lost (IV.B.4)
   h. Nucleus is extruded (IV.B.4)
   i. Protein (hemoglobin) synthesis ceases (IV.B.5)
   j. Cells leave hematopoietic cords and enter sinusoids (IV.B.5)
   k. Cells lack nucleus but retain some ribonucleoprotein precipitable with cresyl blue stain (IV.B.5)
   l. Remaining organelles are broken down by nonlysosomal enzymes (IV.B.5)
   m. Cells are mature (IV.B)

11. Describe each of the six cell types listed in question 9 (IV.B.1–5; 12.III.A.1) in terms of their cell diameter, nuclear morphology (diameter, chromatin pattern, visibility of nucleoli), and cytoplasmic staining properties.

12. How do erythrocyte precursors receive iron to complex with hemoglobin (III.A.3)?

13. Describe the erythron (VII.A) in terms of:
   a. General functions and functional compartments
   b. Circulating erythrocyte number and life span in adults
   c. Erythrocyte number produced and destroyed daily (calculate from b)

14. Describe the hormone erythropoietin (VII.A) in terms of:
   a. Biochemical composition
   b. Site(s) of synthesis
   c. Effect of hypoxia on its synthesis and concentration in blood
   d. Effect on erythroid progenitor cell division
   e. Effect on erythroid precursor differentiation

15. List some vitamins and minerals that are essential to erythropoiesis (VII.A).

16. Beginning with the first recognizable cell type in the granulocytic series, list, in order, the six stages of granulocyte differentiation (V.A.2.a–e).

17. Return to your list of stages in question 16 and indicate at which stage(s) or between which stages the following events occur:
   a. Cells divide (V.A.2.c and d)
   b. Azurophilic granules are formed (V.A.2.a and b)
   c. Azurophilic granules first appear (V.A.2.b)
   d. Specific granules appear (V.A.2.b and c)
   e. Neutrophilic, eosinophilic, and basophilic precursors become discernible (V.A.2.c)
   f. Capacity for mitosis is lost (V.A.2.d)
   g. Cells are mature (V.A.2)
   h. Cells leave hematopoietic cords and enter sinusoids (V.A.2.e)

18. Describe each cell type listed in question 16 in terms of cell diameter, nuclear morphology (shape, chromatin pattern, and visibility of nucleoli), cytoplasmic staining properties, and the types of granules present (V.A.2.a–e).

19. Compare azurophilic granules and specific granules (V.A.2.b; 12.III.B.2.a–c) in terms of:
   a. Diameter
   b. Contents
   c. Staining properties
   d. Order of appearance (V.A.2.a–c)
   e. Changing abundance (increase or decrease) as differentiation and maturation proceed (V.A.2.a–c)

20. List, in order, the hematologic compartments through which a neutrophil passes during the stages between its differentiation and diapedesis. Indicate the approximate time spent in each compartment and its location (VII.B).

21. A decrease in the number of neutrophils in which compartment serves as a potent stimulus of neutrophilopoiesis (VII.B.3)?

22. Name the sites in the body where the following occur (V.B.2; VII.C):
    a. Lymphoblasts divide to form prolymphocytes
    b. Prolymphocytes or their derivatives are programmed to become T lymphocytes
    c. Prolymphocytes or their derivatives are programmed to become B lymphocytes
23. List four stages in the life cycle of monocytes that lead to the formation of macrophages and name the sites in the body where cells at each stage may be found (V.B.1; VII.D).
24. Name the cell type that produces platelets (VI) and describe it in terms of the cell type from which it is derived, its size, the shape of its nucleus, and the amount of DNA it contains compared with most other cells.
25. How are platelets formed (VI)?
26. List, in order, the three overlapping stages of intrauterine hematopoiesis and name the sites in the body where hematopoiesis occurs during each stage (II.A.1–3).
27. Compare primitive erythroblasts, definitive erythroblasts, and erythrocytes in terms of size, site of production, and the presence of a nucleus (II.A.1 and 2.a).
28. During which of the stages listed in answer to question 26 are leukocytes first produced (II.A.2)?
29. In adults whose bone marrow has become injured, diseased, or destroyed, which organs can help to compensate for the loss by resuming hematopoietic functions (II.C)?

SYNOPSIS

I. GENERAL FEATURES OF HEMATOPOIESIS

Hematopoiesis is blood cell production. It involves the proliferation and differentiation of hematopoietic stem cells and may be subdivided, according to the cell type formed, into erythropoiesis, leukopoiesis, granulopoiesis, agranulopoiesis, lymphopoiesis, and thrombopoiesis.

A. Hematopoietic Stem Cells: These are undifferentiated mesodermal derivatives able to divide repeatedly and differentiate into mature blood cells. The nature and structure of the earliest blood cell precursors are debatable. The best available evidence supports the **monophyletic theory** of hematopoiesis, according to which a single **pluripotent stem cell** (formerly called a **hemocytoblast**) can form all mature blood cell types. Hematopoietic stem cells are called **colony-forming cells (CFCs),** or colony-forming units (CFUs), because they form colonies of recognizable blood cell types in culture. Pluripotent CFCs were first demonstrated in spleen cell cultures and are called CFC-S cells. Some CFC-S cells may circulate in a form resembling lymphocytes. CFC-S cells divide only rarely, most likely because each of their progeny can give rise to so many cells. The progeny of a dividing CFC-S cell remain pluripotent or differentiate into one of several **unipotential stem cell** types, which can divide but each of which produces only one mature blood cell type (eg, CFC-E cells form erythrocytes).

B. Hematopoietic Tissues: These tissues are collections of CFCs and their progeny at various stages of maturation, suspended in a reticular connective tissue stroma. Active hematopoiesis shifts its location in overlapping stages during development (II.A.1–3): it occurs first in the extraembryonic mesoderm of the yolk sac; next in the fetal liver, spleen, and thymus; and finally in the bone marrow and lymphoid tissue.

C. Bone Marrow (III.A): Bone marrow (medullary tissue) is the primary hematopoietic tissue from the fifth month of fetal life. All bone marrow begins as active, or **red, marrow.** During growth, development, and aging, portions of the red marrow are replaced by adipocytes to form **yellow marrow.** Yellow marrow can be reactivated by an increased demand for blood cells (eg, during chronic hypoxia and hemorrhage). Yellow marrow does not produce blood cells and thus is not a hematopoietic tissue. Red marrow has a limited distribution in adults. It contains masses of reticular connective tissue **stroma** that support the CFCs and their progeny (the **hematopoietic cords**), separated by vascular **sinusoids** whose walls have openings through which maturing blood cells enter the circulation.

D. Blood Cell Life Span: Blood cells have a limited life span in the circulation, owing to the recognition and removal of worn and damaged erythrocytes by macrophages and to the migration of leukocytes into the surrounding tissues. To keep constant numbers of each cell type in circulation, hematopoiesis must be continuous. Otherwise, a decrease in the number of circulating cells, or **anemia,** results.

E. Regulation of Hematopoiesis: Regulation involves colony-stimulating factors (CSFs), such as erythropoietin, leukopoietin, and thrombopoietin. These hormones act at various steps in hematopoiesis to enhance the proliferation and differentiation of CFCs; however, except for erythropoietin (VII.A), their nature and actions are not clear. This reflects the discovery of a variety of CSFs (eg, GM-CSF, G-CSF, M-CSF, and Steel factor) with overlapping hematopoietic activities.

II. DEVELOPMENT OF HEMATOPOIETIC TISSUES

A. Sites of Intrauterine Hematopoiesis:
1. **Primordial (prehepatic) phase.** During week three of embryonic development, cell clusters called **blood islands** form in the extraembryonic mesoderm of the yolk sac. Cells at the periphery form the endothelium of the primitive blood vessels. By a process called **megaloblastic erythropoiesis,** cells at the center form the first blood cells, called **primitive erythroblasts.** These differ from **definitive erythroblasts** of later stages in that they are larger, contain a unique type of hemoglobin, and retain their nuclei. Leukocytes and platelets do not appear until the next phase.
2. **Hepatosplenothymic phase.** During the second month, hematopoiesis shifts to the liver, spleen, and thymus. Hematopoietic stem cells invade these organs and begin producing a wider variety of blood cell types. The **liver** produces granulocytes, platelets, and red blood cells that may be nucleated (definitive erythroblasts) or enucleate (erythrocytes). Hematopoiesis in the liver declines during the fifth month, but continues at low levels until a few weeks after birth. The **spleen** produces mainly erythrocytes and small numbers of granulocytes and platelets. Just before birth, lymphopoiesis becomes an important splenic function. The **thymus** produces lymphocytes. Lymphocytes produced here become T cells with a variety of specialized functions (14.III.A.2).
3. **Medullolymphatic (definitive) phase.** During the third month, hematopoiesis begins shifting to the bone marrow and lymphoid tissue, where it remains throughout adulthood. **Medullary tissue** (bone marrow) first becomes hematopoietic in the clavicle's diaphysis, between months two and three. As other bones ossify, their marrow becomes active. By the fifth month, bone marrow is the primary hematopoietic tissue, producing platelets and all blood cell types. Additional lymphocytes form in the developing **lymphoid tissues and organs** (eg, thymus, lymph nodes, spleen). Before birth, the lymph nodes also may produce red blood cells.

B. Sites of Postnatal Hematopoiesis: Beginning in infancy, hematopoiesis is restricted to the bone marrow (medullary or myeloid tissue) and the lymphoid tissues.

C. Extramedullary Hematopoiesis in Disease: In adults, erythropoiesis, granulopoiesis, and thrombopoiesis in sites other than bone marrow are abnormal. When bone marrow cannot meet the demand for blood cells, the liver, spleen, or lymph nodes may resume their embryonic hematopoietic activity.

III. GENERAL STRUCTURE OF MATURE HEMATOPOIETIC TISSUE

Mature hematopoietic tissues share a basic architecture supported by a reticular connective tissue scaffolding (stroma) permeated by many sinusoids. The meshwork between the sinusoids contains developing blood cells; as these complete their differentiation, they enter the circulation through openings in the sinusoid walls.

A. Bone Marrow: All marrow begins as red marrow, also called **active,** or **hematogenous,** marrow. During growth, the blood cells are gradually depleted and replaced by adipocytes. In adults, red marrow is restricted to the skull, vertebrae, ribs, sternum, ilia, and the proximal epiphyses of some long bones. The fatty, nonhematopoietic replacement tissue in other bony cavities is termed yellow marrow.

1. **Stroma** consists of adipocytes (as much as 75% of red marrow), macrophages, and reticular connective tissue composed of **adventitial cells** (reticular cells) and the reticular fibers (type-III collagen) they produce. Adventitial cells are highly branched, mesenchymal derivatives resembling fibroblasts. Their processes separate the developing blood cells from the endothelium of sinusoids.

2. **Hematopoietic cords,** which comprise the stromal scaffolding, are crowded with overlapping blood cells of all types and at all stages of differentiation. Nests of similar cells, often the progeny of a single stem cell, occupy different microenvironments in the marrow cords. Abundant sinusoids lie between the cords and have openings in their walls through which maturing blood cells and platelets enter the circulation. In histologic section, the dense packing makes the identification of individual cell types difficult. Differentiating blood cells are therefore commonly studied in smears.

3. **Bone marrow functions.** In addition to being the primary site for hematopoiesis, bone marrow helps destroy old red blood cells. Macrophages in the bone marrow, spleen, and liver break down hemoglobin to form (1) **globin,** which is quickly hydrolyzed; (2) **porphyrin** rings, which are converted to **bilirubin;** and (3) **iron,** which is complexed with and transported by the plasma protein **transferrin** to other bone marrow sites for reuse by developing erythrocytes. Iron is stored in bone marrow macrophages as **ferritin** (iron complexed with the protein **apoferritin**) and **hemosiderin.** In some sections, clusters of developing erythrocytes surround and receive iron from macrophages in groupings called **erythroblastic islands.**

B. **Lymphoid Tissues and Organs:** The thymus, spleen, lymph nodes, and lymphatic aggregations, such as the tonsils and Peyer's patches, contribute to postnatal hematopoiesis by providing sites for lymphocyte proliferation, programming, and differentiation **(lymphopoiesis).** Lymphoid organs and tissues are also assembled on a reticular connective scaffolding and are described in Chapter 14.

IV. ERYTHROPOIESIS

A. **General:** In healthy adults, erythropoiesis (red blood cell formation) occurs exclusively in bone marrow. Erythrocytes derive from CFC-E cells, which in turn derive from CFC-S cells. Erythrocyte differentiation is commonly described by naming cell types at specific stages in the process according to their histologic characteristics (IV.B). Cellular changes that occur during erythroid differentiation include (1) a decrease in cell size; (2) condensation of nuclear chromatin; (3) a decrease in nuclear diameter; (4) an accumulation of hemoglobin in the cytoplasm (increased acidophilia); (5) a decline in ribosome numbers in the cytoplasm (decreased basophilia); and (6) ejection of the nucleus.

B. **Stages of Erythroid Differentiation:** Erythrocyte maturation is commonly divided into six stages (Fig. 13–1). These stages are identified by overall cell diameter, nuclear size and chro-

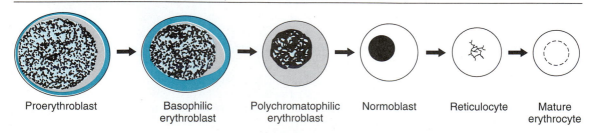

| Proerythroblast | Basophilic erythroblast | Polychromatophilic erythroblast | Normoblast | Reticulocyte | Mature erythrocyte |

Figure 13–1. Erythropoiesis. Schematic diagram of erythrocyte precursor cells at various stages of erythroid development. The proerythroblast derives from a CFU-E cell. Drawings are roughly to scale.

matin pattern, and cytoplasmic staining properties. Cells in transition between these stages are commonly found in bone marrow smears. Cell division occurs throughout the early stages, but cells lose their ability to divide during the normoblast stage. The following discussion begins with the least mature cells; the sixth (final) stage produces the mature erythrocyte (12.III.A.1).

1. **Proerythroblasts** are large (14–19 μm in diameter) and contain a large, centrally located, pale-staining nucleus with one or two large nucleoli. The small amount of cytoplasm (approximately 20% of cell volume) contains polyribosomes actively involved in hemoglobin synthesis. The resulting cytoplasmic basophilia allows these cells to be distinguished from myeloblasts, with which they are most easily confused. Proerythroblasts are capable of multiple mitoses and may be considered unipotential stem cells.

2. **Basophilic erythroblasts** are slightly smaller than proerythroblasts, with a diameter of 13 to 16 μm. They have slightly smaller nuclei with patchy chromatin. Their nucleoli are difficult to distinguish. The cytoplasm is more intensely basophilic, typically staining a deep royal blue. A prominent, clear, juxtanuclear cytocenter often is visible. Basophilic erythroblasts continue hemoglobin synthesis at a high rate and are capable of mitosis.

3. **Polychromatophilic erythroblasts** are smaller yet (12–15 μm in diameter), and more hemoglobin accumulates in their cytoplasm. The conflicting staining affinities of the polyribosomes (basophilic) and hemoglobin (acidophilic) give the cytoplasm a grayish appearance. The nucleus is smaller than in less mature cells, with more condensed chromatin forming a checkerboard pattern. Cells at this stage retain the ability to synthesize hemoglobin and to divide.

4. **Normoblasts** (orthochromatophilic erythroblasts) are easily identified because of their small size (8–10 μm in diameter); an acidophilic cytoplasm with only traces of basophilia; and small, eccentric nuclei with chromatin so condensed that it appears black. Although early normoblasts may divide, erythroid cells lose their ability to do so during this stage, which ends with the extrusion of the **pyknotic** (degenerated, dead) nucleus.

5. **Reticulocytes** are nearly indistinguishable from mature erythrocytes with standard stains; however, when they are stained with the supravital dye cresyl blue, **residual polyribosomes** form a blue-staining, netlike precipitate in the cytoplasm. Reticulocytes complete their maturation to become erythrocytes (12.III.A.1) during their first 24 to 48 hours in circulation. This process involves the ejection or enzymatic digestion of their remaining organelles and assumption of the biconcave shape.

V. LEUKOPOIESIS

Leukopoiesis (white blood cell formation) encompasses both granulopoiesis and agranulopoiesis. Leukopoietic CFCs that have been identified include CFC-GM (forms both granulocytes and macrophages), CFC-G (forms all granulocyte types), CFC-M (forms macrophages), and CFC-EO (forms only eosinophils). All of these CFCs with limited capabilities derive from the pluripotential CFC-S cells.

A. Granulopoiesis:

1. **General.** Granulopoiesis occurs in the bone marrow of healthy adults. The three granulocyte types—neutrophils, basophils, and eosinophils—may all derive from a single precursor (CFC-G). The structural changes that characterize granulopoiesis include (1) a decrease in cell size; (2) condensation of nuclear chromatin; (3) changes in nuclear shape (flattening → indentation → lobulation, a progression resembling the gradual deflation of a balloon); and (4) an accumulation of cytoplasmic granules.

2. **Stages of granulocyte differentiation.** Granulocyte maturation is commonly divided into six stages (Fig. 13–2). These stages are identified by overall cell diameter; size, shape, and chromatin pattern in the nuclei; and type and number of specific granules in the cytoplasm. The specific granules, with their characteristic staining properties, first appear at the myelocyte stage; from this point, the cells are named according to the mature granulocyte type they will form (eg, neutrophilic myelocyte). In the granulocyte series, cell division ceases at the metamyelocyte stage. The following discussion begins with the least mature cells; the sixth (final) stage produces the mature granulocyte (12.III.B.2.a–c).

   a. **Myeloblasts,** the earliest recognizable granulocyte precursors, are approximately 15 μm in diameter and are difficult to distinguish from other stem cells. Each has a large, spher-

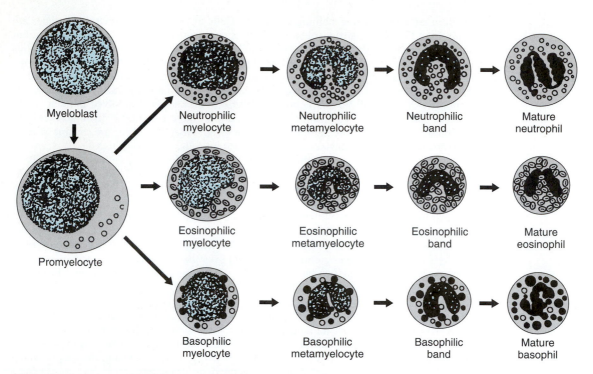

Figure 13–2. Granulopoiesis. Schematic diagram of granulocyte precursor cells at various stages of granulocyte development. Drawings are roughly to scale.

ical, euchromatic nucleus with as many as three smudgy nucleoli. Their cytoplasm lacks granules and is more basophilic than that of their CFC precursors but less basophilic than that of proerythroblasts, with which they are most often confused.

b. Promyelocytes (15 to 24 μm in diameter) are larger than myeloblasts and their chromatin is slightly more condensed. Their otherwise spherical nuclei may be flattened on one side and may contain nucleoli. Their cytoplasm is more basophilic than that of myeloblasts and contains azurophilic granules (0.05 to 0.25 μm in diameter) but not specific granules (12.III.B), which appear during the subsequent stage. Because azurophilic granules are synthesized mainly during this stage, the number per cell decreases during division and maturation. These granules contain lytic enzymes and function as lysosomes.

c. Myelocytes typically are smaller than promyelocytes (10 to 16 μm in diameter). This is the first stage at which enough specific granules accumulate in the cytoplasm to enable distinction among the three immature granulocyte types: **neutrophilic myelocytes, eosinophilic myelocytes,** and **basophilic myelocytes.** Myelocyte nuclei are round to kidney-shaped, with chromatin that is more condensed than during previous stages. Like their precursors, myelocytes can divide.

d. Metamyelocytes. The three metamyelocyte types—**neutrophilic, eosinophilic,** and **basophilic**—are smaller (10 to 12 μm in diameter) and more densely packed with specific granules. The nucleus is deeply indented, often resembling a mask, and its chromatin is more condensed. During this stage, the capacity for mitosis is lost.

e. Band cells. The three band cell types—**neutrophilic, eosinophilic,** and **basophilic**—have horseshoe-shaped nuclei. They range in diameter from 10 to 12 μm. Like the erythroid reticulocytes, these nearly mature cells circulate in small numbers (3 to 5% of circulating leukocytes) but may appear in larger numbers when granulopoiesis is hyperstimulated. During final maturation, the nuclei undergo further chromatin condensation and lobulation. Mature granulocytes (ie, neutrophils, eosinophils, and basophils) also occur in bone marrow.

B. **Agranulopoiesis:** Agranulocytes (monocytes and lymphocytes), like the other blood cell types, derive from CFC-S cells. The morphologic changes during maturation include decreases in overall cell and nuclear diameter and an increase in nuclear chromatin density. The morphologic characteristics of agranulocytes at immature stages are much less distinct than those of erythrocytes and granulocytes.

1. **Monocytopoiesis.** CFC derivatives that give rise to monocytes are called **monoblasts** and are difficult to identify in bone marrow smears. Monoblast derivatives, **promonocytes,** are slightly easier to identify and serve as immediate precursors of monocytes. Promonocytes are larger (10 to 20 μm in diameter) than monocytes and have pale-staining nuclei and basophilic cytoplasm.

2. **Lymphopoiesis.** In adults, lymphopoiesis occurs mainly in lymphoid tissues and organs and to a lesser extent in bone marrow. Before division, the precursor, or **lymphoblast,** is larger than the typical circulating lymphocyte. However, many circulating lymphocytes can respond to antigenic stimulation by blasting (enlarging to assume lymphoblast morphology), which indicates that they are dormant stem cells. Some of these, called **null cells,** are neither T nor B cells and may represent circulating CFC-S cells.

VI. THROMBOPOIESIS

Platelet (thrombocyte) production is carried out in the bone marrow by unusually large cells (100 μm in diameter) called **megakaryocytes.** Immature megakaryocytes, called **megakaryoblasts,** derive from CFC-Megs, which in turn derive from CFC-S cells. Megakaryoblasts undergo successive incomplete mitoses involving repeated DNA replications without cellular or nuclear division. The result of this process, called **endomitosis,** is a single large megakaryocyte with a single, large, multilobed, **polyploid** (as many as 64 n) nucleus. Maturation involves lobulation of the nucleus and development of an elaborate **demarcation membrane system** that subdivides the peripheral cytoplasm, outlining cytoplasmic fragments destined to become platelets. As the demarcation membranes fuse to form the plasma membranes of the platelets, ribbonlike groups of platelets are shed from the megakaryocyte periphery into the marrow sinusoids to enter the circulation.

VII. COMPARTMENTS & THE LIFE CYCLE OF BLOOD CELL TYPES

A. **Erythrocytes:** The total population of mature and developing red blood cells constitutes the widely dispersed but functionally discrete **erythron,** which is subdivided into two compartments. The **circulating compartment** includes all mature erythrocytes in the circulation (approximately 2.5×10^{13}). The **medullary compartment** (erythropoietic pool) includes the bone marrow sites where erythropoiesis occurs. Erythrocytes usually leave the bone marrow to enter the circulation as reticulocytes and undergo final maturation within 24 to 48 hours. Mature erythrocytes circulate for approximately 120 days before they are retired by macrophages (primarily in the spleen, but also in the bone marrow and liver). Approximately 10^{11} erythrocytes are retired daily. The iron in the hemoglobin is conserved and eventually returned to the marrow by transferrin. Iron-free hemoglobin is converted by the liver into bile pigment called **bilirubin.** Red cell replacement is controlled by the glycoprotein hormone **erythropoietin,** which stimulates erythrocyte precursors in the bone marrow to proliferate and differentiate. Erythropoietin is produced by unknown cells in the kidney cortex in response to low oxygen tension in the blood. Other factors affecting erythrocyte production and function include iron, intrinsic factor, vitamin B_{12}, and folic acid.

B. **Granulocytes:** Neutrophils and other granulocytes are continually produced in the bone marrow and, because their numbers remain relatively constant, they also must be continually destroyed. Granulocytes constantly move from the marrow to the circulation to the tissues, where many of them die.

1. The **medullary formation compartment** in the bone marrow comprises the stem cells and is the site of granulopoiesis. Cells spend approximately 7 days in this compartment.

2. The **medullary reserve compartment** in the bone marrow comprises newly formed granulocytes that have yet to enter the circulation. Neutrophils remain here for another 4 days.

3. The **circulating compartment** comprises mature granulocytes circulating in the blood. The number of cells in the circulating compartment remains relatively constant, even though most granulocytes circulate for only a few hours. When the cell number in this compartment decreases as a result of margination or removal of the cells from the blood (eg, by leukopheresis), granulocyte production in the bone marrow is stimulated to replace the missing cells by an unknown mechanism. This stimulation of bone marrow activity appears to be mediated by multiple CSFs, which together are called **leukopoietin.**

4. The **marginating compartment** comprises cells that have entered the circulation but have attached to the walls of blood vessels, become confined by vasoconstriction in some capillary beds, or passed through intercellular junctions between endothelial cells to move out of the blood vessels and into the connective tissues—a process called **diapedesis.** After they have entered the tissues, granulocytes rarely reenter the circulation. However, exchanges between the rest of the marginating compartment and the circulating compartment occur continuously. The total time spent in the circulating and marginating compartments is approximately 6 to 7 hours.

C. **Lymphocytes:** Precursors of both B cells and T cells are produced in the bone marrow. Those destined to become T cells migrate to the thymus, where they are programmed to assume the specialized functions of this lymphocyte class before reentering the circulation and moving to the spleen or lymph nodes for final maturation. Mature T cells return to the circulation for a long period of time; in humans, they have a life span that is measured in years. Precursors destined to become B cells never enter the thymus but are programmed to be B cells in the bone marrow and are subsequently distributed to the spleen, lymph nodes, and other lymphatic aggregations, where they respond to specific antigens. B cells have a life span of at least 6 weeks in humans. In response to antigenic stimulation, they proliferate and differentiate into plasma cells. Lymphopoiesis and lymphocyte function are discussed further in Chapter 14.

D. **Monocytes:** Monocytes form in the bone marrow and remain in circulation for approximately 2 days before leaving the bloodstream by passing between the endothelial cells in the walls of capillaries and venules. They enter the connective tissues to differentiate into macrophages and other mature components of the mononuclear phagocyte system, including the Kupffer cells in the liver and osteoclasts in bone.

E. **Platelets:** Platelets are formed in the bone marrow, most likely in response to increased blood levels of one or more CSFs referred to as thrombopoietin. Platelets have a life span of approximately 10 days in the circulation. Aside from their involvement in clot formation and the eventual removal of clots by sloughing or phagocytosis, the fate of platelets is unknown.

MULTIPLE-CHOICE QUESTIONS

Select the single best answer.

13.1. Which of the following is the most widely held theory of hematopoiesis, according to which all cell types are believed to derive from a single pluripotential stem cell?
(A) Medullolymphatic theory
(B) Monophyletic theory
(C) Monopoietic theory
(D) Polyphyletic theory

13.2. Which of the following bone marrow components generally increases in abundance as hematopoietic activity declines?
(A) Adipocytes
(B) Erythroblastic islands
(C) Hematopoietic cords

(D) Sinusoids
(E) Stem cells

13.3. Which cell type has a nucleus with a checkerboard pattern and grayish-staining cytoplasm because of the presence of roughly equal amounts of basophilic and acidophilic components?
(A) Basophilic erythroblast
(B) Normoblast (orthochromatophilic erythroblast)
(C) Polychromatophilic erythroblast
(D) Proerythroblast
(E) Reticulocyte

13.4. Which of the following cytoplasmic components are the main constituents of the dark precipitate that forms in reticulocytes in re-

sponse to staining with the vital dye cresyl blue?
(A) Golgi complexes
(B) Hemoglobin
(C) Nucleoli
(D) Nuclear fragments
(E) Polyribosomes
(F) SER
(G) Spectrin

13.5. Which of the following terms best describes a cell in the neutrophil lineage with a deeply indented, or mask-shaped, nucleus?
(A) Mature neutrophil
(B) Monocyte
(C) Myeloblast
(D) Neutrophilic band
(E) Neutrophilic metamyelocyte
(F) Neutrophilic myelocyte
(G) Promyelocyte

13.6. Which of the following is true of erythrocytes?
(A) Enter the circulation only after becoming fully mature
(B) Undergo mitosis in the circulation in response to erythropoietin
(C) Are removed from the circulation by macrophages after approximately 120 days
(D) Contain mitochondria and are capable of oxidative phosphorylation
(E) Have precursors that produce hemoglobin on their RER

13.7. Which of the following is true of granulocytes that have entered the marginating compartment?
(A) Undergo mitosis
(B) May cross capillary walls to enter connective tissues
(C) Cannot reenter the circulation
(D) May eventually differentiate into tissue macrophages

13.8. Which of the following is true of monocytes?
(A) Have precursors in bone marrow that are virtually indistinguishable from early granulocyte precursors
(B) Typically remain in the circulation for several weeks
(C) Have no cytoplasmic granules
(D) Undergo no structural or functional changes after leaving the bone marrow

13.9. Which of the following is true of megakaryocytes?
(A) Are multinucleated
(B) Are formed by the fusion of many haploid cells
(C) Serve as precursors to bone marrow macrophages
(D) Are located primarily in the spleen
(E) Contain a network of intracellular membranes

13.10. Which of the following cell types typically contains the largest and most easily visualized nucleolus?
(A) Erythrocyte
(B) Basophilic erythroblast
(C) Orthochromatophilic erythroblast
(D) Polychromatophilic erythroblast
(E) Proerythroblast
(F) Reticulocyte

13.11. Which of the following cell types has reached the first stage in its lineage in which it is incapable of further mitosis?
(A) Erythrocyte
(B) Basophilic erythroblast
(C) Orthochromatophilic erythroblast
(D) Polychromatophilic erythroblast
(E) Proerythroblast
(F) Reticulocyte

13.12. Which of the following is the earliest stage at which specific granulocyte types can be distinguished from one another?
(A) Band form
(B) Mature form
(C) Metamyelocyte
(D) Myeloblast
(E) Myelocyte
(F) Promyelocyte

13.13. Which of the following is the earliest stage at which azurophilic granules first accumulate?
(A) Band form
(B) Mature form
(C) Metamyelocyte
(D) Myeloblast
(E) Myelocyte
(F) Promyelocyte

13.14. Appearance in the peripheral blood of large numbers of which of the following cell types is termed a "shift to the left" and may signal a bacterial infection?
(A) Band form
(B) Mature form
(C) Metamyelocyte
(D) Myeloblast
(E) Myelocyte
(F) Promyelocyte

13.15. Which of the following processes occurs during granulocyte maturation but not during erythrocyte maturation?
(A) Cells eventually lose their capacity for mitosis
(B) Cytoplasmic content of hemoglobin increases
(C) Nuclear euchromatin content increases
(D) Nucleus becomes increasingly lobulated
(E) Nucleus is expelled
(F) Overall cell diameter generally decreases
(G) Overall nuclear size decreases

13.16. In which order do the phases of intrauterine hematopoiesis occur?
  (A) Hepatosplenothymic, medullolymphatic, primordial
  (B) Hepatosplenothymic, primordial, medullolymphatic
  (C) Medullolymphatic, hepatosplenothymic, primordial
  (D) Medullolymphatic, primordial, hepatosplenothymic
  (E) Primordial, hepatosplenothymic, medullolymphatic
  (F) Primordial, medullolymphatic, hepatosplenothymic

13.17. A dramatic reduction in the oxygen-carrying capacity of a patient's blood after extensive erythrocyte lysis during a sickle cell crisis triggers the secretion of which substance from the kidney?
  (A) Aldosterone
  (B) Angiotensinogen
  (C) Apoferritin
  (D) Erythropoietin
  (E) Renin

13.18. Which of the following cell types is capable of further mitosis after leaving the hematopoietic organ in which they were formed?
  (A) Basophil
  (B) Eosinophil
  (C) Erythrocyte
  (D) Lymphocyte
  (E) Monocyte
  (F) Neutrophil
  (G) Platelet

13.19. In which of the following sites are blood cells first formed during embryogenesis?
  (A) Amnion
  (B) Bone marrow
  (C) Liver
  (D) Neural crest
  (E) Spleen
  (F) Thymus
  (G) Yolk sac

13.20. A decrease in the number of neutrophils in which of the following compartments is known to act as a potent stimulus for neutrophilopoiesis?
  (A) Circulating compartment
  (B) Marginating compartment
  (C) Medullary formation compartment
  (D) Medullary reserve compartment

ANSWERS TO MULTIPLE-CHOICE QUESTIONS

13.1. B (I.A)
13.2. A (I.C)
13.3. C (IV.B.3)
13.4. E (IV.B.5)
13.5. E (V.A.2.d)
13.6. C (IV.B.4 and 5; VII.A)
13.7. B (VII.B.4 and D; V.A.2.d)
13.8. A (V.B.1; VII.D; 12.III.B.1.b)
13.9. E (VI)
13.10. E (IV.B.1 and 2)

13.11. C (IV.B.4)
13.12. E (V.A.2.c)
13.13. F (V.A.2.b)
13.14. A (V.A.2.e; 12.III.B.2.a)
13.15. D (IV.B.1–5; V.A.2.a–e)
13.16. E (II.A.1–3)
13.17. D (VII.A)
13.18. D (VII.A–E)
13.19. G (II.A.1)
13.20. A (VII.B.3)

Lymphoid System

14

OBJECTIVES

This chapter should help the student to:

- Know the functions of the lymphoid system.
- Know the names, locations, and functions of the cells, tissues, and organs of the lymphoid system and identify them, as well as their components, in a micrograph.
- Know the distinguishing features of the lymphoid organs.
- Distinguish between central and peripheral lymphoid organs.
- Distinguish between cell-mediated and humoral immunity.
- Describe lymphocyte differentiation from stem cells to T or B memory and effector cells.
- Know the five immunoglobulin classes and their distinguishing features.
- Describe the steps in lymphocyte activation by antigens.
- Describe antigen disposal by cell-mediated and humoral mechanisms.
- Describe the path taken by lymph as it flows through the lymph nodes.
- Describe blood flow through the spleen according to the open and closed theories of circulation.

MAX-Yield™ STUDY QUESTIONS

1. Describe the general structure of lymphoid tissue in terms of the:
   a. Type of connective tissue that makes up the stroma (I.A[1])
   b. Types of cells and fibers that make up the stroma (III.D.1 and 2)
   c. Types of cells suspended within the spaces of the stroma (I.A)
   d. Lymphoid organ in which the composition of the stroma differs from that of all other lymphoid tissues and organs (III.D.1 and 2)
2. Compare cellular and humoral immunity in terms of the:
   a. Type of lymphocyte (B or T) primarily associated with each (I.D.1 and 2)
   b. Requirement for direct lymphocyte contact during antigen disposal (I.D.1 and 2)
   c. Names of the effector cells involved in each type (I.F.4; III.A.1 and 2)
3. Compare the central and peripheral lymphoid organs (I.B) in terms of the names of the organs that comprise each group, and the antigen dependence or independence of lymphocyte proliferation in these organs.
4. Sketch an IgG molecule (Fig. 14–1; II.A.1–6) and label the light and heavy chains, Fc and Fab fragments, constant and variable regions, and cell-binding and antigen-binding regions.
5. List the five types of immunoglobulins (Igs) secreted by plasma cells, and indicate which Ig is described by each of the following characteristics:
   a. Most abundant in blood (II.B.1)
   b. Can cross the placenta (II.B.1)
   c. Secretory form consists of two Igs, protein J, and a transport component (II.B.2)
   d. Predominant Ig in secretions (eg, mucus, tears, saliva; II.B.2)
   e. Usually exists as a pentamer (II.B.3)
   f. Most effective in activating the complement system (II.B.3)
   g. Fc portion has a great affinity for the surface of mast cells and basophils (II.B.4)
   h. Is the primary mediator of allergic reactions (II.B.4)

---

[1] See footnote on page 1.

    i. Is least understood (II.B.5)

    j. Are found on the surface of B lymphocytes (list two; II.B.3 and 5)

6. Indicate the order in which both T and B lymphocytes undergo the following processes after encountering an antigen (I.F.3 and 4):

    a. Differentiation into effector and memory cells

    b. Blast transformation (formation of immunoblasts)

    c. Clonal expansion (proliferation)

7. List the T-lymphocyte effector cells and describe the basic functions of each (III.A.2).

8. Name the B-lymphocyte effector cell and describe its basic function and its most important organelle (III.A.1 and C).

9. How do memory cells make the response to subsequent encounters with a particular antigen (secondary immune response) more effective than the response to the first encounter (primary immune response) (I.F.5; II.B.1 and 3)?

10. Describe the thymus in terms of:

    a. Its primary functions (VI.B.1 and 3)

    b. Its location in the body (VI.A)

    c. Its classification as a lymphoid organ (central or peripheral, encapsulated or unencapsulated; VI.A)

    d. Its embryonic germ layer(s) of origin (VI.B.6)

    e. The embryonic pharyngeal pouch(es) from which it derives (VI.B.6)

    f. The source of lymphocyte precursors that populate it before and after birth (VI.B.6)

    g. The type of reticular cells it contains (III.D.2) and their embryonic germ layer of origin (VI.B.6)

11. List the thymic hormones that may be secreted by the epithelial reticular cells (III.D.2). What is the general effect of these hormones on other lymphoid organs (VI.B.3 and 5)?

12. Compare the cortex and medulla of the thymus (VI.A.1 and 2) in terms of the:

    a. Packing density of the lymphocytes

    b. Number of epithelial reticular cells

    c. Hassall's corpuscles

    d. Type of blood vessels present

    e. Location of the blood–thymus barrier

    f. Site of T-cell programming

    g. Site where involution begins

13. List, in order, the layers through which a substance in the blood must pass to cross the blood–thymus barrier (VI.B.2).

14. What is the most likely function of the blood–thymus barrier (VI.B.2)?

15. What happens to the size and functional activity of the thymus, beginning at puberty and continuing into old age? What is this process called (VI.B.6)?

16. Compare the thymus and other lymphoid organs in terms of:

    a. Embryonic germ layers of origin (III.D.1 and 2; VI.B.6)

    b. Lymphocyte types they produce (III.A; VI.B.1; VII.B.2; VIII.B.2)

    c. Primary type of reticular cells present (III.D.1 and 2)

    d. The presence of lymphoid nodules (IV; Table 14–1)

    e. The presence of a cortex and medulla (Table 14–1)

    f. The presence of sinuses and cords (Table 14–1)

    g. Filtering functions (I.D)

17. Compare normal animals with those thymectomized at birth (VI.B.5) in terms of the:

    a. Number of lymphocytes circulating in the blood and lymph

    b. Ability to mount a delayed hypersensitivity reaction

    c. Ability to reject a foreign graft

    d. Length of survival

    e. Ability to mount a cellular immune response

    f. Ability to mount a humoral immune response

18. Describe lymph nodes in terms of:

    a. Their general functions (VII.B.1–3)

    b. Their location in the body (VII)

    c. Their classification as lymphoid organs (central or peripheral, encapsulated or unencapsulated; I.B)

    d. Their embryonic germ layer(s) of origin (I.F.1; III.D.1)

    e. The sources of the lymphocyte precursors that populate them (VII.A.1 and 3; I.F.1)

    f. The type of reticular cells they contain (III.D.1)

19. Sketch a lymph node (Fig. 14–2) and label the following:

a. Capsule	**h.** Cortex
b. Hilum	**i.** Medulla
c. Trabeculae	**j.** Medullary cords
d. Subcapsular sinus	**k.** Medullary sinuses
e. Peritrabecular sinus	**l.** Paracortical zone
f. Lymphoid nodules	**m.** Afferent lymphatic vessels
g. Germinal centers	**n.** Thymus-dependent region

20. Beginning with the afferent lymphatic vessels and ending with the efferent lymphatic vessels, trace the path of lymph through a lymph node (VII.A.5; Fig. 14–2). What percentage of lymph actually penetrates the nodules (VII.B.1)?

21. Name the cells and structures commonly found in the lumens of lymph node sinuses (VII.A.5).

22. How does the composition of lymph change as it passes through a lymph node? Which substances are removed or added? By which cell types (VII.B.1–3)?

23. For each cell type listed below, give the basic function and name the part(s) of a lymph node in which it can be found. If a cell type is found throughout the node, indicate any sites where it occurs in higher proportions.

a. B lymphocyte (III.A.1; VII.A.1)	**e.** Plasma cell (III.C)
b. T lymphocyte (III.A.2; VII.A.3)	**f.** Follicular dendritic cell (III.E)
c. Lymphoblast (I.C; IV; VII.A.1)	**g.** Macrophage (III.B; VII.A.5)
d. Memory cell (I.F.4; IV)	**h.** Reticular cell (III.D.1; VII.A.1 and 2)

24. Through which blood vessels in a lymph node can lymphocytes directly exit the bloodstream? Where in the lymph node are these vessels found (VII.A.3)?

25. Describe the spleen in terms of:

a. General functions (VIII)	**d.** Germ layer(s) of origin (I.F.1; III.D.1)
b. Location (VIII)	**e.** Lymphocyte source(s) (VIII.A.1.a)
c. Classification (I.B; VIII)	**f.** Reticular cell type (III.D.1)

26. Beginning with the splenic artery and ending with the splenic vein, name the vessels, sinuses, and any other structures through which blood travels as it passes through the spleen, according to both the open and closed theories of splenic circulation (VIII.A.2.a–c; Fig. 14–3).

27. Compare the white and red pulp of the spleen in terms of the:

    a. Relative amount present (VIII.A.1.a and b)

    b. Predominant cell type (VIII.A.1.a and b)

    c. Site of lymphocyte activation (VIII.B.2)

    d. Site of red blood cell destruction by splenic macrophages (VIII.B.3)

    e. Site of highest concentrations of mature, active plasma cells (VIII.B.2)

28. Name the two major components of white pulp (VIII.A.1.a) and compare them in terms of their location, predominant lymphocyte type, and germinal center location.

29. Describe the sequence of events in the life of a B lymphocyte after its encounter with an appropriate antigen in the marginal zone of the white pulp (VIII.A.1.c and B.2).

30. Name the two major components of the red pulp (VIII.A.1.b).

31. Compare red pulp sinusoids with common capillaries (VIII.A.1.b) in terms of:

    a. Luminal diameter

    b. Openings between endothelial cells

    c. Basal lamina

    d. Endothelial cell shape

32. Describe collections of unencapsulated lymphoid tissue (II.B.2; IV; V) in terms of location, predominant lymphocyte type, presence of germinal centers, and predominant type of Ig secreted.

33. Name the three tonsil types in the human mouth and pharynx (IX; Table 14–2) and compare them in terms of location, number, crypts, and epithelial covering.

SYNOPSIS

I. GENERAL FEATURES OF THE LYMPHOID SYSTEM

A. Components: The lymphoid system's major functional components comprise two main cell types: **T** and **B lymphocytes.** Lymphocytes circulate in the blood and lymph and are scattered in loose connective tissue; most are concentrated in clusters called **lymphatic** (or **lymphoid**) **aggregates.** These can be large and encapsulated, forming **lymphoid organs** such as the **thymus, spleen,** and **lymph nodes.** They also form small, partly encapsulated **tonsils.** Still smaller, unencapsulated aggregates often occur in the respiratory, digestive, and urinary tract walls. In addition to lymphocytes, lymphoid tissues typically include a **reticular connective tissue stroma** in whose meshwork lymphocytes, macrophages, and antigen-presenting cells are suspended. Lymphatic vessels and circulation are described in Chapter 11.

B. Classification of Lymphoid Tissues and Organs: In **peripheral lymphoid organs** (lymph nodes, spleen, tonsils) and **unencapsulated lymphatic aggregates** (V), lymphocyte production is antigen-*dependent* and provides committed immunocompetent cells that respond to specific antigens. In **central lymphoid organs** (thymus, bone marrow, bursa of Fabricius [in birds]), lymphocyte production is antigen-*independent* and supplies uncommitted T-lymphocyte (thymus) or B-lymphocyte (bone marrow, bursa) precursors that subsequently move to peripheral organs and tissues. Mounting effective immune responses to new antigens requires ongoing production of uncommitted lymphocytes by the central lymphoid organs.

C. Lymphoid Nodules (Follicles): These occur in all lymphatic aggregates except the thymus. Active (lymphocyte-producing) nodules each have a dark-staining periphery, or **mantle zone,** containing tightly packed small lymphocytes, and a light-staining core, or **germinal center,** containing **immunoblasts** (lymphoblasts; ie, lymphocytes stimulated by antigens to enlarge and proliferate). The lighter staining reflects the increased cytoplasmic volume and decreased nuclear heterochromatin that accompany lymphocyte activation.

D. General Functions of Lymphoid Tissues: All lymphoid tissues and organs produce lymphocytes. Lymph nodes also filter lymph and add antibodies to it, whereas the spleen filters and adds antibodies to blood and removes and destroys old red blood cells. Unencapsulated lymphoid aggregates filter and add antibodies to tissue fluid. The thymus has no significant filtering function but supports the proliferation and programming of T-lymphocyte precursors. The thymus also secretes hormones (eg, thymosin, thymopoietin) that promote the function and maintenance of lymphoid tissues in general and T cells in particular. Lymphoid functions are all directed toward a single objective: **antigen disposal,** which involves the two major mechanisms of cellular and humoral immunity.
1. **Cellular (cell-mediated) immunity.** Activated T lymphocytes differentiate into specialized cell types, some of which (CD8+) contact and kill intruding cells, and some of which (CD4+) release **cytokines,** substances that enhance various aspects of the immune response.
2. **Humoral immunity.** Activated B lymphocytes differentiate into plasma cells that secrete antigen-binding **immunoglobulins (antibodies),** which circulate in the blood and lymph.
3. **Immunologic memory.** Lymphoid function in response to initial exposure to a particular infection protects an organism during subsequent exposure to the same infective agent (I.F.4 and 5).
4. **Specificity.** An ability to respond to one type of infection (eg, chicken pox) does not imply resistance to another (eg, tuberculosis).
5. **Tolerance.** Antigen-disposal mechanisms directed toward the body's own cells (as occurs occasionally in **autoimmunity**) can be disastrous, even fatal. Thus, a key aspect of immune function is the ability to distinguish "self" from "nonself" antigens, and to tolerate the self.

E. Immunoglobulins: There are five major classes of circulating antibodies, or immunoglobulins (Igs): IgM, IgA, IgD, IgG, and IgE (easily recalled with the mnemonic MADGE). All are secreted by plasma cells, but each class has distinguishing features (II.B.1–5). Each Ig binds with great specificity to its antigen to inactivate toxic substances and to mark (opsonize) them for removal by macrophages, neutrophils, and eosinophils.

F. Lymphocyte Programming and Activation: This multistep process is outlined below.

1. Cells of mesodermal origin are programmed in the bone marrow or thymus as B- or T-lymphocyte precursors, respectively.

2. These cells subsequently move to peripheral organs (I.B), where each encounters a specific antigen (I.G) to which it becomes programmed (committed) to respond. The concentration of antigens on the surfaces of **antigen-presenting cells** (III.E), or the delivery of processed antigens to lymphocytes by macrophages (III.B), improves the efficiency of this step over that available from random lymphocyte–antigen collisions.

3. Not all lymphocytes can respond to all antigens. Our ability to respond to a variety of antigens rests in the diversity of antigen-binding capabilities of virgin (preactivated) lymphocytes. It is estimated that lymphocytes able to bind more than a billion different antigens are present prior to any antigenic challenge. When such a challenge occurs, a lymphocyte able to bind the antigen is selectively stimulated to divide (activated). Activated cells enlarge and form lymphoblasts **(blast transformation)** and subsequently undergo a series of divisions **(clonal expansion),** forming a **clone** of cells competent to recognize that antigen. This process is termed **clonal selection.** Many immunocompetent lymphocyte clones may be generated in response to different parts of a single antigen.

4. The products of this initial clonal expansion undergo **differentiation** into two basic cell types: **effector cells,** which immediately begin antigen disposal **(primary immune response),** and **memory cells,** which are held in reserve for subsequent encounters with the antigen **(secondary immune response).** T-lymphocyte derivatives form three main effector cell types (III.A.2), which enter the circulation and search the body for their antigens, providing cellular immunity. B-lymphocyte derivatives form only one effector cell type: plasma cells. These usually remain in the tissue or organ, where they differentiate and secrete into body fluids the Igs that provide humoral immunity.

5. When the same antigen is again encountered, memory cells generated during the initial clonal selection and expansion (either T or B) undergo the same process—blast transformation, clonal expansion, and differentiation—that occurs during the primary response, but more rapidly (with a shorter lag time between exposure and response) and more effectively (owing to the increased number of responsive cells, and the greater affinity of the antibodies) than before.

G. Antigens: These are foreign (nonself) substances that are able to elicit an immune response (cellular, humoral, or both). They can be entire cells (eg, bacteria, tumor cells) or large molecules (eg, proteins, polysaccharides, nucleoproteins). Their antigenicity is determined by several factors: larger and more complex (eg, branched or folded) molecules are more potent antigens than smaller, simpler ones; proteins are more antigenic than carbohydrates; and lipids are nonantigenic unless complexed with a more potent antigen. Particularly potent antigens are said to be immunodominant. The site of entry of an antigen into the body also can affect its antigenicity. The specific part of an antigen that elicits the immune response (and to which the antibodies bind) is called an antigenic determinant, or epitope; it can consist of a monosaccharide or as few as four to six amino acids. Thus a bacterium can have many antigenic determinants and elicit many cellular and humoral responses.

II. IMMUNOGLOBULINS

These antibodies are proteins secreted by plasma cells into body fluids (blood, lymph, tissue fluid, saliva, tears, milk, mucus) in response to antigenic stimulation. They bind with high affinity to the antigenic determinants that elicited their production and make up most of the blood's gamma-globulins (12.II.B.1).

A. Immunoglobulin Structure: Familiarity with the Y-shaped structure of Igs and the positions of their components (Fig. 14–1) facilitates understanding of the lymphoid system.

1. **Heavy and light chains.** Each IgG has two heavy chains (50 kDa each) and two light chains (23 kDa each). The heavy chains form the stem and part of each arm of the Y. The light chains lie in the arms, parallel to the heavy chains.

2. **Constant and variable domains.** Each chain (heavy or light) includes a region that is constant from one IgG to another and a region of variable structure that determines the anti-

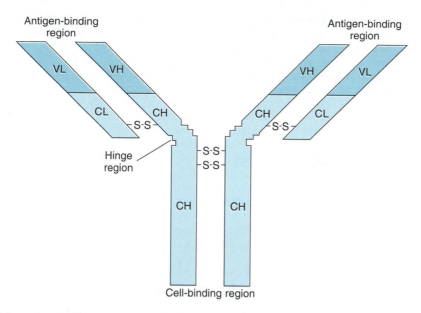

Figure 14–1. Schematic diagram of immunoglobulin structure. The dark-colored regions show the variable (V) domains of the heavy (H) and light (L) chains. The light-colored regions show the constant (C) domains. Interchain linkages by disulfide bonds are indicated by —S—S—. The hinge region harbors the papain cleavage site. Papain digestion results in the production of two Fab fragments, each with an antigen-binding region, and an Fc fragment with a cell-binding region.

body's binding specificity. The variable domains occupy the distal ends of the arms, and the constant regions are in the stem and proximal parts of the arms.

3. **Fc and Fab regions.** The proteolytic enzyme papain cleaves each Ig into three fragments at the branch point of the Y (hinge region). The single crystallizable fragment (Fc region) includes part of the constant domain occupying the stem. It is crystallizable because only a pure preparation of a single protein crystallizes and because even in a mixture of antibodies with different binding affinities, the stem structure is constant. There are two antigen-binding fragments (Fab regions), which include the entire light chain and variable and constant portions of the heavy chain. Because the combined variable regions of the light and heavy chains determine antigen-binding specificity, these fragments retain the original IgG's binding specificity. Because they vary from one antibody to another, Fab fragments from a mixture of IgGs are not crystallizable.

4. **Carboxyl and amino termini.** The carboxyl termini are the constant portion's free ends, and the amino termini are the free ends of the light and heavy chains' variable portions.

5. **Antigen-binding and cell-binding regions.** The amino-terminal region of the variable portions of each arm of the Y is the antigen-binding site. Thus, each Ig has two antigen-binding sites. The cell-binding region is the carboxyl terminus at the base of each heavy chain. Thus, the Fc fragment harbors the cell-binding region and differs among the immunoglobulin types (II.B).

6. **Disulfide bonds.** *Inter*chain disulfide bonds link the heavy chains to each other and to the light chains near the hinge region. *Intra*chain disulfide bonds occur at various sites along both light and heavy chains.

B. **Characteristics of Immunoglobulin Types:** Human Igs comprise five major groups based on the nature of their Fc regions. These differences determine how the Igs interact with one another and with various cells involved in immune responses:

1. **IgG.** The most abundant type in blood (75% of serum Ig), IgG occurs mainly as a monomer. When it binds to its antigen, its Fc region extends away from the antigen and is accessible to Fc receptors on cells (eg, neutrophils, macrophages), making IgG very effective at promoting antigen disposal by phagocytosis. IgG appears later than IgM after an initial antigenic

challenge and is a bit less effective in complement activation, but it shows greater antigen-binding specificity. It constitutes most of the secondary humoral immune response and can remain active in blood for many weeks (six times as long as IgM). IgG can cross the placenta to confer passive immunity on the fetus; it is also found in human milk.

2. **IgA.** This secretory antibody is the main Ig in body secretions (saliva, tears, mucus, colostrum, milk, semen, vaginal fluid), but makes up only 0.2% of serum Ig. Secretory IgA includes two IgA monomers linked through their Fc regions by **protein J** to form dimers. This renders IgA more soluble and less likely to be sequestered by binding to Fc receptors on cells. Another protein, the **secretory,** or **transport, component** is produced by mucosal epithelial cells. This protein is carried on mucosal cell surfaces and allows these cells to pick up IgA–protein J complexes from plasma cells in the connective tissue underlying epithelia and transport them from the cells' basal surface to the lumen where they are released into the secretions.

3. **IgM.** Although it constitutes only 10% of serum Ig, IgM is the major Ig in the primary immune response. Secreted soon after a new antigenic challenge, it is larger and less antigen-specific than IgG. It occurs as a monomer, along with IgD, on the surface of virgin B lymphocytes. When antigen binds to these surface antibodies, B cells are activated. They begin differentiating into plasma cells and secreting a soluble IgM. Secreted IgM forms pentamers, with its Fc regions joined at the core of the macromolecule and its antigen-binding regions directed outward. IgM is highly effective in complement activation (II.C.2).

4. **IgE.** Normally, IgE occurs as a monomer in very small amounts in the serum. Its Fc portion binds avidly to cell-surface Fc receptors on mast cells and basophils, leaving its antigen-binding sites extending away from the cell surface. Antigens binding to IgE cross-link the receptors and stimulate the release of histamine, heparin, and leukotrienes (eg, slow-reacting substance of anaphylaxis [SRS-A] and eosinophil chemotactic factor of anaphylaxis [ECF-A]) from the cytoplasmic granules. Antigens that bind to IgE or stimulate its production are termed **allergens,** and IgE plays a major role in allergic reactions and parasitic infections.

5. **IgD.** The least understood immunoglobulin, IgD may function as an embryonic or fetal Ig. It is rarely secreted, and its plasma concentration is low (0.2% of serum Ig). It occurs chiefly as an antigen receptor on the surface of B lymphocytes along with IgM.

C. **General Mechanisms of Immunoglobulin Action:**
1. **Opsonization.** Foreign cells and molecules to which antibodies have bound are more easily recognized as intruders by antigen-disposing cells (macrophages, cytotoxic T cells, neutrophils, eosinophils), largely through the display of their Fc regions. Antibody-labeled antigens are thus opsonized (marked for disposal). IgG, IgM, and some components of the complement system act as **opsonins.**
2. **Complement activation.** The complement system is a complex of plasma enzymes that catalyze a cascade of reactions when activated (both IgG and IgM can initiate the cascade). Effects of complement activation include (1) increased blood flow to the affected area (inflammation); (2) **chemotaxis** of the inflammatory cells (eosinophils, basophils, neutrophils, cytotoxic T cells); (3) opsonization; and (4) lysis of the invading cells (ie, components of the system act together to puncture the plasma membrane of the invading cells).
3. **Formation of antigen–antibody complexes.** Antigenic molecules in body fluids precipitate when antibodies bind and cross-link them into large macromolecular complexes. In the process, the antigens may be inactivated (ie, their toxicity is diminished or eliminated). The antigen–antibody complexes subsequently undergo phagocytosis by macrophages, neutrophils, or eosinophils.

III. CELLS OF THE LYMPHOID SYSTEM

A. **Lymphocytes:** These are the principal cells of the lymphoid system. Their ability to recognize and respond to foreign cells and substances is the basis for initiating an immune response, but lymphocytes are not phagocytic. The functional classes of lymphocytes differ in cell-surface composition and in their response to antigenic challenges, but they are indistinguishable with standard histologic stains. (The appearance of lymphocytes is described in Table 12–1; lymphocyte precursor origin in bone marrow is described in 13.VII.C.) Bone marrow-derived precursors enter the circulation and populate central lymphoid organs. Those in the thymus become T-lymphocyte precursors. B-lymphocyte programming apparently occurs in specific bone marrow microenvironments.

1. **B lymphocytes (B cells)** are primarily responsible for humoral immunity (I.D.2) and carry IgM and IgD on their membranes as antigen receptors. When antigens bind to these Igs, B lymphocytes undergo blast transformation and clonal expansion (I.F.3). Most of the resulting daughter cells differentiate into **plasma cells** (III.C); others become memory cells that react to the same antigen in subsequent encounters. B cells require assistance from helper (CD4+) T cells to respond to antigens incapable of cross-linking B-cell surface antigen receptors; these antigens are called **T-dependent (thymus-dependent) antigens.**

2. **T lymphocytes (T cells)** are primarily responsible for cell-mediated immunity (I.D.1). They carry antibody-like antigen receptors (but not Igs) on their surfaces. When antigens bind to these receptors, T lymphocytes undergo blast transformation and proliferation and produce both effector and memory cells (I.F.4); they require the aid of macrophages or other types of antigen-presenting cells (III.E) for an optimal response. This reflects the need for an antigen to be complexed with **major histocompatibility complex (MHC)** molecules for T-cell activation. Two major T-lymphocyte effector cell types are distinguishable based on characteristic cell-surface molecules (ie, CD4 and CD8) and their different roles in immunity:

   a. **Helper T cells** carry the CD4 marker on their surface and are thus said to be CD4-positive or **CD4+** T cells. They aid B lymphocytes in mounting a humoral immune response to T-dependent antigens, in part by secreting cytokines (eg, interleukins and interferon) when activated. They are activated by antigen complexed with **MHC class II** molecules on the surface of antigen-presenting cells. Because proteins in the coat of the human immunodeficiency virus (HIV) bind selectively to the CD4 protein, CD4+ cells are important targets of HIV infection. Moreover, the disposal of HIV-infected CD4+ cells by the immune system is a major factor in the immunodeficiency characterizing acquired immunodeficiency syndrome (AIDS).

   b. **Cytotoxic T cells** carry the CD8 marker on their surface and are thus **CD8+** T cells. They recognize, adhere to, and kill—by cell lysis—invading bacteria, virus-infected cells, transplanted cells, and tumor cells. These cells play a principal role in graft rejection. Their killing activity requires activation by their specific antigen.

   c. There appears to be a third type of T effector, called **suppressor T cells** (also CD8+), based on evidence that some interactions between T and B cells actually inhibit B-cell activity. Whether this is true inhibition or redirection of the immune response is not yet clear. CD8+ cells are activated by antigen complexed with **MHC class I** molecules on the surface of antigen-presenting cells. Activated CD8+ cells and macrophages also release cytokines (eg, blastogenic factor, migration-inhibiting factor, proliferation-inhibiting factor) that control B- and T-cell proliferation and macrophage activity. T-lymphocyte precursors programmed in the thymus enter the circulation and populate T-dependent regions of the lymph nodes (paracortical zone) and spleen (periarterial lymphatic sheaths). T-lymphocyte effector cells reenter the circulation more readily than do B-lymphocyte effectors (plasma cells).

3. **Natural killer (NK)** cells are circulating lymphocytes that cannot be classed as either T or B cells (ie, they lack both T and B surface antigens). Like cytotoxic T cells, they attack and lyse invading cells (eg, tumor cells and virus-infected cells) through direct cell–cell contact. However, NK cell-mediated killing appears to be independent of antigenic activation (ie, it is natural or innate). The mechanism whereby these cells target nonself cells for destruction is not entirely clear, but may involve IgG. They also enhance immune responses by secreting the cytokine interferon.

B. **Macrophages:** These typically are monocyte derivatives (ie, components of the mononuclear phagocyte system). Others may differentiate *in situ* from mesenchymal precursors. They are large, often migratory phagocytic cells (5.II.E.2.b). In both cellular and humoral immunity, they phagocytose complex antigens and enhance their antigenicity by breaking them into myriad antigenic determinants and by complexing them with MHC molecules (III.A.2) on their surface for presentation to lymphocytes. They also phagocytose antigen–antibody complexes. Macrophages interact with T lymphocytes chiefly through direct cell contact, presenting the MHC-complexed antigens on their surface. Macrophages line vascular sinuses, are distributed among the lymphocytes of lymphoid organs and tissues, and are dispersed in loose connective tissues.

C. Plasma Cells: These differentiated B-lymphocyte effector cells secrete the Igs primarily responsible for humoral immunity. Their morphology includes a "clock face" nucleus and abundant RER typical of protein-secreting cells (5.II.E.2.c). Plasma cells, found in all lymphoid tissues, occur in high concentration in the medullary cords of lymph nodes, the red pulp cords in the spleen, and the lamina propria under mucosal and glandular epithelia. They are rare in the thymus, occurring only in the medulla. Each plasma cell secretes only one class of Ig that binds only one antigen.

D. Reticular Cells: The long processes of these typically stellate cells form a mesh in which lymphocytes, plasma cells, and other tissue components are suspended. Lymphoid organs contain either of two major reticular cell types:
1. **Mesenchymal reticular cells.** Reticular cells of lymph nodes, spleen, tonsils, and bone marrow are of mesodermal origin. Each has a pale central nucleus with a prominent nucleolus and pale, sparse cytoplasm that contains RER, a Golgi complex, free ribosomes, lysosomes, glycogen granules, and intermediate filaments composed of vimentin. They produce a reticular fiber network (5.II.B and III.B) on which they are suspended and which they partly surround with their long filopodia. Other functions ascribed to these cells and their derivatives include (1) phagocytosing antigenic organisms, inert foreign matter, dead cells, and cell debris; (2) trapping antigens on their surfaces and subsequently stimulating adjacent lymphocytes; and (3) acting as hematopoietic (lymphoid and myeloid) stem cells.
2. **Epithelial reticular cells.** Reticular cells of the thymus are of endodermal origin (from the lining of the third pharyngeal pouch). Like the mesoderm-derived reticular cells, these may be stellate, but they do not secrete reticular fibers. Rather, they form their reticular mesh by attaching to one another at the tips of their long cell processes by means of **desmosomes.** Their intermediate filaments consist of cytokeratins. The cells have pale, oval nuclei with prominent nucleoli; the cytoplasm contains a Golgi complex, RER, and ribosomes. They also contain small (0.1 mm), dense granules believed to be secretory granules containing thymic hormones (eg, serum thymic factor, thymic humoral factor, thymopoietin, thymosin). In the thymic medulla, these cells assume many shapes; some become flattened to form tight concentric bodies called **Hassall's corpuscles** (VI.A.2). In the cortex, they are mainly stellate and help form the **blood–thymus barrier** (VI.B.2).

E. Antigen-presenting Cells: These cells, often mesenchymal reticular cell derivatives, bind antigen–antibody complexes on their surfaces for long periods, sometimes without phagocytosing them. In other cases, these cells do phagocytose bound antigens, complex them with MHC molecules, and subsequently present them on their surfaces. In both ways, they collect and concentrate antigens for presentation to, and stimulation of, lymphocytes. Because they are generally longer-lived than macrophages, these antigen-presenting cells can hold antigen–antibody complexes on their surfaces long after an infection has been eliminated, thereby enhancing immunologic memory. Furthermore, the continued presence of the antigen can assist in refining antibody-binding specificity over time as more specific lymphocyte clones are selected. With HIV infection, long-term presentation can be a threat because viruses on antigen-presenting cell surfaces, even when coated by antibodies, retain considerable infectivity. In this case, the antigen-presenting cells harbor a chronic source of infection and may even assist in evolving immunoresistant virus isotypes. Antigen-presenting cells appear in the lymph nodes as follicular dendritic cells of the cortex and dendritic cells of the paracortical zone; in the spleen, they are the dendritic cells of the marginal zone; in the skin (18.II.C), they are Langerhans cells; and in the liver (16.IV.D.2), they are Kupffer cells. Macrophages (III.B) also have important antigen-presenting functions and are often included in this classification.

IV. LYMPHOID NODULES

These spherical collections of lymphocytes constitute the primary functional subunits of all encapsulated and unencapsulated lymphoid aggregates except the thymus. B lymphocytes predominate, but smaller numbers of helper T cells often are present. **Primary nodules** lack germinal centers and contain only small lymphocytes. They are present prenatally and in the absence of antigens (eg, in animals housed in sterile surroundings). **Secondary nodules** appear after birth. These are primary nodules activated by antigen exposure; their size and number are proportionate to the degree of antigenic stimulation. Structurally, they have a narrow, dark-staining halo of small lymphocytes surrounding a larger, lighter-staining **germinal center** that contains mainly lymphoblasts. The dark periphery often shows a **cap,** a localized crescent-shaped thickening of the mantle zone where memory cells (I.F.4) collect. The size of the germinal center decreases when antigenic stimuli are removed. Sections through a secondary nodule's periphery may resemble primary nodules, but the presence of primary nodules is doubtful if nearby nodules have germinal centers.

V. UNENCAPSULATED LYMPHATIC AGGREGATES

These are lymphoid nodules occurring singly or in small clusters. The classic example is **Peyer's patches,** clusters of lymphoid nodules in the lamina propria of the small intestine (ileum; 15.VII.C.3). Nodule clusters also occur in the appendix, and solitary nodules are scattered beneath the epithelium in the digestive, respiratory, urinary, and genital passage walls. These occur especially at branch points (eg, in the respiratory tree), where two or more organs join (eg, gastroesophageal junction), and where transitions in epithelial linings occur. Nodules may be covered by a layer of flattened reticular cells, but they lack the collagenous capsule that surrounds lymphoid organs.

VI. THYMUS

This is the only discrete central lymphoid organ in humans. It produces only T-lymphocyte precursors and has no lymphoid nodules. Its reticular cells derive from endoderm and produce no reticular fibers. The thymus is the only organ containing **Hassall's corpuscles.** Its age-dependent structural atrophy, or **involution** (VI.B.6), is also unique among lymphoid organs. Structural features that allow the rapid identification of lymphoid organs are shown in Table 14–1.

A. Structure: The thymus lies in the mediastinum anterior to the large vessels emerging from the heart. Its two lobes are joined and covered by a thin loose connective tissue capsule that penetrates the lobes as septa, dividing the lobes into incomplete lobules. Each lobule has a peripheral dark-staining cortex, adjoining the capsule and septa, and a central light-staining medulla. The septa penetrate only to the corticomedullary junction; thus, each lobule's medulla is continuous with that of adjacent lobules.

1. **Cortex.** Small lymphocytes predominate in this dark-staining periphery of each lobule. The dark color reflects the tight packing of lymphocyte nuclei, which are suspended in a meshwork of long, epithelial reticular cell processes. The reticular cells, which are stellate and less numerous than in the medulla, form a boundary between the cortex and the connective tissue of the capsule and septa. They also ensheathe the cortical capillaries, the only blood vessels found in the cortex. The cortex is the site of T-lymphocyte precursor proliferation and of the blood–thymus barrier (VI.B.2).

2. **Medulla.** In effect, each thymic lobe has one medulla that extends into the core of each lobule. The light staining reflects the presence of more epithelial reticular cells and fewer lymphocytes than in the cortex. Medullary reticular cells assume many shapes and sizes; some have granules containing thymic hormones. The lymphocytes, which are more mature than in the cortex, enter the circulation from the medulla to populate the T-dependent areas of other lymphoid organs. The spherical **Hassall's corpuscles** (30 to 150 μm in diameter),

Table 14–1. Distinguishing structural features of the lymphoid organs.

Key Features	Thymus	Lymph Nodes	Spleen	Tonsils
Cortex and medulla	Yes	Yes	No	No
Lymphoid nodules	No	Yes	Yes	Yes
Cords and sinuses	No	Yes	Yes	No
Unique structures	Hassall's corpuscles	Cortical nodules, subcapsular sinus	Central arteries	Epithelial covering

whose function is unknown, consist of concentric layers of flattened epithelial reticular cells. With age, cells in the core of the corpuscles may die and calcify.

B. **Functions:**

1. **T-lymphocyte production** is the primary function of the thymus. T-lymphocyte precursors formed in the bone marrow populate the thymic cortex. The cortical environment influences these thymic lymphocytes (**thymocytes**) to proliferate and acquire T-lymphocyte characteristics. Thymic programming involves two important developments: (1) the acquisition of antigen-recognition capacity by the reordering of antigen-receptor genes; and (2) the acquisition of T-cell characteristics through expression of the cell-surface molecules and signal-transduction complexes that regulate interactions with antigenic and accessory cells, and the homing capabilities that direct virgin T cells to secondary sites for further programming. Most cortical thymocytes undergo cell death and fragmentation (**apoptosis**) followed by phagocytosis by macrophages. This may eliminate cells prematurely activated or targeted toward "self" antigens. Maturing survivors move to the medulla, where they enter the circulation through postcapillary venules or efferent lymphatic vessels. They populate the T-dependent regions of secondary lymphoid organs (eg, lymph nodes, spleen). Here they further differentiate into functional T lymphocytes. Most thymocytes cannot respond to antigens. Therefore, thymocytes—especially those in the cortex—should be considered distinct from, but precursors to, the T lymphocytes that carry out the cellular immune response.

2. **Blood supply and blood–thymus barrier.** The arterial blood supply enters the thymus through the capsule, penetrating the organ with the septa. Branches of septal vessels extend along the border between the cortex and medulla, feeding capillaries that penetrate both regions. The cortical capillaries arch through the cortex and empty into postcapillary venules in the medulla, as do the medullary capillaries. The venous drainage follows the arterial course in reverse. The thymus contains continuous (nonfenestrated) capillaries surrounded by a thick basal lamina. In the cortex, capillary endothelial cells may penetrate the basal lamina and contact epithelial reticular cell processes ensheathing the cortical capillaries. This three-layered structure (nonfenestrated capillary endothelium, thick basal lamina, and reticular cell sheath) forms the blood–thymus barrier. This barrier, found only in the cortex, separates proliferating thymocytes from the blood. Together with the disposition of the blood vessels (directing blood flow toward the medulla and away from the cortex), the barrier limits the antigenic material to which the cortical thymocytes are exposed. This helps maintain a supply of uncommitted (naïve) stem cells for later programming during encounters with new antigens.

3. **Hormone production.** Epithelial reticular cells of the thymic medulla have cytoplasmic granules thought to contain thymic hormones (eg, **thymopoietin, thymosin**). These humoral factors have trophic effects on the entire lymphoid system and promote thymocyte proliferation and T-cell differentiation.

4. **Effects of exogenous hormones.** Adrenocorticosteroids and ACTH slow thymocyte proliferation and reduce the thickness of the thymic cortex. Androgens and estrogens accelerate thymic involution; castration delays it. Growth hormone stimulates thymic growth.

5. **Effects of thymectomy.** Destruction or removal of the thymus at birth results in complete failure of T-lymphocyte production. It reduces the number of circulating lymphocytes and causes T-dependent regions of the spleen and lymph nodes to remain unpopulated. Affected newborns cannot generate a cell-mediated immune response and, consequently are incapable of graft rejection and delayed hypersensitivity. Nor can they generate a T-dependent humoral immune response, and the lack of thymic hormones causes general atrophy of other lymphoid organs. By 3 to 4 months after experimental postnatal thymectomy, a laboratory animal weakens, loses weight, and finally dies. Thymectomy in adult animals has less dramatic effects because many thymocytes have already left the thymus. Functional T lymphocytes are already distributed in the tissues and T-dependent regions of the secondary lymphoid organs. The number of circulating lymphocytes is reduced, however, and the response to new and unusual antigens is compromised. Grafting thymic tissue into thymectomized animals at any age reverses the effects of thymectomy. The graft is repopulated with thymocyte precursors from the host bone marrow, and thymic function is restored.

6. **Histogenesis and involution.** The thymus arises from the ventral portion of the paired third pharyngeal pouches, whose endodermal lining gives rise to the epithelial reticular cells.

Pouch ectoderm and neural crest-derived mesenchyme also contribute importantly to thymic structure. After 6 weeks of gestation, the thymic rudiments detach from the pharyngeal wall and migrate to the mediastinum, where they partially fuse to form the two lobes of the thymus. The thymus is populated by hematopoietic stem cells of mesodermal origin from the liver and bone marrow during hepatosplenothymic and medullary hematopoiesis, respectively; the stem cells divide and fill the cortex with thymocytes. The thymus increases in size until puberty, but it reaches its maximum size (relative to body weight) shortly after birth. At puberty, involution begins. The cortex thins as thymocyte proliferation slows and more cells leave the thymus. The relative area of the medulla increases, and the Hassall's corpuscles enlarge, sometimes calcifying. Even in adults, the thymus can produce large numbers of thymocytes when needed. In the elderly, much of the active thymic tissue is replaced by connective and adipose tissue.

VII. LYMPH NODES

These are the smallest but most numerous encapsulated lymphoid organs. Scattered in groups along lymphatic vessels in the neck, axilla, groin, thorax, and abdomen, they act as in-line filters of the lymph, removing antigens and cellular debris and adding Igs.

A. **Structure:** Lymph nodes are bean-shaped, with convex and concave surfaces (Fig. 14–2). The parenchyma consists of a peripheral cortex, adjacent to the convex surface, and a central medulla, lying near the depression (hilum) in the concave surface. The connective tissue capsule gives off **trabeculae** that penetrate between the cortical nodules and subdivide the cortex. Blood vessels enter and leave through the hilum.
1. **Cortex.** The cortex stains darkly owing to the tight packing of lymphocytes. These are suspended in a reticular connective tissue network and arranged as a layer of typical secondary lymphoid nodules (containing primarily B cells) with germinal centers. The cortex also contains reticular cells, antigen-presenting **follicular dendritic cells,** macrophages, a few plasma cells, and some helper T cells.
2. **Medulla.** Lighter staining than the cortex, the medulla consists of cords of lymphoid tissue **(medullary cords)** separated by **medullary sinuses.** The lymphocytes are mainly small, less numerous than in the cortex, and concentrated in the cords. The cords are rich in reticular cells and fibers and contain many plasma cells that have migrated from the cortex.
3. **Paracortical zone.** This T-dependent region lies between the cortical lymphoid nodules and the medulla. It contains mainly T lymphocytes suspended in a reticular connective tissue network. B lymphocytes, plasma cells, macrophages, and antigen-presenting **interdigitating dendritic cells** also may be present. This zone is also characterized by many **high-endothelial (postcapillary) venules (HEVs).** T lymphocytes home to HEVs in a two-stage process. **L-selectins** on the lymphocyte surface adhere loosely to receptors on the endothelial cell surface in HEVs. This association promotes tighter lymphocyte binding through **integrins,** initiating diapedesis. Immobilized T cells leave the blood to enter the paracortical zone by passing between the cuboidal endothelial cells.
4. **Lymphatic vessels.** Lymphatic vessels associated with lymph nodes are of two types. Both contain valves to ensure unidirectional lymph flow through the node. **Afferent lymphatic vessels** deliver lymph by penetrating the capsule at several points on the convex surface. **Efferent lymphatic vessels** carry filtered lymph away from the node, exiting through the hilum on the concave surface.
5. **Sinuses and lymph flow.** Lymph node sinuses filter lymph and direct its flow. Partly lined with reticular cells and many macrophages, they are not simply open spaces, but are traversed by a mesh of reticular cells and fibers, macrophages, and follicular dendritic cells. The complex sieving action slows lymph flow to facilitate antigen removal. Lymph is delivered by the afferent vessels to the cuplike **subcapsular sinus** between the capsule and the cortical parenchyma. From here, it flows directly into the **peritrabecular sinuses** surrounding the trabeculae. It subsequently flows through the anastomotic network of **medullary sinuses** that converge on the efferent lymphatic vessels exiting through the hilum.

B. **Functions:**
1. **Filtration of lymph.** Cellular debris and antigens carried by incoming lymph are removed by the macrophages and follicular dendritic cells of the sinuses (similar cells are found in

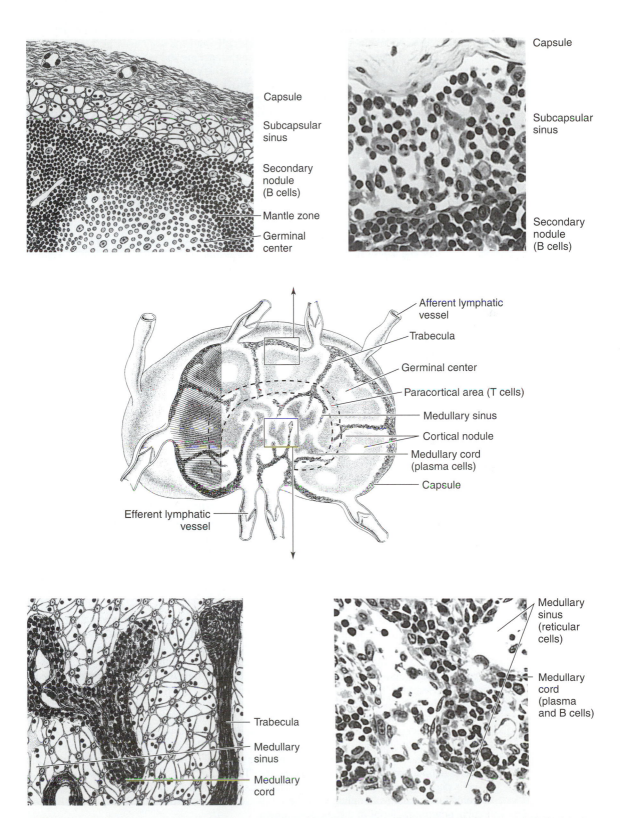

Figure 14–2. Schematic diagram of lymph node structure. The rectangles in the center drawing are magnified in the upper and lower drawings. Section VI.B of the synopsis contains a description. (Reproduced, with permission, from Junqueira LC, Cameiro J, Kelley RO: *Basic Histology,* 9th ed. Stamford, CT: Appleton & Lange, 1998.)

the cortical nodules and medullary cords). Lymphocytes carried by the lymph may flow through the nodes, contacting antigen-presenting cells and macrophages in the sinuses, or leave the sinuses and enter the parenchyma. Lymph reaching the efferent lymphatic vessels has been cleared of more than 90% of antigens and cellular debris. Less than 1% of the lymph passing through a node penetrates the nodules.

2. **Lymphocyte production (lymphopoiesis).** Stimulated by antigens removed from the lymph, **T lymphocytes** undergo blast transformation and clonal expansion and subsequently differentiate into effector and memory cells that recognize and respond to a specific antigen. T-lymphocyte effector cells seek and destroy the antigen, entering the sinuses and leaving the node through efferent vessels. The cells reenter the blood where the lymphatic vascular system empties into the venous system. Similarly stimulated, **B lymphocytes** move to the cortical nodules' germinal centers to undergo the blast transformation that yields memory and effector (plasma) cells. Differentiated plasma cells migrate to the medullary cords. Memory B cells either return to the nodule's peripheral mantle zone or leave the node by entering the sinuses.

3. **Immunoglobulin production.** Most plasma cells remain in the medullary cords, secreting Igs into the lymph flowing through the medullary sinuses and exiting through the efferent lymphatic vessels. These Igs reach the blood as the lymph empties into the venous system in the neck.

VIII. SPLEEN

This largest lymphoid organ lies in the abdominal cavity's upper left quadrant. Its functions include lymphopoiesis, Ig production, and the filtration of cellular debris and antigens from the blood. Because it is the blood's immunologic filter, the spleen's blood supply and circulation are especially important. Unlike other lymphoid organs, the spleen lacks a definitive cortex and medulla. The parenchyma (splenic pulp) lacks true lobules; however, the dense connective tissue capsule, which contains a small amount of smooth muscle, gives rise to trabeculae that divide the splenic pulp into incomplete compartments.

A. Structure:

1. **Splenic pulp** consists of many erythrocytes, leukocytes, and macrophages, and a variety of blood vessels, all suspended within a meshwork of mesenchymal reticular cells and fibers. Unstained slices of splenic pulp exhibit many whitish islands of lymphoid tissue (white pulp) embedded in a sea of dark red, erythrocyte-rich tissue (red pulp).

   a. **White pulp** consists of the lymphoid tissue surrounding each of the many central arteries (VIII.A.2.a); it has two major components. The sleeve of lymphoid tissue around each central artery is called a **periarterial lymphatic sheath (PALS)** (Fig. 14–3). These sheaths contain mainly T lymphocytes and constitute the spleen's T-dependent regions. Surrounding each PALS, or appended to one side, is the second component, the **peripheral white pulp (PWP)**. PWP contains mainly B lymphocytes and usually includes a secondary lymphoid nodule with a germinal center.

   b. **Red pulp** makes up most of the spleen and also has two major components: the red pulp cords and the splenic sinusoids between them. The **red pulp (Billroth's) cords** are irregular reticular connective tissue sheets that branch and anastomose to surround the sinuses. They vary in thickness according to the distention of the sinusoids. In addition to reticular cells and fibers, the cords contain many cell types, including all of the blood's formed elements, dendritic cells, macrophages, plasma cells, and lymphocytes. **Splenic sinusoids** differ from capillaries: the lumen is wider and more irregular; there are 2- to 3-μm spaces between the lining endothelial cells; and there is a sparse, discontinuous basal lamina composed largely of reticular fiber bands that run roughly perpendicular to the vessel's length. The overall arrangement resembles a barrel, with the endothelial cells (elongated on the sinusoid's long axis) representing the wooden staves and the basal lamina bands representing the hoops. The slitlike spaces between the endothelial cells permit an extensive exchange of fluids, solutes, and flexible cells between the sinusoids and cords. Macrophages in the cords extend their processes through the slits and phagocytose material in the sinusoid lumen.

   c. The **marginal zone** forms a moat of blood sinuses and loose lymphoid tissue between the white and red pulp. Blood-borne antigens delivered to the **marginal sinuses** are phagocy-

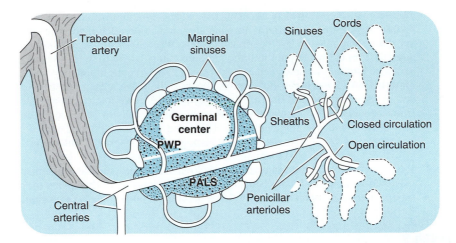

Figure 14–3. Schematic diagram of the arterial supply to the splenic sinusoids. Both open and closed theories of circulation are depicted. A T lymphocyte–rich periarterial lymphatic sheath (PALS) surrounds each central artery. The B lymphocyte–rich peripheral white pulp (PWP), with its characteristic germinal center, is separated from the red pulp cords and sinuses by a moatlike arrangement of marginal zone sinuses. The sheaths of sheathed arterioles are composed mainly of macrophages.

tosed by the many macrophages and trapped by interdigitating dendritic cells in the zone. Its rich blood supply, cellular composition, and location make the marginal zone important in concentrating blood-borne antigens for presentation to the splenic lymphocytes.

  2. Splenic circulation (see Fig. 14–3).

    a. Arterial supply. The spleen receives blood from the **splenic artery** (a branch of the celiac trunk off the abdominal aorta). Near the hilum, the splenic artery branches to form several **trabecular arteries.** These enter the spleen through the trabeculae and branch to enter the parenchyma as the numerous **central arteries** around which the white pulp is organized. After passing through the white pulp, the arteries give off many **penicillar arterioles,** which in turn give off many capillaries and **sheathed arterioles.** Near their termination, the sheathed arterioles have localized wall thickenings consisting of macrophages. Capillaries arising from the central artery loop back to feed the marginal sinuses. Others, including those arising from the penicillar and sheathed arterioles, feed the red pulp sinuses.

    b. Open and closed theories of splenic circulation. How blood reaches the sinusoid lumens is not clear. The **closed theory** holds that the capillary walls are continuous with the sinusoid walls and that the capillaries empty directly into the sinusoid lumens. The **open theory** holds that the capillaries end abruptly in the red pulp cords, leaking blood that reaches sinusoid lumens by percolating through the cords and passing through openings in the sinusoid walls. Current evidence favors the open theory.

    c. Venous drainage. From the sinusoids, blood flows into red pulp veins that converge on and empty into trabecular veins; these are unusual in that they lack a distinct tunica media. At the hilum, trabecular veins empty into the **splenic vein,** which joins the inferior mesenteric vein and empties into the hepatic portal vein before it enters the liver.

B. Functions:

  1. Filtration of blood. Antigens carried by capillaries to the marginal sinuses are removed by macrophages and dendritic cells; they are concentrated and processed for presentation to lymphocytes in the white pulp. Other macrophages lie in red pulp cords and around sheathed arterioles. Antigenic material in the sinusoids can be removed by macrophage processes extending into the lumen; in the cords, such materials are cleared by macrophages and dendritic cells.

  2. Lymphocyte production (lymphopoiesis). Both T and B lymphocytes are activated in the spleen. Lymphocyte–antigen interactions are more intense in white pulp, particularly near the marginal zone, but may occur in red pulp. T-lymphocyte effector cells formed in the PALS migrate to the sinusoids to enter the circulation. B lymphocytes stimulated in the marginal zone move to PWP germinal centers, where they divide. Plasma cells thus formed migrate from the white pulp into the red pulp cords, where they remain, producing Igs that percolate into the sinusoids and exit the spleen in the venous blood.

3. **Destruction of worn red blood cells** occurs in both the spleen and bone marrow. At the end of their average 120-day life span, erythrocytes have lost terminal sialic acid residues from their glycocalyx, making them recognizable as cells to be retired. They also become less flexible and can fragment. Old RBCs and fragments trapped in red pulp cords are phagocytosed by macrophages. The hemoglobin is degraded into several components (13.III.A.3) and delivered to the liver by means of the portal vein.

4. **Extramedullary hematopoiesis.** In pathologic conditions such as leukemia, in which bone marrow function (medullary hematopoiesis) is compromised, the spleen may resume erythropoietic or granulopoietic activity. Liver and lymph nodes may resume similar functions.

IX. TONSILS

These partially encapsulated lymphoid aggregates contain many lymphoid nodules; they underlie the mucous membranes (epithelial lining) of the mouth and pharynx. Together with the diffuse subepithelial lymphoid tissue that connects them to form a ring, they guard the common entrance to the digestive and respiratory tracts. The three types—**palatine, pharyngeal,** and **lingual tonsils**—differ in number, type of epithelial covering, number and type of epithelial invaginations (or **crypts**), and the presence or absence of a partial capsule (Table 14–2).

Table 14–2. Comparison of the tonsils.

	Palantine Tonsils	Pharyngeal Tonsil	Lingual Tonsils
Location	Lateral walls of oral pharynx, below level of soft palate	Posterior of naso-pharynx in midline, above level of soft palate	Posterior third of tongue (floor of pharynx)
Number per individual	2	1	Small and numerous
Number of crypts per tonsil	10–20	No crypts, but surface is pleated	One crypt per tonsil
Epithelial covering	Nonkeratinized stratified squamous	Ciliated pseudostratified columnar	Lightly keratinized stratified squamous
Capsule	Thick partial capsule of dense connective tissue	Thin partial connective tissue capsule	No definitive capsule

MULTIPLE-CHOICE QUESTIONS

Select the single best answer.

14.1. Which of the following is true of epithelial reticular cells of the thymus?
(A) Are derived from embryonic mesoderm
(B) Are structural components of the blood–thymus barrier
(C) Are found exclusively in the thymic cortex
(D) Contain tonofilaments composed of vimentin
(E) Synthesize and secrete reticular fibers

14.2. Which of the following functions is carried out by all lymphoid tissues and organs?
(A) Filtration of lymph
(B) Filtration of blood
(C) Extramedullary erythropoiesis
(D) Production of lymphocytes
(E) Destruction of old erythrocytes

14.3. Which of the following is true of the thymic cortex?
(A) Is the site of the blood–thymus barrier
(B) Among blood vessels, contains only sinusoids

(C) Contains Hassall's corpuscles
(D) Lacks reticular cells
(E) Is the location of thymic plasma cells

14.4. Which of the following reflects the order in which lymph passes through the sinuses of lymph nodes?
(A) Medullary → peritrabecular → subcapsular
(B) Medullary → subcapsular → peritrabecular
(C) Peritrabecular → subcapsular → medullary
(D) Subcapsular → medullary → peritrabecular
(E) Subcapsular → peritrabecular → medullary

14.5. Which of the following are B-lymphocyte–derived cells that contain abundant RER and are located in the red pulp cords (Billroth's cords) in the spleen?
(A) CD4+ lymphocytes
(B) CD8+ lymphocytes
(C) Macrophages
(D) Mast cells
(E) NK cells
(F) Plasma cells
(G) Reticular cells

14.6. Which of the following is a "thymus-dependent" region of a peripheral lymphoid organ?
(A) Medullary cords of lymph nodes
(B) Hassall's corpuscles of the thymus
(C) Germinal centers of Peyer's patches
(D) Periarterial lymphatic sheaths of the spleen
(E) Crypts of the pharyngeal tonsils

14.7. Which of the letters in Figure 14–4 best corresponds to the component of an IgE molecule that binds directly to the surface of a mast cell?

14.8. Which of the letters in Figure 14–4 best corresponds to the antigen-binding region on the light chain of an IgG molecule?

14.9. Which of the following are nonphagocytic cells in the lymph nodes that bind antigen on their surfaces and present them to lymphocytes for recognition and stimulation?
(A) Dendritic cells
(B) Epithelial reticular cells
(C) Macrophages
(D) NK cells
(E) Plasma cells

14.10. Which of the following cells are the primary cellular component of the stroma of the splenic lymphoid nodules?
(A) B lymphocytes
(B) Dendritic cells
(C) Erythrocytes
(D) Macrophages
(E) NK cells
(F) Reticular cells
(G) T lymphocytes

14.11. Which of the following immunoglobulin types is most important in conferring passive immunity on the fetus and on the newborn infant?
(A) IgA
(B) IgD
(C) IgE
(D) IgG
(E) IgM

14.12. Which of the following immunoglobulin types is most effective in fixing complement and typically occurs in the serum as a pentamer?
(A) IgA
(B) IgD
(C) IgE
(D) IgG
(E) IgM

14.13. Which of the following is partly encapsulated and covered by nonkeratinized stratified squamous epithelium?
(A) Appendix
(B) Lingual tonsil
(C) Palatine tonsil
(D) Peyer's patch
(E) Pharyngeal tonsil

14.14. Which of the following is an encapsulated peripheral lymphoid organ that serves as the primary immunologic filter of the blood?
(A) Bone marrow
(B) Kidney
(C) Lymph node
(D) Spleen
(E) Thymus

14.15. Which of the following lymphoid organs has a cortex, a medulla, and a subcapsular sinus?
(A) Appendix
(B) Lymph node
(C) Palatine tonsil
(D) Spleen
(E) Thymus

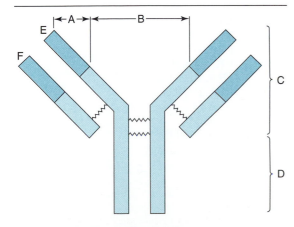

Figure 14–4.

14.16. Which of the following processes best accounts for the pale staining of the germinal centers of lymphoid nodules?
(**A**) Antibody production
(**B**) Antigen disposal
(**C**) Blast transformation
(**D**) Clonal expansion
(**E**) Complement activation

14.17. Which of the following lymphoid organs receives afferent lymphatic vessels?
(**A**) Appendix
(**B**) Lymph node
(**C**) Palatine tonsil
(**D**) Spleen
(**E**) Thymus

14.18. Which of the following gives rise to both memory and effector cells and is primarily associated with humoral immunity?
(**A**) B lymphocyte
(**B**) NK cell
(**C**) Macrophage
(**D**) Plasma cell
(**E**) T lymphocyte

14.19. Which of the following structures would be most heavily labeled by an immunohistochemical method targeting the CD8 surface antigen?
(**A**) Germinal center
(**B**) Marginal sinus
(**C**) Paracortical zone
(**D**) Peyer's patch
(**E**) Sheathed arterioles
(**F**) Splenic cords
(**G**) Subcapsular sinus

14.20. Which of the following cells carry the CD4 marker cells on their surface and are thus important targets of HIV infection?
(**A**) B lymphocytes
(**B**) Cytotoxic T cells
(**C**) Epithelial reticular cells
(**D**) Helper T cells
(**E**) Natural killer (NK) cells
(**F**) Plasma cells
(**G**) Suppressor or T cells

14.21. Which of the following is true of secondary (peripheral) lymphoid organs?
(**A**) Are capable of antigen-independent lymphopoiesis
(**B**) Include the human thymus
(**C**) Are defined as the initial sites of hematopoiesis in the embryo
(**D**) Include the bursa of Fabricius in birds
(**E**) All contain lymphoid nodules
(**F**) All lack connective tissue capsules
(**G**) All contain epithelial reticular cells

14.22. Which of the following cells contain secretory granules believed to contain thymic hormones?
(**A**) B lymphocytes
(**B**) Epithelial reticular cells
(**C**) Helper T lymphocytes
(**D**) Langerhans cells
(**E**) Mast cells
(**F**) Plasma cells
(**G**) Thymocytes

ANSWERS TO MULTIPLE-CHOICE QUESTIONS

14.1. B (III.D.2)
14.2. D (I.D)
14.3. A (III.D.2; VI.A.1 and 2)
14.4. E (VII.A.5; Fig. 14–2)
14.5. F (III.C; VIII.B.2)
14.6. D (III.A.2.c; VIII.A.1.a)
14.7. D (II.A.3,5 and B.4; Fig. 14–1)
14.8. F (II.A.1 and 5; Fig. 14–1)
14.9. A (III.E; VII.A.1)
14.10. F (VIII.A.1)
14.11. D (II.B.1)
14.12. E (II.B.3)
14.13. C (IX; Table 14–2)
14.14. D (VIII)
14.15. B (VII; Table 14–1)
14.16. C (I.F.3; IV)
14.17. B (VII.A.4 and 5)
14.18. B (I.F.4; III.A.1 and 2)
14.19. C (III.A.2; VII.A.3)
14.20. D (III.A.2)
14.21. E (I.B and C)
14.22. B (VI.A.2 and B.3)

Digestive Tract

15

OBJECTIVES

This chapter should help the student to:

- Name the parts of the digestive tract and the primary function of each.
- Describe the structure of the tongue, teeth, and gingiva.
- Describe the development of the teeth.
- Compare the digestive tract organs in terms of the four layers comprising their walls and relate any structural variations to differences in organ function.
- Know the distinguishing regional structure of each digestive tract component.
- Name the secretory product(s), the distinguishing structural features, and (where appropriate) the staining properties of each secretory cell type in digestive tract mucosa.
- List the features of the small intestine that promote nutrient absorption and trace the steps in this process.
- Identify the organ, region, cell types present, and type of section (ie, transverse or longitudinal) in a micrograph of any part of the digestive tract.

MAX-Yield™ STUDY QUESTIONS

1. List the organs of the digestive tract in the order in which food traverses them (I.A[1]). Describe what happens to the food in each (I.C.1–3).
2. Sketch a cross-section of a generalized tubular organ of the digestive tract that shows the layered structure of its walls (Fig. 15–1), and indicate the location of the following:
   - **a.** Lumen
   - **b.** Mucosa
   - **c.** Submucosa
   - **d.** Muscularis externa
   - **e.** Serosa
   - **f.** Epithelium
   - **g.** Lamina propria
   - **h.** Muscularis mucosae
   - **i.** Meissner's (submucosal) plexus
   - **j.** Auerbach's (myenteric) plexus
   - **k.** Mesothelium
   - **l.** Attachment of the mesentery
3. Describe the oral cavity in terms of its epithelial lining, the muscle type in its walls, and the structural difference between the hard and soft palates (II.A).
4. Describe the tongue in terms of its predominant tissue and the epithelium that covers it (II.C).
5. Name the four types of lingual papillae and compare them in terms of their characteristic shape, distribution of taste buds, and relative abundance (II.C.1–4).
6. List the four types of teeth (by shape) found in humans (III.A).
7. Compare the "dental formula" for permanent and deciduous teeth (III.B).
8. Sketch a tooth and its surrounding structures in sagittal (midline longitudinal) section (Fig. 15–2) and label the following:
   - **a.** Gingiva
   - **b.** Alveolar bone
   - **c.** Crown
   - **d.** Neck
   - **e.** Root
   - **f.** Apical foramen
   - **g.** Enamel
   - **h.** Cementum
   - **i.** Dentin
   - **j.** Pulp
   - **k.** Periodontal ligament
   - **l.** Epithelial attachment (of Gottlieb; III.D.3)

---

[1] See footnote on page 1.

9. Compare dentin, enamel, and cementum (III.C.5–7) in terms of:
   a. Hardness
   b. Porosity
   c. Collagen content
   d. Cell responsible for synthesis
   e. Capacity for replacement
10. Describe tooth pulp in terms of its predominant tissue, major cell types, blood supply, and innervation (III.C.4 and 5.b; Fig. 15–2).
11. Describe the periodontal ligament in terms of its composition, location, functions, the structures to which it attaches, and the effects of dietary vitamin C and protein deficiency (III.D.1; Fig. 15–2).
12. Compare ameloblasts, odontoblasts, and cementoblasts in terms of their embryonic tissue of origin (III.E), the layer of tooth structure formed by each (III.E), and their survival into adulthood (III.C.5.c, 6.c, and 7).
13. Beginning with the dental laminae, list, in order, the named stages of crown development (Fig. 15–3).
14. Sketch a developing tooth in the bell stage (Fig. 15–3) and label the following:
   a. Ameloblasts
   b. Odontoblasts
   c. Enamel organ
   d. Inner enamel epithelium
   e. Outer enamel epithelium
   f. Stellate reticulum
   g. Dental papilla
   h. Cervical loop
15. By the time a deciduous tooth is shed, prior to the eruption of the permanent tooth, only the crown remains. What happens to the root (III.E.4)?
16. Describe the oral pharynx (IV) in terms of its epithelial lining and the type of muscle in its walls.
17. Compare the esophageal wall closest to the pharynx with that closest to the stomach (IV; Table 15–1) in terms of:
   a. Epithelium
   b. Location of the mucus-secreting glands
   c. Tissue type of the muscularis externa
   d. Outer covering (serosa versus adventitia)
18. Sketch the stomach's outline and show the boundaries of the cardia, fundus, body, and pylorus (Fig. 15–4).
19. Name the epithelium covering the stomach's luminal surface and lining the gastric pits (VI.B.1).
20. Compare the mucosal glands in the stomach's four regions in terms of their major and minor secretory products and the depth of their gastric pits (VI.C.1–4).
21. Name four secretory cell types found in the gastric glands (VI.B.3–6) and compare them in terms of their secretory product(s), staining properties, distribution, and organelles.
22. How does the stomach's muscularis externa differ from that of the other tubular organs of the digestive tract (I.B.3)?
23. Name, in the order that food passes them, the three segments of the small intestine; also name the type of epithelium lining each (VII).
24. List three features of the wall of the small intestine that increase the surface area exposed to the lumen and thus promote absorption of nutrients (VII.A, B.1 and 3).
25. Name five important cell types found in the epithelium lining the intestinal lumen and the crypts of Lieberkühn (VII.B.3–8) and compare them in terms of their primary function (including secretory products), distribution, and distinguishing structural or staining properties.
26. Compare the duodenum, jejunum, and ileum in terms of the presence of submucosal glands, amount of lymphoid tissue, and number of goblet cells (VII.C.1–3).
27. Compare the absorption of lipids, amino acids, and monosaccharides from the small intestine lumen (VII.B.3.a–d) in terms of the mechanism by which they enter the absorptive cells, modifications (if any) carried out in the absorptive cell (eg, packaging), and their selective uptake by blood versus lymph.
28. List the functions of the large intestine (VIII).
29. Indicate how the large and small intestines differ in terms of the presence of plicae circulares and villi (VIII.A), number of goblet cells (VIII.A), amount of lymphoid tissue (VIII.A), and the outer layer of their muscularis externa (VIII.C).
30. Describe how the appendix (IX) differs from most of the large intestine in terms of its overall diameter, the depth of its crypts, the abundance of lymphoid follicles, and the muscularis externa.

31. Compare the rugae of the stomach (VI.A) and the circular folds (plicae circulares) of the small intestine (VII.A) in terms of their permanence.
32. List the segments of the digestive tract that are covered (I.B.4):
  a. Primarily by serosa
  b. Primarily by adventitia
  c. By both serosa and adventitia
33. Describe the trends in wall structure from the duodenum through the colon (VII.C; VIII) in terms of the increase or decrease in:
  a. Number of goblet cells
  b. Number of enteroendocrine cells
  c. Amount of lymphoid tissue
  d. Number of villi
  e. Number of circular folds

SYNOPSIS

I. GENERAL FEATURES OF THE DIGESTIVE TRACT

  A. Components: The digestive tract is a series of organs forming a long muscular tube whose continuous lumen opens to the exterior at both ends. The organs include the oral cavity, oral pharynx, esophagus, stomach, small intestine (duodenum, jejunum, ileum), large intestine (cecum and appendix; ascending, transverse, descending, and sigmoid colon), rectum, and anal canal.

  B. General Structural Features: Each organ's wall has four concentric layers (Fig. 15–1): the mucosa, submucosa, muscularis externa, and serosa or adventitia. (To master digestive tract histology, first learn the general composition and location of each layer and then focus on distinguishing features of each organ; Table 15–1.) Distinguishing structural features make more sense when considered in relation to their functions (I.C).
  1. Mucosa. This layer borders the lumen and has three parts. The **epithelium** derives from endoderm throughout the tract, except in the oral cavity and anal canal, where it derives from

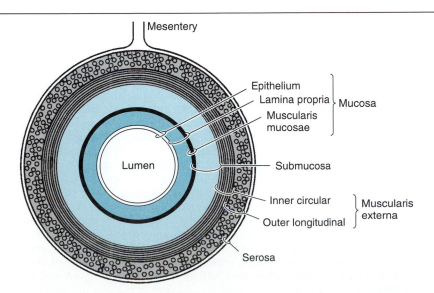

Figure 15–1. Simplified schematic diagram of the layers in the walls of the digestive tract.

Table 15–1. Distinguishing features of the walls of the digestive tract.

| Organ | Region | Mucosa | |
		Epithelium	Lamina Propria
Esophagus	Upper third	Nonkeratinized stratified squamous	Mucous glands
	Middle third		
	Lower third		Cardiac glands (mostly mucous, with shallow pits)
Stomach	Cardia	Simple columnar	
	Fundus and body		Gastric glands (fundic type with shallow pits)
	Pylorus		Gastric glands (pyloric type, mostly mucous, with deep pits)
Small intestine	Duodenum	Simple columnar; striated border; goblet cells	Crypts of Lieberkühn, lymphoid nodules; lamina propria forms core of villi
	Jejunum		
	Ileum		Peyer's patches
Colon and rectum	Appendix	Simple columnar; shorter microvilli; abundant goblet cells	Many lymphoid nodules
	Cecum		No villi; few crypts
	Ascending and descending colon		
	Transverse and sigmoid colon		
	Rectum	Rectal columns	
Anal canal	Anorectal area	Nonkeratinized stratified squamous	Large venous plexus (hemorrhoidal plexus)
	Anocutaneous area	Keratinized stratified squamous	No distinguishing features
	Cutaneous area		Pilosebaceous follicles and apocrine sweat glands

invaginating ectoderm. The epithelium is stratified squamous in the oral cavity, oral pharynx, esophagus, and anal canal; it is simple columnar in the stomach, intestines, and rectum. The **lamina propria** is the loose connective tissue layer containing blood and lymphatic vessels beneath the epithelium. The **muscularis mucosae** is a thin, smooth muscle layer bordering the submucosa.

2. **Submucosa.** This dense, irregular connective tissue layer contains blood and lymphatic vessels and the submucosal (Meissner's) plexus of nerves. Some organs are characterized by glands and lymphoid nodules in this layer.

3. **Muscularis externa.** This component consists of two layers (inner circular and outer longitudinal) of smooth muscle throughout most of the tract. Between them lies the myenteric (Auerbach's) plexus. The muscle around the oral cavity is skeletal; where it is absent (eg, hard palate, gingiva), the submucosa binds tightly to bone. In the upper esophagus, this layer contains skeletal muscle, which is replaced by smooth muscle in the lower portion. The stomach's muscularis externa has three layers: outer longitudinal, middle circular, and inner oblique. The colon's outer longitudinal layer is gathered into three bands called the teniae coli. Smooth and skeletal muscles encircling the anal canal form involuntary and voluntary sphincters, respectively.

4. **Serosa and adventitia.** The tract's outer covering differs according to location. The esophagus and rectum are surrounded and held in place by a connective tissue adventitia similar to that around blood vessels. Intraperitoneal organs (stomach, jejunum, ileum, transverse and sigmoid colon) are suspended by mesenteries and covered by a serosa (ie, a thin layer of loose connective tissue covered by simple squamous epithelium, or mesothelium). Retroperitoneal organs (duodenum, ascending and descending colon) are bound to the posterior abdominal wall by adventitia and are covered on their anterior surfaces by serosa.

Table 15–1. (Cont'd)

Muscularis Mucosae	Submucosa	Muscularis Externa	Adventitia/Serosa
Thick	Esophageal glands (mostly mucous, some serous)	Skeletal muscle	Adventitia
Thin		Skeletal and smooth muscle	
		Smooth muscle	
Two layers, with radial projections into the lamina propria	No distinguishing features; no permanent folds	Three layers of smooth muscle (inner oblique, middle circular, outer longitudinal)	Serosa
Elevated by plicae circulares	Brunner's glands	Two layers of smooth muscle (inner circular, outer teniae coli)	Adventitia and serosa
	Circular folds formed by submucosa (plicae circulares)		Serosa
Discontinuous	Lymphoid nodules	Very thin	Serosa
	No circular folds	Two layers of smooth muscle (inner circular, outer teniae coli)	Adventitia and serosa
	Semicircular folds		Serosa
			Adventitia
No distinguishing features	Circular folds formed by submucosa	Internal sphincter formed by thickened inner circular layer of smooth muscle External sphincter formed by outer layer of skeletal muscle	Adventitia

C. General Functional Features: The primary functions of the digestive tract include the absorption of nutrients and water and the excretion of wastes and toxins.

  1. Digestion. Enzymatic degradation of food is a prerequisite for absorption. Enzymes act mainly at food surfaces, and chewing exposes more surface area. Lip, cheek, and tongue muscles position food between the teeth. Saliva dissolves water-soluble particles and contains enzymes that attack carbohydrates (16.II.A and C). Taste buds (24.IV.A) check for contaminants, toxins, and nutrients. The tongue moves chewed food back into the oral pharynx and closes the epiglottis to protect the airway. Skeletal muscle in the walls of the oral pharynx and upper third of the esophagus aids the tongue in swallowing and moves food down the esophagus, where smooth muscle takes over. The esophagus adds mucus to reduce friction but mainly moves material to the stomach. Glands in the stomach wall add acid (HCl), a protease (pepsin), and mucus to the mixture (now called chyme). Smooth muscles in the stomach wall mix and pulverize the chyme and move it to the small intestine (duodenum), where pancreatic enzymes and bile are added. These enzymes hydrolyze nutrients to an absorbable form. Bile's detergent action disperses water-insoluble lipid into tiny droplets, increasing the surface area available to pancreatic lipases. The epithelial cells lining the small intestine (**enterocytes**) have additional enzymes on their luminal surfaces to complete the hydrolysis of certain nutrients.

  2. Absorption. This primary function of the digestive tract occurs mainly in the intestines: the small intestines absorb nutrients, and the large intestines absorb water. To maximize the absorptive surface, the small intestine's lining has multiple permanent folds, including plicae circulares (VII.A) and villi (VII.B.1). Intestines are lined by absorptive cells (enterocytes; VII.B.3) whose apical microvilli further increase the surface area. These cells absorb and transfer amino acids and sugars to capillaries in the lamina propria, whose blood carries them to the liver for further processing. Enterocytes assemble chylomicrons from absorbed lipids and transfer them to lymphatic capillaries (lacteals) in the lamina propria. Here, lipids reach the blood through the lymphatic vascular system.

3. **Excretion.** Metabolic wastes are excreted by the liver as bile and emptied into the duodenal lumen by the bile duct. Smooth muscles in the small intestine's walls move undigested material and waste products to the large intestine (colon). Here, more mucus is added and most of the water is extracted. This process concentrates and solidifies the intestinal contents, forming feces. This material is further dehydrated, stored in the rectum, and finally expelled through the anal canal.

4. **Endocrine function.** Individual cells with characteristics of the diffuse neuroendocrine system (DNES) (4.VI.C.2) are scattered among the epithelial cells lining the tract's mucosal glands and crypts. These **enteroendocrine cells** were formerly called argentaffin, argyrophilic, and enterochromaffin cells, owing to their affinity for silver and chromium stains. They secrete hormones and amines (eg, serotonin, secretin, gastrin, somatostatin, cholecystokinin, glucagon) that regulate such local gastrointestinal functions as gut motility and the secretion of acid, enzymes, and hormones by other cell types.

5. **Innervation.** Distributed in and along the tract's walls are the myenteric (Auerbach's) and submucosal (Meissner's) autonomic nerve plexuses. These include postsynaptic sympathetic fibers, presynaptic and postsynaptic parasympathetic fibers, parasympathetic ganglion cell bodies, and some visceral sensory fibers. After voluntary swallowing, these autonomic plexuses coordinate **peristalsis**—wavelike contractions of the muscularis externa that propel ingested material through the tract. They also control the muscularis mucosa, which maintains contact between the mucosa and the tract's contents and helps empty mucosal glands. These plexuses also modulate the secretory activity of certain DNES-like cells. In general, sympathetic action inhibits, and parasympathetic action stimulates, gut motility.

6. **Blood supply.** Mesenteric branches of the abdominal aorta ramify further in the mesenteries to form a series of arcades. Small arteries penetrate the tract walls to feed capillaries of the lamina propria. Only small veins accompany branches of the mesenteric arteries. The larger veins draining these organs diverge from the arterial path and empty either directly or through tributaries into the **hepatic portal vein,** which branches within the liver to feed the **hepatic sinusoids** (16.IV.C.3). Amino acids, sugars, small fatty acids, and any toxins absorbed in the intestine thus travel directly to the liver to be metabolized, stored, or detoxified before reaching the general circulation.

7. **Protection.** The extensive absorptive surface of the digestive tract increases the risk of infection. The risk is reduced by immunoreactive cells, including IgA-secreting plasma cells, in the lamina propria and submucosa. Other defenses include lysozyme secreted by Paneth's cells, digestive enzymes in the lumen, the layer of mucus covering the epithelium, and the tight junctions between absorptive cells. Toxic substances that do reach the blood are carried directly to the liver for detoxification in the SER of the hepatocytes.

II. ORAL CAVITY

The digestive tract's upper end is bounded anteriorly by the teeth and lips, posteriorly by the oral pharynx, laterally by the teeth and cheeks, superiorly by the hard and soft palate, and inferiorly by the tongue and floor of the mouth.

A. **Wall Structure:** The **mucosa** includes the lining epithelium and the underlying lamina propria. Nonkeratinized stratified squamous epithelium (mucous membrane) covers all internal surfaces of the oral cavity and pharynx except the teeth. The lamina propria is a vascular connective tissue with papillae similar to those in the dermis (18.I.B.2). The papillary capillaries nourish the epithelium. The oral cavity has no muscularis mucosae. The **submucosa** is more fibrous than the lamina propria; it contains many blood vessels and small salivary glands. The oral cavity also lacks a standard muscularis externa. Skeletal muscle underlies the submucosa in the lips, cheeks, tongue, floor of the mouth, oral pharynx, and soft palate, including its downward extension, the **uvula.** Bone underlies the thin submucosa of the hard palate and gums (gingiva).

B. **Lips:** Here, a transition occurs from nonkeratinized mucous membrane to the keratinized stratified squamous epithelium of the skin. The thin keratinized layer covering the lips' **vermilion border** allows the reddish color of blood in vessels of the lamina propria to show through. Hair follicles, keratin, and additional pigment help distinguish the outer lip surface from the inner lip surface in tissue sections.

C. Tongue: This component of the oral cavity is a mass of skeletal muscle covered by a mucosa. The mucosa is bound tightly to the muscle by the lamina propria, which penetrates the muscle. The tongue has little or no submucosa. The muscle is arranged in bundles of many sizes; these are separated by connective tissue and cross each other in three planes. This gives the tongue the flexibility required for speech, positioning food, chewing, and swallowing. The mucosa differs on the dorsal (upper) and ventral (lower) surfaces. The ventral surface has a thin, nonkeratinized stratified squamous epithelium underlain with a lamina propria. The epithelium covering the dorsal surface is partly keratinized. The anterior two thirds of the dorsal surface is separated from the posterior third by a V-shaped groove. Behind this, the epithelium invaginates to form the crypts of the lingual tonsils (14.IX). Cryptless patches of lymphoid tissue in the lamina propria cause surface bulges in this region. The anterior two thirds of the dorsal surface has many papillae, which are mucosal projections. There are four types of papillae.

1. **Filiform papillae** are the most numerous. They are sharp, often partly keratinized, conical projections lacking taste buds.
2. **Fungiform papillae** resemble mushrooms. Each papilla has taste buds (24.IV.A) on its expanded upper surface but not on its narrow stalk. Fungiform papillae occur singly and are scattered among the filiform papillae.
3. **Foliate papillae** occur in rows separated by furrows into which serous glands in the lamina propria drain. The furrow walls (sides of the papillae) harbor many taste buds.
4. **Circumvallate papillae** are the largest and least numerous; only 7 to 12 occur near the V-shaped groove at the back of the tongue. Each is surrounded by a ringlike mucosal ridge from which it is separated by a circular furrow, whose walls contain taste buds on both sides. As with the foliates, ducts from serous (von Ebner's) glands empty into the furrow and wash chemical stimuli from the taste buds, allowing new tastes to be sensed.

III. TEETH & ASSOCIATED STRUCTURES

Note to medical and dental students: Although the teeth are the focus of dental histology, detailed coverage of tooth development and histology varies widely among medical school curricula. Few questions on teeth show up on the USMLE. Thus, medical students may wish to restrict their review of this topic to the major headings and learn the meaning of the bold-faced words, leaving further study to their dental colleagues (for whom this book is also intended).

A. Tooth Shape: Humans have four types of teeth, each with a distinctive crown and root structure. The structure and location of each type suit its functions. **Incisors** are located directly behind the lips. Each has a single root and a chisel-shaped crown for cutting. **Canines (cuspids)** lie lateral to the incisors. Each has a single root and a conical crown for grasping and tearing. **Premolars (bicuspids)** lie posterolateral to the canines. Each has two roots and a squat ovoid crown with a flat upper surface for crushing. Their location near the front of the mouth allows them to aid in grasping. **Molars (tricuspids)** lie behind the premolars. Each has three roots and a rounded, boxlike crown with a flat upper surface for crushing and grinding. Their location near the angle of the jaw allows them to exert greater crushing force than the premolars.

B. Permanent and Deciduous Teeth: Human adults (barring loss from decay, trauma, or other causes) normally have 32 permanent teeth arranged in two arches (**maxillary,** or upper, and **mandibular,** or lower). Each arch has two bilaterally symmetric quadrants. The eight teeth in each quadrant define the adult "dental formula": two incisors, one canine, two premolars, and three molars. **Deciduous (baby) teeth** develop first and are normally replaced by permanent teeth. The arrangement of the 20 deciduous teeth is like that of the permanent teeth, but there are no molars. The dental formula for the five deciduous teeth in each quadrant is two incisors, one canine, and two premolars.

C. Tooth Structure: Each tooth has the following parts (Fig. 15–2), which lie above, at, or below the gum line:
1. **Crown (corona).** Projecting above the gum, this is the only part covered by enamel.
2. **Root (radix).** This part projects below the gum into the bony socket (**alveolus**) that anchors the tooth. A tooth can have one to three roots, which are covered by **cementum.** A small

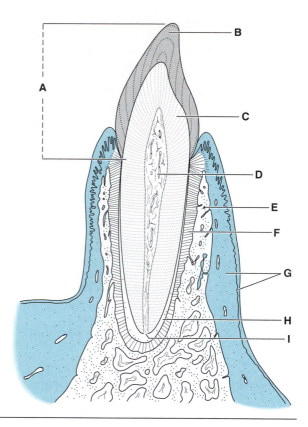

Figure 15–2. Schematic diagram of an incisor. Labeled components include the crown (A), enamel (B), dentin (C), pulp (D), alveolar bone (E), periodontal ligament (F), gingiva (G), cementum (H), and apical foramen (I). (Revised, with permission, from Leeson TS, Leeson CR: *Histology,* 2nd ed. Philadelphia, PA: WB Saunders, 1970.)

opening at the root's apex **(apical foramen)** provides vessels and nerves access to the pulp cavity.

3. **Neck (cervix).** Lying at the crown–root junction, at or just below the gum line, this is defined as the point where the enamel and cementum meet.

4. **Pulp cavity.** This part lies at the tooth's core, mainly in the root but extending into the crown. It is filled with pulp, a loose, vascular connective tissue. Vessels and nerves enter by means of the apical foramen. Some nerve (pain) fibers lose their myelin after entering the cavity; they may extend for short distances into the **dentinal tubules** (III.C.5.b).

5. **Dentin.** A layer of bonelike tissue, dentin envelops the pulp in both the crown and root.

   a. **Composition.** Hydroxyapatite crystals (8.III.A.2.b) make up 70% of dentin's dry weight, placing it between bone and enamel in hardness. Organic components include type-I collagen and glycosaminoglycans.

   b. **Organization.** The dentin and pulp cavity are separated by a single layer of columnar cells called **odontoblasts.** These have basal nuclei, a well-developed Golgi complex, an RER, and many ribosomes. A long, branched, tapered odontoblast process **(Tomes' fiber)** extends from each cell's apical (dentinal) surface and penetrates the dentin's width in a dentinal tubule. Transverse sections of dentin have a honeycomblike appearance (dentin forms the comb; Tomes' fibers and tissue fluid, the honey). An unmyelinated nerve fiber often lies in the dentinal tubule.

   c. **Histogenesis.** Dentin's organic matrix, termed **predentin,** is secreted by odontoblasts from their apices. As the predentin is deposited and the dentin layer thickens, the cells retreat, leaving in place a thin cell process that gradually elongates to form a Tomes' fiber. Mineralization begins when the cells release, into the predentin, membrane-limited **matrix vesicles** containing fine hydroxyapatite crystals. The crystals act as nucleation sites for further mineral deposition. The crystals grow by accruing more mineral from the tissue fluid.

6. **Enamel.** A thick layer of calcified material covering the dentin of the crown, enamel is not a true tissue when mature, because it lacks cells or cell processes.

a. **Composition.** Because mineral salts (mainly hydroxyapatite) make up 95% of enamel, it is the body's hardest substance. Unlike bone and dentin, its organic components (preenamel) do not include collagen, but rather two unique classes of proteins known as **amelogenins** and **enamelins.**

b. **Organization.** Enamel is arranged as tightly packed hydroxyapatite columns (**enamel rods or prisms**) bound together by **interrod enamel.**

c. **Production.** During tooth formation, enamel is produced by tall columnar cells called **ameloblasts.** Each has a basal nucleus, a well-developed Golgi complex, an RER, and a short apical cell process (**Tomes' process**). This process extends into the enamel matrix and contains secretory vesicles filled with preenamel. As the organic material is secreted from the ameloblast's apical surface, the cell recedes. Unlike the Tomes' fibers of odontoblasts, the Tomes' processes recede along with the ameloblasts, leaving behind a solid rod of organic preenamel. Calcification begins at each rod's periphery and proceeds toward its core. Ameloblasts do not accompany the tooth during eruption; instead, they degenerate. Enamel is thus irreplaceable.

7. **Cementum.** This bonelike tissue covering the dentin of the root is thicker at its apex than at its neck. It contains **cementocytes,** which, like osteocytes, lie in lacunae, communicate through canaliculi, and produce the matrix. Cementum is an active tissue that can undergo either enhanced production or resorption, depending on the stresses to which it is subjected; thus, it helps keep the root in close contact with the socket wall.

D. **Associated Structures:**

1. **Periodontal ligament.** The collagen fibers of this dense connective tissue sling surround the tooth's root, inserting into both cementum and alveolar bone. This ligament serves as the alveolar periosteum, binds the root to the socket wall, suspends the tooth, and permits slight movement. Its pressure-sensitive nerve endings warn against biting too hard and prevent the resorption of alveolar bone that would otherwise accompany direct transmission of pressure to the socket walls. Because its matrix undergoes rapid and continual turnover, it contains soluble collagen and glycosaminoglycans and is particularly susceptible to nutritional deficiencies. Vitamin C or protein deficiencies may cause it to degenerate, resulting in the loosening or loss of teeth.

2. **Alveolar bone** is simply the bone of the mandible and maxilla that lines the alveoli (sockets) and to which the teeth attach by periodontal ligaments. Even in adults it consists of primary (woven) bone (8.III.C.1).

3. **Gingiva (gums).** The oral mucosa covering the mandibular and maxillary arches in which the teeth are anchored consists of nonkeratinized stratified squamous epithelium. It includes an underlying lamina propria, whose long papillae interdigitate with epithelial ridges. The lamina propria binds tightly to the epithelium by hemidesmosomes and to the periosteum of the underlying bone by interwoven collagen fibers. The gingival epithelium forms a cuff around the crown's base and is separated from the tooth by a narrow gingival crevice. At the base of the crevice, the gingiva forms a basal lamina-like thickening, the **cuticle,** that encircles the tooth and attaches to the enamel. This is the **epithelial attachment of Gottlieb.**

E. **Tooth Development:** Beginning during week 6 of gestation, tooth development involves a cascade of epithelial–mesenchymal interactions and proceeds through a series of morphologic stages. This complex process is more easily understood by monitoring the changes in epithelium and mesenchyme that occur during each stage and by focusing on the tooth components formed by each tissue. The oral epithelium derives from oral ectoderm and gives rise to the ameloblasts that form the enamel. The mesenchyme is the ectomesenchyme underlying the oral epithelium. This embryonic connective tissue derives from the neural crest and gives rise to the odontoblasts and cementoblasts, which form dentin and cementum, respectively. It also forms the dental pulp. Mesenchyme around the developing tooth forms the periodontal ligament and alveolar bone.

1. **Crown development** (Fig. 15–3) is completed shortly before eruption. It begins in oral ectodermal ridges called **dental laminae** with the formation of **epithelial tooth buds.** The buds form a cap that envelops a papilla of ectomesenchyme. A wave of interactions between the epithelial cap and papillary mesenchyme begins at the top of the crown and progresses toward the cervical loop (see Fig. 15–3). Briefly stated, ectomesenchymal clusters induce epithelial tooth buds in the dental lamina, prompting the proliferation and condensation of the papillary mesenchyme. This process, in turn, induces formation of the **inner enamel ep-**

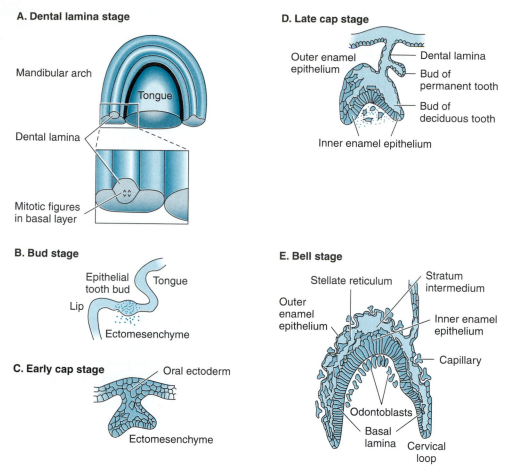

A. Dental lamina stage

Mandibular arch

Tongue

Dental lamina

Mitotic figures
in basal layer

B. Bud stage

Epithelial
tooth bud — Tongue

Lip

Ectomesenchyme

C. Early cap stage — Oral ectoderm

Ectomesenchyme

D. Late cap stage

Outer enamel
epithelium — Dental lamina
— Bud of
permanent tooth
— Bud of
deciduous tooth

Inner enamel epithelium

E. Bell stage

Stellate reticulum — Stratum
intermedium

Outer
enamel
epithelium — Inner enamel
epithelium

— Capillary

Odontoblasts

Basal
lamina   Cervical
loop

Figure 15–3. Stages in crown development. **A. Dental lamina stage.** Localized bands of proliferating cells in the basal layer of the stratified oral epithelium, peripheral to the developing tongue, form two (one per jaw) horseshoe-shaped epithelial ridges, or dental laminae, over the mesenchyme of the future mandibular and maxillary arches. **B. Bud stage.** Stimulated by local clusters of neural crest–derived mesenchyme cells, proliferation increases in the base of each dental lamina at the 10 sites of future deciduous teeth. These epithelial tooth buds enlarge and bulge into the underlying mesenchyme. **C. Early cap stage.** With further proliferation, the deep bud surfaces invaginate and widen to form solid caps over mesenchymal clusters. In the cap's core, the cell density decreases as internal cells become stellate and the interstices accumulate tissue fluid. The peripheral cells, which contact the basal lamina, form a simple epithelial shell and continue to divide, increasing the cap's size. A stalk of dental lamina connects each cap to the oral epithelium. The mesenchyme under the cap proliferates and condenses, indenting the cap's base. **D. Late cap stage.** Mesenchyme within the indentation grows to form the dental papilla, further indenting the cap's base. The epithelial cells over the papilla (inner enamel epithelium) become columnar, whereas those forming the rest of the shell (outer enamel epithelium) remain low cuboidal. The stellate cells and fluid inside the shell make up the stellate reticulum. Between the stellate reticulum and the inner enamel epithelium lies a layer of epithelial cells called the stratum intermedium. Together, the inner and outer epithelia, stratum intermedium, and stellate reticulum constitute the enamel organ. The outer enamel epithelium is continuous with the narrowing stalk of the dental lamina; this gives rise to another tooth bud that will subsequently form a permanent tooth (IV.F). **E. Bell stage.** As the cap grows, the indentation deepens, the inner enamel epithelium expands around the enlarging papilla, and the developing tooth becomes bell-shaped. Mesenchyme cells near the inner enamel epithelium condense, differentiate into a layer of columnar odontoblasts, and begin forming predentin. Mesenchyme in the papilla's core forms the dental pulp. Columnar cells of the inner enamel epithelium differentiate into ameloblasts and begin producing enamel soon after the dentin begins to calcify. After the enamel layer is complete, the ameloblasts shorten and become inactive. The ringlike junction of the inner and outer enamel epithelium at the rim of the bell is termed the cervical loop. Capillaries indent the outer enamel epithelium, and it loses its connection with the oral epithelium as the dental lamina degenerates. (Revised and redrawn, with permission, from Warshawsky H: The teeth. In: Weiss L (ed.), *Histology: Cell and Tissue Biology,* 5th ed. New York: Elsevier, 1983.)

ithelium (see Fig. 15–3), causing the papillary mesenchyme cells to become odontoblasts. The inner enamel epithelial cells are induced to become ameloblasts, which cause the odontoblasts to produce predentin. Calcified predentin induces the ameloblasts to produce enamel. (See Figure 15–3 for a detailed description of the process.)

2. **Root development.** After the crown is formed, the cervical loop grows rootward, enclosing the dental papilla. The inner and outer enamel epithelia fuse around the root, forming **Hertwig's root sheath,** whose inner layer induces odontoblast differentiation in the adjacent papillary mesenchyme. Once the predentin around the root calcifies, the root sheath degenerates. This brings surrounding mesenchymal cells into contact with the dentin, inducing them to become cementoblasts. Cementum secreted by these cells onto the root surface traps the ends of fibers produced by nearby fibroblasts. The fibroblasts remodel these fibers to form the periodontal ligament.

3. **Eruption.** As the root elongates, alveolar bone limits its downward growth, forcing the crown upward. Tissue between the crown and gingival surface degenerates, allowing the crown to erupt into the oral cavity. Ameloblasts covering the crown degenerate. No enamel forms after eruption.

4. **Development of permanent teeth.** In the late cap stage, a secondary (permanent) tooth bud arises from the labial (lip) surface of each dental lamina stalk (see Fig. 15–3). Dental lamina tissue from each second premolar burrows backward, successively budding off three permanent molar buds. Permanent tooth buds remain dormant until activated after birth; subsequently, they undergo the same developmental steps as deciduous teeth. As each permanent tooth enlarges, it induces osteoclast-mediated resorption of the alveolar bone that separates the bony crypt in which it lies from the baby tooth's socket. Continued growth of the permanent tooth leads to resorption of the baby tooth's root until only the crown is left, held only by its cuticle to the gingiva. After this is lost, the permanent tooth erupts into the oral cavity.

IV. PHARYNX

A short, broad, muscular tube behind the tongue and soft palate, the pharynx is shared by the respiratory and digestive tracts. Its superior portion, the **respiratory pharynx,** lies above the soft palate, communicates with the nasal cavity, and is lined by respiratory epithelium. The inferior portion, the **oral pharynx (oropharynx),** lies below the level of the soft palate. It communicates with the oral cavity and is lined by nonkeratinized stratified squamous epithelium. Its walls contain the palatine and pharyngeal tonsils (14.IX), many small subepithelial mucous glands, and skeletal muscle arranged as circular pharyngeal constrictors and longitudinal pharyngeal muscles. The pharynx also communicates with both the esophagus and the larynx. During swallowing, the back of the tongue helps close the epiglottis (17.V.A) to direct food away from the larynx and into the esophagus.

V. ESOPHAGUS

This long, narrow, muscular tube transports food from the pharynx to the stomach. Its mucosa includes nonkeratinized stratified squamous epithelium, a lamina propria that interdigitates with the scalloped basal border of the epithelium, and a muscularis mucosae. Mucus-secreting **esophageal glands** characterize its submucosa and help distinguish esophagus from vagina (23.VIII) in histologic sections. The muscularis externa of the esophagus comprises skeletal muscle in the upper third, a mixture of skeletal and smooth muscle in the middle third, and smooth muscle in the lower third. The outer surface is covered by adventitia, except for the short serosa-covered segment in the abdominal cavity between the diaphragm and stomach. Mucus-secreting esophageal cardiac glands are found in the lamina propria of the region near the stomach.

VI. STOMACH

This dilated portion of the digestive tract temporarily holds ingested food, adding mucus, acid, and the digestive enzyme **pepsin.** Its contractions blend these components into a viscous mixture called **chyme.** The chyme is subsequently divided into parcels for further digestion and absorption by the intestines.

A. **General Structure:** The stomach wall comprises the same layers as the rest of the tract. The complex mucosa contains numerous **gastric glands,** a two- or three-layered muscularis mucosae that helps empty the glands, and an intervening lamina propria. When the stomach is empty and contracted, the mucosa and underlying submucosa are thrown into irregular, temporary folds called **rugae,** that flatten when it is full. The smooth muscle of the muscularis externa comprises three layers: outer longitudinal, middle circular, and inner oblique. The stomach has four major regions: **cardia, fundus, body,** and **pylorus** (Fig. 15–4).

B. **Gastric Mucosa:** The stomach's simple columnar epithelial lining is perforated by many small holes called **foveolae gastricae.** The foveolae are the openings of epithelial invaginations, the **gastric pits,** which penetrate the lamina propria to various depths. The pits serve as ducts for the branched tubular gastric glands. Each gland has three regions: an isthmus at the bottom of the pit, a straight neck that penetrates deeper into the lamina propria (perpendicular to the surface), and a coiled base that penetrates deeper still and ends blindly just above the muscularis mucosae. The mucosa is characterized by the following epithelial cell types.
   1. **Surface mucous cells** form the simple columnar epithelium lining the stomach, the gastric pits, and much of the isthmus of each gastric gland. They secrete a neutral mucus that protects the stomach's surface from the acidic gastric fluid.
   2. **Undifferentiated cells** are low columnar cells with basal ovoid nuclei scattered in the neck of the gastric glands. After dividing in the neck, some move upward to replace pit and surface mucous cells. Others move deeper into the glands and differentiate into the other cell types listed below. Surface mucous cells turn over more rapidly than do the other cell types.
   3. **Mucous neck cells** occur singly or in clusters between the parietal cells in the neck of the gland. They differ from the surface mucous cells by secreting acidic mucus.
   4. **Parietal (oxyntic) cells** secrete HCl and intrinsic factor.
      a. **Structure and location.** These cells lie mainly between mucous neck cells in the neck of the gland. They are large, and round to pyramidal, with one or two central nuclei and a pale, acidophilic cytoplasm. The many mitochondria indicate that their secretory activity is energy-dependent. Each cell has a circular invagination of its apical plasma membrane that is visible only with the electron microscope. When the cells are stimulated to produce HCl, the many **tubulovesicles** in the apical cytoplasm fuse with the invaginated plasma membrane to form a deeper, more highly branched invagination termed the **intracellular canaliculus.**
      b. **Function.** HCl production involves the active transport of H^+ and Cl^- ions across canalicular membranes into the lumen. The Cl^- derives from blood-borne chloride. H^+ formation involves a two-step process in which CO_2 is converted by carbonic anhydrase to carbonic acid, which dissociates into H^+ and bicarbonate. Acid production is greatly enhanced by histamine and gastrin produced by enteroendocrine cells in gastric glands (and elsewhere). **Intrinsic factor** is a glycoprotein required for vitamin B_{12} absorption. B_{12} deficiency leads to a disorder of erythropoiesis called **pernicious anemia.** Parietal cell secretion is stimulated by cholinergic nerve endings.
   5. **Chief (zymogenic) cells** secrete pepsinogen and some lipase.
      a. **Structure and location.** These cells predominate in the base of gastric glands and are smaller than parietal cells. They are basophilic owing to the RER's ribosomes. They also contain membrane-limited pepsinogen-filled zymogen granules.
      b. **Function.** The RER synthesizes pepsinogen and lipase, which are packaged in granules by the Golgi complex and stored in the cytoplasm for secretion. **Pepsinogen** is an inactive proenzyme, or zymogen, that is converted to the active protease **pepsin** when exposed to acid in the stomach lumen. Gastric lipase has only weak lipolytic activity.
   6. **Enteroendocrine cells.** In the stomach, these cells (I.C.4) occur mainly in the base of gastric glands. They produce various endocrine and paracrine amines (eg, histamine, serotonin) and peptide hormones (eg, gastrin). They are considered DNES components.

C. **Regional Differences:**
   1. **Cardia.** A narrow collarlike region, the cardia surrounds the entry of the esophagus. Here, the lamina propria contains simple or branched tubular cardiac glands similar to those in the terminal esophagus. These glands often have shallow crypts and coiled bases with wide lumens. Although they produce mainly mucus and lysozyme, some parietal cells may be present.

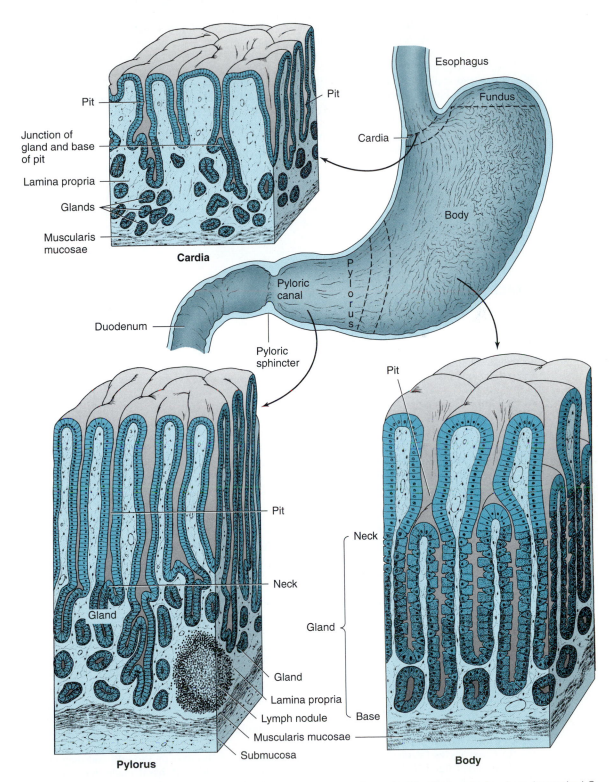

Figure 15–4. Regions of the stomach and their histologic structure. (Reproduced, with permission, from Junqueira LC, Carneiro J, Kelley RO: *Basic Histology,* 9th ed. Stamford, CT: Appleton & Lange, 1998.)

2. **Fundus and body.** The glands in these regions are similar in structure and function. The body is the stomach's largest region, extending from the cardia to the pylorus. The fundus is a smaller, roughly hemispherical region extending above the cardia. Gastric glands—termed **fundic glands** in both regions—are characterized by shallow pits and long glands. The pits extend approximately one third of the distance from the mucosal surface to the base of the glands. Fundic glands contain abundant **parietal** and **chief cells.** Parietal cells are concentrated in the neck and upper base; chief cells predominate in the lower base. **Serotonin** (5-hydroxytryptamine)-secreting cells are found at the base.

3. **Pylorus.** This region comprises the distal 4 to 5 cm of the stomach leading to the small intestine. **Pyloric glands** have deep pits and short glands (mnemonic: P for both pylorus and pits). The pits extend one half to two thirds of the distance from the mucosal surface to the base of the glands. Large pale-staining mucus-secreting cells with basal nuclei predominate. Parietal cells and especially chief cells are scarce here. Gastrin-secreting cells **(G cells)** lie in bases of these glands. At the pylorus–small intestine junction, a thickened band of the muscularis externa's middle circular layer, the pyloric sphincter, controls the passage of chyme.

VII. SMALL INTESTINE

The small intestine, which includes the **duodenum, jejunum,** and **ileum,** receives chyme from the stomach, bile from the liver, and digestive enzymes from the pancreas. Here, nutrients are hydrolyzed to an absorbable form; they are absorbed and transferred to blood and lymphatic capillaries. Undigested material is moved to the large intestine by peristalsis. The word "small" refers to diameter: the small intestine is longer and narrower than the large intestine.

A. **General Structure:** The small intestine's walls are made of the same layers that are found in the rest of the tract (I.B; Fig. 15–1). The two-layered muscularis externa (I.B.3) exhibits typical organization, as does the submucosa (I.B.2), except in the duodenum, where distinctive submucosal (Brunner's) glands (VII.C.1) are present. A series of permanent folds, the **plicae circulares (valves of Kerckring),** composed of both submucosa and mucosa, extend into the lumen and increase the surface area approximately threefold. The main distinguishing features of the small intestine (when viewed through the microscope) are the composition and organization of the mucosa.

B. **Mucosa of the Small Intestine:** This is a simple columnar epithelium with goblet cells, underlain by a lamina propria and separated from the submucosa by a muscularis mucosae.

1. **Villi.** The presence of these epithelium-covered, fingerlike mucosal projections into the lumen is the most diagnostic feature of small intestine structure. The lamina propria core of each consists of loose connective tissue (5.III.A.1) and contains a central, blind-ended lymphatic capillary **(lacteal),** as well as blood capillaries. Smooth muscle fibers run lengthwise in the villus core; however, the muscularis mucosae per se does not extend into the villi. Rhythmic contractions (shortening) of the villi speed up during digestion and help propel nutrients in the blood and lymphatic capillaries to the general circulation. The villi increase the mucosal surface area approximately 10-fold and thus enhance absorption; their shape and abundance differ by region (VII.C).

2. **Intestinal glands (crypts of Lieberkühn).** These simple (often coiled) tubular glands extend into the lamina propria below the bases of the villi. They are lined by absorptive, goblet, Paneth, enteroendocrine, and undifferentiated cells. Their secretions enter the lumen by means of small openings between the villi. Similar glands are seen in the large intestine, where they contain many more goblet cells.

3. **Enterocytes (absorptive cells)** are the predominant cells covering the villi and they occur in small numbers in the crypts. These tall columnar cells with basal nuclei have densely packed, glycocalyx-covered microvilli extending from their apical surfaces into the lumen. The approximately 3000 microvilli per cell give the cell–lumen border a striped appearance, referred to as a **striated border.** Enterocytes attach to neighboring cells by junctional complexes (4.IV.B), including tight junctions near the lumen. Although their structure is comparatively simple, these cells perform several complex and important functions.

    a. Digestion. Disaccharidases and dipeptidases bind to the luminal surface of the microvilli and complete the hydrolysis of nutrients begun by pancreatic enzymes in the lumen. The resulting monosaccharides and amino acids are more readily absorbed.

    b. Absorption. Apical microvilli increase the absorptive surface area approximately 20-fold and thus enhance absorption. Amino acids and monosaccharides cross the apical plasma membrane by facilitated diffusion, whereas the products of lipid hydrolysis (fatty acids and monoglycerides) cross passively. Larger molecules may enter through pinocytotic vesicles (**caveolae**) that form at the bases of the microvilli.

    c. Lipid processing and chylomicron assembly. Absorbed monoglycerides and fatty acids collect in the SER, where they are resynthesized into triglycerides and subsequently assembled into chylomicrons—small lipid spheres with a thin protein coat. Chylomicrons are packed in vesicles by the Golgi complex and move to the basolateral plasma membrane for exocytosis; from here, most enter the lymphatic capillaries.

    d. Transport of smaller nutrients. Amino acids, monosaccharides, and short-chain fatty acids cross the cytoplasm and then the basolateral cell membrane to reach the lamina propria, where they enter the blood and lymphatic capillaries.

4. **Goblet cells** lie between the absorptive cells, gradually increasing in number from duodenum to ileum. The acid glycoprotein (mucus) they secrete onto the mucosal surface lubricates the digestive tract's walls, protecting them from pancreatic enzymes and impeding bacterial invasion (see 4.VI.C.3 for details of mucus-secreting cell structure).

5. **M cells.** These flat membranous epithelial cells overlie solitary lymphoid nodules and Peyer's patches (14.V) of the intestinal lamina propria. Their apical (luminal) surfaces have small folds rather than microvilli. The cells help initiate immune responses by endocytosing antigens from the lumen and passing them to lymphoid cells in underlying nodules.

6. **Paneth's cells** lie in the bases of the crypts and produce a protein–polysaccharide complex. In addition to RER and Golgi complexes, they have large acidophilic secretory granules containing lysozyme, an antibacterial enzyme that helps control intestinal flora.

7. **Enteroendocrine cells.** Most known enteroendocrine cells (I.C.4) are found in the crypts of the small intestine. Here they produce hormones and amines, such as **secretin,** which increases pancreatic and biliary bicarbonate and water secretion; **cholecystokinin,** which increases pancreatic enzyme secretion and gallbladder contraction; **gastric inhibitory peptide,** which decreases gastric acid production; and **motilin,** which increases gut motility.

8. **Undifferentiated cells.** Mucosal epithelial cells undergo continual turnover. Replacement occurs through the mitosis of undifferentiated (stem) cells located near the base of the crypts. Evidence of this mitotic activity includes the presence of highly condensed, dark-staining chromosomes (ie, **mitotic figures**) in the walls of the crypts. Products of these divisions differentiate into all of the cell types described above; by a mechanism that is still unclear, they move toward the crypt base or toward the tips of the villi, from which they are finally sloughed into the lumen.

C. Regional Differences:

1. **Duodenum.** The major distinguishing feature of this C-shaped first part of the small intestine is the presence of **duodenal (Brunner's) glands** in the submucosa. The mucous cells of these glands produce an alkaline secretion (pH 8.1–9.3) that enters the lumen through the crypts. It protects the duodenal lining from the acidity of the chyme and raises the luminal pH to optimize pancreatic enzyme activity. Unlike the jejunum and ileum, the duodenum is retroperitoneal (I.B.4). It is also the entry point for the bile and pancreatic ducts, which penetrate the full thickness of the duodenal wall. It typically has fingerlike or leaflike villi and relatively few goblet cells. Because the arms of the C cradle the pancreas in situ, some pancreatic tissue may accompany duodenal sections, providing another clue for identification.

2. **Jejunum.** An intraperitoneal organ, the jejunum has long, leaflike villi, many plicae circulares, and an intermediate number of goblet cells. The key to its identification, however, is that although it has villi (and is thus part of the small intestine), it contains neither Brunner's glands nor Peyer's patches.

3. **Ileum.** This intraperitoneal organ has fewer villi, which are short and broad-tipped (club-like), and abundant goblet cells. Its lamina propria typically contains many lymphoid nodule clusters (Peyer's patches; 14.V). These may be large enough to produce a visible bulge on the luminal surface and extend into the submucosa.

VIII. LARGE INTESTINE (COLON)

The large intestine comprises the cecum; the ascending, transverse, descending, and sigmoid colon; and the rectum. It converts undigested material received from the small intestine into feces by removing water and adding mucus. The colon is shorter and wider than the small intestine. Its walls differ from the small intestine at both the gross and microscopic levels.

A. **Mucosa:** The colon's lining has no folds, except in the rectum, where vertical folds, called the **rectal columns (of Morgagni),** occur at the rectoanal junction. No villi are present. The epithelium is simple columnar with abundant goblet cells. The interposed absorptive cells have irregular short microvilli. Water absorption by these cells is passive; it follows the active transport of sodium out of their basal surfaces. The mucosa has many deep crypts, containing many goblet cells and few enteroendocrine cells. The lamina propria has more lymphoid cells and nodules than does that of the small intestine. Nodules may extend into the submucosa.

B. **Submucosa:** This layer is generally unremarkable except in the lower rectum, where it contains portions of the **hemorrhoidal plexus** of veins, which extends into the lamina propria. The absence of valves in the veins within and draining the plexus, coupled with the great abdominal pressure changes to which they are subjected (eg, during coughing and straining), often causes these veins to become varicosed, resulting in the formation of hemorrhoids.

C. **Muscularis Externa:** In the colon, this component is unique in that the outer longitudinal layer of smooth muscle is gathered into three thick longitudinal bands called **teniae coli.** A thin layer of longitudinal smooth muscle often exists between the bands. The inner circular muscle layer resembles that of the small intestine.

D. **Adventitia and Serosa:** The outer covering on the various parts of the colon varies, depending on whether they are intraperitoneal (cecum, transverse, sigmoid) or retroperitoneal (ascending, descending) (I.B.4; see Table 15–1). The rectum passes vertically through the pelvis, surrounded by adventitia. The colon's serosa is characterized by the presence of many teardrop-shaped adipose-filled outpocketings termed **appendices epiploicae.**

IX. APPENDIX (VERMIFORM APPENDIX)

This is a narrow fingerlike evagination of the inferior end of the cecum. Histologically, it resembles the colon except that it has a smaller lumen, fewer and shorter crypts, many more lymphoid nodules, and no teniae coli.

X. ANAL CANAL

In humans, this canal is approximately 4-cm long and connects the rectum and the anal opening. The mucosa of the first 2 cm has typical colonic epithelium with very short crypts. This is replaced by stratified squamous epithelium, which continues to the anal opening. The lamina propria contains extensions of the hemorrhoidal plexus, and the submucosa under the stratified epithelium contains sebaceous glands and large circumanal apocrine sweat glands (18.VIII.B). The muscularis in this region has a thickened inner circular layer of smooth muscle that forms the involuntary internal anal sphincter. Distal to this, the canal is encircled by the voluntary external anal sphincter, which is composed of skeletal muscle from the pelvic diaphragm.

MULTIPLE-CHOICE QUESTIONS

Select the single best answer.

15.1. Which of the following is the dental formula for adult (permanent) teeth?
(A) 1 incisor, 1 canine, 3 premolars, 3 molars
(B) 2 incisors, 1 canine, 2 premolars, 3 molars
(C) 2 incisors, 2 canines, 2 premolars, 2 molars
(D) 2 incisors, 1 canine, 3 premolars, 2 molars
(E) 3 incisors, 1 canine, 2 premolars, 3 molars

15.2. The ectomesenchyme located in the region of developing teeth is derived from which of the following embryonic tissues?
(A) Oral endoderm
(B) Mesoderm
(C) Oral ectoderm
(D) Neural crest
(E) Neural ectoderm

15.3. A cascade of interactions between which two tissues is primarily responsible for tooth formation?
(A) Blood and ectomesenchyme
(B) Blood and epithelium
(C) Ectomesenchyme and epithelium
(D) Ectomesenchyme and nerve
(E) Epithelium and nerve

15.4. Which of the letters in Figure 15–5 corresponds to a tissue derived from oral ectoderm?

15.5. Which of the letters in Figure 15–5 corresponds to the periodontal ligament?

15.6. Which of the letters in Figure 15–5 corresponds to a tissue formed by odontoblasts?

15.7. Which of the letters in Figure 15–6 corresponds to the ameloblasts?

15.8. Which of the letters in Figure 15–6 corresponds to the boundary between the epithelial and mesenchymal derivatives, whose interactions lead to tooth development?

15.9. Which of the letters in Figure 15–6 corresponds to the cells that form dentin?

15.10. Which of the letters in Figure 15–6 corresponds to the cells that form Tomes' processes?

15.11. Which of the following types of lingual papillae lack taste buds?
(A) Circumvallate
(B) Filiform
(C) Foliate
(D) Fungiform

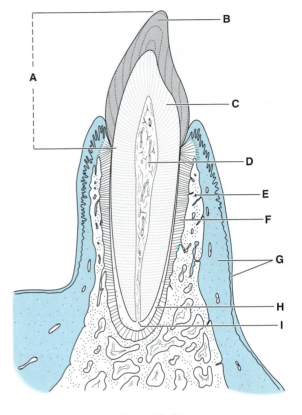

Figure 15–5.

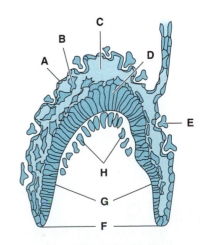

Figure 15–6.

15.12. Which of the following organs is shown in Figure 15–7?
(**A**) Appendix
(**B**) Cardiac stomach
(**C**) Colon
(**D**) Lower esophagus
(**E**) Pyloric stomach
(**F**) Rectum
(**G**) Upper esophagus

15.13. Which of the following organs is shown in Figure 15–8?
(**A**) Colon
(**B**) Cardiac stomach
(**C**) Duodenum
(**D**) Fundic stomach
(**E**) Pyloric stomach
(**F**) Rectum
(**G**) Upper esophagus

15.14. Which of the following organs is shown in Figure 15–9?
(**A**) Anal canal
(**B**) Colon
(**C**) Duodenum
(**D**) Fundic stomach
(**E**) Ileum
(**F**) Jejunum
(**G**) Pyloric stomach

15.15. Which of the following organs is shown in Figure 15–10?
(**A**) Anal canal
(**B**) Cardiac stomach
(**C**) Colon
(**D**) Duodenum
(**E**) Fundic stomach
(**F**) Ileum
(**G**) Pyloric stomach

15.16. Which of the following is true of the wall of the stomach?
(**A**) Simple squamous epithelium covers its outer surface
(**B**) Forms temporary folds called plicae circulares when the stomach is empty
(**C**) Stratified squamous epithelium covers its inner surface
(**D**) Three layers of skeletal muscle comprise its muscularis externa
(**E**) Lacks a definitive submucosa

15.17. Which of the following structures acts primarily to promote the absorption of nutrients by increasing the internal surface area of the small intestine?
(**A**) Goblet cells
(**B**) Microvilli
(**C**) Peyer's patches
(**D**) Pyloric sphincter
(**E**) Teniae coli

15.18. Which of the following is true of the absorptive cells of the small intestine?

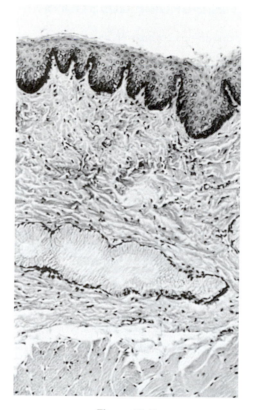

Figure 15–7.

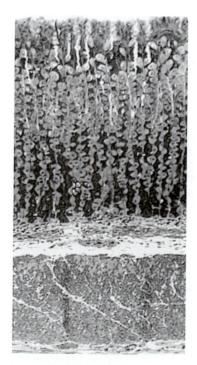

Figure 15–8.

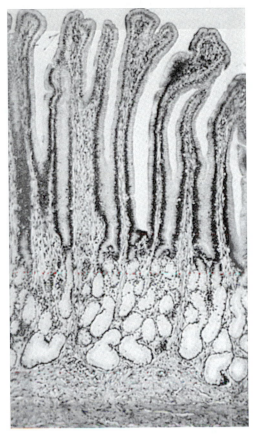

Figure 15–9.

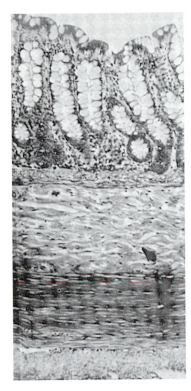

Figure 15–10.

(A) Are also called enteroendocrine cells
(B) Many microvilli cover their basal surfaces
(C) Absorb lipids by active transport
(D) Use absorbed lipids to synthesize triglycerides
(E) Undergo mitosis at the tips of the villi and are sloughed into the crypts

15.19. Which of the following is true of Brunner's glands?
(A) Are characteristic components of the jejunal wall
(B) Produce a serous secretion that is rich in digestive enzymes
(C) Lie in the submucosal layer
(D) Empty their secretions into the lacteals
(E) Are composed of collections of goblet cells

15.20. Antibiotic therapies that slow the replacement of the cells lining the small intestine may cause the loss of what tissue type?
(A) Ciliated pseudostratified columnar epithelium
(B) Pseudostratified columnar epithelium with stereocilia

(C) Simple columnar epithelium
(D) Simple cuboidal epithelium
(E) Stratified squamous, nonkeratinized epithelium
(F) Stratified cuboidal epithelium
(G) Transitional epithelium

15.21. Capillaries normally responsible for nourishing duodenal mucosal epithelial cells, which are the first vessels degraded by a bleeding duodenal ulcer, are embedded in which tissue type?
(A) Dense regular connective tissue
(B) Dense irregular connective tissue
(C) Glandular epithelia of Brunner's glands
(D) Loose connective tissue
(E) Simple columnar epithelium

15.22. Which of the following is true of the muscularis mucosa of the upper esophagus?
(A) Is composed of skeletal muscle
(B) Is innervated by postganglionic autonomic fibers
(C) Is located within the submucosa
(D) Possesses intermediate filaments composed of cytokeratin
(E) Extends into leaflike villi
(F) Is penetrated by Brunner's glands
(G) Contains the submucosal plexus

15.23. Pernicious anemia can result from insufficient intrinsic factor secretion by which of the following?
  (A) Bone marrow reticular cells
  (B) Bone marrow reticulocytes
  (C) Brunner's glands
  (D) Enteroendocrine cells
  (E) Polychromatophilic erythroblasts
  (F) Parietal cells
  (G) Kidneys

15.24. Which of the following is true of the appendix?
  (A) Is a blind-ended evagination of the ileum
  (B) Has few lymphoid nodules in its walls
  (C) Has relatively shallow intestinal glands
  (D) Longitudinal smooth muscle gathers into its teniae coli
  (E) Contains short villi

15.25. Which of the following is true of the mucosal glands of the pylorus?
  (A) Resemble those in the body of the stomach
  (B) Contain abundant parietal cells in their basal half
  (C) Secrete more pepsinogen than do fundic glands
  (D) Have shallower pits than those in the fundic region
  (E) Secrete mainly mucus

15.26. Skeletal muscle may occur in the muscularis externa at the junction between which of the following pairs of organs?
  (A) Cecum and appendix
  (B) Duodenum and jejunum
  (C) Esophagus and stomach
  (D) Ileum and cecum
  (E) Jejunum and ileum
  (F) Pharynx and esophagus
  (G) Stomach and duodenum

15.27. Diarrhea may result if which of the following organs fails to carry out its primary role in absorbing water from the feces?
  (A) Anal canal
  (B) Cecum
  (C) Colon
  (D) Esophagus
  (E) Pharynx
  (F) Small intestine
  (G) Stomach

15.28. Permanent circular folds (plicae circulares) are a characteristic structural feature of the walls of which of the following organs?
  (A) Anal canal
  (B) Cecum
  (C) Colon
  (D) Esophagus
  (E) Pharynx
  (F) Small intestine
  (G) Stomach

15.29. Which of the following organs characteristically contains the most goblet cells in the mucosa covering its villi?
  (A) Anal canal
  (B) Cecum
  (C) Colon
  (D) Duodenum
  (E) Ileum
  (F) Jejunum
  (G) Stomach

15.30. The teniae coli of the large intestine represent an organ-specific specialization of which of the following layers of the intestinal tract wall?
  (A) Epithelium
  (B) Glycocalyx
  (C) Lamina propria
  (D) Muscularis mucosa
  (E) Muscularis externa
  (F) Serosa
  (G) Submucosa

ANSWERS TO MULTIPLE-CHOICE QUESTIONS

15.1. B (III.B)
15.2. D (III.E)
15.3. C (III.E and E.1)
15.4. B (III.E and III.E.1; Fig. 15–2)
15.5. F (III.D; Fig. 15–2)
15.6. C (III.C.5.c; Fig. 15–2)
15.7. D (III.C.6.c and E.1; Fig. 15–3)
15.8. G (III.E.1; Fig. 15–3)
15.9. H (III.C.5.c; Fig. 15–3)
15.10. D (III.C.6.c; Fig. 15–3)

15.11. B (II.C)
15.12. G (V; note the skeletal muscle in the muscularis)
15.13. D (VI.C.2; Fig. 15–4)
15.14. G (VI.C.3; Fig. 15–4)
15.15. C (VIII.A; note the abundant goblet cells and lack of villi)
15.16. A (I.B.3 and 4; VI.A and B.1)
15.17. B (VII.A, B.1, and 3.b)
15.18. D (VII.B.3.a–d)

15.19. C (VII.C.1)
15.20. C (VII.B)
15.21. D (I.B.1)
15.22. B (I.B.1 and C.5; V)
15.23. D (VI.B.4 and 4.b)
15.24. C (IX)
15.25. E (VI.C.1–3)

15.26. F (IV and V)
15.27. C (VIII.A)
15.28. F (VII.A)
15.29. E (VII.B.1 and 4; VIII.A; the colon has more goblet cells but lacks villi)
15.30. E (VIII.C)

16

Glands Associated
with the Digestive Tract

OBJECTIVES

This chapter should help the student to:

- List the accessory glands attached to the digestive tract by their ducts and describe their roles in digestion.
- Compare mucous and serous secretory cells in terms of their structure, staining, and secretions.
- Distinguish between the major salivary glands based on the content and distribution of serous and mucous cells.
- Relate the ultrastructure of the pancreatic acinar cell to its function.
- Describe the liver's double blood supply.
- Relate the hepatocyte's complex ultrastructure to its many functions.
- Describe the boundaries and contents of the classic liver lobule, the portal lobule, and the hepatic acinus. Understand the functions that gave rise to these overlapping views of liver organization.
- Describe the structure, function, and location of the gallbladder.
- Identify the gallbladder in a micrograph and distinguish it from the small intestine.
- Identify and distinguish among digestive glands in micrographs; distinguish between adenomeres and ducts; and identify the different types of associated ducts, cells, and other substructures.

MAX-Yield™ STUDY QUESTIONS

1. Compare serous cells with mucous cells in terms of their secretory product, staining properties, and the type of alveoli they typically comprise (acinar or tubular) (II.B.1[1]).
2. Name the three types of paired salivary glands (II.C–E) and compare them in terms of:
   a. Contribution to salivary volume
   b. Proportion of serous and mucous cells found in each gland
   c. Presence of serous demilunes
   d. Composition of their secretions
   e. Presence of striated ducts (I.F.1.b)
3. Describe the function of striated ducts (I.F.1.b).
4. Compare the saliva produced in response to sympathetic versus parasympathetic stimulation (II.G).
5. Name the endocrine component of the pancreas (III.A).
6. Sketch a cross-section of a pancreatic acinus (Fig. 16–1) and label the following:
   a. Acinar cells
   b. Nucleus
   c. RER
   d. Golgi complex
   e. Zymogen granules
   f. Basal lamina
   g. Lumen
   h. Centroacinar cell
   i. Intercalated duct
7. Describe two types of pancreatic exocrine secretion (III.B.1 and 2) in terms of their composition and role in digestion, the cells primarily responsible for their secretion, and the enteroendocrine hormone that stimulates their release.
8. Into which digestive tract segment are pancreatic exocrine secretions delivered (III.A)?

---

[1] See footnote on page 1.

9. Compare the exocrine pancreas with the parotid gland in terms of the chief secretory cell type (serous or mucous) (I.D), and the presence of centroacinar cells (III.B.2), islets of Langerhans (III.A), and striated intralobular ducts (I.F.1.b).

10. Name the two vessels that provide the liver's blood supply (IV.A, C.1 and 2) and compare the blood they carry in terms of its origin (the vessels giving rise to those entering the liver), its contribution (%) to liver blood volume, and its oxygen, nutrient, and bilirubin content.

11. In which vessels in the liver does the dual blood supply first become mixed (IV.C.3)?

12. Sketch a cross-section of three adjacent classic liver lobules and include and label the following (Figs. 16–2 and 16–3):

   a. Portal triads
   b. Branches of the hepatic artery
   c. Branches of the hepatic portal vein
   d. Bile ducts
   e. Central veins
   f. Hepatic sinusoids
   g. Space(s) of Disse
   h. Hepatocytes
   i. Endothelial cells
   j. Kupffer cells
   k. A portal lobule (outline)
   l. A hepatic acinus (outline)
   m. The direction of (use arrows):
      (1) Blood flow
      (2) Bile flow
      (3) Lymph flow

13. Name the three principal components of a portal triad (IV.E.1.a). Which has the largest lumen?

14. Name the two major cell types that border the hepatic sinusoids (IV.C.3).

15. Name the three major cell types that border the space of Disse (IV.C.3).

16. Name the cells that border the bile canaliculi (IV.D.1.a).

17. Name several proteins synthesized and secreted by hepatocytes (IV.D.1.b). Are these endocrine or exocrine products?

18. Name the hepatocyte organelles or inclusions involved in the following functions:

   a. Protein synthesis and secretion (2.III.C.1.a and b)
   b. Drug detoxification and inactivation (IV.D.1.b)
   c. Synthesis of bilirubin glucuronide (IV.F.1)
   d. Synthesis of bile acids (IV.F.1)
   e. Storage of lipid (2.III.H)
   f. Storage of glycogen (2.III.H)

19. Name the liver's exocrine secretory product; describe its composition and function (IV.D.1.b and F.1).

20. List, in order, the named structures through which bile flows after secretion by the hepatocytes and name the segment of the digestive tract into which it is delivered (IV.F.2).

21. How is the flow of bile into the digestive tract regulated (IV.F.2)?

22. List the functions of the gallbladder (V).

23. Compare the walls of the gallbladder with those of the small intestine in terms of the lining epithelium, mucosal branching, submucosa, and fiber orientation in the muscularis (V.A–C).

24. Describe the process by which bile fills the gallbladder (IV.F.2).

25. Name the enteroendocrine hormone that stimulates the contraction of smooth muscle in the gallbladder wall, thereby ejecting bile into the duct system (V).

SYNOPSIS

I. GENERAL FEATURES OF THE GLANDS ASSOCIATED WITH THE DIGESTIVE TRACT

   A. Components of the System: The **salivary glands** (parotid, submandibular, and sublingual), **pancreas,** and **liver** are digestive glands situated outside the digestive tract. The main function of the gallbladder, also discussed in this chapter, is to store bile.

   B. Embryonic Origin and Association with the Tract: Each component arises as an outpocketing of the embryonic gut tube and retains its connection with the tract's lumen through a duct. The duct-lining cells and exocrine secretory cells are epithelial and of endodermal origin. The connective tissue that forms the organ capsules and divides the glands into lobes and lobules derives from mesoderm.

C. Exocrine and Endocrine Functions: The presence of ducts indicates that these are exocrine glands; however, the pancreas and liver also have important endocrine functions.

D. Serous and Mucous Exocrine Secretory Cells: In general, the secretory cells of the liver, pancreas, and parotid gland are exclusively serous, whereas the submandibular and sublingual glands contain a mixture of serous and mucous cells (4.VI.C.3 and 4).

E. Glandular Subunits: Exocrine glands are structurally and functionally subdivided by septa, platelike invaginations of their connective tissue capsules. This arrangement applies mainly to the pancreas and salivary glands; the more complex subdivision of the liver is considered separately (IV.E).
1. **Lobes** are the largest subunits and are separated by connective tissue septa.
2. **Lobules** are subunits of lobes and are separated by thin extensions of the septa.
3. **Adenomeres** are secretory subunits of lobules. They consist of all of the secretory cells that release their products into a single intralobular duct.
4. **Acini (or alveoli)** are smaller secretory subunits. Each acinus is a spheric collection of secretory cells surrounding the blind-ended termination of a single intercalated duct (Fig. 16–1). An adenomere may include one or more acini.

F. Exocrine Ducts: Digestive gland ducts (see Fig. 16–1) are classified by their location and identified by their location, size, and epithelial lining.
1. **Intralobular ducts.** Several small ducts may occur in each lobule. They are usually surrounded by a thin layer of connective tissue and lined by simple cuboidal epithelial cells with central or basal nuclei. They transport the secretions from the adenomeres to the interlobular excretory ducts. Two intralobular ducts deserve special mention.
   a. **Intercalated ducts** are the smallest ducts. Their narrow lumen is continuous with that of the acini. They typically have a simple squamous or low cuboidal epithelial lining. They transport secretions from the acini to larger intralobular ducts.
   b. **Striated ducts** differ from standard intralobular ducts in the appearance of their lining epithelial cells. The nucleus is displaced toward the cell apex by extensive basal plasma membrane infoldings, giving the cell's base a radially striped (striated) appearance at

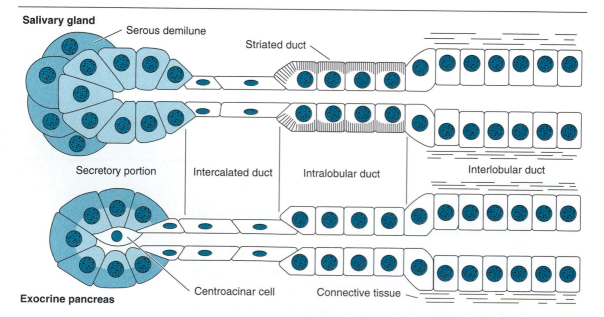

Figure 16–1. Schematic diagram of the secretory units and ducts of the salivary gland and exocrine pancreas. The salivary gland depicted has mucous alveolus with a serous demilune. Other salivary gland secretory acini may be serous, resembling the pancreatic acinus shown, but without the centroacinar cell. Note that the intralobular ducts of salivary glands have striations in the basal region of their lining cells that displace their nuclei apically.

high magnification. In EMs, these infoldings are seen to interdigitate with numerous mitochondria. This arrangement is characteristic of sites of energy-dependent ion transport activity, such as that performed by Na^+/K^+-ATPase (4.VI.A.1). Striated ducts are typical of salivary glands but not of the pancreas.

2. **Interlobular ducts,** located in the septa between the lobes and lobules, are larger than the intralobular ducts and are quickly distinguished by the greater amount of connective tissue surrounding them. Their lining is typically simple cuboidal to simple columnar to stratified columnar epithelium. Generally, larger ducts have a taller epithelial lining. These ducts include the large ducts within the glands and the still larger excretory ducts that exit the glands. They transport secretory products from the intralobular ducts to the lumen of the digestive tract.

II. SALIVARY GLANDS

A. **General Structure and Function:** Three major pairs of glands—the **parotid, submandibular,** and **sublingual**—surround the oral cavity. Each lobule contains many adenomeres that empty their secretions (saliva) into the oral cavity through a series of intercalated, striated, and interlobular ducts. Saliva moistens the food, lubricates the digestive tract, and begins the enzymatic digestion of carbohydrates. The glands also excrete certain salts and protect against bacterial invasion by releasing **lysozyme** and **IgA** into the saliva.

B. **Cell Types:**
1. **Serous and mucous cells** are the predominant secretory cells of salivary adenomeres. The key to identifying the three major salivary glands in section lies in knowing the different staining properties of these cells, their organization, and the proportion of each type found in each gland. Serous cells are relatively small and more basophilic. They produce a protein-rich, watery secretion and usually form acinar (spheric) alveoli. Mucous cells are larger, more acidophilic, and may have a foamy appearance. They produce a thick glycosaminoglycan-rich secretion (mucus) and typically form tubular alveoli.
2. **Myoepithelial cells** are contractile cells between the basal lamina and the epithelial cells of acini and ducts. Those around serous acini are called **basket cells,** owing to their stellate shape and long cell processes embracing the acinus. Myoepithelial cells of ducts are spindle-shaped and oriented lengthwise. Both types contain abundant actin microfilaments and myosin and help propel secretory products toward the oral cavity.
3. **Other cells.** IgA-secreting plasma cells and other cells typically found in areolar tissue are scattered in the connective tissue around the adenomeres.

C. **Parotid Glands:** These branched acinar glands contain almost exclusively serous secretory cells. The granules in these cells are PAS-positive (owing to their polysaccharide content) and are rich in protein. Parotid secretions, approximately 25% of the total salivary volume, contain amylase, maltase, sialomucin, and enzyme-resistant secretory IgA (14.II.B.2).

D. **Submandibular (Submaxillary) Glands:** These branched tubuloacinar glands, which produce approximately 70% of the salivary volume, contain both serous and mucous (mostly serous) adenomeres. The serous acini consist of small basophilic cells with PAS-positive cytoplasm and basal membrane infoldings. Their serous secretions contain sialomucin and have weak amylase activity. The mucous adenomeres may be capped by **serous demilunes** (see Fig. 16–1) composed of several **lysozyme**-secreting serous cells.

E. **Sublingual Glands:** These branched tubuloalveolar glands contain both mucous and serous cells (mostly mucous). Only mucous adenomeres are present, but many are capped by serous demilunes. These glands produce approximately 5% of the salivary volume.

F. **Modulation of Salivary Composition by Duct Epithelium:** The saliva produced by the secretory cells (**primary saliva**) is isosmotic with blood; saliva reaching the oral cavity, however, is typically hypotonic. The ion-transporting cells lining the striated and excretory ducts absorb sodium from, and add potassium and water to, the saliva. Like the cells lining the kidney's distal tubules (19.II.B.4), they respond to aldosterone and help regulate electrolyte balance.

G. Modulation of Salivary Volume by Autonomic Innervation: **Parasympathetic** stimulation (eg, sitting down to a meal) provokes the secretion of copious watery saliva containing less than average amounts of organic material. **Sympathetic** stimulation (eg, fear and stress) yields low-volume viscous saliva that is rich in organic material (causing what is sometimes called "cotton mouth").

III. PANCREAS

A. General Structure and Function: The pancreas is a serous, compound-acinar gland. Microscopically, it resembles the parotid gland, except that it lacks striated ducts (see Fig. 16–1) and contains islets of Langerhans. (These endocrine cell clusters [21.III] constitute the endocrine pancreas; only the exocrine pancreas is discussed here.) The pancreatic lobules contain serous adenomeres that secrete a variety of digestive enzymes into a branched duct system that empties into the duodenum.

B. Cell Types:

1. **Pancreatic acinar cells.** Each acinus consists of several pyramid-shaped, enzyme-secreting cells whose apices border on a small lumen and whose bases abut a basal lamina. The base of each cell is basophilic, owing to the ribosomes of the enzyme-synthesizing RER. Acinar cells synthesize myriad enzymes that can hydrolyze proteins (proteases such as trypsin, chymotrypsin, and elastase), lipids (lipases such as triacylglycerol lipase and phospholipase A_2), carbohydrates (amylase), and nucleic acids (RNase and DNase). Nascent enzymes are packaged and concentrated by the juxtanuclear Golgi complex and stored in the acidophilic apical region as membrane-bound **zymogen granules.** Here they await exocytosis upon stimulation by **cholecystokinin,** which is produced by the small intestine's enteroendocrine cells, or by parasympathetic stimulation from the vagus nerve. The enzymes in the granules (zymogens) are inactive before release. One such zymogen, trypsinogen, is enzymatically converted to the active protease trypsin in the small intestine by enterokinase, an enzyme secreted by enterocytes.

2. **Centroacinar cells.** Unique to the exocrine pancreas, each centroacinar cell has a condensed nucleus and a clear cytoplasm. One or two may be seen in the lumen of each acinus at the origin of the intercalated duct (see Fig. 16–1). These and other duct-lining cells produce a watery, bicarbonate-rich fluid in response to **secretin,** another enteroendocrine product of the small intestine's mucosa. This fluid helps to adjust the acidic chyme (15.I.C.1) to neutral pH, optimizing pancreatic enzyme activity.

IV. LIVER

A. General Structure: The liver is the body's largest gland. It is partly covered by a thin capsule **(Glisson's capsule)** and has a sparse, delicate, reticular connective tissue stroma accompanying the blood vessels as they penetrate the parenchyma. Its predominant cell type is the **hepatocyte.** These cells are arranged in one- or two-cell-thick plates that are separated by the hepatic sinusoids (Fig. 16–2). The liver has a dual blood supply: it receives blood from both the portal vein and the hepatic artery. It also has three drainage systems: the hepatic veins, lymphatic vessels, and bile ducts.

B. General Functions: The liver has several important functions, most of which are carried out by hepatocytes. Its primary role in digestion involves the enzymatic processing (metabolism) of nutrients absorbed by the intestines to provide the chemical building blocks and fuel needed to support life. Some hepatocyte enzymes modify and detoxify potentially dangerous chemicals and drugs. Hepatocytes synthesize many important proteins (eg, albumin, prothrombin, fibrinogen, lipoproteins) and secrete them into the blood, which is by definition an **endocrine** function. Moreover, they synthesize bile from the wastes of erythrocyte destruction and secrete it into the biliary tract (IV.F.2), which is an **exocrine** function. The liver also stores glucose, fats, and vitamin A. Because the hepatocytes carry out most of the liver's functions, knowledge of their structure and functions (detailed in IV.D.1.a and b) is a prerequisite for understanding liver function.

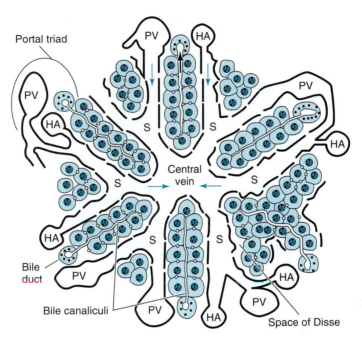

Figure 16–2. Schematic diagram of a classic liver lobule. Branches of the hepatic artery (HA) and hepatic portal vein (PV) empty blood into hepatic sinusoids (S), through which it flows toward the central vein. The endothelial lining of the sinusoids is discontinuous and is separated from the radial plates of hepatocytes by the space of Disse. Bile canaliculi receive bile from the hepatocytes that border them and convey it toward the bile ducts in the portal triads. The arrows show that blood and bile flow in opposite directions.

C. **Blood Supply:** The liver's complex structure is organized around its many vessels and sinusoids. Thus, it is best to begin detailed study with an examination of the liver's dual blood supply.

1. **Hepatic portal vein.** This large vein is formed by the junction of mesenteric and splenic veins. Mesenteric veins deliver oxygen-poor, nutrient-rich blood from capillaries in the intestinal walls. The splenic vein delivers by-products of red blood cell destruction from the splenic sinusoids (14.VIII.B.3). The hepatic portal vein supplies approximately 75% of the liver's blood volume. Entering through the **hilum** on the liver's inferior surface, it branches repeatedly to form the **portal venules** that empty their blood into the hepatic sinusoids.

2. **Hepatic artery.** This smaller vessel, a branch of the celiac artery, enters the liver alongside the portal vein. It follows the latter's branching pattern and empties oxygen-rich blood into the same sinusoids. It supplies approximately 25% of the liver's blood volume.

3. **Hepatic sinusoids.** Serving as the liver's blood capillaries, these lie between the radially oriented hepatocyte plates and receive blood from branches of both the portal vein and the hepatic artery (see Fig. 16–2). The mixed arterial and venous blood flows through the sinusoids into the central veins. The sinusoid lumen is separated from the free surface of the hepatocyte plates by a discontinuous wall composed of typical endothelial cells and **Kupffer cells.** Between the endothelium and the hepatocytes is the narrow **space of Disse.** Blood plasma enters this space through openings between the endothelial cells that are too small for blood cells to pass. Blood-borne substances thus directly contact the microvillus-covered hepatocyte surface. The cells absorb nutrients, oxygen, and toxins from, and release endocrine secretions into, these spaces. Fluid in the spaces of Disse flows toward lymphatic vessels in the portal triads (ie, in a direction opposite the blood flow). Thus, the spaces of Disse also serve as the liver's lymphatic capillaries.

4. **Central veins.** So named because each lies at the center of a classic liver lobule (IV.E.1), these veins receive blood from the sinusoids and deliver it to larger **sublobular veins,** which merge to form even larger hepatic veins. Central veins can be distinguished from other hepatic vessels by their position at the hub of the radially arranged hepatocyte plates and sinusoids and by the paucity of surrounding connective tissue (see Fig. 16–2).

5. Hepatic veins. These collect oxygen- and nutrient-poor blood from the sublobular veins. They converge to form larger veins that exit the liver's upper surface and empty into the inferior vena cava.

D. **Cell Types:**
   1. **Hepatocytes.** These are the primary structural and functional subunits of the liver.
      a. **Structure and organization.** Hepatocytes can have one or two nuclei; they contain every type of membrane-bound organelle (2.III), as well as glycogen granules and lipid droplets. Organized into one- or two-cell-thick plates that are separated by sinusoids, the polygonal hepatocytes are cubelike, with six major surfaces. Four of these typically abut on similar surfaces of adjacent hepatocytes. The abutting hepatocyte plasma membranes are attached by desmosomes and are closely apposed, except where they form the walls of a **bile canaliculus.** These small, tubular, plasma membrane-bound gaps between adjacent hepatocytes receive bile (IV.F.1). They are sealed with continuous occluding junctions to prevent bile leakage into the sinusoids. Short microvilli project into each canaliculus. The other two hepatocyte surfaces generally face the sinusoids on either side of the plate and are covered with short microvilli that project into the space of Disse.
      b. **Function.** The metabolism of absorbed nutrients may involve their further degradation, storage of excess as glycogen granules or lipid droplets, or the use of one type of nutrient to synthesize another (eg, amino acids to make glucose). Unlike most endocrine glands, the liver's endocrine functions are not limited to hormone (eg, somatomedin) secretion, but include the release of processed nutrients (eg, glucose, lipoproteins) and the production of plasma proteins (eg, albumin, fibrinogen). The liver's main exocrine function is **bile** production. Metabolic wastes, by-products of red blood cell destruction, and toxic substances removed from the blood are enzymatically inactivated (detoxified) in the SER and released into the bile canaliculi.
   2. **Kupffer cells.** Monocyte-derived members of the mononuclear phagocyte system, these are interspersed among the sinusoidal endothelial cells and on their luminal surfaces. They contain ovoid nuclei, many mitochondria, a well-developed Golgi complex, scattered lysosomes, phagosomes, and RER. They are more easily distinguished from the endothelial cells in standard H & E preparations when they have phagocytosed colored particles (eg, India ink) prior to fixation.
   3. **Fat-storing (Ito) cells.** These stellate cells lie in the space of Disse. They accumulate fat and store vitamin A as retinyl esters (eg, retinyl acetate).

E. **Liver Lobules:** The relationship between hepatic structure and function is best demonstrated by three models of liver substructure: the classic lobule, the portal lobule, and the hepatic acinus of Rappaport (Fig. 16–3).
   1. **Classic liver lobule.** This model is based on the direction of blood flow. In sections, liver substructure exhibits a pattern of interlocking hexagons; each of these is a classic lobule. Pig lobules are bounded by a connective tissue sheath. Human lobule boundaries are indistinct, but can be estimated by noting the portal triads at the lobule periphery, the central vein at its center, and the alternating hepatocyte plates and sinusoids between them.
      a. **Portal triad.** One triad occupies a potential space (**portal space**) at each of the six corners of the lobule. Each triad contains three main elements surrounded by connective tissue: a **portal venule** (a branch of the portal vein), a **hepatic arteriole** (a branch of the hepatic artery), and a **bile ductule** (tributary of a bile duct). A lymphatic vessel also may be seen.
      b. **Central vein.** A single vein marks the center of each lobule. It is easily distinguished from those in the portal triad by its lack of a connective tissue sheath.
      c. **Hepatocyte plates and hepatic sinusoids.** Many such plates radiate from the central vein toward the lobule periphery (like the spokes of a wheel). The plates are separated by hepatic sinusoids, which receive blood from the vessels in the triads, converging on the lobule center to empty directly into the central vein.
   2. **Portal lobule.** This model is based mainly on the direction of bile flow, which is counter to that of blood. From this perspective, the liver parenchyma is divided into interlocking triangles, each with a portal triad at its center and a central vein at each of its three corners. Bile produced by the hepatocytes enters the membrane-bound bile canaliculi between them and

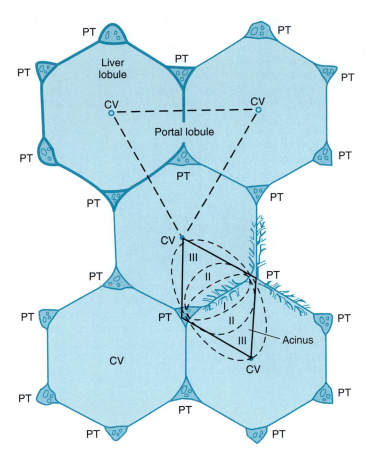

Figure 16–3. Schematic drawing of the territories of the classic liver lobules, hepatic acini, and portal lobules. The classic lobule has a central vein (CV) and is outlined by lines that connect the portal triads (PT; *solid lines*). The centers of the portal lobules are located in the portal triads and are outlined by lines that connect the central veins (*upper triangle*). They constitute the portion of the liver from which bile flows toward a portal triad. The hepatic acinus is the region irrigated by one distributing vein (*diamond-shaped figure*). Zones of the hepatic acinus are indicated by roman numerals I, II, and III. (Revised, with permission, from Leeson TS, Leeson CR: *Histology,* 2nd ed. Philadelphia, PA: WB Saunders, 1970.)

flows within the hepatocyte plates toward the bile duct in the portal triad. Liver lymph in the spaces of Disse flows in the same direction as bile, toward lymphatic vessels in the triad.

    3. Hepatic acinus (of Rappaport). This model is more abstract; it is based on changes in oxygen, nutrient, and toxin content as blood flowing through the sinusoids is acted on by hepatocytes. Each diamond-shaped acinus contains two central veins and two portal triads that define its four corners. The diamond is divided into two triangles by a line connecting the portal triads. Along this line run terminal branches of the portal and hepatic vessels that deliver blood to the sinusoids. Each triangle is divisible into three zones, according to their distance from the terminal distributing vessels. Zone I, for example, is closer to these vessels, whereas zone III is closer to the central vein. Blood in zone-I sinusoids has higher oxygen, nutrient, and toxin concentrations than in the other zones. As the blood flows toward the central vein, these substances are gradually removed by hepatocytes. Zone-I hepatocytes thus have a higher metabolic rate and larger glycogen and lipid stores. They are also more susceptible to damage by blood-borne toxins, and their energy stores are the first to be depleted during fasting. This model helps explain regional histopathologic differences in patients with liver damage.

F. Biliary System: Bile synthesis and secretion are the liver's main exocrine functions.
    1. Bile. Bile consists of bile acids, phospholipids, cholesterol, bilirubin, water, and elec-

trolytes. This composition gives it detergent properties that aid in digesting dietary fat. Approximately 90% of bile comes from recycled substances that were added to the intestinal contents in the duodenum and subsequently reabsorbed into the portal circulation by the epithelial lining of the distal part of the intestine. Hepatocytes merely reabsorb them from the sinusoids and transport them back to the bile canaliculi. Approximately 10% of bile is synthesized de novo in the hepatocyte's SER.

    a. Bile acids. Cholic acid is synthesized from cholesterol and conjugated with glycine or taurine to form glycocholic and taurocholic acid, respectively.

    b. Bilirubin is a water-insoluble by-product of the hemoglobin catabolism that accompanies the disposal of worn erythrocytes by cells of the mononuclear phagocyte system in the spleen, liver, and bone marrow (13.III.A.3; 14.VIII.B.3). It is carried by the blood to the hepatocytes, which conjugate it with glucuronic acid to form **bilirubin glucuronide.** This now water-soluble substance is secreted, with other bile components, into the bile canaliculi.

  2. Biliary tract. Bile in the canaliculi flows toward the portal triads (ie, counter to the blood flow in the sinusoids). At the lobule periphery, the canaliculi empty into short, narrow **bile ductules** (also called **cholangioles,** or **Hering's canals**), which are lined by cuboidal cells with clear cytoplasm. The ductules deliver the bile to **bile ducts** in the portal triads (in cross-section, the nuclei of the duct-lining cells resemble strings of beads). The bile ducts empty into successively larger ducts, ending in a single **hepatic duct** that joins the cystic duct from the gallbladder to form the **common bile duct (ductus choledochus).** This common duct empties the bile into the duodenum. Where the common bile duct penetrates the duodenal wall, it is encircled by a thick layer of smooth muscle, the **sphincter of Oddi.** Although the liver produces bile continuously, the sphincter opens fully only when a particularly fatty meal enters the duodenum. When the sphincter is closed, bile backs up the common duct, through the cystic duct, and into the gallbladder.

V. GALLBLADDER

The gallbladder **(cholecyst)** is a small, blind-ending sac attached to the liver's underside. It stores and concentrates bile and releases it in response to **cholecystokinin.** Its layered walls resemble the digestive tract, although it characteristically lacks a submucosa.

A. Mucosa: This layer consists of simple columnar epithelial cells with abundant apical microvilli overlying a typical lamina propria. The cells secrete mucus, and a sodium pump in their basal membranes facilitates water absorption from stored bile. The many mucosal folds are branched, and thus differ from intestinal villi. Near the cystic duct, the mucosa invaginates deeply into the lamina propria and even into the underlying muscularis. These invaginations form glands with large lumens whose continuity with the principal lumen of the organ may not be apparent in cross-section. These large sinuses also help to distinguish gallbladder from intestines. The cells lining these sinuses contribute most of the mucus to the stored bile.

B. Muscularis: This is a layer of interwoven smooth muscle fibers under the lamina propria. It contracts and empties the gallbladder in response to cholecystokinin released by enteroendocrine cells in the intestinal mucosa when dietary fat enters the intestinal lumen.

C. Adventitia and Serosa: As for all retroperitoneal organs, the gallbladder's outer layer consists of both an adventitia that attaches it to the liver and a serosa that covers its free (peritoneal) surface.

MULTIPLE-CHOICE QUESTIONS

Select the single best answer.

16.1. Which of the following is the type of gland shown in Figure 16–4?
(A) Liver
(B) Pancreas
(C) Parotid gland
(D) Sublingual gland
(E) Submandibular gland

16.2. Which of the following is the type of gland shown in Figure 16–5?
(A) Liver
(B) Pancreas
(C) Parotid gland
(D) Sublingual gland
(E) Submandibular gland

16.3. Which of the following is the type of gland shown in Figure 16–6?
(A) Liver
(B) Pancreas
(C) Parotid gland
(D) Sublingual gland
(E) Submandibular gland

16.4. Which of the following components of saliva initiates the digestion of polysaccharides?
(A) Amylase
(B) Lysozyme

(C) Potassium
(D) Secretory IgA
(E) Sialomucin

16.5. Which of the following is true of serous demilunes?
(A) Produce IgA
(B) Border directly on salivary intercalated ducts
(C) Produce mucinogen
(D) Produce lysozyme
(E) Occur only in parotid glands

16.6. Which of the following is true of myoepithelial cells?
(A) Are usually found outside the acinar basal lamina
(B) Cannot contract
(C) Contain actin and myosin in their cytoplasm
(D) Are found around the bile canaliculi
(E) Are responsible for emptying the gallbladder

16.7. Which of the following is typically found in the connective tissue of a portal triad?
(A) Bile canaliculus
(B) Branch of the central vein
(C) Branch of the hepatic artery
(D) Hepatic acinus
(E) Lymphatic capillary

16.8. Which of the following is true of Kupffer cells?
(A) Are nonphagocytic
(B) Contain lysosomes
(C) Derive from circulating neutrophils
(D) Resemble hepatocytes in H & E–stained sections
(E) Border on the bile canaliculi

16.9. Which of the following is true of the gallbladder?
(A) Dilutes bile
(B) Absorbs bile
(C) Secretes mucus
(D) Has a thick submucosa
(E) Is covered entirely by serosa

16.10. Which of the following is true of the cells lining striated ducts?
(A) Have basal plasma membranes containing numerous infoldings
(B) Are characterized by a paucity of mitochondria
(C) Are characterized by basally located nuclei
(D) Are characterized by abundant, long apical microvilli
(E) Are surrounded by striated muscle fibers

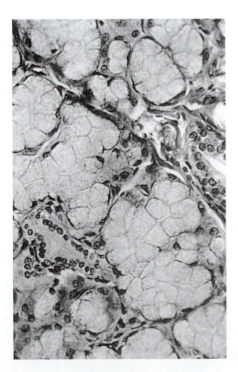

Figure 16–4.

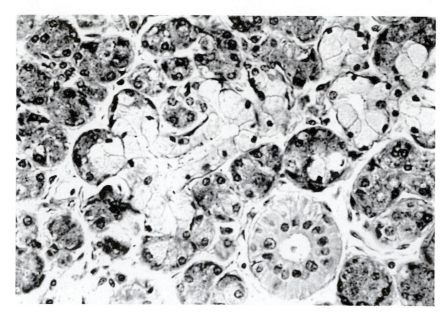

Figure 16–5.

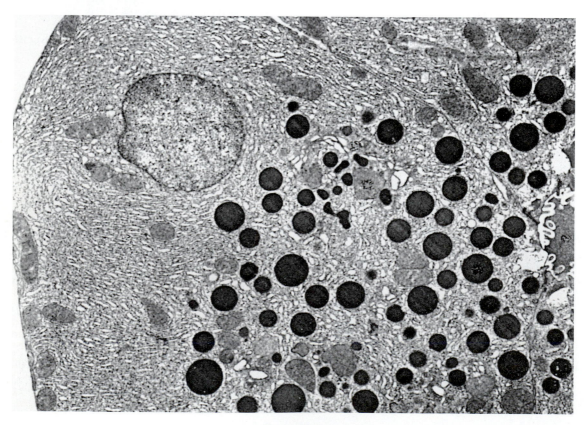

Figure 16–6.

16.11. Which of the following glands lacks serous acini but contains serous demilunes?
 (A) Liver
 (B) Parotid gland
 (C) Pancreas
 (D) Sublingual gland
 (E) Submandibular gland

16.12. Which of the following glands contains both mucous and serous adenomeres?
 (A) Liver
 (B) Parotid gland
 (C) Pancreas
 (D) Sublingual gland
 (E) Submandibular gland

16.13. Which of the following is true of the hepatic sinusoids?
 (A) Are bordered directly by hepatocytes
 (B) Contents flow toward the portal triads
 (C) Are surrounded by the space of Disse
 (D) Lumen is entirely sealed by junctional complexes
 (E) Contents empty into a cholangiole (canal of Hering)

16.14. Which of the following is true of the space of Disse?
 (A) Is bordered directly by hepatocytes
 (B) Contents flow toward the central vein
 (C) Is surrounded by the hepatic sinusoid
 (D) Lumen is entirely sealed by junctional complexes
 (E) Contents empty into a cholangiole (canal of Hering)

16.15. Which of the following is true of the bile canaliculi?
 (A) Are bordered directly by endothelial cells
 (B) Contents flow toward the central vein
 (C) Are surrounded by the hepatic sinusoids
 (D) Lumens are entirely sealed by junctional complexes
 (E) Normally contain blood plasma

16.16. Which of the following is a unique structural feature of the exocrine pancreas?
 (A) Centroacinar cells
 (B) Insulin-secreting β-cells
 (C) Predominantly mucous secretory cells
 (D) Predominantly serous secretory cells
 (E) Serous demilunes
 (F) Striated interlobular ducts
 (G) Striated intralobular ducts

16.17. Which of the following is true of the hepatic portal vein and its branches?
 (A) Carry oxygen-rich blood
 (B) Carry nutrient-poor blood
 (C) Carry blood rich in bilirubin
 (D) Receive blood from the central veins
 (E) Receive blood from the hepatic sinusoids
 (F) Receive blood from the cholangioles

16.18. Which of the following is true of cholecystokinin?
 (A) Stimulates the production of a watery, protein-poor, bicarbonate-rich secretion by pancreatic duct-lining cells
 (B) Stimulates the release of zymogens by pancreatic duct-lining cells
 (C) Stimulates relaxation of smooth muscle in the gallbladder
 (D) Is produced by enteroendocrine cells in the intestinal mucosa
 (E) Is produced by pancreatic duct-lining cells

16.19. Which of the following is true of pancreatic zymogens?
 (A) Are packaged for secretion in the SER
 (B) Are synthesized on free polyribosomes
 (C) Are inactive until they reach the duodenal lumen
 (D) Are stored in the basal cytoplasm of pancreatic acinar cells
 (E) Are produced by the pancreatic duct-lining cells

16.20. Which of the following glands is shown in Figure 16–7?
 (A) Sublingual gland
 (B) Submandibular gland
 (C) Pancreas
 (D) Parotid gland
 (E) Liver

Figure 16–7.

16.21. Which of the following is true about the embryonic origins of hepatocytes, pancreatic acinar cells, and cells lining the pancreatic ducts?
(**A**) All derive from endoderm
(**B**) All derive from neural crest
(**C**) All derive from mesoderm
(**D**) All derive from ectoderm
(**E**) Derive from different germ layers

16.22. Which of the following structures is located at the center of a classic liver lobule?
(**A**) Bile duct
(**B**) Branch of the hepatic artery
(**C**) Branch of the hepatic portal vein
(**D**) Central vein
(**E**) Cholangiole
(**F**) Portal triad
(**G**) Zone I

16.23. Which of the following is an endocrine secretory product of the liver?
(**A**) Albumin
(**B**) Bile acids
(**C**) Bilirubin glucuronide
(**D**) Cholecystokinin

16.24. Which of the following occurs in response to parasympathetic stimulation of the salivary glands?
(**A**) Secretory volume increases
(**B**) Cell division in secretory acini increases
(**C**) Secretory viscosity increases
(**D**) Organic content in saliva increases
(**E**) Cell division in interlobular ducts increases

ANSWERS TO MULTIPLE-CHOICE QUESTIONS

16.1. D (II.E; note that the adenomeres are almost exclusively mucous)
16.2. E (II.D; note the presence of both serous acini and mucous acini containing serous demilunes; a striated duct is also shown)
16.3. B (III.B.1; note the abundant RER in the basal cytoplasm and the zymogen granules in the apical region)
16.4. A (II.A and C–E)
16.5. D (II.D)
16.6. C (II.B.2)
16.7. C (IV.E.1.a)
16.8. B (IV.D.2)
16.9. C (V.A–C)
16.10. A (I.F.1.b)

16.11. D (II.E)
16.12. E (II.D)
16.13. C (IV.C.3)
16.14. A (IV.C.3)
16.15. D (V.D.1.a)
16.16. A (III.B.2)
16.17. C (IV.C.1 and F.1.b; 14.VIII.B.3)
16.18. D (III.B.1)
16.19. C (III.B.1)
16.20. C (III.A and B; Fig. 16–1)
16.21. A (I.B)
16.22. D (IV.E.1.b; Fig. 16–3)
16.23. A (IV.D.1.b)
16.24. A (II.G)

INTEGRATIVE MULTIPLE-CHOICE QUESTIONS: THE DIGESTIVE SYSTEM

Select the single best answer.

DS.1. Which of the following organs is shown in Figure DS–1?
(**A**) Colon
(**B**) Duodenum
(**C**) Fundic stomach
(**D**) Gallbladder
(**E**) Ileum
(**F**) Jejunum
(**G**) Pyloric stomach

DS.2. Which of the following organs is shown in Figure DS–2?
(**A**) Colon
(**B**) Duodenum
(**C**) Fundic stomach
(**D**) Gallbladder
(**E**) Ileum
(**F**) Jejunum
(**G**) Pyloric stomach

DS.3. Which of the following organs is shown in Figure DS–3?
(**A**) Colon
(**B**) Duodenum

(**C**) Fundic stomach
(**D**) Gallbladder
(**E**) Ileum
(**F**) Jejunum
(**G**) Pyloric stomach

DS.4. Which of the following cell types is known to exhibit both exocrine and endocrine secretory activity?
(**A**) Enterocyte
(**B**) Enteroendocrine
(**C**) Gastric chief
(**D**) Hepatocyte
(**E**) Myoepithelial
(**F**) Pancreatic acinar
(**G**) Parietal

DS.5. Which of the following organs has both adventitial and serosal coverings?
(**A**) Appendix
(**B**) Ascending colon
(**C**) Esophagus
(**D**) Ileum
(**E**) Jejunum
(**F**) Rectum
(**G**) Stomach

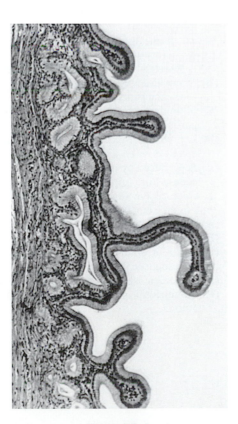

Figure DS–1.

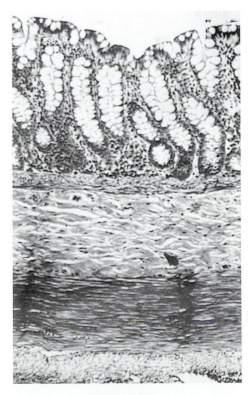

Figure DS–2.

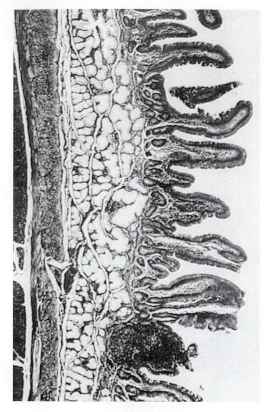

Figure DS–3.

DS.6. Which of the following cell types secretes lysozyme?
(A) Ameloblast
(B) Centroacinar
(C) Hepatocyte
(D) Kupffer
(E) M cell
(F) Odontoblast
(G) Paneth's cell

DS.7. Which of the following is the source of hepatic endothelial cells?
(A) Ectoderm (other than neural crest)
(B) Neural crest
(C) Mesoderm
(D) Endoderm

DS.8. Which of the following is the source of the epithelium lining the esophagus?
(A) Ectoderm (other than neural crest)
(B) Neural crest

(C) Mesoderm
(D) Endoderm

DS.9. Which of the following is the source of Kupffer cells?
(A) Ectoderm (other than neural crest)
(B) Neural crest
(C) Mesoderm
(D) Endoderm

DS.10. Which of the following is the source of the ganglion cells in the myenteric plexus?
(A) Ectoderm (other than neural crest)
(B) Neural crest
(C) Mesoderm
(D) Endoderm

DS.11. Which of the following is the source of goblet cells?
(A) Ectoderm (other than neural crest)
(B) Neural crest
(C) Mesoderm
(D) Endoderm

DS.12. Hepatocytes in which location are the first to be destroyed by hepatotoxic substances in the blood?
(A) Zone I of hepatic acinus
(B) Portal space
(C) Zone II of hepatic acinus
(D) Portal triad
(E) Zone III of hepatic acinus
(F) Portal lobule

DS.13. Which of the following is true of the hormone cholecystokinin?
(A) Is secreted by the gallbladder (cholecyst)
(B) Stimulates the secretory activity of cells lining the pancreatic ducts
(C) Is produced by the liver in response to a particularly fatty meal
(D) Is produced by cells derived from neural crest
(E) Is produced by gastric parietal cells in response to gastrin

DS.14. In which of the following locations are mitotic figures most likely to be found?
(A) Neck of the stomach's fundic glands
(B) Tips of the villi in the ileum
(C) Layer of epithelial cells in contact with the esophageal lumen
(D) Pancreatic islets of Langerhans
(E) Submucosal plexus in the wall of the colon

ANSWERS TO INTEGRATIVE MULTIPLE-CHOICE QUESTIONS

DS.1. D (16.V.A; note branched mucosal folds, sinuses, and lack of submucosa)

DS.2. A (15.VIII.A; note the abundant goblet cells and lack of villi)

DS.3. B (15.VII.A and C.1; note the villi and submucosal mucous [Brunner's] glands)

DS.4. D (16.IV.B)

DS.5. B (Table 15–1)

DS.6. G (15.VII.B.6)

DS.7. A (Table 4–1; 13.II.A.1)

DS.8. D (15.I.B.1)

DS.9. C (13.II.A.1 and B.1; 16.IV.D.2)

DS.10. B (9.I.E; 15.I.B.3)

DS.11. D (15.I.B.1)

DS.12. A (16.IV.E.3; Fig. 16–3)

DS.13. D (4.VI.C.2; 15.I.C.4 and VII.B.7)

DS.14. A (15.VI.B.2)

17

Respiratory System

OBJECTIVES

This chapter should help the student to:

- Name the three divisions of the respiratory system and the components of each.
- Compare the right and left lungs.
- Describe the respiratory tract walls in terms of the arrangement, composition, and function of their layers and cells.
- Distinguish between respiratory tract components based on differences in wall structure.
- Describe the structure of the interalveolar septum.
- Describe the blood–air barrier's structure and function. Identify its components in electron micrographs.
- Compare sympathetic and parasympathetic effects on bronchial smooth muscle.
- Describe the structure, function, and location of the pleura.
- Identify the organ, tissues, and cell types present and distinguish among the various components of the respiratory system in micrographs of respiratory tract or lung tissue.

MAX-Yield™ STUDY QUESTIONS

1. List the components of the ventilating mechanism involved in inhaling, exhaling, or both (I.A.1[1]).
2. Name the respiratory tree's two basic portions and the functions of each (I.A.2 and 3).
3. List, in order, the respiratory tract segments through which inspired air passes (I.A.2 and 3).
4. Name three ways in which inspired air is conditioned in the respiratory tract (en route to the alveoli) to optimize gaseous exchange (I.A.2.). Name the structure(s) associated with each type of conditioning (I.B.1.a; II.A and B).
5. Compare the right and left lungs in terms of the number of primary and secondary bronchi each receives (VII.A and B), the number of lobes in each (VII.B), and the angle at which the primary bronchi enter (VII.A).
6. Indicate whether the following components of the respiratory tract wall increase or decrease from the nose to the alveoli (I.B.1; Table 17–1):
   - a. Diameter of lumen
   - b. Thickness of walls
   - c. Height of epithelium
   - d. Number of cilia
   - e. Number of goblet cells
   - f. Number of glands
   - g. Amount of elastic tissue
   - h. Amount of smooth muscle
   - i. Amount of bone
   - j. Amount of cartilage
   - k. Size of individual cartilages
   - l. Number of alveoli
7. Indicate the level(s) of the respiratory tree (Table 17–1) at which the following transitions occur:
   - a. Conducting portion to the respiratory portion
   - b. Ciliated pseudostratified columnar to nonkeratinized stratified squamous (and back to ciliated pseudostratified columnar) epithelium
   - c. Ciliated pseudostratified columnar to simple ciliated columnar epithelium
   - d. Simple columnar to simple cuboidal epithelium
   - e. Simple cuboidal to simple squamous epithelium

---

[1] See footnote on page 1.

8. At which level(s) of the respiratory tree (Table 17–1) are the following initially lost?
   a. Goblet cells
   b. Cilia
   c. Glands
   d. Cartilage
   e. Smooth muscle
   f. Lymphatic capillaries (IX.B)

9. At which level(s) of the respiratory tree are the following found?
   a. Vibrissae (II.A)
   b. Swell bodies (II.B)
   c. Elastic cartilage (V.A & B)
   d. C-shaped cartilages (VI)
   e. Platelike cartilage islands (VII.B)
   f. Anastomoses (IX.A.2)
   g. Clara cells (VII.E)
   h. First appearance of alveoli (VII.F; Table 17–1)
   i. Pulmonary surfactant (VIII.C)
   j. Type-I cells (VIII.B.1)
   k. Type-II cells (VIII.B.2)

10. What is the function of each of the following in the respiratory system?
    a. Conchae (II.B)
    b. Vibrissae (II.A)
    c. Swell bodies (II.B)
    d. Epiglottis (V.A)
    e. Cilia (I.B.1.a)
    f. Small granule cell (I.B.1.a)
    g. Clara cells (VII.E)
    h. Alveolar pores (VIII.A.2)
    i. Pulmonary surfactant (VIII.C)
    j. Type-I cells (VIII.B.1)
    k. Type-II cells (VIII.B.2)
    l. Alveolar macrophages (VIII.B.3)

11. Sketch a tissue section through three alveoli (Fig. 17–2; VIII.A–C) and include and label the following:
    a. Alveolar sac
    b. Type-I cells
    c. Type-II cells
    d. Alveolar macrophages
    e. Alveolar septum
    f. Interstitium
    g. Alveolar pore
    h. Capillaries
    i. Endothelial cells
    j. Fused basal laminae
    k. Surfactant
    l. Blood–air barrier

12. Compare the effects of sympathetic and parasympathetic stimulation on bronchial and vascular smooth muscle (X).

13. Describe the pleura in terms of structure, function, and location (XI).

SYNOPSIS

I. GENERAL FEATURES OF THE RESPIRATORY SYSTEM

A. **Components and Basic Functions of the Respiratory System:** The respiratory system comprises the lungs, airways (ie, pharynx, larynx, trachea, bronchi), and associated structures. Specialized for gaseous exchange between blood and air, including oxygen uptake and carbon dioxide release, it is functionally divisible into three major parts: its conducting and respiratory portions and the ventilating mechanism.

1. **Ventilating mechanism.** This mechanism creates the pressure differences that move air into (inspiration) and out of (expiration) the lungs. It includes the **diaphragm, rib cage, intercostal muscles, abdominal muscles,** and the lungs' **elastic connective tissue.** Inspiration (inhalation) is active and involves muscle contraction. The intercostal muscles lift the ribs while the diaphragm and abdominal muscles lower the thoracic cavity floor. This enlarges the cavity, creating a vacuum that draws air in through the airways. The air expands the airways, inflates the lungs, and stretches the elastic connective tissue. Expiration (exhalation) is more passive: relaxing the muscles allows the elastic fibers to retract, contracting the lungs and forcing air out.

2. **Conducting portion.** The walls of this system of tubes are specialized to carry air to and from the site of gas exchange without collapsing under the pressures created by the ventilating mechanism. This portion also conditions the air; it warms, moistens, and cleans it to enhance gas exchange. The components of this portion include the **nasal cavity** (II), **nasopharynx** (IV), **larynx** (V), **trachea** (VI), **bronchi** (VII), **bronchioles** (VII.D), and **terminal bronchioles** (VII.E).

3. **Respiratory portion.** This portion is distinguished by the presence of **alveoli** (VIII), small saccular structures whose thin walls allow gas exchange between air and blood. Alveoli oc-

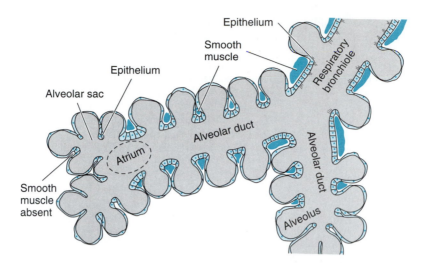

Figure 17–1. Schematic diagram of the respiratory portion of the respiratory tract. The major distinguishing feature of this portion of the tract is the presence of alveoli. Alveolar ducts can be distinguished from alveolar sacs by the presence of smooth muscle in the walls of the former and its absence in the walls of the latter. Details of alveoli and alveolar septa are shown in Figure 17–2.

cur in clusters at the end of the bronchial tree. These clusters extend (like rooms from a hallway; Fig. 17–1) from the walls of **respiratory bronchioles** (VII.F), **alveolar ducts** (VII.G), and **atria** and **alveolar sacs** (VII.H).

B. **Wall Structure:** Like the digestive tract, the respiratory tract has layered walls whose lining epithelium derives from endoderm. The wall layers include an epithelium, a lamina propria that contains both mucous glands and cartilage that prevents the tract from collapsing under pressure, smooth muscle that regulates the luminal diameter, and an adventitia that contains collagen and elastic fibers. Each layer undergoes gradual changes as the wall's overall thickness decreases from the nasal cavity to the alveoli (Table 17–1).

1. **Respiratory epithelium.** The epithelium that lines most of the tract is ciliated pseudostratified columnar with goblet cells. As the respiratory tract branches and its luminal diameter decreases, the epithelium drops in height and loses first goblet cells and then cilia as it approaches the alveoli.

   a. **Epithelial cell types. Ciliated columnar cells** predominate in the tract. Each has approximately 300 motile cilia (2.III.J; 4.IV.A.1) on its apical surface and associated basal bodies in the apical cytoplasm. **Mucous goblet cells** are the next most numerous. They secrete the mucus that covers the epithelium and traps and removes bacteria and other particles from inspired air. Cilia projecting from columnar cells sweep the contaminated mucus toward the mouth for disposal. **Brush cells** are columnar but lack cilia; they often have abundant apical microvilli. Two types are present: one resembles an immature cell and may replace dead ciliated or goblet cells; the other has basal nerve endings and appears to be a receptor. **Basal cells** are small round cells that lie on the basal lamina but do not reach the lumen. They may be stem cells that replace other cell types. **Small granule cells** resemble basal cells, but they contain many small cytoplasmic granules and exhibit DNES activity (4.VI.C.2).

   b. **Metaplasia** refers to changes in epithelial organization or type in response to changes in the physical or chemical environment. For example, a smoker's respiratory epithelium typically develops more goblet cells in response to high pollutant levels and fewer ciliated cells in response to carbon monoxide. These changes, which are reversible, frequently cause congestion of the smaller airways.

2. **Lamina propria.** Consisting of loose connective tissue, the lamina propria contains mucous glands in the upper tract (from the nasal cavity to the bronchi). Its elastic fiber content increases toward the alveoli. Skeletal connective tissue support begins as cartilage and bone in

Table 17–1. Distinguishing features of the respiratory tract components.

Tract Components			Wall Layers				
			Epithelium	Lamina Propria	Glands	Skeletal Connective Tissue	Muscle
Conducting portion (no alveoli)	Nasal cavity	Vestibule	Keratinized stratified squamous to nonkeratinized stratified squamous to respiratory epithelium	Hair follicles	Bowman's glands	Bone/cartilage	None
		Fossa		Venous sinuses			
	Nasopharynx		Respiratory epithelium	Lymphoid nodules	Mucous and serous glands (mostly mucous)	None	Skeletal
	Larynx		Respiratory epithelium and nonkeratinized stratified squamous over true vocal cords	Gradual decrease in thickness and increase in number of elastic fibers		Large hyaline cartilage rings and small elastic cartilages	Skeletal in vocalis muscle and around cartilages
	Trachea		Respiratory epithelium			C-shaped hyaline cartilage rings	Smooth in trachealis muscle
	Bronchi: primary, secondary, tertiary		Gradual decrease in height, cilia, and goblet cells		Complete rings in primary, plates of cartilage in smaller bronchi	Several layers of circular smooth muscle	
	Bronchioles		Simple columnar with cilia and goblet cells		None	None	Decreasing numbers of smooth muscle cells
	Terminal bronchioles		Simple cuboidal, some cells ciliated, goblet cells rare				
Respiratory portion (alveoli present)	Respiratory bronchioles		Low cuboidal, with few cilia; no goblet cells				Few smooth muscle cells
	Alveolar ducts and sacs		Some low cuboidal cells; no cilia.				
	Alveoli		Mostly simple squamous (type I); some low cuboidal (type II) in septa	Interstitium rich in capillaries and elastic fibers			No muscle

the nasal cavity, becomes cartilage (only) in the larynx, and subsequently decreases gradually, disappearing at the level of the bronchioles.

3. **Smooth muscle** begins in the trachea, where it joins the open ends of the C-shaped tracheal cartilages (VI). In the bronchi, many layers of smooth muscle encircle the walls in a spiral. The muscle layer's thickness gradually decreases until it disappears at the level of the alveolar ducts.

II. NASAL CAVITY

This cavity is divided by the **nasal septum** into two bilaterally symmetric cavities that open to the exterior through the **nares** (nostrils). Each consists of two chambers—a **vestibule** and a **nasal fossa**—which differ in position, size, and wall structure.

A. **Vestibule:** The smaller, wider, and more anterior chamber of each cavity, the vestibule lies just behind the nares. The medial septum and lateral walls are supported by cartilage, and the epithelial lining is a continuation of the epidermis (18.I.B.1) covering the nose. Just inside, the epithelium is keratinized; here, the lining contains many sebaceous and sweat glands, as well as thick short hairs called **vibrissae,** which filter large particles from inspired air. Deeper in the vestibule, the epithelium changes from keratinized to nonkeratinized stratified squamous and then to respiratory epithelium (I.B.1) just before entering the nasal fossa.

B. Nasal Fossa: This is the larger, narrower, and more posterior chamber in each cavity. Here the septum and lateral walls are lined by respiratory epithelium. They are supported by bone and contain mucous glands and venous sinuses in the lamina propria. Three curved bony shelves, termed **conchae,** or **turbinate bones,** project into each fossa from its lateral wall. These help warm and moisten the air by increasing the mucosal surface area and forming a system of baffles that cause turbulence and slow the air flow through the cavity. Alternating sides every 20 to 30 minutes, venous plexuses **(swell bodies)** in the conchal mucosa engorge with blood. The swelling restricts air flow, directing it through the other side, and thus helps to limit the drying of mucosal surfaces. Arterial vessels in the fossa walls create a countercurrent system that warms air by directing blood flow from posterior to anterior (opposite to the flow of inspired air) in a series of arches. Specialized olfactory epithelium (24.IV.B) is present in the roof of each fossa.

III. PARANASAL SINUSES

These are cavities in the frontal, maxillary, ethmoid, and sphenoid bones around the nose and eyes. Their thin respiratory epithelial lining has few goblet cells and binds tightly to the surrounding bones' periosteum by a lamina propria containing a few small mucous glands. Mucus produced here drains into the nasal fossa through small openings under the conchae.

IV. NASOPHARYNX

This upper part of the pharynx (15.IV) is a broad cavity overlying the soft palate. It is continuous anteriorly with the nasal fossae and inferiorly with the oral part of the pharynx **(oropharynx).** The walls, lined by respiratory epithelium, are supported by bone and skeletal muscle.

V. LARYNX

A bilaterally symmetric tube, the larynx lies in the neck between the base of the oropharynx and the trachea. During swallowing, its opening is covered by the epiglottis. Its walls, supported by several laryngeal cartilages in the lamina propria, contain skeletal muscle and house the vocal apparatus.

A. Epiglottis: This tissue flap extends toward the oropharynx from the front of the larynx. Its superior surface is covered by nonkeratinized stratified squamous epithelium and its inferior surface by respiratory epithelium. The lamina propria contains a few mucous glands and an elastic cartilage plate. During swallowing, the tongue's backward motion forces the epiglottis over the laryngeal opening, directing food away from the airway and into the esophagus. After swallowing, the elastic cartilage helps to reopen and maintain the airway.

B. Laryngeal Cartilages: Several cartilages frame the laryngeal lumen and serve as attachments for the skeletal muscles that control the vocal apparatus. The larger thyroid and cricoid and most of the paired arytenoid cartilages are hyaline, whereas the smaller ones (paired cuneiform and corniculate, epiglottic, and tips of the arytenoids) are elastic.

C. Vocal Apparatus: The broad part of the larynx, below the epiglottis and surrounded by the thyroid cartilage, contains two bilaterally symmetric pairs of mucosal folds. **False vocal cords (vestibular folds)** are the upper pair of laryngeal folds. They are covered by respiratory epithelium and contain glands whose ducts open mainly into the cleft separating them from the lower pair of folds, or **true vocal cords.** These are covered by stratified squamous epithelium. Each contains a large elastic fiber bundle running front to back, called the **vocal ligament,** and a parallel skeletal muscle bundle, called the **vocalis muscle.** Air forced through the larynx by the ventilating mechanism (I.A.1) vibrates the true cords. The vocalis muscle regulates cord tension. Other muscles control the shape and position of the laryngeal lumen. In this way, the laryngeal muscles control the pitch (frequency) and other aspects of the sounds produced by the vibrating cords. The cords also assist the epiglottis in preventing foreign objects from reaching the lungs; they close to build up pressure when coughing is required to dislodge materials blocking the airway.

VI. TRACHEA

This 10-cm tube extends from the larynx to the primary bronchi. It is lined by respiratory epithelium, and its lamina propria contains mixed seromucous glands that open onto its lumen. Its most characteristic features are the 16 to 20 C-shaped cartilage rings whose open ends are directed posteriorly. The opening is bridged by a fibroelastic ligament, which prevents overdistension, and by smooth muscle bundles **(trachealis muscle),** which constrict the lumen and increase the force of air flow during coughing, forced expiration, and speech.

VII. BRONCHIAL TREE

This tree begins where the trachea branches to form two primary bronchi, one of which penetrates the hilum of each lung. The hilum is also where arteries and nerves enter and veins and lymphatic vessels exit the organ. These structures, together with the dense connective tissue that binds them, form the **pulmonary root.** The bronchial tree branches extensively within the lungs. Changes in the bronchial tree's wall structure as it progresses toward the alveoli (see Table 17–1) occur gradually and not at sharp boundaries.

A. Primary Bronchi: One primary bronchus enters each lung. Each resembles the trachea, but the cartilage rings and spiral bands of smooth muscle completely encircle the lumen. The path of the right primary bronchus is more vertical than that of the left. As a result, foreign objects reaching the bronchial tree are more likely to lodge in the right side.

B. Secondary Bronchi: These **lobar bronchi** branch directly from the primary bronchi; each supplies one pulmonary lobe. Because the right lung has three lobes and the left only two, the right primary bronchus gives rise to three secondary bronchi and the left gives rise to two. They resemble the primary bronchi except that their supporting cartilages (and those of the smaller bronchi) form irregular plates, or islands, rather than rings.

C. Tertiary Bronchi: Arising directly from the secondary bronchi, each of these **segmental bronchi** supplies one **bronchopulmonary segment (pulmonary lobule).** Although each lung has 10 such segments, the different number of secondary bronchi causes the tertiary branching pattern to differ in the right and left lungs. Except for a decrease in overall diameter, tertiary bronchi resemble secondary bronchi. Tertiary bronchi may branch several times to form successively smaller branches, which are considered bronchi if their walls contain cartilage and glands.

D. Bronchioles: These are branches of the smallest bronchi, from which they differ only by lacking cartilage and glands in their walls. Large bronchioles are lined by typical respiratory epithelium; as they branch, their epithelial height and complexity decrease until they are simple ciliated columnar or cuboidal. Each bronchiole gives rise to five to seven terminal bronchioles.

E. Terminal Bronchioles: The smallest components of the conducting portion of the respiratory system, these are lined by ciliated cuboidal or columnar epithelium and have few or no goblet cells. (The elimination of goblet cells before cilia in the bronchial tree's lower reaches helps prevent individuals from drowning in mucus). The lining here also includes dome-shaped cilia-free **Clara cells,** whose cytoplasm contains glycogen granules, lateral and apical Golgi complexes, elongated mitochondria, and a few secretory granules. The function of these cells is unclear. Each terminal bronchiole can branch to form two or more respiratory bronchioles.

F. Respiratory Bronchioles: These make up the first part of the respiratory division. Their cuboidal epithelial lining resembles that of the terminal bronchioles but is interrupted by thin-walled saccular evaginations called **alveoli.** The number of alveoli increases as the respiratory bronchioles proceed distally. As the alveoli increase in number, the cilia decrease until they disappear. Goblet cells are absent.

G. Alveolar Ducts: These are the distal extensions of the respiratory bronchioles. Here the alveoli are so dense that the wall consists almost entirely of these sacs, and the lining has been reduced to small knobs of smooth muscle covered by cilia-free simple cuboidal cells. The knobs appear to project into the duct's elongated lumen, each resting atop a thin septum that separates adjacent alveoli. The alveolar duct thus resembles a hallway with so many doorways leading to small rooms (alveoli) that the hallway (alveolar duct) appears almost to lack walls (see Fig. 17–1).

H. Atria and Alveolar Sacs: Atria are the distal terminations of alveolar ducts (see Fig. 17–1). The arrangement is comparable to a hallway (alveolar duct) leading to a rounded foyer (atrium). The foyer has small doorways leading to some rooms (alveoli) but also has two or more larger doorways leading into shorter, dead-end hallways (alveolar sacs). The shorter hallways lead to more rooms (alveoli). Put simply, the difference between atria and alveolar sacs is that the atria open into alveolar ducts, alveoli, and alveolar sacs, whereas the alveolar sacs open only into alveoli and atria. Although these distinctions can be made fairly easily in sections cut longitudinally through the entire system of passageways, beginning with the alveolar duct, such perfect cuts are rare in slides of lung tissue. More often, the various components are cut in oblique or cross-section and only the openings to the alveoli are seen, making it hard to distinguish between the sacs and atria. In such cases, the best clue is the size of the knobs projecting into the passageways. Those projecting into alveolar sacs lack smooth muscle and are thus smaller than those projecting into either the atria or the alveolar ducts.

VIII. ALVEOLI

Found only in the respiratory portion (which their presence defines), these small (approximately 200-μm) sacs open into a respiratory bronchiole, an alveolar duct, an atrium, or an alveolar sac. They are separated by thin walls, termed interalveolar (or alveolar) septa (Fig. 17–2).

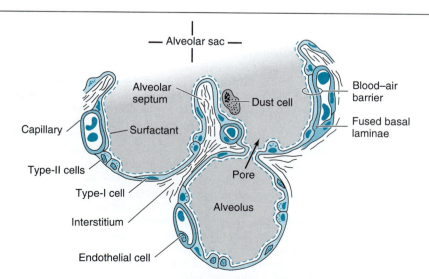

Figure 17–2. Schematic diagram of three alveoli and alveolar septa. The lower alveolus is shown in cross-section and communicates with another alveolar sac (*not shown*).

A. **Interalveolar Septa:** The structure of these septa, which are specialized for gas exchange, is critical to respiratory function. The septa consist of two simple squamous epithelial layers, with the **interstitium** sandwiched between them. The interstitium consists of continuous (nonfenestrated) capillaries embedded in an elastic connective tissue that comprises elastic and collagen fibers, ground substance, fibroblasts, mast cells, macrophages, leukocytes, and interstitial cells that contract in response to epinephrine and histamine. This elastic tissue is an important component of the ventilating mechanism. Gas exchange occurs between the air in the alveolar lumen and the blood in the interstitial capillaries.

1. **Blood–air barrier.** This term refers to the structures that oxygen and CO_2 must cross to be exchanged. Varying from 0.1 to 1.5 µm in thickness, it comprises four layers: (1) the film of pulmonary surfactant (VIII.C) on the alveolar surface; (2) the cytoplasm of the squamous epithelial (type-I alveolar) cells (VII.B.1); (3) the fused basal laminae between the type-I alveolar and capillary endothelial cells; and (4) the cytoplasm of the squamous endothelial cells lining the interstitial capillaries.

2. **Alveolar pores.** Each septum may be interrupted by one or more pores that vary from 10 to 15 µm in diameter. These connect adjacent alveoli, equalize pressure, and allow collateral air flow. They maximize the use of available alveoli when some small airways are blocked.

B. **Alveolar Cell Types:**

1. **Type-I cells.** Also called type-I alveolar cells, type-I pneumocytes, and squamous alveolar cells, these squamous epithelial cells make up 97% of the alveolar surfaces. They are specialized to serve as very thin (often only 25 nm in width) gas-permeable components of the blood–air barrier. Their organelles (eg, Golgi complex, endoplasmic reticulum, mitochondria) cluster around the nucleus. Much of the cytoplasm is thus unobstructed by organelles, except for the abundant small pinocytotic vesicles that assist in the turnover of pulmonary surfactant and the removal of small particles from the alveolar surfaces. Type-I cells attach to neighboring epithelial cells by desmosomes and occluding junctions. The latter limit pleural effusion—leakage of tissue fluid into the alveolar lumen. Type-I cells can be distinguished from the nearby capillary endothelial cells by their position bordering the alveolar lumen and by their slightly more rounded nuclei.

2. **Type-II cells.** These cells, which are also called type-II alveolar cells, type-II pneumocytes, great alveolar cells, and alveolar septal cells, cover the remaining 3% of the alveolar surface. They are interspersed among the type-I cells, to which they attach by desmosomes and occluding junctions. Type-II cells are roughly cuboidal with round nuclei; they often occur in small groups at the angles where alveolar septal walls converge. In EMs, they contain many mitochondria and a well-developed Golgi complex, but they are mainly characterized by the presence of large (0.2-µm), membrane-limited **lamellar (multilamellar) bodies.** These exhibit many closely apposed parallel membranes (lamellae) and contain phospholipids, glycosaminoglycans, and proteins. Type-II cells are secretory cells. Their secretory product, **pulmonary surfactant,** is assembled and stored in the lamellar bodies, which also carry it to the apical cytoplasm. There, the bodies fuse with the plasma membrane and release surfactant onto the alveolar surface.

3. **Alveolar macrophages (dust cells).** These large monocyte-derived representatives of the mononuclear phagocyte system lie both on the surface of alveolar septa and in the interstitium. Macrophages remove debris that escapes the mucus and cilia in the system's conducting portion. They also phagocytose blood cells entering the alveoli as a result of heart failure. Alveolar macrophages, when stained positively for iron pigment (hemosiderin), are thus designated "heart failure cells."

C. **Pulmonary Surfactant:** Continuously synthesized and secreted by type-II alveolar cells onto the alveolar surfaces, pulmonary surfactant is removed by alveolar macrophages and by type-I and type-II alveolar cells. Its composition and continuous turnover allow it to serve two major functions. It reduces surface tension in the alveoli and has some bactericidal effects, which include cleaning the alveolar surface and preventing bacterial invasion of the many septal capillaries. The surfactant forms a thin two-layered film over the entire alveolar surface. The film consists of an aqueous basal layer (hypophase) composed mainly of protein, which is covered by a monomolecular film of phospholipid (mainly **dipalmitoyl lecithin**) whose fatty acid tails extend into the lumen. By reducing surface tension, the surfactant helps prevent alveolar collapse during expiration. It thus eases breathing by decreasing the force required to reopen the alveoli during the next inspiration. Because surfactant secretion begins in the last weeks of fetal development, premature infants often suffer **respiratory distress syndrome,** evidenced by labored breathing caused by the lack of surfactant. Surfactant secretion can be induced by administering glucocorticoids, significantly improving the infant's condition and chances for survival.

D. Alveolar Lining Regeneration: Daily turnover of approximately 1% of the type-II cells, whose mitotic progeny form both type-I and type-II cells, allows for normal alveolar lining renewal. When these lining cells are destroyed by the inhalation of toxic gases, replacements for both types of cells are similarly derived from surviving type-II cells.

IX. PULMONARY CIRCULATION

A. Blood Supply: The lungs have a dual blood supply: the functional (pulmonary) circulation and the systemic (nutrient) circulation. The two systems enter separately but communicate through extensive anastomoses near the capillary beds.

1. **Functional circulation** is provided by the pulmonary arteries and veins. **Pulmonary arteries** arise from the heart's right ventricle as large-diameter elastic arteries. The pulmonary arteries branch and enter the lung at the pulmonary root (VII). They follow the bronchial tree's branching pattern to carry oxygen-poor blood to the lungs' capillary beds for oxygenation. Smaller branches (less than 1 mm in diameter) are of the muscular type, with a definitive internal elastic lamina. Pulmonary arteries have a thin intima and thinner media than do other arteries of equal size. **Pulmonary veins** collect oxygenated blood from the lungs' capillaries and return it to the heart's left atrium for distribution through the aorta and its branches. The larger pulmonary vein branches accompany the bronchi. Smaller branches travel unaccompanied in the connective tissue septa separating the bronchopulmonary segments (VII.C). The thin intima of these vessels differs from that of other veins in that it lacks valves and contains a rich elastic fiber network in its subendothelial layer. Whereas the media is absent in vessels smaller than 100 μm, in larger vessels, it contains both smooth muscle and elastic fibers. The adventitia is thicker than that of pulmonary arteries.

2. **Systemic circulation** is provided by the bronchial arteries and veins. **Bronchial arteries** are typical muscular arteries (see Table 17–1) arising from the aorta or intercostal arteries. They are always smaller than accompanying branches of the pulmonary arteries. The bronchial arteries enter at the pulmonary root and follow the bronchial tree's branching pattern to the respiratory bronchiole level. Here, they anastomose with branches of the pulmonary artery. Branches of the bronchial arteries carry oxygen-rich blood to capillaries in the bronchi, bronchioles, interstitium, and pleura. The blood collects in submucosal venous plexuses in various parts of the bronchial tree before entering the bronchial veins. **Bronchial veins** are typical small veins that carry blood from the submucosal bronchial venous plexuses and accompany the bronchial tree. Those following the larger bronchi empty into the azygous, hemiazygous, or posterior intercostal veins. Those associated with the smaller portions of the bronchial tree empty directly into branches of the pulmonary veins.

B. Lymphatic Drainage: The lungs' lymphatic vessels form superficial and deep networks, both of which drain to the lymph nodes near the hilum. Deep vessels have few valves; they accompany either the bronchial tree or the pulmonary veins in the intersegmental connective tissue. Superficial vessels, which have many valves, lie in the visceral pleura. Lymph in the superficial network travels to the hilar nodes through vessels that either traverse the pleural surface or penetrate the lung surface and empty into intersegmental vessels. Lymphatic vessels are notably absent from interalveolar septa; at this level, the rich capillary network is responsible for draining excess interstitial fluid.

X. INNERVATION

Autonomic motor and general sensory nerves penetrate the pulmonary root, accompanying the blood vessels and the bronchial tree. Sensory nerves, carrying poorly localized pain sensations, monitor airway irritants and participate in the cough reflex. Parasympathetic motor fibers (branches of the vagus nerve) stimulate bronchial constriction, and sympathetic fibers cause bronchial dilation. Sympathomimetic drugs such as isoproterenol stimulate bronchodilation during asthma attacks.

XI. PLEURA

This serous membrane has two layers, one covering the lungs (**visceral pleural**) and the other covering the thoracic cavity's inner wall (**parietal pleura**). Like the peritoneum and the pericardium, the pleura comprises a thin squamous mesothelium attached to the organ or wall by a thin connective tissue layer containing collagen and elastic fibers. Bordered by the mesothelial cells, the narrow pleural cavity lies between the parietal and visceral pleurae. This cavity normally contains only a thin film of lubricating fluid, which (together with the smooth mesothelial surfaces) reduces friction between the lung surfaces and thoracic walls during respiratory movements. Certain diseases and wounds allow air or fluid to enter the pleural cavity, increasing its size and restricting respiratory movement. Small amounts of air and fluids can be absorbed. Larger amounts may precipitate lung collapse and require medical intervention.

MULTIPLE-CHOICE QUESTIONS

Select the single best answer.

17.1. Which of the arrows in Figure 17–3 is located in a respiratory bronchiole?

17.2. Which of the following components of the respiratory tract wall increase(s) in amount from trachea to alveoli?
(**A**) Cilia
(**B**) Elastic fibers
(**C**) Smooth muscle
(**D**) Cartilage
(**E**) Goblet cells

17.3. Which of the following is true of the visceral pleura?
(**A**) Can be described as an adventitia
(**B**) Is devoid of lymphatic vessels
(**C**) Includes a layer of simple squamous epithelium
(**D**) Lacks elastic fibers
(**E**) Covers the outer wall of the pleural cavity

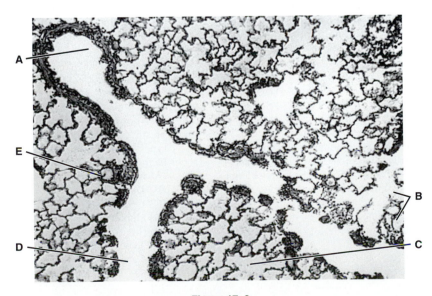

Figure 17–3.

17.4. Which of the following is (are) located at the transition between the conducting and respiratory portions of the respiratory tract?
(A) Bronchioles
(B) Epiglottis
(C) Nasopharynx
(D) Primary bronchi
(E) Respiratory bronchioles
(F) Secondary bronchi
(G) Tertiary bronchi

17.5. Which of the following is true of pulmonary surfactant?
(A) Is secreted by type-I cells
(B) Consists of a monolayer of phospholipid overlying an aqueous hypophase
(C) Prevents alveolar collapse by increasing surface tension
(D) May promote the growth of bacteria
(E) Excessive production in the lungs of premature infants causes respiratory distress syndrome

17.6. Which of the following is true of Clara cells?
(A) Are found in primary bronchi
(B) Are ciliated
(C) Are typically shorter than surrounding epithelial cells
(D) Contain secretory granules

17.7. C-shaped hyaline cartilages are a characteristic feature of the walls of which of the following structures?
(A) Alveolar ducts
(B) Bronchi
(C) Bronchioles
(D) Larynx
(E) Respiratory bronchioles
(F) Terminal bronchioles
(G) Trachea

17.8. Which of the following structures is (are) partly lined by nonkeratinized stratified squamous epithelium?
(A) Bronchi
(B) Bronchioles
(C) Larynx
(D) Nasopharynx
(E) Respiratory bronchioles
(F) Terminal bronchioles
(G) Trachea

17.9. The pulmonary (functional) and bronchial (nutrient) arterial systems enter the lungs separately through the hilus but anastomose into a single system at which of the following levels?
(A) Alveolar ducts
(B) Bronchi
(C) Bronchioles
(D) Larynx

(E) Respiratory bronchioles
(F) Terminal bronchioles
(G) Trachea

17.10. Which of the following have an epithelial lining in which cilia are typically present but goblet cells are absent?
(A) Alveolar ducts
(B) Bronchi
(C) Bronchioles
(D) Larynx
(E) Respiratory bronchioles
(F) Terminal bronchioles
(G) Trachea

17.11. Which of the following is a distinctive structural characteristic of the left lung?
(A) Three secondary bronchi
(B) Ten bronchopulmonary segments
(C) Two lobes
(D) Two primary bronchi
(E) Vertical primary bronchus

17.12. Which of the following structural characteristics enables the distinction between terminal and respiratory bronchioles?
(A) Alveoli
(B) Cilia
(C) Cuboidal epithelial lining
(D) Mucous glands in the lamina propria
(E) Smooth muscle

17.13. Which of the following provides immediate relief from the uncontrolled and excessive bronchial smooth muscle contraction suffered by asthmatic patients?
(A) Glucocorticoid injections
(B) Sympathomimetic drugs
(C) Local degranulation of mast cells
(D) Tracheotomy
(E) Vagal stimulation

17.14. Which of the following is a protective mechanism in the respiratory tract but is absent from the digestive tract?
(A) Cilia
(B) Goblet cells
(C) Lymphoid nodules
(D) Secretory IgA
(E) Zonula occludens

17.15. The presence of iron deposits in the cytoplasm of which of the following lung cell types is a diagnostic feature of congestive heart failure?
(A) Clara cell
(B) Dust cell
(C) Erythrocyte
(D) Goblet cell
(E) Small granule cell
(F) Type-I cell
(G) Type-II cell

ANSWERS TO MULTIPLE-CHOICE QUESTIONS

17.1. D (VII.F; Fig. 17–1; note alveoli in otherwise bronchiolar wall)

17.2. B (I.B.2; Table 17–1)

17.3. C (IX.B; XI)

17.4. E (I.A.2 and 3; VII.E and F)

17.5. B (VIII.C)

17.6. D (VII.E)

17.7. G (VI; Table 17–1)

17.8. C (V.C; Table 17–1)

17.9. E (IX.A.2)

17.10. E (VII.A–H; Table 17–1)

17.11. C (VII.A–C)

17.12. A (I.A.2 and 3; VII.E and F; Table 17–1)

17.13. B (X)

17.14. A (I.B.1–3; 15.I.C.7)

17.15. B (VIII.B.3)

Skin

18

OBJECTIVES

This chapter should help the student to:

- List the skin's functions and relate them to its structure.
- Name the skin's two major layers. For each, name the basic tissue type that predominates and describe the arrangement and distinguishing features of its constituent layers.
- Name the four cell types of the epidermis and describe their structure, function, and location.
- Relate the steps in cell renewal and keratinization to the epidermal layers.
- Compare thick and thin skin.
- Describe melanin granule synthesis and turnover.
- Identify and describe the components of hair follicles and nail complexes. Briefly describe nail and hair growth.
- Describe the skin's blood and nerve supply in terms of structure, location, and specialized functions.
- Name and compare three types of glands in skin in terms of structure, function, and location.
- Identify skin type, layers, cell types, hair follicles, and glands in a micrograph of a section of skin.

MAX-Yield™ STUDY QUESTIONS

1. Name the two major layers of the skin (I.B[1]) and compare them in terms of thickness (I.B.3; Fig. 18–1), vascularity (I.B.1 and 3), and embryonic germ layer of origin (I.B.1 and 3).
2. Describe the hypodermis in terms of its structure, function, and location (I.B.4).
3. List the four cell types in the epidermis and compare them in terms of their number, location, and primary function (I; II.A, B.2, C, and D).
4. Compare thick and thin skin in terms of epidermal layers, numbers of hair follicles, sweat glands and sebaceous glands, and location (Table 18–1).
5. Beginning at the surface, list the five layers of the epidermis (II.A.1–5). Compare the keratinocytes in each layer in terms of their cell shape, capacity for cell division, staining properties, and the visibility of their nuclei and other organelles.
6. Compare the structures binding the keratinocytes and melanocytes in the stratum basale to the underlying basal lamina (II.A.1 and B.2).
7. Name the embryonic cell type that forms melanocytes (II.B.2).
8. Name the enzyme in melanocyte granules that is primarily responsible for melanin production (II.B.3).
9. Describe how melanin granules enter keratinocytes (II.B.4).
10. Name two mechanisms that darken the skin after ultraviolet light exposure (II.B.6).
11. Describe Langerhans' cells (II.C) in terms of:
    a. Shape
    b. Location
    c. Staining properties
    d. Granules
    e. Immune function
12. Describe Merkel's cells (II.D) in terms of the type of skin (thick or thin) in which they are most abundant, their association with nerve endings, and two possible functions.

---

[1] See footnote on page 1.

13. Name the two layers of the dermis (III.A and B) and compare them in terms of their primary tissue type, thickness, and location in relation to the epidermis and hypodermis.
14. Is hair growth continuous or discontinuous (V.D)?
15. Explain the expression "growth in mosaic" as it applies to hair (V.D).
16. Sketch a hair follicle in *longitudinal* section (V.B.1–4; Fig. 18–3). Include and label the following:
    - **a.** Hair bulb
    - **b.** Dermal papilla
    - **c.** Germinal matrix
    - **d.** Melanocytes
    - **e.** Hair shaft
    - **f.** Internal root sheath
    - **g.** External root sheath
    - **h.** Glassy membrane
    - **i.** Connective tissue sheath
    - **j.** Arrector pili muscle
17. Sketch a *cross-section* through a hair follicle above the bulb (V.B.1–4; Fig. 18–3). Include and label the following:
    - **a.** Medulla
    - **b.** Cortex
    - **c.** Cuticle
    - **d.** Internal root sheath
    - **e.** External root sheath
    - **f.** Glassy membrane
    - **g.** Connective tissue sheath
18. Above which level in the hair follicle is the internal root sheath absent (V.B.3.a)?
19. Compare the hardness of keratin in hairs with that of keratin in the epidermis (V.C).
20. Sketch a longitudinal section of a fingertip through the nail (VI.B; Fig. 18–4). Include and label the following:
    - **a.** Root of the nail
    - **b.** Eponychium
    - **c.** Nail plate
    - **d.** Nail matrix
    - **e.** Nail bed
    - **f.** Hyponychium
21. Describe sebaceous glands in terms of:
    - **a.** Class, as determined by adenomere shape (tubular or acinar; VII.A)
    - **b.** Association with hair follicles (VII.A)
    - **c.** Sites in which they occur without hair follicles (VII.A)
    - **d.** Mode of secretion (merocrine, apocrine, or holocrine; VII.B)
    - **e.** Composition of secretion (VII.B)
22. Name the two sweat gland types in humans (VIII.A and B) and compare them in terms of:
    - **a.** Class, as determined by adenomere shape (tubular or acinar)
    - **b.** Distribution in skin
    - **c.** Mode of secretion (merocrine, apocrine, or holocrine)
    - **d.** Composition of secretion
    - **e.** Innervation
23. Name the type of epithelium that commonly lines the ducts of sweat glands (VIII.A.2.a).
24. Sketch a vertical section of skin (Fig. 18–1); show the boundaries of the epidermis, papillary dermis, reticular dermis, and hypodermis. Indicate the location of the two arterial plexuses, the papillary capillaries, and the three venous plexuses (IV.A–D).
25. Name the types of sensory receptors found in the epidermis (I.B.1; II.D), dermal papillae (III.A), and reticular dermis (III.B; 24.II.A–G).
26. Name the components of the skin that play a role in protection from dehydration (II.A.3), abrasion (I.B.2; II.A.1; III.A), infection (II.A.3, C; III.A; 5.III.A.1), and ultraviolet radiation (II.B.5 and 6); regulation of blood pressure (IV.D) and body temperature (IV.D; VIII.A.3); sensory reception (I.B.1 and 3; II.D; 24.II.A–G); and excretion (VIII.A.3).

SYNOPSIS

I. GENERAL FEATURES OF THE SKIN

    A. General Functions: The skin is the largest and heaviest organ. It protects against microorganisms, toxic substances, dehydration, ultraviolet radiation, impact, and friction. It acts as a sensory receptor and has roles in excretion, vitamin D metabolism, and the regulation of blood pressure and body temperature.

B. General Organization: Human skin (the integument) comprises two types. **Thick skin,** limited to the palms and soles, lacks hair and has abundant sweat glands. **Thin skin** has hairs and covers the rest of the body (Table 18–1). Thick or thin, the skin consists of two distinct but tightly attached layers—the **epidermis** and **dermis**—which are underlain by the hypodermis (Fig. 18–1).

1. **Epidermis.** This outer (superficial) layer of skin, composed of **keratinized stratified squamous epithelium,** derives from embryonic surface ectoderm. It is avascular, receiving nourishment from vessels in the underlying dermis. Its only innervation is by unencapsulated (free) nerve endings (24.II.A). The epidermal layer is further divided into five layers; these include, in order from superficial to deep, the **stratum corneum, stratum lucidum, stratum granulosum, stratum spinosum,** and **stratum basale** (II.A.1–5). The width of these layers differs in thick and thin skin (see Table 18–1).

2. **Dermal–epidermal junction.** The stratum basale is underlain by a basement membrane connecting the epidermis and dermis. The junction has the appearance of zigzagging interdigitations between upward projections of the dermis (**dermal papillae**) and downward projections of the epidermis (**epidermal ridges).**

3. **Dermis.** This inner (deeper) layer is a vascular connective tissue of mesodermal origin. It is further divisible into a superficial **papillary layer** and a deeper **reticular layer.** The papillary layer contains extensive capillary networks, which nourish the epidermis. The reticular layer contains many arteriovenous anastomoses that help regulate blood pressure and body temperature. The dermis is richly supplied with free nerve endings, a variety of encapsulated sensory receptors (24.II.B–G), and autonomic fibers that control the vascular smooth muscle. Even in thick skin, the dermis is much thicker than the overlying epidermis.

4. **Hypodermis.** Although not a part of the skin, this layer of mesoderm-derived loose connective and adipose tissue under the dermis flexibly binds the skin to deeper structures. Its thickness varies, depending on nutritional status, activity level, body region, and gender. It is also called **subcutaneous fascia** and, where it is thick enough, the **panniculus adiposus** (6.II.B.1).

C. Structures Associated with the Skin: Glands (sebaceous and sweat), hairs, and nails arise from epidermal downgrowths into the dermis during embryonic development. These structures, which are mainly of epithelial origin, require epithelial–mesenchymal interactions between the epidermis and dermis for their formation and maintenance (V–VIII).

D. Wound Healing: Lacerations (cuts), abrasions, and burns create defects in skin structure of variable size and depth. Such defects are initially filled with granulation tissue comprising proliferating fibroblasts that secrete matrix materials (mainly type-III collagen). The granulation tissue is subsequently remodeled and most of the type-III fibers are replaced by type-I collagen. The extent to which healing rather than scarring occurs depends on the width and depth of the

Table 18–1. Comparison of thick and thin skin.

	Thick Skin	Thin Skin
Location	Palms and soles	Entire body (except palms and soles)
Total thickness	Thicker (0.8–1.4 mm)	Thinner (0.07–1.12 mm)
Epidermis		
Stratum corneum	Thicker (15–> 40 layers)	Thinner (10–20 layers)
Stratum lucidum	Present (a few layers)	Usually absent
Stratum granulosum	A few layers	Single, often discontinuous layer
Stratum basale	More Merkel's cells	Fewer Merkel's cells
Dermatoglyphics	Present (eg, fingerprints)	Absent
Dermis		
Hair follicles	Absent	Present (except in glans penis, labia minora, clitoris, and lips)
Sebaceous glands	Fewer	More (associated with hairs)
Eccrine sweat glands	More	Fewer
Meissner's corpuscles	More (in dermal papillae)	Fewer
Elastic fibers	Fewer	More

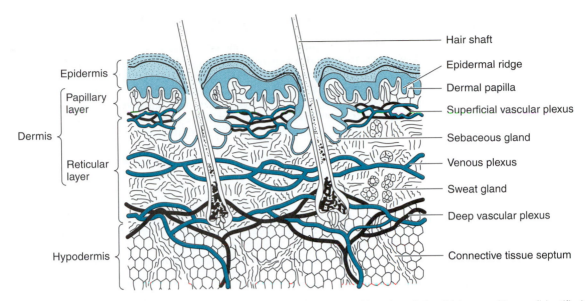

Figure 18–1. Schematic diagram of a vertical section through thin skin. Note the relative thickness of layers (identified at left) and the locations of the dermal vascular plexuses and papillary capillaries.

wound. Small, shallow wounds are typically recovered by keratinocytes arising from the stratum basale. In wounds that remove the epidermis from a larger area, keratinocytes in the external root sheath of the hair follicles divide and migrate to cover the granulation tissue. Eventually, normal epidermal and dermal architecture may be restored. Larger and deeper wounds, in which the hair follicles are lost or destroyed, may never be completely recovered by normal epidermis and, thus, may leave a dense connective tissue scar at the site.

II. EPIDERMIS

The epidermis contains two major cell populations (keratinocytes and melanocytes) and two minor cell populations (Langerhans' and Merkel's cells).

A. **Keratinizing System:** **Keratinocytes** comprise most of the epidermis. They participate in the continuous turnover (renewal) of the skin surface, which involves four overlapping stages: cell renewal (**mitosis**), cell differentiation (**keratinization**), cell death (**apoptosis**), and the sloughing of dead cells from the surface (**exfoliation**) (Fig. 18–2). The entire process takes 15 to 30 days and occurs in waves. Cells produced by a mitotic pulse in the basal layer undergo keratinization in synchrony. Each pulse pushes the cell layers produced earlier toward the surface in waves. Layers from several waves, each at a different depth and step in the process, give

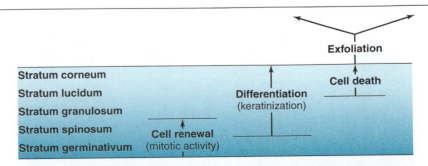

Figure 18–2. Schematic diagram of the keratinizing system of the epidermis.

sectioned epidermis a stratified appearance. Five epidermal layers are distinguishable by keratinocyte shape, staining properties, contents, and position.

1. **Stratum basale (stratum germinativum).** This single layer of columnar basophilic keratinocytes rests on the basal lamina between epidermis and dermis. These cells give rise to the keratinocytes in all other layers. They attach to their neighbors by **desmosomes** and to the basal lamina by **hemidesmosomes** (4.IV.C.2). **Cytokeratin** intermediate filaments (**tonofilaments;** 2.III.I.3.a, 4.IV.B.3) help to stabilize both junctions. Cytokeratin content increases as cells approach the stratum corneum, where it constitutes approximately 50% of their protein. The basophilia of the cells in this layer results from cytokeratin synthesis by their ribosomes.

2. **Stratum spinosum.** This stratum comprises several layers of large keratinocytes overlying the stratum basale. The cells are cuboidal or polygonal in the deeper layers and slightly flattened in the upper layer. **Tonofibrils** (tonofilament bundles) fill the cytoplasm, extend into the numerous cell processes that give these cells their spiny appearance, and insert into the desmosomes that attach the tips of these processes to those of adjacent cells. The mitotic rate here is lower than in the stratum basale. Mitosis occurs only in the **malpighian layer,** which includes the stratum basale and stratum spinosum.

3. **Stratum granulosum.** This layer lies above the stratum spinosum and, in thick skin, consists of three to five layers of flattened polygonal (often diamond-shaped) cells filled with membraneless **keratohyalin granules.** The intense basophilia of these granules stems from their possession of a histidine-rich precursor of the protein **filaggrin.** Cells in this layer also contain small ovoid or rodlike **lamellar granules;** these fuse with the plasma membrane and release their contents (glycosaminoglycans and phospholipids) into the intercellular spaces. This helps to seal deeper layers from the exterior, providing protection from dehydration.

4. **Stratum lucidum.** Apparent only in thick skin, this layer overlies the stratum granulosum. It is a narrow, acidophilic, translucent band of flattened keratinocytes whose organelles, nuclei, and intercellular borders are not visible. The cytoplasm contains dense cytokeratin aggregates embedded in an amorphous electron-dense matrix derived from the keratohyalin granules. This intracellular mixture of intermediate filaments and matrix constitutes immature keratin, sometimes called **eleidin.**

5. **Stratum corneum.** This surface layer consists of many layers of dead, platelike enucleate keratinocytes with thickened plasma membranes. These cells represent the final stage of keratinization and are filled with mature **keratin,** a birefringent scleroprotein consisting of at least six polypeptides. Keratin's substructure includes tonofilament subunits formed by three intertwined polypeptide chains. Nine of these subunits coil to form each 10-nm thick intermediate filament; as they aggregate end to end, the tonofilament increases in length. Tonofilaments are embedded in and bound by the amorphous matrix from keratohyalin granules. Dead cells are continuously exfoliated from the surface and replaced by cells from the deeper layers.

B. **Pigmentation System:** Skin color is conferred mainly by the pigments melanin and carotene but is also affected by epidermal thickness, dermal blood vessels, and the color of the blood in those vessels.

1. **Melanins** contribute to skin, eye, and hair color. Synthesized by **melanocytes,** they include the dark brown **eumelanin** (found in the epidermis, iris, and brown and black hair) and the cysteine-rich **pheomelanin** (found in red hair).

2. **Melanocytes** derive from the neural crest and migrate to the epidermis during embryogenesis. They are scattered among the keratinocytes of the stratum basale but lack desmosomes. They have round cell bodies, central nuclei, and long cytoplasmic processes that pass between the cells of the strata basale and spinosum to terminate in small indentations on keratinocyte surfaces. Melanocytes make up 10 to 25% of this layer's cells but do not participate in keratinization. Their cytoplasm contains many mitochondria, a well-developed Golgi complex, short RER cisternae, and membrane-bound **melanosomes,** in which melanin is synthesized. There is no difference in melanocyte numbers per unit area of skin in dark- and light-skinned races; differences in pigmentation reflect differences in the rates of melanin synthesis, accumulation, and degradation.

3. **Melanin synthesis** involves the **tyrosinase**-mediated enzymatic conversion of **tyrosine** to **DOPA** (3,4-dihydroxyphenylalanine) and of DOPA to **dopaquinone.** Additional steps con-

vert dopaquinone to melanin. In melanin granule formation, tyrosinase is synthesized on the RER and transported to the Golgi complex. Membrane-limited tyrosinase-filled vesicles called **melanosomes** pinch off from the Golgi complex, accumulate in the Golgi region, and develop through four stages to mature melanin granules. **Stage-I** melanosomes are round vesicles characterized by tyrosinase activity associated with fine granular to filamentous material in the vesicle periphery; melanin is not yet present. **Stage-II** melanosomes are ovoid and contain parallel filaments; tyrosinase activity triggers melanin deposition on these filaments. **Stage-III** melanosomes have the same structure as those in stage II, but melanin deposition has partly obscured the filaments. **Stage-IV** melanosomes (mature **melanin granules**) are 1-μm long, 0.4-μm wide, and completely filled with melanin. Their ultrastructure is no longer visible.

4. **Fate of mature melanin granules.** Mature granules move into the tips of the melanocytes' long processes. They are subsequently injected into the keratinocytes of the strata basale and spinosum in a process termed **cytocrine secretion.** (Keratinocytes act as melanin depots and usually contain more melanin than do melanocytes). Melanin granules accumulate over the nuclei of the dividing keratinocytes, protecting the DNA from the damaging effects of the sun. Keratinocytes carry melanin to the skin surface. During keratinization, the granules, along with the keratinocytes' nuclei and organelles, are digested by lysosomes.

5. **Melanin function.** Some of the sun's rays break apart molecules in the skin, forming highly reactive free radicals. DNA cleavage by ionizing radiation or the recombination of normal DNA with other free radicals can alter its structure, causing cell death or neoplastic transformation. The euchromatic DNA of dividing cells in the Malpighian layer is more susceptible to these effects. Although melanin's dark color allows it to absorb some rays directly, its major protective effect is its ability to absorb free radicals.

6. **Factors affecting melanin synthesis.** Melanogenesis increases or decreases in response to various factors. **Increased exposure to ultraviolet rays** both darkens existing melanin and speeds tyrosinase synthesis, increasing the amount and rate of melanin production. **Melanocyte-stimulating hormones** (α- and β-MSH) from the pituitary gland (20.III.C) enhance pigmentation but do not exist in free form in humans. Pituitary ACTH, however, contains a peptide sequence identical to α-MSH that influences human pigmentation. **Addison's disease** involves an insufficient production of cortisol by the adrenal cortex (21.II.A.4.b). The pituitary releases excess ACTH in an attempt to stimulate the adrenals. The resulting stimulation of tyrosinase activity in melanocytes causes hyperpigmentation. **Albinism,** in which no melanin pigment is produced, is caused by a genetic defect in tyrosinase synthesis and a consequent absence of tyrosinase activity. Melanocytes of affected individuals typically contain melanosomes that have not progressed beyond stage II. **Hydroquinone,** the active ingredient in some over-the-counter treatments for "age spots," inhibits melanin synthesis.

C. **Langerhans' Cells:** These star-shaped cells lack tonofilaments and occur mainly in the stratum spinosum (400–1000 cells/mm^2 of skin surface). They stain selectively with gold chloride and contain many rodlike or racket-shaped cytoplasmic granules **(Birbeck's granules).** They are antigen-presenting cells (14.III.E) that process and present to the lymphocytes antigenic material that penetrates the skin's surface. Of mesodermal origin, they arise in bone marrow and may belong to the mononuclear phagocyte system. Langerhans' cells also occur in oral and vaginal epithelia and in the thymus.

D. **Merkel's Cells:** Scattered in the stratum basale, these cells are more numerous in thick skin (see Table 18–1). They resemble basal keratinocytes but have a clearer cytoplasm containing many small dense granules. Free nerve endings form an expansion **(Merkel's disk)** that covers the basal surface of each Merkel's cell (24.II.B). This arrangement suggests that the cells function as mechanoreceptors, but other evidence suggests DNES-related functions.

III. DERMIS

The dermis comprises two layers of vascular connective tissue that blend at their common border. It contains hair follicles (in thin skin; see Table 18–1; V) and sebaceous and sweat glands (VII; VIII).

A. Papillary Layer: This layer of loose connective tissue, rich in elastic fibers, lies directly beneath the epidermal basement membrane. Its projections—**dermal papillae**—interdigitate with the epidermal ridges, increasing the area of contact. Special collagen fibers called **anchoring fibrils** extend into the epidermal basal lamina to reinforce the dermal–epidermal junction. The papillary layer contains immunoprotective cells (5.III.A.1), a rich capillary network (IV.B), and abundant free nerve endings (24.II.A), some of which penetrate the epidermis. Many dermal papillae contain encapsulated touch receptors called Meissner's corpuscles (24.II.D).

B. Reticular Layer: Deep to the papillary layer, this is a thicker layer of dense irregular connective tissue. Richly vascularized, this layer contains many **arteriovenous anastomoses** (IV.D). The reticular layer also contains a rich supply of nerves in both free and encapsulated endings (eg, Pacinian corpuscles; 24.II.A–G).

IV. BLOOD SUPPLY TO THE SKIN

Although the epidermis is avascular, the skin receives an extensive vascular supply through the dermal blood vessels (see Fig. 18–1), which can hold approximately 4.5% of the body's total blood volume.

A. Arterial Plexuses: One of the two arterial plexuses that provide the skin's blood supply lies at the border between the papillary and reticular layers of the dermis. The other lies between the dermis and hypodermis. Both give rise to arterioles that feed the papillary capillaries.

B. Papillary Capillaries: The dermal papillae, which surround the epidermal ridges, contain a rich capillary network that provides oxygen and nutrients to the avascular epidermis.

C. Venous Plexuses: The capillary bed in each papilla drains, by a single venule, into one of three venous plexuses. Two of these lie in the same position as the arterial plexuses; the other lies between them in the middle of the reticular dermis.

D. Arteriovenous Anastomoses (Shunts): Within the dermal plexuses, there are many direct connections, or anastomoses, between the arteries and veins. Postganglionic autonomic fibers control the opening and closing of these shunts, helping to control blood pressure and body temperature by regulating the amount of blood in the papillary capillaries. When the shunts are closed, more blood flows through the papillary capillaries; when they are open, they direct blood away from the capillaries, increasing blood volume in the larger vessels and thereby increasing blood pressure. Opening of the shunts also reduces the loss of body heat through the skin.

V. HAIR

Hair occurs only in thin skin; its color, size, shape, and distribution vary according to race, age, sex, and body region. The structures that form hairs and maintain their growth are called **hair follicles.**

A. Follicle and Hair Development:
  1. Follicles. Early in the third month of human development, local epidermal thickenings form at the sites of future hairs: first on the eyebrows, chin, and upper lip and subsequently over the rest of the thin skin. Cells at the base of each thickening invade the dermis, and a small **dermal papilla** invades the leading edge of the epidermal downgrowth. Interactions between the papilla and the invaginating epidermis induce hair follicle differentiation. Hair begins to form in the **hair bulb** at the base of the follicle through the keratinization of the bulb's epithelial cells. These cells are pushed toward the surface by mitosis in the **germinal matrix** (hair bulb epithelium). Some epithelial cells in the walls of the developing follicle divide, forming bulges that differentiate into sebaceous glands (VIII).
  2. Hairs. By the fifth or sixth month, the fetus is covered with fine hairs (**lanugo**). Just before birth, most of this hair is shed, except for that on the scalp and eyebrows and eyelashes. A few months after birth, the remaining lanugo has been replaced by coarser mature **terminal hairs;** the rest of the body is covered with a coat of fine short hairs, called **vellus.** At puberty, coarse terminal hairs replace the vellus in specific body areas. In males, terminal hairs

develop in the axilla and pubic region, on the face, and, to some extent, over the rest of the body. In females, they develop mainly in the axilla and pubic regions.

B. Follicle and Hair Structure: Hair follicles extend from the surface deep into the dermis or hypodermis. The follicle's broad base, or hair bulb, consists of a cap of rapidly dividing epithelial cells (the germinal matrix) overlying a dermal papilla that harbors the nerve and blood supply. Cells from the germinal matrix keratinize, forming the concentric layers of the hair shaft as they move toward the surface. Near the surface, distinct layers ensheathe the canal that contains the hair shaft.

1. **Germinal matrix.** This cluster of epithelial cells capping the dermal papilla forms four indistinct concentric zones around the papilla. The zone closest to the papilla resembles the stratum basale in both structure and function. It contains both columnar epidermal cells and the melanocytes that give the hair its color. This germinal layer gives rise to the poorly keratinized cells of the hair shaft's **medulla** and to the cells in the other three zones of the germinal matrix. Around the hair bulb's base, this layer is continuous with the external root sheath that surrounds the entire bulb and shaft; near the surface, it is continuous with the stratum basale. Cells in the next layer form the **cuticle.** The most peripheral layer of the germinal matrix forms the poorly keratinized cells of the internal root sheath.

2. **Hair shaft layers.** These three concentric layers are formed by the germinal matrix (V.B.1). The cell borders are indistinct, however, and cross-sections through hair follicles near the skin surface often do not show these layers' cellular nature. In addition, the hair may be dislodged by tissue processing, leaving only the space (**follicular canal**) originally occupied by the shaft. The **medulla** forms the shaft's thin core of poorly keratinized and often vacuolated cells. The **cortex** surrounds the medulla and consists of several layers of well-keratinized polygonal cells. The **cuticle** is the shaft's outermost layer. Within the bulb, its cells are cuboidal; farther up the shaft they become tall columnar, fill with keratin, and finally change their orientation to become a few layers of flattened, highly keratinized cells. These cells form the hard, shinglelike outer covering of the hair.

3. **Root sheaths.** The concentric sheaths around the hair shaft are more clearly distinguished in the area between the bulb and the skin surface.
   a. **Internal root sheath.** This layer is closest to the hair shaft; it extends from the bulb only to the level of the sebaceous gland ducts. At this point, the soft keratin-filled cells are shed into the follicular canal. There are three component layers: the **cuticle of the internal root sheath** is a layer of flat cells separated from the hair shaft cuticle only by the follicular canal; the middle layer is **Huxley's layer,** which comprises one to three layers of low cuboidal cells; and the outermost layer is **Henle's layer,** a translucent layer of flattened to cuboidal cells resembling the epidermal stratum lucidum.
   b. **External root sheath.** This layer surrounds the internal root sheath and is continuous with the epidermis. Above the level of the sebaceous glands, it includes all of the epidermal layers. Below this level, it retains only the granulosum, spinosum, and basale. The granulosum is also lost near the follicle's base, where the spinosum and basale become continuous with the germinal matrix.
   c. **Glassy membrane.** This is the thickened basal lamina underlying the stratum basale of the external root sheath and separating it from the surrounding connective tissue sheath.
   d. **Connective tissue sheath.** This layer of condensed connective tissue surrounds the entire follicle, including the bulb. It extends along the follicle to the surface, where it blends into the looser papillary dermis.

4. **Associated structures.** Found near the neck of the root sheath, **sebaceous glands** (VII) always accompany hairs. They empty their secretions through a short duct into the follicular canal. **Arrector pili muscles** are small bundles of smooth muscle that originate in the papillary dermis and extend obliquely toward the hair follicle to insert into the connective tissue sheath below the sebaceous glands. When they contract, the hairs stand upright, giving the appearance of gooseflesh. Their contraction also compresses the sebaceous glands, pushing their secretions into the neck of the follicular canal and out onto the skin's surface.

C. Keratinization of Hair: Hair and epidermis differ in their keratinization. For example, the keratin of the hair's cortex and cuticle is harder than that of the epidermis; keratinized hair cells remain tightly attached to one another, whereas those of the skin are continuously sloughed; keratinization of the hair is intermittent and is restricted to the bulb, whereas that of the skin is continuous and occurs over the entire surface; and keratinized cells of the epidermis are identical, whereas those in hair differ in structure and function depending on their position in the hair.

D. Hair Growth: Hair grows in cycles, which involve repeated growing **(anagen),** regression **(catagen),** and resting **(telogen)** phases. During anagen, the proliferation and differentiation of cells in the germinal matrix cause the hair to elongate. During catagen and telogen, the germinal matrix becomes inactive and may atrophy. The hair detaches from the bulb, moving upward as the external root sheath retracts toward the surface. Eventually, the hair is shed. During the next anagen phase, the lower part of the external root sheath grows downward again, either forming a new germinal matrix over the old papilla or stimulating the formation of a new papilla. The bulb reforms, and the cycle continues. Hair growth cycles do not occur synchronously over the entire body surface. Rather, they occur in patches, a pattern called **growth in mosaic.** Hormones, especially androgens, influence the pattern of terminal hair distribution and growth rate.

VI. NAILS

These plates of highly keratinized cells are analogous to, but harder than, the stratum corneum.

A. Nail Development: Nail development resembles that of hair but involves the formation of plates rather than cylinders. At the end of the third month of gestation, a narrow plate of epidermis on the terminal phalanges' dorsal surface invades the underlying dermis of each finger and toe. This invasion continues proximally, forming a furrow called the **nail groove.** Epithelial cells beneath the groove proliferate to form the **nail matrix,** whose composition and function resemble the hair's germinal matrix. Proliferation in the nail matrix pushes the upper cells toward the surface. These cells differentiate and become highly keratinized, thus forming the **nail plate.** The plate is gradually pushed out of the groove by further cell proliferation and differentiation in the nail matrix. The growing plate slides distally on the dorsal surface of the digit. The epidermis over which it slides is the **nail bed.**

B. Nail Complex Structure: The nail plate (or nail) has two parts: the **nail body** (the visible part of the nail) and the **nail root** (the part hidden in the nail groove). The nail and its supporting structure are surrounded by papillary dermis. The **nail matrix** is a thickened region of epidermis that contains dividing cells in the layer directly contacting the dermis and keratinizing cells between this basal layer and the nail plate. The nail matrix surrounds the root and extends beyond the nail groove. The **nail bed** lies beneath the nail body, distal to the nail matrix. It consists of only the deeper epidermal strata, for which the nail serves as a stratum corneum. The **eponychium** (or **cuticle**) is a thick keratinized layer extending from the upper surface of the nail groove over the most proximal part of the nail body. The **hyponychium** is a local thickening of the stratum corneum underlying the free (distal) end of the nail. The **lunula** is the whitish, opaque, crescent-shaped region on the proximal nail body, adjacent to the nail groove. Its distal border corresponds roughly to that of the underlying nail matrix.

VII. SEBACEOUS GLANDS

A. Structure and Location: These exocrine glands occur in all thin skin, most often in association with the hair follicles into which their ducts empty, but are most numerous in the skin of the face, forehead, and scalp. In hairless skin, they open directly onto the surface. Their acinar secretory sections contain many large lipid-filled cells that appear pale-staining and foamy.

B. Function: The acinar cells of sebaceous glands fill with lipid droplets containing a mixture of triglycerides, waxes, squalene, and cholesterol and its esters. Their nuclei become pyknotic and the cells eventually burst, releasing their contents and other cell debris (together termed **sebum**) into the ducts. The entire cell is shed, a type of secretion known as **holocrine secretion.** The oily sebum moves through the ducts and into the hair follicle. It covers the hair and moves out onto the surface. Here, it lubricates the skin and may have some antibacterial or antifungal effects. The secretory activity of these glands, which accelerates during puberty, is continuous and is increased by androgens.

VIII. SWEAT GLANDS

Two types of sweat glands, **eccrine** (or merocrine) and **apocrine,** occur in human skin. Both develop as epidermal invaginations into the dermis, and they differ mainly in their size, distribution, and secretory products.

A. Eccrine Sweat Glands:

1. **Distribution.** These are the most numerous sweat glands in humans; approximately 3 million are present in each individual. They occur over most of the body, except for the glans penis, glans clitoridis, and the vermilion border of the lips. They are most abundant in thick skin, such as the palms, where there are approximately 3000 per square inch.

2. **Structure.** These simple coiled tubular glands have slightly coiled **ducts** lined with simple to stratified cuboidal epithelium; the cells that line these ducts are smaller and stain darker than those in the secretory components of the glands. Each duct opens directly onto the skin surface. The highly coiled **secretory components** of the sweat glands lie in deep reticular dermis or shallow hypodermis. Surrounding connective tissue condenses to form a sheath around the basal lamina, and many myoepithelial cells lie between the basal lamina and the secretory cells. The secretions are released by means of exocytosis **(merocrine secretion).** Secretory cells, which are larger and stain lighter than the duct-lining cells, comprise two types. **Dark (mucoid) cells** are pyramidal and line most of the gland's secretory segment; their bases do not reach the basal lamina. They contain rodlike mitochondria, a well-developed Golgi complex, RER, many free ribosomes, and dark glycoprotein-containing granules. **Clear cells,** also pyramidal, lack secretory granules, contain abundant glycogen, and surround the inner layer of dark cells. Their basal plasma membranes do contact the basal lamina and are highly folded, reflecting a role in ion and water transport.

3. **Secretory product.** Eccrine sweat is a watery secretion whose main components (besides water) include NaCl, urea, ammonia, and uric acid. The glands thus assist in excreting the by-products of protein metabolism. Evaporation of eccrine sweat from the skin surface reduces body temperature by cooling the blood in the papillary capillaries.

B. Apocrine Sweat Glands:

1. **Distribution.** Less numerous than the eccrine type, these glands occur mainly in the axilla, pubic and anal regions, and the areolae of the breasts.

2. **Structure.** Apocrine sweat glands are also simple coiled tubular glands but are generally larger and have wider lumens than eccrine glands. The coiled ducts are lined with cuboidal epithelium and open into hair follicles. Coiled and embedded in the dermis, each secretory portion has a wide lumen lined by cuboidal to columnar cells. Myoepithelial cells lie between the secretory cells and the basal lamina.

3. **Secretory product.** Apocrine sweat is a viscous, odorless fluid that, once secreted, acquires a distinctive odor as a result of bacterial degradation. The term "apocrine" derives from early evidence that the secretory cells of these glands released their apical cytoplasm along with the secretory product. However, recent evidence does not support apical shedding.

MULTIPLE-CHOICE QUESTIONS

Select the single best answer.

18.1. Which of the following is true of the arrector pili muscles?
(A) Are composed of bundles of myoepithelial cells
(B) Help push sebum onto the skin surface
(C) Have no effect on hair position in humans
(D) Attach directly to the external root sheath by desmosomes
(E) Contain capillary networks that nourish growing hairs

18.2. Which of the following skin-associated regions contains the epithelial cells with the lowest mitotic rate?
(A) Germinal matrix of hair bulbs
(B) Nail matrix
(C) Stratum germinativum
(D) Stratum granulosum
(E) Stratum malpighian
(F) Stratum spinosum

18.3. Which of the following is true of hair growth?
(A) Is unaffected by circulating androgens
(B) Occurs continuously
(C) Involves cell division in the cuticle
(D) Occurs in patches
(E) Is most active during the telogen phase

18.4. Which of the skin's many functions are partly or completely provided by components of the dermis?
(A) Protection from dehydration
(B) Regulation of blood pressure
(C) Protection from ultraviolet radiation
(D) Synthesis of keratin
(E) Synthesis of melanin

18.5. Which of the following are stellate cells that stain positively with gold chloride and contain rodlike or racket-shaped Birbeck's granules in their cytoplasm?
(A) Keratinocytes
(B) Langerhans' cells
(C) Melanocytes
(D) Merkel's cells

18.6. Which of the following cell types is found mainly in the stratum basale, is characterized by a clear cytoplasm containing small cytoplasmic granules, and appears to carry out mechanoreceptive and/or DNES-related functions?
(A) Keratinocyte
(B) Langerhans' cell
(C) Melanocyte
(D) Merkel's cell

18.7. Which of the following cells are neural crest derivatives that are found mainly in the stratum basale, have long cell processes that extend into the stratum spinosum, and contain lamellar granules that exhibit intense tyrosinase activity?
(A) Keratinocytes
(B) Langerhans' cells
(C) Melanocytes
(D) Merkel's cells

18.8. Which of the following cells are of mesodermal origin, derive from bone marrow precursors, and function as antigen-presenting cells in the skin and other organs?
(A) Keratinocytes
(B) Langerhans' cells
(C) Melanocytes
(D) Merkel's cells

18.9. Which of the following epidermal layers is most likely to contain columnar keratinocytes?
(A) Stratum basale
(B) Stratum corneum
(C) Stratum granulosum
(D) Stratum lucidum
(E) Stratum spinosum

18.10. Which of the following epidermal layers contains enucleate squamous keratinocytes and is the most superficial layer?
(A) Stratum basale
(B) Stratum corneum
(C) Stratum granulosum
(D) Stratum lucidum
(E) Stratum spinosum

18.11. Which of the following epidermal layers is most likely to contain both desmosomes and hemidesmosomes?
(A) Stratum basale
(B) Stratum corneum
(C) Stratum granulosum
(D) Stratum lucidum
(E) Stratum spinosum

18.12. Keratohyaline granules are particularly abundant and mitotic activity is rare in which of the following epidermal layers?
(A) Stratum basale
(B) Stratum corneum
(C) Stratum granulosum
(D) Stratum lucidum
(E) Stratum spinosum

18.13. Which of the following epidermal layers is the most superficial layer in which keratinocyte mitosis occurs continuously and in which cells are attached to their neighbors during interphase by abundant desmosomes?
(A) Stratum basale
(B) Stratum corneum
(C) Stratum granulosum
(D) Stratum lucidum
(E) Stratum spinosum

18.14. Which of the labels in Figure 18–3 corresponds to the layer that contains the highest concentration of type-IV collagen?

18.15. Which of the labels in Figure 18–3 corresponds to the layer that is notably absent from the hair follicle at levels superficial to the entry of the sebaceous gland ducts?

18.16. Which of the labels in Figure 18–3 corresponds to the region containing the capillaries that nourish the germinal matrix and support hair growth?

18.17. Which of the labels in Figure 18–3 corresponds to the hair follicle layer that is analogous to the skin's Malpighian layer and also is responsible for reepithelialization of areas where abrasions or burns have removed small patches of epidermis?

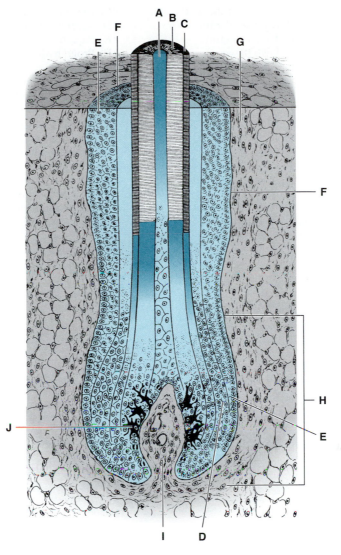

Figure 18–3.

18.18. Which of the labels in Figure 18–4 corresponds to the part of the nail complex that is analogous to the hair follicle's germinal matrix?

18.19. Which of the labels in Figure 18–4 corresponds to the part of the nail complex that is analogous to the skin's stratum corneum?

18.20. Which of the following is (are) often absent in thick skin but typically present in thin skin?

   (**A**) Arrector pili muscles
   (**B**) Meissner's corpuscles
   (**C**) Stratum basale
   (**D**) Stratum corneum
   (**E**) Stratum lucidum
   (**F**) Stratum spinosum
   (**G**) Sweat glands

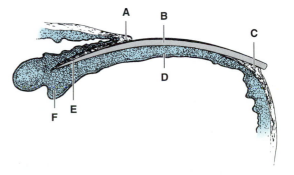

Figure 18–4.

18.21. Which of the following is typical of the reticular dermis but not of the papillary dermis?
(A) Capillaries that nourish the epidermis
(B) Dense irregular connective tissue
(C) Loose (areolar) connective tissue
(D) Meissner's corpuscles
(E) Sweat gland ducts
(F) Type-I collagen
(G) Arteriovenous anastomoses

18.22. Which of the following is characteristic of sweat glands but not of sebaceous glands?
(A) Adenomeres are typically acinar in shape
(B) Adenomeres contain clear cells and dark cells
(C) Adenomeres have a "foamy" appearance
(D) Ducts empty mainly into hair follicles
(E) Mode of secretion is holocrine
(F) Rate of secretion is controlled by circulating androgens
(G) Secretory product is oily to waxy in nature

18.23. Albinism can result from an inherited defect in the gene encoding which of the following enzymes?
(A) Catalase
(B) Collagenase
(C) Lysyl oxidase
(D) Na$^+$/K$^+$-ATPase
(E) Papain
(F) Prolyl hydroxylase
(G) Tyrosinase

18.24. The border between which of the following is a significant site of arteriovenous anastomoses in the skin?
(A) Dermis and epidermis
(B) Epidermis and basal lamina
(C) Papillary dermis and reticular dermis
(D) Stratum basale and stratum spinosum
(E) Stratum spinosum and stratum granulosum

18.25. Which of the following is a factor produced or activated in the dermis that stimulates calcium uptake in the digestive tract?
(A) Calcitonin
(B) Calsequestrin
(C) Intrinsic factor
(D) Melanin
(E) Tannin
(F) Vitamin A
(G) Vitamin D

ANSWERS TO MULTIPLE-CHOICE QUESTIONS

18.1. B (V.B.4)
18.2. D (II.A.1–3; V.B.1; VI.A and B)
18.3. D (V.D)
18.4. B (I.B.1 and 3; III.B; IV.D)
18.5. B (II.C)
18.6. D (II.D)
18.7. C (II.B.2)
18.8. B (II.C)
18.9. A (II.A.1–5)
18.10. B (II.A.1–5)
18.11. A (II.A.1–5)
18.12. C (II.A.1–5)
18.13. E (II.A.1–5)
18.14. F (V.B.3.c; 4.IV.C.1.a)
18.15. D (V.B.3.a)
18.16. I (V.A.1 and B)
18.17. E (I.D; II.A.2; V.B.3.b)
18.18. F (V.B.1; VI.A)
18.19. B (VI.B)
18.20. A (V.B.4; Table 18–1)
18.21. B (III.A and B; Fig. 18–1)
18.22. B (VII.A and B; VIII.A and B)
18.23. G (II.B.6)
18.24. C (III.A and B; Fig. 18–1)
18.25. G (I.A)

19

Urinary System

OBJECTIVES

This chapter should help the student to:

- List the organs of the urinary system and describe the role of each in the system's functions.
- Identify the structures and regions visible in a frontal section of a kidney and describe their functions.
- Describe the structure, function, and location of each component of a nephron and identify these components in histologic sections.
- Describe the function of the juxtaglomerular apparatus and identify its components.
- Trace the flow of blood through the kidney and identify renal vascular elements in histologic sections.
- Trace the flow of urinary filtrate from Bowman's space to the exterior, naming in order the tubules and components of the urinary tract and describing any changes in filtrate composition and epithelial lining that occur in each component.
- Describe the kidney's endocrine functions and the hormonal regulation of renal function.

MAX-Yield™ STUDY QUESTIONS

1. Name the organs of the urinary system (I.A.1 and 2^1) and describe their roles in the system's functions (I.B).
2. Sketch a frontal section of the kidney (Fig. 19–1) and label the following:
   a. Capsule (II.A)
   b. Cortex (II.A.3)
   c. Medulla (II.A.4)
   d. Medullary pyramids (II.A.4)
   e. Medullary rays (II.A.5)
   f. Renal columns (Fig. 19–1)
   g. Major calyx (I.A.1.a)
   h. Minor calyx (I.A.1.a)
   i. Renal papilla (II.A.4)
   j. Renal pelvis (I.A.1.a)
   k. Renal sinus (I.A.1.a)
   l. Hilum (I.A.1.a)
   m. Renal artery and vein (I.A.2.a)
   n. Interlobar artery and vein (II.E)
   o. Arcuate artery and vein (II.E)
   p. Interlobular artery and vein (II.E)
   q. Renal lobe (II.A.6)
   r. Renal lobule (II.A.7)
3. Sketch a nephron, label its major components, and show which components lie in the cortex and which in the medulla (II.B.1–5; Fig. 19–5).
4. Sketch a renal corpuscle and label the following:
   a. Glomerulus (II.B.1.a)
   b. Visceral and parietal layers of Bowman's capsule (II.B.1.b)
   c. Urinary space (II.B.1.b)
   d. Afferent and efferent arterioles (II.B.1.f)
   e. Vascular pole (II.B.1.d)
   f. Urinary pole (II.B.1.e)
   g. Proximal convoluted tubule (II.B.2)
   h. Mesangial cells (II.B.1.a)
5. Sketch the ultrastructure of a portion of the glomerular filtration barrier (Fig. 19–2) and label the following:

---

[1] See footnote on page 1.

    a. Glomerular capillary lumen   **e.** Pedicels

    b. Glomerular capillary endothelial cell   **f.** Filtration slits (slit pores)

    c. Endothelial fenestrae   **g.** Diaphragms covering filtration slits

    d. Fused basal laminae   **h.** Urinary space

6. Describe a role undertaken by mesangial cells in maintaining filtration barrier integrity (II.B.1.f).

7. Compare proximal and distal convoluted tubules in terms of:

    a. Location in the kidney (II.B.2 and 4)

    b. Epithelial lining (height, microvilli, mitochondria, staining, lateral and basal plasma membrane infoldings; II.B.2 and 4)

    c. Luminal diameter (II.B.2 and 4)

    d. Substances absorbed from or secreted into filtrate (II.B.2 and 4)

8. What is the general function of the loop of Henle (II.B.3.b)?

9. Compare the ascending and descending loop of Henle in terms of the type of convoluted tubules its thick portions resemble, the epithelial lining of its thin portions, and permeability to water (II.B.2, 3.b.[1],[2], and 4).

10. Compare cortical and juxtamedullary nephrons in terms of their numbers, their roles (active versus passive) in establishing medullary hypertonicity, and the length of their loops (II.B.5).

11. Compare collecting tubules and ducts (II.C.1 and 2) with convoluted tubules (II.B.2 and 4) in terms of location and epithelia (epithelial height variation, cytoplasmic staining intensity, and intercellular border visibility).

12. Describe the juxtaglomerular apparatus (II.D) in terms of its components and their locations. Which cells secrete renin?

13. Trace the flow of blood from the renal arteries to the glomerular capillaries (II.E).

14. Compare the paths taken to the renal vein by blood leaving the cortical and the juxtamedullary glomeruli. (*Hint:* one path includes peritubular capillaries and stellate veins; the other includes the vasa recta; II.E.)

15. Compare the vasa recta (II.E) with the thin loops of Henle (II.B.3) from juxtamedullary nephrons in terms of location, contents, epithelial type and thickness, and countercurrent function.

16. Compare aldosterone (II.B.4) and antidiuretic hormone (ADH; II.C.2) in terms of:

    a. Site of synthesis and secretion

    b. Stimulus for secretion

    c. Site of action in the kidney

    d. Role in kidney function

17. Beginning with the stimulus for the production of renin, diagram the cascade that produces angiotensin II and explain its role in renal function (II.D).

18. Trace the flow of fluid from the urinary (Bowman's) space to a minor calyx, naming, in order, the tubules through which the fluid flows. Describe any changes in fluid composition (substances added or removed), volume, and osmolarity (tonicity) that occur in each tubule segment (II.F; Fig. 19–3).

19. Describe the structural features of the walls of the renal calyces (III), renal pelvis (III), ureters (IV), and urinary bladder (V).

20. Name—in order, from bladder to exterior—the parts of the male urethra and the epithelial lining of each part (VI.A.1–3).

21. Compare male and female urethras in terms of length, function, and epithelial lining (VI.A and B).

22. Compare the internal (V) and external (VI.A.2) urinary sphincters in terms of location and muscle type.

SYNOPSIS

I. GENERAL FEATURES OF THE URINARY SYSTEM

    A.   **Components of the System:**   The urinary system comprises the kidneys and the urinary tract.

        1. Kidneys. These paired, bean-shaped, retroperitoneal organs are located in the posterior wall of the abdominal cavity.

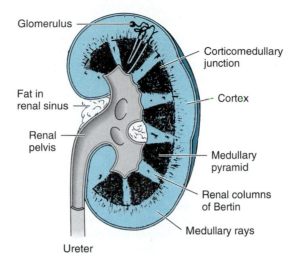

Figure 19–1. General organization of the kidney.

a. **Structural and functional subdivisions.** In frontal section (Fig. 19–1), each kidney shows a dark-staining outer **cortex** and a light-staining inner **medulla** that partly surrounds the **renal hilum.** Extensions of the medulla into the cortex are called **medullary rays,** and extensions of the cortex into the medullar region are called **renal columns (of Bertin).** The hilum consists of the **renal sinus** and its contents, which include the larger renal blood vessels, the **renal pelvis,** and adipose tissue. Each human kidney consists of several pyramid-shaped subunits—**renal lobes**—whose bases lie in the cortex and whose apices lie in the medulla. The apices are cupped by **minor calyces** that collect and empty the urine from each lobe into the larger **major calyces.** These in turn empty into the single, funnel-shaped renal pelvis, which is continuous with the **ureter.** Each lobe consists of numerous **renal lobules,** each containing hundreds of **nephrons.** These largely tubular structures filter the blood, modify the filtrate, and empty into a series of **collecting tubules and ducts** that converge on the medulla to empty urine into the minor calyces.

b. **Blood supply.** Because they are blood-filtering organs, the kidneys' blood supply is crucial to their function. A pair of **renal arteries**—one per kidney—arise from the aorta in the upper abdomen. Each undergoes successive branching to feed specialized capillary beds in both the cortex (**glomeruli** and **peritubular capillaries**) and medulla (**vasa recta**). Knowing the renal artery's branching pattern within each kidney aids in understanding how the blood reaches the capillaries that play integral roles in renal function. In addition, the structure, route, and location of the branches provide clues to the arrangement of the structural and functional subdivisions of the kidney.

2. **Urinary tract.** The **ureters, urinary bladder,** and **urethra** are described mainly in terms of their wall structure. Except for portions of the urethra, the lumen of the tract is characteristically lined with transitional epithelium.

B. **General Functions of the System:** The kidneys filter the blood; reabsorb nutrients and excrete metabolic wastes and foreign substances; regulate the ion, salt, and water concentrations of the fluids that bathe the body's tissues; and produce renin and erythropoietin. The collection of raw filtrate from the blood in the glomerular capillaries is only the first step in urine production. It is followed by the reabsorption of important ions, small proteins, nutrients, and most of the water from the filtrate. These are returned to the blood in the peritubular capillaries and vasa recta in precise proportions. The portion of the filtrate that is not reabsorbed constitutes the urine; it is carried by the ureters from the kidneys, temporarily stored in the urinary bladder, and released through the urethra.

II. KIDNEYS

A. General Organization: The kidneys, which measure approximately 11 X 6 cm, are bean-shaped, retroperitoneal organs encapsulated by dense connective tissue and surrounded by adipose tissue. Several components can be distinguished without the aid of a microscope.

1. **Renal sinus.** This medial concavity of each kidney contains the renal pelvis, the entering and exiting blood vessels and nerves, and adipose tissue.
2. **Hilum.** This region comprises the renal sinus and its contents.
3. **Cortex.** This dark-staining outer region underlies the capsule. It contains the renal corpuscles, proximal and distal convoluted tubules, peritubular capillaries, and medullary rays.
4. **Medulla.** This light-staining inner region partly surrounds the renal sinus. It comprises 8 to 18 conical **medullary pyramids** whose bases abut the cortex and whose apices (renal papillae) point toward the renal sinus. It also contains the collecting ducts, loops of Henle, and vasa recta. Each **renal papilla,** perforated by openings of the collecting ducts, is cradled by a minor calyx into which the ducts empty. Several minor calyces empty into a major calyx. The major calyces empty into the renal pelvis, which in turn drains into the ureter.
5. **Medullary rays.** These extensions of medullary tissue, which penetrate the cortex, comprise clusters of collecting tubules and ducts. One medullary ray occupies the center of each renal lobule.
6. **Renal lobes.** Each human kidney has 8 to 18 lobes. Each lobe (a medullary pyramid [II.A.4] and its associated cortex) contains numerous renal lobules.
7. **Renal lobules.** Each lobule consists of a central medullary ray and all of the nephrons emptying into its collecting tubules. The borders between adjacent renal lobules are marked by interlobular arteries and veins.

B. Nephrons: Nephrons are the functional subunits of the kidney. Each includes a renal corpuscle, a proximal convoluted tubule, a loop of Henle, and a distal convoluted tubule.

1. **Renal corpuscle.** This blood-filtering unit of the nephron consists of a glomerulus covered by a Bowman's capsule; together, these structures form the filtration barrier. Each corpuscle has both a urinary and a vascular pole.
   a. **Glomerulus.** This small tuft of capillaries has fenestrae covered by thin diaphragms. Modified smooth muscle cells called **mesangial cells** lie between the capillary loops.
   b. **Bowman's capsule** is a double-walled epithelial chamber. Its inner wall, or **visceral layer,** consists of **podocytes.** These cells have long **primary processes,** from which arise interdigitating foot processes **(pedicels)** that grasp the glomerular capillaries like fingers around a broom handle and adhere tightly to the fused capillary–podocyte basal lamina. The outer wall—the **parietal layer**—is simple squamous epithelium. The chamber between the visceral and parietal layers is the **urinary** (or **Bowman's**) **space.**
   c. **Filtration barrier.** The structures separating the capillary lumen from the urinary space (Fig. 19–2) include (1) the diaphragm-covered capillary fenestrations; (2) the fused basal laminae of the capillary endothelial cells and podocytes; and (3) the diaphragm-covered filtration slits between the interdigitating pedicels.
   d. **Vascular pole.** This side of the corpuscle is where the afferent arterioles feeding the glomerular capillaries enter and the efferent arterioles draining them exit. It lies opposite the urinary pole.
   e. **Urinary pole.** This side of the corpuscle is where the proximal convoluted tubule exits.
   f. **Filtration mechanism.** Blood is delivered to the glomerulus by the **afferent arteriole.** Arterial pressure forces fluid from the blood through the filtration barrier and into the urinary space. Each component of the barrier (fenestrae, diaphragms, basal lamina, filtration slits) aids in limiting the passage of blood components by size, thus preventing blood cells and large proteins from entering the urinary space. Molecules trapped in the basal lamina are periodically removed by the mesangial cells. A reduced volume of blood leaves the glomerulus by means of the narrower **efferent arteriole,** and the raw filtrate in the urinary space enters the proximal convoluted tubule for further processing.
2. **Proximal convoluted tubule.** This epithelial tube begins at the renal corpuscle's urinary pole. Its simple low-columnar-to-cuboidal lining cells have abundant long microvilli, which form a **brush border** that partly obscures the lumen and increases the surface area available for absorption. The lining cells absorb approximately two thirds of the sodium from the filtrate; water follows passively, reducing the filtrate volume by approximately the same pro-

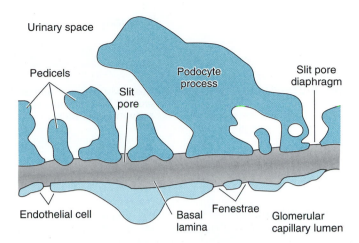

Figure 19–2. Schematic diagram of the glomerular filtration barrier. Fluid from the capillary lumen passes through fenestrae in the capillary wall, subsequently moves through the fused basal laminae of the capillary endothelial cells and podocytes, and finally proceeds through diagram-covered filtration slits between the pedicels of the podocytes to enter the urinary space. The basal lamina exhibits a central lamina densa sandwiched between two less dense laminae rarae.

portion. All of the glucose (unless present in great excess), amino acids, acetoacetate, and vitamins are reabsorbed by facilitated transport (2.II.C.1.b), and small proteins are reabsorbed by pinocytosis. The many mitochondria required for the energy-intensive absorptive function interdigitate with basal membrane infoldings and make the lining cells acidophilic. The convoluted part of the proximal tubule lies in the cortex and empties into its straight portion (also called the thick descending limb of the loop of Henle), which has the same epithelium and function. Together, the convoluted and straight portions of the proximal tubule measure approximately 14 mm, making this the longest portion of the nephron in the cortex and the most often encountered tubule type in cortical sections.

3. **Loop of Henle.**
   a. **Structure.** Henle's loop, a U-shaped epithelial tube, includes thick and thin descending limbs and thin and thick ascending limbs (Fig. 19–3). It extends from the proximal convoluted tubule in the cortex, dips into the medulla, and returns to the cortex, where it empties into the distal convoluted tubule. The abrupt transition from thick to thin in both arms of the U reflects changes from low columnar or cuboidal to squamous epithelium and from squamous back to cuboidal epithelium. The luminal diameter changes less than the external diameter.
   b. **Function.** A prerequisite for forming hypertonic urine, the loop acts as a **countercurrent multiplier** to establish an osmotic gradient in the interstitial fluid of the medulla. **Hypertonic** (concentrated) and **hypotonic** (dilute) are relative terms. The point of reference assumed is the tonicity of normal tissue fluid or blood (**isotonic**). The medullary interstitium, for example, is approximately isotonic near the corticomedullary junction and gradually becomes most hypertonic near the tips of the medullary papillae. The descending and ascending portions of the loop of Henle play important roles in establishing and maintaining this osmotic gradient.
   (1) **Descending part.** This first segment of Henle's loop plays a passive role in making the medullary interstitium hypertonic and helps maintain the gradient. The filtrate delivered to the descending part of the thin loop by the descending thick loop is isotonic, but the removal of salt, nutrients, and water in the proximal tubule reduces the volume from its raw state in the urinary space. The descending loop is permeable to both water and salt, although it is more permeable to water. As the fluid in its lumen passes deeper into the hypertonic medulla, it loses water to the interstitium and becomes more hypertonic. As water is lost, the filtrate volume decreases. The filtrate's tonicity equilibrates with the hypertonic interstitium, peaking at the bottom of the U as it enters the ascending portion.

A. Flow of Fluid

B. Changes in Fluid Composition

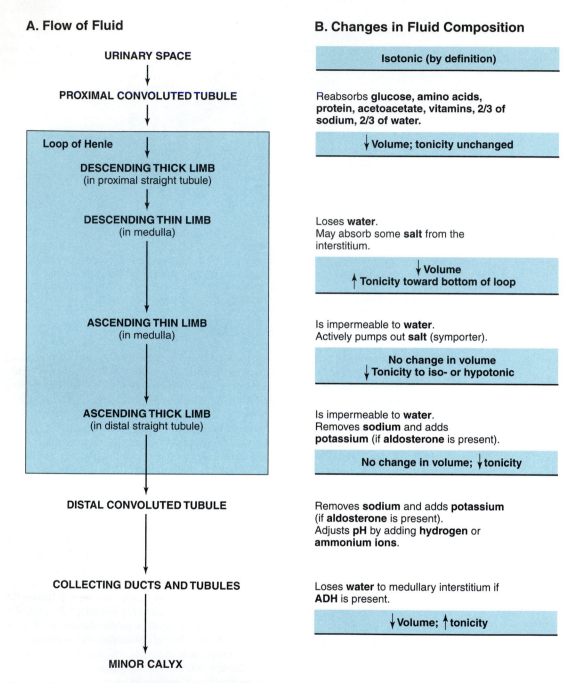

Figure 19–3. A. Flow of fluid from the urinary space to the minor calyx. **B.** Summary of the changes in fluid composition (*shaded*) and tonicity that occur in each tubule segment, owing to the actions carried out by that segment (*unshaded*).

(2) **Ascending part.** This segment of Henle's loop has a more active role in setting up the gradient and, in particular, making the medullary interstitium hypertonic. The cells lining the thick ascending component resemble those in the distal convoluted tubule (II.B.4). They contain a $Na^+/K^+/Cl^-$ pump **(symporter)** that constantly pumps these ions (in a 1:1:2 ratio) from the filtrate into the interstitial fluid around the tubules. This symporter thus increases the salt concentration (and tonicity, or osmolarity) in the interstitium and is responsible for approximately 20% of the reabsorption of these ions

from the filtrate. Because this part of the loop is impermeable to water, water in the filtrate cannot follow the salt into the interstitium and dilute it. As the reduced volume of filtrate ascends toward the distal convoluted tubule in the cortex, the removal of salt (but not water) by the cells lining this part of the loop causes the fluid in its lumen to gradually become isotonic or hypotonic. The importance of the osmotic gradient to hypertonic urine production becomes clearer in light of the events occurring in the collecting ducts as they pass through the medulla en route to the calyces (II.C.2).

4. **Distal convoluted tubule.** This last segment of the nephron lies in the cortex. Its epithelial lining is low cuboidal and lacks a brush border, making its lumen appear wider. The lining cells' nuclei are displaced apically by abundant mitochondria interdigitating with folds of the basal plasma membrane—a common feature of ion-transporting cells. These cells are more basophilic than those lining the proximal tubules. The lateral cell boundaries are indistinct as a result of extensive lateral membrane interdigitations with their neighbors. The distal tubule epithelium forms a disk of tightly packed columnar cells called a **macula densa** at the point near the renal corpuscle's vascular pole where it contacts an afferent arteriole. This disk may monitor NaCl concentration or the flow rate of the tubular fluid (II.D). The distal convoluted tubule makes final adjustments of the salt, water, and acid balance. The presence of **aldosterone** (from the adrenal cortex; 21.A.3.a) causes the lining cells to remove more sodium from and add potassium to the fluid. **Atrial natriuretic factor** (11.III.B.2.a) increases sodium excretion. Elevated levels of parathyroid hormone cause the lining cells to reabsorb more calcium and release more phosphate. The cells may further adjust the pH by secreting hydrogen and ammonium ions into the lumen.

5. **Cortical and juxtamedullary nephrons.** Renal corpuscles are found throughout the cortex. Although most belong to the cortical nephrons, the 15% closest to the medulla belong to the juxtamedullary nephrons. The latter group has short, thick descending limbs and longer thin limbs that extend deeper into the medulla. The juxtamedullary nephrons bear the primary responsibility for setting up the medulla's osmotic gradient.

C. Collecting Tubules and Ducts:

1. **Structure.** These differ from the nephrons in their embryonic origin and are easily distinguished from proximal and distal tubules in sections. Their blocklike lining cells have distinct intercellular borders; they are cuboidal in the smaller tubules and columnar in the larger (eg, papillary) ducts of the medulla. Because their cytoplasm stains poorly, the lining cells appear clear or white.

2. **Function.** Cortical collecting tubules receive a reduced volume of hypotonic or isotonic urine from the nephrons and empty it into larger collecting ducts. These leave the cortex in medullary rays and enter the medulla, increasing in size until they open into a minor calyx through the tip of a papilla. The medullary collecting ducts play the final role in forming hypertonic urine. Under the influence of pituitary **antidiuretic hormone** [(**ADH**) or vasopressin; 20.IV.A.1], they become permeable to water. As they pass through the osmotic gradient of the medulla, water diffuses passively from their lumens into the hypertonic medullary interstitium, causing the osmolarity of the fluid in their lumens to equilibrate with the interstitium; only a small volume of hypertonic urine is released. Without ADH, the collecting ducts remain impermeable to water, and a larger amount of hypotonic or isotonic urine is produced. Without the hypertonic medullary interstitium established by the ascending portion of the loop, the excess water would not leave the ducts, even in the presence of ADH. This mechanism for concentrating the urine is inhibited by drugs such as **furosemide,** which inactivates the ascending limb's symporter (II.B.3.b.[2]), reducing the hypertonicity of the medullary interstitium.

D. Juxtaglomerular Apparatus: Located near each renal corpuscle's vascular pole, at the point of contact between a distal convoluted tubule and an afferent arteriole, this apparatus includes **juxtaglomerular (JG) cells, a macula densa,** and **polkissen (extraglomerular mesangial cells).** The JG cells, modified smooth muscle cells in the afferent arteriole's wall, exhibit secretory ultrastructure and numerous PAS-positive cytoplasmic granules. Although the macula densa's influence on the JG cells is poorly understood, subnormal blood volume, pressure, or sodium causes the JG cells to secrete **renin.** This enzyme cleaves plasma **angiotensinogen** to produce **angiotensin I,** which is converted to active **angiotensin II** by enzymes in the lungs. Angiotensin II, a vasoconstrictor, increases blood pressure and stimulates aldosterone production by the adrenal cortex, thereby increasing chloride and sodium reabsorption by the distal tubule. The sodium and chloride enter the blood in the peritubular capillaries. Although the distal tubules are impermeable to water, the increased tonicity of blood leaving the kidneys draws water into the blood as it passes through other tissues, increasing blood volume and pressure. Increased blood pressure distends the afferent arterioles, stretching the JG cells and halting renin secretion. The function of polkissen is unknown.

E. Blood Supply and Circulation: The arteries are accompanied by similarly named veins. Each kidney receives a **renal artery**—a branch from the abdominal aorta. **Anterior** and **posterior** branches arise from the renal artery before reaching the hilum. **Interlobar arteries** arise from the anterior and posterior branches in the hilum and penetrate the medulla between the pyramids. **Arcuate arteries** arise from the interlobar arteries and course along the arched border between the cortex and medulla. **Interlobular arteries** arise at right angles from the arcuate arteries; they penetrate the cortex between the medullary rays and lie between neighboring renal lobules. Many **afferent arterioles** arise from each interlobular artery, each of which supplies a glomerulus (II.B.1.a). An **efferent arteriole** carries blood away from the glomerulus. Efferent arterioles of **cortical nephrons** branch to form abundant **peritubular capillaries** that carry absorbed products away from the proximal and distal tubules and converge to form the **stellate veins** of the peripheral cortex. These drain into the interlobular veins. Efferent arterioles of **juxtamedullary nephrons** give rise to numerous straight capillary loops—**vasa recta**—that descend into the medulla. Vasa recta arise mainly from the efferent arterioles of the juxtamedullary nephrons; some arise from the arcuate artery. The descending vasa recta carry isotonic blood into the medulla. This blood loses water and picks up salt as it passes deeper into the medulla. Unlike the loop of Henle, the ascending vasa recta are permeable to salt and water. As blood ascends through the gradient, its tonicity equilibrates with the interstitium. The blood leaving the medulla is thus isotonic or slightly hypertonic. The passive salt and water exchange between the vasa recta and the interstitium is known as the **countercurrent exchange** mechanism. This mechanism is important in removing water lost to the filtrate during its descent into the medulla and thus in maintaining the osmotic gradient set up by the countercurrent multiplier of Henle's loop. Blood in the ascending vasa recta drains into interlobular veins and exits through veins accompanying the larger arteries.

F. Summary of Renal Function: An organized arterial system carries blood to the glomeruli of the renal corpuscles. Each corpuscle acts as both filter and funnel, collecting raw filtrate and directing it to the proximal convoluted tubule, where glucose, amino acids, acetoacetate, small proteins, vitamins, sodium, and water are reabsorbed (see Fig. 19–3). The remaining fluid enters the loop of Henle, which sets up a hypertonic osmotic gradient in the medulla. The fluid subsequently leaves the loop and enters the distal convoluted tubule. Here, aided by the juxtaglomerular apparatus and aldosterone, the salt, ion, and water balance between the blood and urine is adjusted. Next, the fluid exits the nephron through the collecting ducts, which pass back through the medulla. ADH renders the medullary collecting ducts permeable to water, allowing the water to flow out of the collecting duct lumens and into the medullary interstitium. This results in the release of a reduced volume of hypertonic urine into the minor calyx.

III. RENAL CALYCES & RENAL PELVIS

The walls of each renal calyx and pelvis consist of mucosa, muscularis, and adventitia; no submucosa is present. The mucosa, which consists of typical urinary (transitional) epithelium (4.III.B.8), attaches to an underlying helical meshwork of smooth muscle (muscularis) by a connective tissue lamina propria of variable density. The epithelium forms an osmotic barrier that protects the surrounding tissues from the hypertonic urine and the urine from dilution. The adventitia blends into the adipose tissue in the renal sinus.

IV. URETERS

These carry urine from the renal pelvis to the urinary bladder. Although the ureter's lumen is narrower than that of the renal pelvis, the wall structure, including the transitional epithelial lining, is similar. The ureter wall thickens and the muscle cells change from a helical to a longitudinal array near the bladder before fanning out in the bladder wall to form the bladder's superficial and deep trigones.

V. URINARY BLADDER

This distensible muscular sac, lined by transitional epithelium over a dense lamina propria, has walls like those of the ureter, pelvis, and calyces but with a thicker muscularis. The smooth muscle fibers run in many directions and are not organized in layers except near the urethral orifice, where they form an involuntary **internal sphincter.**

VI. URETHRA

The urethra differs in length, epithelium, and function in males and females.

A. Male Urethra: Longer than the female urethra, this conducts both urine and seminal fluid; it has three main parts.
  1. **Prostatic segment.** This most proximal part of the male urethra exits the neck of the urinary bladder. It is surrounded by the prostate gland (22.IV.B) and is lined by transitional epithelium. This part receives the prostatic and ejaculatory ducts and empties into the membranous segment.
  2. **Membranous segment.** This is the shortest segment and is encircled by the skeletal muscle of the urogenital diaphragm, whose fibers form a voluntary **external sphincter.** It is lined by pseudostratified columnar epithelium and empties into the cavernous segment.
  3. **Cavernous segment.** This section passes through the **corpus spongiosum** of the penis (22.V.A.2). Within the glans, near the tip of the penis, the urethral lumen widens to form the **fossa navicularis,** where the epithelium changes from pseudostratified columnar to stratified squamous. The urethra opens at the end of the penis through the **urethral meatus.** Many **glands of Littre** empty mucous secretions into the lumen all along the urethra; these glands are more numerous in the pendulous part.

B. Female Urethra: Shorter than the male urethra, this counterpart carries only urine. It is lined by stratified squamous epithelium, with patches that are pseudostratified columnar. Midway along its path from bladder to exterior, it is surrounded by a voluntary external sphincter formed by the urogenital diaphragm.

MULTIPLE-CHOICE QUESTIONS

Select the single best answer.

19.1. Which of the following structures is indicated by the letter A in Figure 19–4?
(A) Afferent arteriole
(B) Bowman's space
(C) Brush border
(D) Efferent arteriole
(E) Macula densa
(F) Parietal layer of Bowman's capsule
(G) Podocyte

19.2. Which of the following best describes the composition of a renal lobule?
(A) Renal pyramid and associated cortex
(B) Medullary ray and all nephrons that empty into it

(C) Renal pyramid and all nephrons that empty into it
(D) Interlobular artery and all nephrons it supplies
(E) Renal corpuscle and all associated renal tubules

19.3. Blood in the arcuate arteries subsequently flows into which of the following vascular channels?
(A) Afferent arteriole(s)
(B) Efferent arterioles
(C) Glomerular capillaries
(D) Interlobar arteries
(E) Interlobular arteries
(F) Peritubular capillaries
(G) Stellate veins

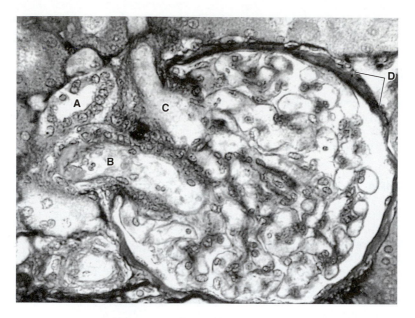

Figure 19–4.

19.4. Collections of cortical tissue between the medullary pyramids are called:
(**A**) Interlobular cortex
(**B**) Juxtamedullary nephrons
(**C**) Medullary rays
(**D**) Renal columns of Bertin
(**E**) Renal lobes

19.5. Which of the following vessels are typically seen at the border between the renal cortex and medulla?
(**A**) Arcuate arteries and veins
(**B**) Interlobar arteries and veins
(**C**) Interlobular arteries and veins
(**D**) Stellate veins
(**E**) Vasa recta

19.6. Which of the labels in Figure 19–5 corresponds to the location of podocytes?

19.7. Which of the labels in Figure 19–5 corresponds to the location of the brush border?

19.8. Which of the labels in Figure 19–5 corresponds to the location of the most hypertonic filtrate?

19.9. Which of the labels in Figure 19–5 corresponds to the nephron component that contains the macula densa in its wall?

19.10. Which of the labels in Figure 19–5 corresponds to the location of the filtration barrier?

19.11. Which of the labels in Figure 19–6 corresponds to the cell body of a podocyte?

19.12. Which of the labels in Figure 19–6 corresponds to an endothelial cell of the glomerular capillary?

19.13. Which of the labels in Figure 19–6 corresponds to the fused basal lamina of the glomerular filtration barrier?

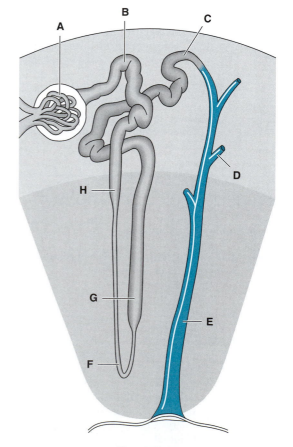

Figure 19–5.

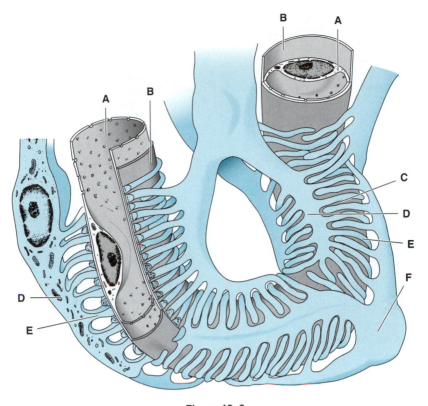

Figure 19–6.

19.14. Which of the labels in Figure 19–6 corresponds to a pedicel?

19.15. Which of the labels in Figure 19–6 corresponds to a filtration slit?

19.16. Which of the diagrams in Figure 19–7 best represents the morphology of an epithelial cell from the proximal convoluted tubule?

19.17. Which of the diagrams in Figure 19–7 best represents the morphology of an epithelial cell from the thin loop of Henle?

19.18. Which of the diagrams in Figure 19–7 best represents the morphology of an epithelial cell from the distal convoluted tubule?

19.19. Which of the diagrams in Figure 19–7 best represents the morphology of an epithelial cell from a collecting duct or tubule?

19.20. Which of the following is the modified smooth muscle cell responsible for the secretion of renin?

    (A) Endothelial cell

    (B) Juxtaglomerular cell

    (C) Mesangial cell

    (D) Podocyte

    (E) Polkissen (extraglomerular mesangial cell)

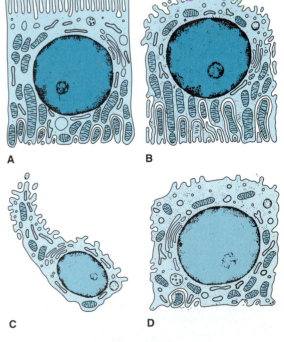

Figure 19–7.

19.21. The basal lamina of the glomerular filtration barrier is cleared of occasional build-up of cellular debris and some antigen–antibody complexes by the phagocytic activity of which of the following cell types?
(A) Endothelial cell
(B) Juxtaglomerular cell
(C) Mesangial cell
(D) Podocyte
(E) Polkissen (extraglomerular mesangial cell)

19.22. Which of the following cell types comprises the visceral layer of Bowman's capsule?
(A) Endothelial cell
(B) Juxtaglomerular cell
(C) Mesangial cell
(D) Podocyte
(E) Polkissen (extraglomerular mesangial cell)

19.23. Which of the following types of epithelium lines the urinary bladder?
(A) Pseudostratified columnar
(B) Simple columnar
(C) Simple cuboidal
(D) Simple squamous
(E) Stratified cuboidal
(F) Stratified squamous
(G) Transitional

19.24. Which of the following types of epithelium comprises the parietal layer of Bowman's capsule?
(A) Pseudostratified columnar
(B) Simple columnar
(C) Simple cuboidal
(D) Simple squamous
(E) Stratified cuboidal
(F) Stratified squamous
(G) Transitional

19.25. Which of the following types of epithelium lines the thick ascending limb of the loop of Henle?
(A) Pseudostratified columnar
(B) Simple columnar
(C) Simple cuboidal
(D) Simple squamous
(E) Stratified cuboidal
(F) Stratified squamous
(G) Transitional

19.26. Which of the following types of epithelium lines the prostatic urethra?
(A) Pseudostratified columnar
(B) Simple columnar
(C) Simple cuboidal
(D) Simple squamous
(E) Stratified cuboidal
(F) Stratified squamous
(G) Transitional

19.27. Which of the following types of epithelium lines the thin loop of Henle?
(A) Pseudostratified columnar
(B) Simple columnar

(C) Simple cuboidal
(D) Simple squamous
(E) Stratified cuboidal
(F) Stratified squamous
(G) Transitional

19.28. Which of the following types of epithelium lines the proximal convoluted tubule?
(A) Pseudostratified columnar
(B) Simple columnar
(C) Simple cuboidal
(D) Simple squamous
(E) Stratified cuboidal
(F) Stratified squamous
(G) Transitional

19.29. Which of the following is characteristic of the proximal but not of the distal convoluted tubules?
(A) Convoluted portion lies in the renal cortex
(B) Epithelial lining is similar to that of the thin loop of Henle
(C) Exhibits a brush border
(D) Lining cells are basophilic
(E) Contains a macula densa
(F) Secretes glucose and amino acids into the filtrate
(G) Is aldosterone's site of action

19.30. Which of the following is characteristic of the renal cortex but not of the renal medulla?
(A) Borders on the renal hilum
(B) Contains the medullary rays
(C) Contains the papillary ducts
(D) Contains the renal pyramids
(E) Contains most of the kidney's vasa recta
(F) Has a hypertonic interstitium
(G) Is the site of action for antidiuretic hormone (ADH)

19.31. Which of the following is characteristic of the thin loops of Henle but not of the vasa recta?
(A) Ascending portion is impermeable to water
(B) Is a component of the juxtaglomerular apparatus
(C) Has a countercurrent exchange system
(D) Contents may be isotonic or hypertonic
(E) Epithelial lining is simple squamous
(F) Is permeable to water regulated by ADH
(G) Secretes angiotensinogen

19.32. Which of the following is characteristic of the cortical, but not the juxtamedullary, nephrons?
(A) Associated efferent arterioles supply the vasa recta
(B) Have long thin loops of Henle that penetrate deep into the renal medulla
(C) Have distal convoluted tubules that empty directly into collecting tubules in the renal cortex
(D) Have proximal convoluted tubules located in the renal cortex

(E) Have distal convoluted tubules located in the renal medulla

(F) Loops of Henle have long, thick descending limbs

(G) Are primarily responsible for establishing the medulla's hypertonic gradient

19.33. Which of the following is characteristic of the internal, but not of the external, urinary sphincter?

(A) Comprises part of the urogenital diaphragm

(B) Consists of smooth muscle

(C) Contracts voluntarily

(D) Surrounds the female urethra midway along its length

(E) Surrounds the membranous urethra

19.34. The efferent arterioles of the juxtamedullary renal corpuscles are considered portal vessels because they connect glomerular capillaries with which of the following vessels?

(A) Afferent arterioles

(B) Arcuate veins

(C) Hepatic sinusoids

(D) Other glomerular capillaries

(E) Peritubular capillaries

(F) Stellate veins

(G) Vasa recta

19.35. Voluntary neural control of urine flow in males is managed by which of the following mechanisms?

(A) Innervation of an internal sphincter around the prostatic urethra as it exits the bladder

(B) Innervation of an external sphincter around the membranous urethra as it exits the pelvic cavity

(C) Regulation of glomerular filtration rate through innervation of the afferent arterioles

(D) Neural control of ADH release by the hypothalamus

(E) Neural control of aldosterone production by the adrenal gland

ANSWERS TO MULTIPLE-CHOICE QUESTIONS

19.1. E (II.B.4 and D)
19.2. B (II.A.7)
19.3. E (II.E)
19.4. D (Fig. 19–1)
19.5. A (II.E)
19.6. A (II.B.1.b)
19.7. B and H (II.B.2)
19.8. F (II.B.3.b.[1])
19.9. C (II.B.4)
19.10. A (II.B.1)
19.11. F (II.B.1.b)
19.12. A (II.B.1.a; Fig. 19–2)
19.13. B (II.B.1.c; Fig. 19–2)
19.14. E (II.B.1.b; Fig. 19–2)
19.15. C (II.B.1.b; Fig. 19–2)
19.16. A (II.B.2)
19.17. C (II.B.3.a)
19.18. B (II.B.4)
19.19. D (II.C.1)
19.20. B (II.D)
19.21. C (II.B.1.f)
19.22. D (II.B.1.b)
19.23. G (V)
19.24. D (II.B.1.b)
19.25. C (II.B.4)
19.26. G (VI.A.1)
19.27. D (II.B.3.a)
19.28. C (II.B.2)
19.29. C (II.B.2 and 4)
19.30. B (I.A.1.a; II.A.3 and 4)
19.31. A (II.B.3.a and b; II.E)
19.32. F (II.B.5)
19.33. B (V; VI.A.2)
19.34. G (II.B.5; II.E; 11.II.D)
19.35. B (V; VI.A.2)

20 Pituitary Gland & Hypothalamus

OBJECTIVES

This chapter should help the student to:

- Describe the location and embryonic origins of the pituitary gland.
- Name the divisions of the pituitary gland.
- Name the cell types in each pituitary division and indicate any characteristic staining properties.
- List the pituitary hormones, indicating for each the division and cell type of origin, and the target of the hormone.
- Describe the role of the hypothalamus in controlling pituitary gland function.
- Describe the blood supply to the pituitary gland and its role in pituitary function.
- Explain the role of negative feedback in pituitary gland function.
- Distinguish between the neurohypophysis and the adenohypophysis and identify their subdivisions, sinusoids, and cell types in a micrograph of the pituitary gland.

MAX-Yield™ STUDY QUESTIONS

1. Describe endocrine gland characteristics in terms of embryonic origin (I.B[1]), secretory cell arrangement (I.C), abundance of blood capillaries (I.C), mode of secretion release and transport (I.D), and typical secretory product (I.D).
2. Describe hormones in terms of:
   a. Chemical composition (two basic types; I.D.1 and 2)
   b. Relative distance and the route taken between the secretory cell and the target site (I.D)
   c. Relative amount needed to elicit a response from the target cell (I.D)
   d. General function (I.D)
3. Describe the location of the pituitary gland (hypophysis) and its relations to the hypothalamus, sella turcica, and optic chiasm (II).
4. Name the pituitary gland's two major divisions (Table 20–1) and compare them in terms of:
   a. Embryonic origin
   b. Microscopic structure
   c. Connections with the hypothalamus (vascular versus neural; III.F; IV.D; Fig. 20–2)
   d. Major hormones released (III.A.2.a and b; IV.A.1; Table 20–1)
5. Sketch the adult pituitary gland and hypothalamus and label the following (Fig. 20–2):
   a. Adenohypophysis
   b. Neurohypophysis
   c. Hypothalamus
   d. Anterior lobe
   e. Posterior lobe
   f. Optic chiasm
   g. Pars distalis
   h. Pars tuberalis
   i. Pars intermedia
   j. Rathke's cysts
   k. Median eminence
   l. Infundibulum
   m. Pars nervosa
   n. Supraoptic nucleus
   o. Paraventricular nucleus
   p. Primary capillary plexus
   q. Secondary capillary plexus
   r. Hypophyseal portal veins
6. Name the three major subdivisions of the adenohypophysis (Table 20–1).
7. Name the major parenchymal cell types in the adenohypophysis, based on their staining properties (III.A.1, 2.a and b).

---

[1] See footnote on page 1.

8. Compare pituitary acidophils and basophils (III.A.2.a and b; Table 20–2) in terms of:
    a. Staining properties, including:
        (1) Affinity for acidic dyes, such as eosin and orange G
        (2) Affinity for basic dyes, such as hematoxylin and aniline blue
        (3) Positive staining with the PAS reaction
    b. Hormones secreted by each
9. Name the primary target organs of (and their response to) the following hormones (Table 20–2):
    a. FSH   **e.** Growth hormone
    b. LH   **f.** Prolactin
    c. ICSH   **g.** ACTH
    d. TSH
10. Beginning with neural stimulation of the hypothalamus, trace the events leading to the secretion of the thyroid hormones T_3 and T_4 by the thyroid gland. Name, in order, the neural, endocrine, and vascular components involved (III.A.2.b and E; Table 20–2).
11. Name the major subdivisions of the neurohypophysis (Table 20–1).
12. Where are the cell bodies of the pars nervosa's unmyelinated axons found (IV.A)?
13. Describe Herring bodies in terms of their contents and location (IV.A).
14. List the contents of the neurosecretory vesicles in the axons of the pars nervosa (IV.A.1–3).
15. Describe pituicytes in terms of structure and function (IV.C).
16. Name the two major hormones released by the neurohypophysis, their target organs, and the effects they elicit (IV.A.1; Table 20–2).

SYNOPSIS

I. GENERAL FEATURES OF THE ENDOCRINE SYSTEM

    A. Components of the System: The endocrine system comprises several organs (eg, adenohypophysis, thyroid gland, adrenal gland), islands of endocrine tissue in exocrine glands (eg, islets of Langerhans), and some isolated endocrine cells (eg, cells with DNES functions in the digestive tract mucosa).

    B. Origin: Endocrine glands are ductless glands that develop as invaginations of epithelial surfaces, such as oral ectoderm or gut endoderm, and subsequently pinch off, losing contact with the parent epithelium.

    C. Microscopic Structure: Endocrine glands typically contain secretory cells arranged as cords, clumps, or follicles in direct contact with abundant capillaries or sinusoids.

    D. Secretions: Endocrine cells release their merocrine secretions—typically hormones—into the bloodstream. Other products released into the bloodstream rather than into ducts (eg, enzymes, serum albumin) are also considered endocrine secretions. **Hormones** are molecules with specific regulatory effects on a target cell, tissue, or organ that is often located far from the gland. Hormones elicit specific and dramatic effects at very low concentrations. They directly or indirectly affect all tissues and many are essential to maintaining the internal steady-state environment. They regulate carbohydrate, protein, and lipid metabolism; mineral and water balance in body fluids; growth; sex-related differences in body shape and sexual function; and behavior, temperament, and emotions.
        1. Peptide hormones. These proteins, glycoproteins, or short peptides bind to specific receptors on target cell surfaces. They often stimulate the production of intracellular second messengers, such as cyclic AMP, in the target cells (2.II.C.2).
        2. Steroid hormones. These lipid-soluble hormones easily cross target cell plasma membranes to directly affect cell function. They bind to specific binding proteins in the cytoplasm and the nucleus. The nuclear receptors subsequently bind to DNA and directly affect gene transcription.

    E. Neuroendocrine System: The complex, interrelated functions of cells, tissues, and organs are controlled and coordinated by two overlapping systems: the nervous system (Chapter 9) and the endocrine system (Chapters 20–23). Increasingly, these are considered parts of a single neuroendocrine system. Once called the master gland because of its ability to control other glands, the **pituitary gland (hypophysis)** currently is seen as a focal connection between the endocrine and nervous systems. The secretory activities of its two parts, the adenohypophysis and the neurohypophysis, are both controlled by a nearby part of the brain, the **hypothalamus.** Hypothalamic activity is controlled by neural connections with other parts of the nervous system and by negative feedback from the hormones produced by the pituitary gland's target cells. Pituitary-related diseases present chiefly as the effects of hypersecretion or hyposecretion of pituitary hormones; they may be caused by lesions of the pituitary gland, its target organs, or the hypothalamus.

II. LOCATION, GENERAL ORGANIZATION, & EMBRYONIC ORIGINS OF THE PITUITARY GLAND

The pituitary gland is suspended by a stalk from the hypothalamus at the base of the diencephalon. It rests in a depression in the sphenoid bone called the sella turcica, behind the optic chiasm. Its two major divisions, the anterior **adenohypophysis** and the posterior **neurohypophysis,** differ in embryonic origin (Fig. 20–1), structure, and function (Tables 20–1 and 20–2).

III. ADENOHYPOPHYSIS

Each secretory cell in the adenohypophysis synthesizes and stores one of the following hormones: follicle-stimulating hormone (FSH), thyrotropin (thyroid-stimulating hormone; TSH), luteinizing hormone (LH), adrenocorticotropic hormone (ACTH), growth hormone (GH), or prolactin. These hormones control the secretory activities of many other glands. Their release is regulated by specific releasing or inhibiting hormones produced by the hypothalamus and delivered to the adenohypophysis by the blood in the hypophyseal portal system (III.D).

A. **Pars Distalis:**
1. **Chromophobes.** These cells stain poorly and appear clear or white in tissue sections. Together, the three chromophobe subpopulations make up approximately 50% of the pars anterior's epithelial cells. They include (1) undifferentiated nonsecretory cells, which may be stem cells; (2) partly degranulated chromophils, which contain sparse granules; and (3) follicular cells, the predominant chromophobe type, which form a stromal network supporting the chromophils; these stellate cells may be phagocytic.
2. **Chromophils.** These hormone-secreting cells stain intensely, owing to the abundant cytoplasmic secretory granules in which hormones are stored. A specific cell type exists for each hormone. Usually larger than chromophobes, chromophils comprise two classes:
   a. **Acidophils** secrete simple proteins and stain intensely with eosin and orange G but respond negatively to the PAS reaction. More abundant in the gland periphery, they are

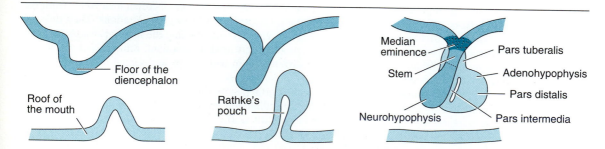

Figure 20–1. Schematic diagram of the development of the pituitary gland (hypophysis). The ectoderm of the roof of the developing oral cavity (*light color*) gives rise to the adenohypophysis. The neurohypophysis is formed by a downgrowth of neural ectoderm from the floor of the developing diencephalon. (Reproduced, with permission, from Junqueira LC, Carneiro J, Kelley RO: *Basic Histology,* 9th ed. Stamford, CT: Appleton & Lange, 1998.)

Table 20–1. General organization and embryonic origins of the pituitary.

Major Divisions	Embryonic Origin	General Structure	Subdivisions
Adenohypophysis (See section III for more detail)	Upward evagination of ectoderm lining primitive oral cavity (Fig. 20–1); contacts and fuses with neurohypophyseal downgrowth	Glandular epithelial cell cords separated by sinusoids of secondary capillary plexus; not directly innervated by hypothalamus, only by autonomic nerves from carotid plexus	**Pars distalis**—largest pituitary subdivision (Fig. 20–2) **Pars tuberalis**—superior extension of distalis; forms partial sleeve around infundibulum of neurohypophysis **Pars intermedia**—narrow band of tissue bordering pars nervosa of neurohypophysis
Neurohypophysis (See section IV for more detail)	Downgrowth of neural ectoderm of the hypothalamus; considered part of the brain (Fig. 20–1)	Contains abundant axons whose cell bodies are located mainly in supraoptic and paraventricular nuclei of the hypothalamus	**Median eminence** of tuber cinereum forms floor of hypothalamus (Fig. 20–2) **Infundibular stem (neural stalk)** carries axons from hypothalamus to pars nervosa and contains capillary loops of primary capillary plexus **Pars nervosa (infundibular process)**—expanded lobe of neurohypophysis; contains axon terminals and capillaries

usually smaller than basophils and have larger and more numerous granules. The acidophils include two major hormone-secreting cell types: **somatotrophs** produce **growth hormone (GH, somatotropin),** and **mammotrophs** produce **prolactin.** (A mnemonic for hormones secreted by acidophils is *GPA: growth hormone, prolactin, acidophils.*)

   b. **Basophils** secrete glycoproteins, stain with hematoxylin and other basic dyes, and respond positively to the PAS reaction. More abundant in the core of the gland, they are usually larger than acidophils, with fewer and smaller granules. The three major hormone-producing basophils produce four major hormones. (A mnemonic for hormones produced by basophils is *B-FLAT: Basophils, FSH, LH, ACTH, TSH.*) Each of the two **gonadotrophs** produces a different gonadotropin. One produces **follicle-stimulating hormone (FSH);** the other produces **luteinizing hormone (LH;** called **interstitial cell–stimulating hormone [ICSH]** in males). **Corticotrophs** produce **adrenocorticotropin (ACTH). Thyrotrophs** produce **thyrotropin (thyroid-stimulating hormone, or TSH).**

B. **Pars Tuberalis:** This funnel-shaped superior extension of the pars distalis surrounds the infundibular stem (Fig. 20–2). It resembles the pars distalis but contains mostly gonadotrophs. The pars tuberalis contains many capillaries of the primary capillary plexus (III.D.1).

C. **Pars Intermedia:** This is a band or wedge of adenohypophysis between the pars distalis and pars nervosa; it is rudimentary in humans. It contains **Rathke's cysts**—small, irregular, colloid-containing cavities lined with cuboidal epithelium that are the remnants of Rathke's pouch. It also contains scattered clumps and cords of basophilic cells, or melanotrophs, which secrete melanocyte-stimulating hormone (β-MSH).

D. **Blood Supply and Hypophyseal Portal System:**
   1. **Primary capillary plexus.** This profusion of capillaries lies in the upper infundibular stalk and lower median eminence; it extends into the pars tuberalis (see Fig. 20–2). The plexus receives blood from the anterior and posterior superior hypophyseal arteries (from the circle of Willis) and drains into the hypophyseal portal veins.
   2. **Hypophyseal portal veins.** These small veins and venules lie mainly in the middle and lower infundibular stalk and in parts of the pars tuberalis. They receive blood from the primary capillary plexus and carry it directly to the secondary capillary plexus in the pars distalis (see Fig. 20–2). Vessels carrying blood directly from one capillary plexus to another without returning to the general circulation are defined as portal vessels.
   3. **Secondary capillary plexus.** This rich plexus of fenestrated capillaries located throughout the pars distalis also penetrates the pars tuberalis and pars intermedia (see Fig. 20–2). Some connections also exist between this capillary bed and that in the pars nervosa. The sinusoids between the clumps and cords of cells in the pars distalis belong to this plexus, which receives venous blood directly from the hypophyseal portal veins and arterial blood from the

Table 20–2. Source, target, and effect of the major pituitary hormones.

Region	Cell Type		Hormone Secreted	Primary Target	Direct Effects	Secondary Targets and Indirect Effects
Pars distalis and tuberalis	Acidophils	Somatotrophs	Growth hormone (GH), somatotropic hormone (STH), somatotropin	Liver, epiphyseal cartilage	Increase somatomedin secretion	Stimulate most cells (especially epiphyseal chondrocytes) to increase growth rate; cause increased breakdown and decreased synthesis of triglycerides in adipocytes and other cells; increase somatostatin secretion by hypothalamus and decrease GHRH levels
		Mammotrophs	Prolactin, lactogenic hormone, luteotropic hormone (LTH)	Mammary gland	Stimulate milk secretion	Increase dopamine secretion by hypothalamus, inhibiting prolactin secretion; decrease prolactin-releasing factor (PRF) secretion
				Brain	Increase maternal behavior	
				Corpus luteum	Maintain progesterone secretion	
	Basophils	Gonadotrophs	Follicle-stimulating hormone (FSH)	Ovarian follicles	Promote follicle development	Inhibit secretion of gonadotropin-releasing hormone (GnRH) by hypothalamic neurons
				Seminiferous tubules	Stimulate spermatogenesis	
			Luteinizing hormone (LH), interstitial cell–stimulating hormone (ICSH)	Ovarian follicles	Stimulate follicle maturation, ovulation	
				Corpus luteum	Stimulate CL development, progesterone secretion	
				Interstitial cells of the testis	Maintain cells, stimulate testosterone secretion	
		Corticotrophs	Adrenocorticotropic hormone (ACTH), corticotropin	Zona fasciculata of adrenal cortex	Stimulate synthesis and secretion of glucocorticoids	Inhibit secretion of corticotropin-releasing hormone (CRH) by hypothalamic neurons; maintain aldosterone production by zona glomerulosa of adrenal cortex
				Zona reticularis of adrenal cortex	Stimulate synthesis and secretion of adrenal androgens	
		Thyrotrophs	Thyroid-stimulating hormone (TSH)	Thyroid follicular cells	Stimulate synthesis and release of thyroxine (T_4) and triiodothyronine (T_3)	Increase metabolic rate in most cells; inhibits production of TSH by pituitary thyrotrophs
Pars intermedia	Basophilic cells	Melanotrophs	Melanocyte-stimulating hormone (MSH)	Melanocytes	Increase melanin production, darken existing melanin	
Pars nervosa	Hypothalamic neuron cell bodies (primarily in the supraoptic nucleus)		Antidiuretic hormone (ADH) or arginine vasopressin	Collecting ducts of kidneys	Absorption of water, production of hypertonic urine	Increase or maintain blood volume, decrease salinity and solute concentration in blood; increase blood pressure
				Vascular smooth muscle	Vasoconstriction	
	Hypothalamic neuron cell bodies (in both supraoptic and paraventricular nuclei)		Oxytocin	Myoepithelial cells of mammary gland	Cell contraction	Milk ejection
				Uterine smooth muscle	Cell contraction	Induction of labor

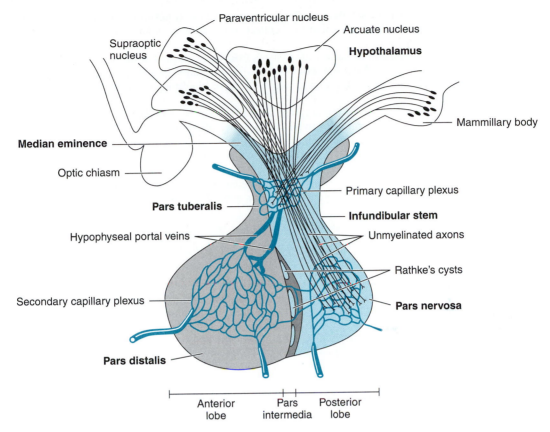

Figure 20–2. Schematic diagram of the subdivisions, blood supply, and innervation of the pituitary gland and hypothalamus. The adenohypophysis (*gray, left*) lies anterior to the neurohypophysis (*right*). For simplicity, only a few of the many nuclei of the hypothalamus *(top)* are shown. Major subdivisions are shown in boldface.

anterior inferior hypophyseal arteries. It is drained by the inferior hypophyseal veins into the internal jugulars.

E. **Hypothalamic Releasing and Inhibiting Hormones:** These small peptides are synthesized in the neuron (neurosecretory) cell bodies of the hypothalamic nuclei and are released by their axon terminals into the primary capillary plexus. They pass through the hypophyseal portal venules and into the secondary capillary plexus, from which they diffuse into the adenohypophysis to stimulate or inhibit the hormone release by acidophils and basophils.

1. **Releasing hormones. Corticotropin-releasing hormone (CRH)** is a 41-amino-acid peptide synthesized in the paraventricular nucleus; it stimulates corticotrophs to release ACTH. **Gonadotropin-releasing hormone (GnRH),** a 10-amino-acid peptide synthesized in the preoptic and arcuate nuclei, stimulates gonadotrophs to release FSH and LH. **Thyrotropin-releasing hormone (TRH)** is a 3-amino-acid peptide that stimulates thyrotrophs to release TSH (thyrotropin).

2. **Inhibiting hormones. Somatostatin (GHIH;** growth hormone–inhibiting hormone) is a 14-amino-acid peptide synthesized in the suprachiasmatic nuclei that inhibits somatotrophs from releasing growth hormone (GH, somatotropin). It also inhibits the secretion of glucagon, insulin, and other hormones associated with the gastrointestinal tract. **Dopamine** (a prolactin-inhibiting hormone [**PIH**]) is a neurotransmitter synthesized in the arcuate nuclei that inhibits prolactin release from mammotrophs.

F. **Summary of Adenohypophyseal Hormone Production:**

1. Neurons of the hypothalamic nuclei synthesize releasing or inhibiting hormones and package them in neurosecretory vesicles.

2. The neurons transport the vesicles down axons in the **tuberoinfundibular** and **hypothala-mohypophysial** tracts, where they collect in the axon terminals surrounding capillaries of the primary plexus.

3. Neural stimulation or hormonal feedback from target organs of the adenohypophysis causes the nerves of the tuberoinfundibular and hypothalamohypophyseal tracts to fire an action potential that releases the appropriate releasing or inhibiting hormone from their axon terminals.

4. The releasing or inhibiting hormone enters the primary capillary plexus and flows through the hypophyseal portal veins to the secondary capillary plexus.

5. Here, the hormone diffuses out of the capillary lumen by means of the fenestrae and stimulates or inhibits the release of stored adenohypophyseal hormones from the acidophils or basophils.

6. The adenohypophyseal hormones enter the capillaries of the secondary plexus; they leave the adenohypophysis through the anterior inferior hypophyseal veins to enter the general circulation.

IV. NEUROHYPOPHYSIS

The subdivisions of the neurohypophysis (outlined in Table 20–1) all exhibit similar microscopic structure. For brevity, the pars nervosa is used here to represent the neurohypophysis. The neurohypophysis has three major structural components: axons, capillaries, and pituicytes.

A. Axons of Neurosecretory Cells: The neurohypophysis stains poorly. It contains many unmyelinated axons whose cell bodies (soma) lie mainly in the **supraoptic** and **paraventricular nuclei** (see Fig. 20–2) of the hypothalamus. Axons passing from these nuclei to the pars nervosa together comprise the hypothalamohypophyseal tract. These axons contain neurosecretory granules and have large granule-filled dilations called **Herring bodies.** The neurosecretory materials in these granules, synthesized and packaged in the cell bodies, include the following products:

1. **Neurohypophyseal hormones.** The hypothalamic neurons terminating in the neurohypophysis release oxytocin and antidiuretic hormone around the capillaries in this part of the pituitary gland. **Oxytocin** is a 9-amino-acid peptide synthesized mainly by cells of the paraventricular nucleus. It stimulates milk ejection by the mammary glands and also stimulates uterine smooth muscle contraction during copulation and childbirth. **Antidiuretic hormone (ADH, arginine vasopressin)** is a 9-amino-acid peptide synthesized mainly by cells in the supraoptic nucleus. It stimulates water resorption by the renal medullary collecting ducts (19.II.C.2) and the contraction of vascular smooth muscle.

2. **Neurophysins** are binding proteins that complex with neurohypophyseal hormones.

3. **ATP (adenosine triphosphate)** acts as a chemical energy source for neurosecretion.

B. Fenestrated Capillary Plexus: Surrounding the axon terminals in the pars nervosa, these capillaries deliver neurosecretory products to the general circulation.

C. Pituicytes: These are highly branched glial cells whose processes surround and support the unmyelinated axons. Their nuclei are typically larger and more euchromatic than those of the many fibroblasts in this tissue.

D. Summary of Neurohypophyseal Hormone Production: Neurons of the supraoptic and paraventricular nuclei of the hypothalamus synthesize ADH and oxytocin, respectively. The neurons package these hormones with neurophysins and ATP in neurosecretory vesicles. The vesicles are transported by the neurons down axons in the hypothalamohypophyseal tract to axon terminals among the capillaries of the pars nervosa. In response to appropriate stimulation, these neurosecretory cells propagate an action potential along their axons, causing exocytosis of the vesicle contents at the axon terminals. The released hormones enter the capillaries of the pars nervosa and leave the pituitary gland to enter the general circulation by means of the posterior inferior hypophyseal veins.

MULTIPLE-CHOICE QUESTIONS

Select the single best answer.

20.1. Which part of the pituitary gland is indicated by the letter B in Figure 20–3?
(A) Infundibulum
(B) Median eminence
(C) Pars distalis
(D) Pars intermedia
(E) Pars nervosa
(F) Pars tuberalis
(G) Rathke's pouch

20.2. Which of the following is true of the primary capillary plexus?
(A) Empties directly into the secondary capillary plexus
(B) Carries releasing hormones to the pars nervosa
(C) Is located in the pars distalis
(D) Picks up pituitary hormones and delivers them to the general circulation
(E) Is an important component of the hypophyseal portal system

20.3. Which of the following is true of releasing and inhibiting hormones?
(A) Are produced only in supraoptic and paraventricular nuclei
(B) Are released by axon terminals located near the primary capillary plexus

(C) Stimulate the synthesis and release of ADH and oxytocin
(D) Are produced by glial cells in the hypothalamus
(E) Are stored in Herring bodies

20.4. Which of the following is produced by cells whose cell bodies lie in hypothalamic nuclei?
(A) ADH
(B) ACTH
(C) FSH
(D) LH/ICSH
(E) TSH

20.5. Which of the following is a gonadotropin?
(A) ADH
(B) ACTH
(C) FSH
(D) TRH
(E) TSH

20.6. Which of the following is a pituitary hormone that directly targets the renal collecting ducts?
(A) ADH
(B) ACTH
(C) FSH
(D) TRH
(E) TSH

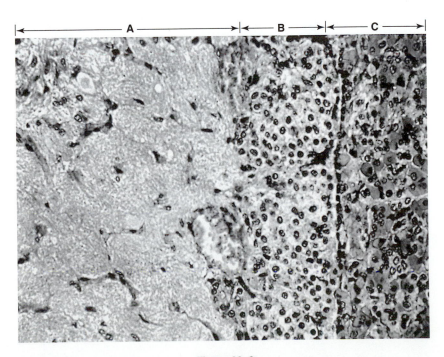

Figure 20–3.

20.7. Which of the following hormones is carried to its target(s) by the hypophyseal portal system?
(A) ADH
(B) ACTH
(C) FSH
(D) TRH
(E) TSH

20.8. The targets of which hormone are limited mainly to mammary gland myoepithelial cells and uterine smooth muscle?
(A) Antidiuretic hormone
(B) Dopamine
(C) Oxytocin
(D) Prolactin
(E) Somatostatin
(F) Somatotropin

20.9. Which of the following hormones is produced primarily in the paraventricular nucleus of the hypothalamus?
(A) Antidiuretic hormone
(B) Dopamine
(C) Oxytocin
(D) Prolactin
(E) Somatostatin
(F) Somatotropin

20.10. The secretory cells of the mammary glands are the primary targets of which of the following hormones?
(A) Antidiuretic hormone
(B) Dopamine
(C) Oxytocin
(D) Prolactin
(E) Somatostatin
(F) Somatotropin

20.11. Which of the labels in Figure 20–4 corresponds to the primary capillary plexus?

20.12. Which of the labels in Figure 20–4 corresponds to the hypothalamic region?

20.13. Which of the labels in Figure 20–4 corresponds to the infundibular stem?

20.14. Which of the labels in Figure 20–4 corresponds to the posterior lobe?

20.15. Which of the labels in Figure 20–4 corresponds to the secondary capillary plexus?

20.16. Which of the labels in Figure 20–4 corresponds to the hypophyseal portal vein?

20.17. Which of the labels in Figure 20–4 corresponds to Rathke's cysts?

20.18. Which of the labels in Figure 20–4 corresponds to the pars distalis?

20.19. Which of the labels in Figure 20–4 corresponds to the pars nervosa?

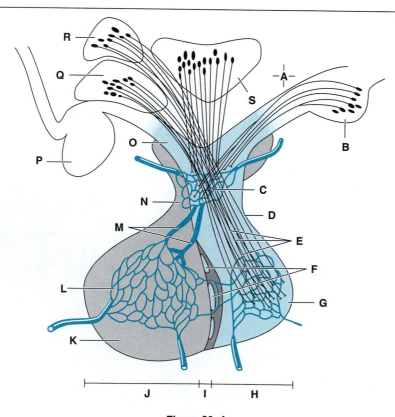

Figure 20–4.

20.20. Which of the labels in Figure 20–4 corresponds to the pars intermedia?

20.21. Which of the following is true of the neurohypophysis?
(A) Contains chromophobes
(B) Contains chromophils
(C) Includes the pars intermedia
(D) Includes the pars tuberalis
(E) Contains axons of hypothalamic neurons
(F) Secretes ACTH
(G) Hormone production is regulated by releasing factors

20.22. Which of the following is true of the adenohypophysis?
(A) Is innervated only by autonomic fibers from the carotid plexus
(B) Stains poorly with standard techniques (eg, H & E)
(C) Contains pituicytes
(D) Contains Herring bodies
(E) Releases oxytocin
(F) Includes the pars nervosa

20.23. Which of the following represents the tissue type that surrounds Rathke's cysts?
(A) Basophils
(B) Follicle cells
(C) Loose (areolar) connective tissue
(D) Melanotrophs
(E) Neuroepithelium
(F) Pituicytes
(G) Simple cuboidal epithelium

20.24. Which of the following is true of the fibroblasts of a normal neurohypophysis?
(A) Do not secrete collagen
(B) Do not secrete glycosaminoglycans
(C) Contain Herring bodies
(D) Derive from neural crest
(E) Myelinate the axons of hypothalamic neurons

20.25. Hypothalamic neurons that produce oxytocin release most of their secretions through axons terminating in or near which of the following sites?
(A) Circle of Willis
(B) Hypophyseal portal veins
(C) Median eminence
(D) Pars nervosa
(E) Pars tuberalis
(F) Primary capillary plexus
(G) Secondary capillary plexus

20.26. Which of the following is true of hormones?
(A) Have very specific effects on their target cells
(B) Are produced mainly by exocrine glands
(C) Must be present at high concentrations to elicit an effect
(D) Must be near their cell of origin to achieve maximum effect
(E) Must consist of short-chain or long-chain polypeptides

20.27. Hypersecretion of dopamine by neurons of the arcuate nucleus of the hypothalamus is likely to elicit which of the following effects?
(A) Hyposecretion of prolactin
(B) Precocious development of the mammary glands
(C) Enhancement of somatotropin production
(D) Hypersecretion of oxytocin
(E) Increased numbers of pituitary basophils

20.28. Pituitary basophils are characteristically PAS-positive because of which of the following structural characteristics?
(A) Granules containing growth hormone
(B) Extensive brush borders
(C) Abundant RER
(D) Abundant SER
(E) Granules containing glycoproteins
(F) Abundant mitochondria
(G) Abundant glycogen granules

ANSWERS TO MULTIPLE-CHOICE QUESTIONS

20.1. D (III.C; Figs. 20–1 and 20–2)
20.2. E (III.D.1–3 and F; Fig. 20–2)
20.3. B (III.E.1 and 2; IV.A.1.b)
20.4. A (III.A.2.a, b, E.1, and F; IV.A.1)
20.5. C (III.A.2.b)
20.6. A (IV.A.1; Table 20–2)
20.7. D (III.E.1)
20.8. C (III.E.1 and 2; IV.A.1)
20.9. C (IV.A.1)
20.10. D (III.A.2.a and E.1; Table 20–2)
20.11. C (III.D.1; Fig. 20–2)
20.12. A (I.E; Fig. 20–2)

20.13. D (Table 20–1; Fig. 20–2)
20.14. G or H (Table 20–1; Fig. 20–2)
20.15. L (III.D.3; Fig. 20–2)
20.16. M (III.D.2; Fig. 20–2)
20.17. F (III.C; Fig. 20–2)
20.18. K or J (III.A; Tables 20–1 and 20–2; Fig. 20–2)
20.19. G or H (Tables 20–1 and 20–2; Fig. 20–2)
20.20. I (III.C; Fig. 20–2)
20.21. E (III.A–F; IV.A–D; Tables 20–1 and 20–2; Fig. 20–2)

20.22. A (III.A–F; IV.A–D; Tables 20–1 and 20–2; Fig. 20–2)

20.23. G (III.C)

20.24. A (IV.C; 5.I.E; Fig. 20–1)

20.25. D (IV.A; Fig. 20–2)

20.26. A (I.D)

20.27. A (III.E.2)

20.28. E (III.A.2.b; 1.VII.E.1)

Adrenals, Islets of Langerhans, Thyroid, Parathyroids, & Pineal Body

21

OBJECTIVES

This chapter should help the student to:

- Describe the structure, function, and location of the islets of Langerhans; the pineal body; and the adrenal, thyroid, and parathyroid glands.
- Describe the embryonic origin of the adrenal cortex, adrenal medulla, and thyroid and parathyroid glands.
- Describe the nerve and blood supply to the adrenal glands, islets of Langerhans, thyroid gland, and pineal body.
- Name the hormones produced by the adrenal cortex, adrenal medulla, islets of Langerhans, thyroid and parathyroid glands, and pineal body; for each hormone, identify the cell type responsible for its secretion, neural and endocrine regulating factors, and main targets and effects.
- Identify the capsule, cortex, zona glomerulosa, zona fasciculata, zona reticularis, medulla, chromaffin cells, and ganglion cells in a micrograph of the adrenal gland.
- Identify the islets of Langerhans in a micrograph of the pancreas.
- Trace the steps in the synthesis, storage, and secretion of hormones by the thyroid's follicular cells.
- Identify the follicles, follicular cells, basement membrane, colloid, capillaries, and parafollicular cells in a micrograph of the thyroid gland.
- Identify the capsule, chief cells, and oxyphil cells in a micrograph of a parathyroid gland section.
- Identify pinealocytes, astroglial cells, and brain sand (corpora arenacea) in a micrograph of the pineal body.

MAX-Yield™ STUDY QUESTIONS

1. Compare the adrenal cortex and adrenal medulla in terms of their embryonic origins (II.A.1 and B.1[1]).
2. Trace the path taken by the blood supplying cells in the adrenal capsule, cortex, and medulla (II.C.1–4).
3. Name the layers of the adrenal cortex, beginning with the layer closest to the capsule (II.A.2.a–c). Compare them in terms of the structure and arrangement of their secretory cells and the classes and examples of hormones secreted by each layer.
4. Compare glucocorticoids, mineralocorticoids, and adrenal androgens (II.A.3.a–c) in terms of sites of synthesis, target organs and effects, and factors that stimulate or inhibit their production.
5. Describe the fetal (provisional) cortex (II.A.5) in terms of:
   a. Size and location
   b. Number of layers
   c. Age when involution begins
   d. Function

---

[1] See footnote on page 1.

6. Name the two major cell types in the adrenal medulla (II.B.2.a and b).
7. Name the two major catecholamines secreted by the adrenal medulla (II.B.3).
8. Describe the factors leading to and the consequences of increased catecholamine secretion by the adrenal medulla (II.B.2.a, b, and 3).
9. Give the location of the islets of Langerhans (III).
10. Name four major hormone-secreting cells in the islets of Langerhans (III.A–D) and compare them in terms of their location (peripheral or central), abundance, and secretory granules and the hormones they produce. Compare the hormones produced in terms of the factors that stimulate their secretion, their targets, and their effects.
11. Describe the thyroid gland in terms of its location, number of lobes, and embryonic origin (IV) and the hormones it secretes (IV.A.2 and C).
12. Sketch a section through three thyroid follicles and the intervening tissue (Fig. 21–1). Label the follicle (epithelial) cells, basal lamina, colloid, capillaries, and parafollicular cells.
13. On the drawing for question 12, show the site of:
    a. Thyroglobulin storage (IV.B.2.a)
    b. Tyrosine residue iodination (IV.B.2.c)
    c. Iodide oxidation (IV.B.2.b)
    d. Thyroglobulin synthesis (IV.B.2.a)
    e. Calcitonin synthesis (IV.C)
    f. The iodide pump (IV.B.2.b)
    g. Cleavage of T_3 and T_4 from iodinated thyroglobulin (IV.B.2.d)
    h. Release of T_3 and T_4 from follicular cells (IV.B.2.d)
14. Describe the effects of TSH on thyroid follicles in terms of follicle cell size and shape (IV.B.1), follicle diameter (IV.A), thyroglobulin synthesis (IV.B.2), iodide uptake (IV.B.2.b), pinocytosis of colloid (IV.B.2.d), and thyroxine production (IV.B.2).
15. List the steps in the synthesis, storage, and iodination of thyroglobulin, naming all intracellular and extracellular components involved (IV.B.2.a–c).
16. Beginning with the uptake of iodinated thyroglobulin, list the steps leading to the release of T_3 and T_4 from the follicular cell and name the organelles involved in each step (IV.B.2.d).
17. Describe parafollicular cells (IV.C) in terms of staining properties, location, hormones secreted and factor(s) that stimulate their secretion, and the targets and effects of their hormones.
18. Describe the parathyroid glands in terms of their number (V), dimensions (V.A.1), location (V), embryonic origin (V), and the major hormone they produce (V.A.2).
19. Name the two major parenchymal cells in the human parathyroid glands and compare them in terms of their relative number, diameter, staining properties, secretory products, and the number of mitochondria they contain (V.A and B).
20. Describe parathyroid hormone in terms of:
    a. The cell that secretes it (V.A.2)
    b. The stimulus required for its secretion (V.A.2)
    c. Its effect on blood calcium (V.A.2)
    d. Its effect on blood phosphate (V.A.2)
    e. Its three main targets and its effects on each (V.A.2)
    f. The effect of high blood calcium levels on its secretion (V.A.2)
    g. Its opposing hormone (IV.C)
21. Describe the pineal body (VI) in terms of:
    a. Shape and dimensions
    b. Location
    c. Covering tissue
    d. Principal cell types
    e. Calcified bodies
    f. Principal hormone
22. Compare pinealocytes (VI.A) with astroglial cells (VI.B) in terms of:
    a. Size, shape, and nuclear staining properties
    b. Cytoplasmic processes
    c. Location relative to blood vessels

SYNOPSIS

I. GENERAL FEATURES OF ENDOCRINE SECRETORY CELLS: STRUCTURE–FUNCTION RELATIONSHIPS

Hormone structure predicts the ultrastructure of the cell that produces it. For example, cells that secrete steroid hormones contain abundant smooth endoplasmic reticulum (SER), whereas those that secrete peptide hormones contain abundant rough endoplasmic reticulum (RER) (see also 20.I.D).

II. ADRENAL (SUPRARENAL) GLANDS

These glands form a cap over each kidney and comprise the adrenal cortex and medulla, which are distinguished by embryonic origin, structure, and function.

A. Adrenal Cortex:
1. **Embryonic origin.** The adrenal cortex derives from coelomic intermediate mesoderm.
2. **Structure in adults.** Cells of the adrenal cortex have a characteristic steroid-synthesizing cell structure (4.VI.C.5). The cortex comprises three layers: the zonae glomerulosa, fasciculata, and reticularis.
   a. **Zona glomerulosa.** This outermost cortical layer lies directly under the capsule and constitutes 15% of adrenal volume. Its cells form arched clusters (glomeruli) surrounded by capillaries. The cells of this layer secrete **mineralocorticoids.**
   b. **Zona fasciculata.** This middle layer of the cortex constitutes 65% of adrenal volume. Its cells form straight cords (fascicles) perpendicular to the organ surface; these cells produce **glucocorticoids** and some **adrenal androgens.**
   c. **Zona reticularis.** This innermost layer of the adrenal cortex constitutes 7% of adrenal volume. Its cells are arranged in irregular cords that form a network (reticulum). Its cells also are smaller and more acidophilic than those in the fasciculata and contain fewer lipid droplets, more mitochondria, and many lipofuscin granules. The reticularis and fasciculata may constitute a single functional zone, with the reticularis producing most of the glucocorticoids and adrenal androgens. The fasciculata may be a reserve zone activated by prolonged stimulation.
3. **Normal function.** The adrenal cortex produces three types of steroid hormones. The mnemonic "salt, sugar, sex" (used by many students for simplicity) suggests that each layer of the cortex produces only one of the three adrenocortical hormone types. Although mineralocorticoids (the "salt" steroids) are produced only in the glomerulosa, the mnemonic may be too simplistic in its association of glucocorticoids (the "sugar" steroids) with the fasciculata and adrenal androgens (the "sex" steroids) with the reticularis. Nevertheless, it can be a useful mnemonic device, and the simplistic view it represents may assist the student in answering some multiple-choice questions on course and board exams.
   a. **Mineralocorticoids,** mainly **aldosterone,** are produced by the zona glomerulosa in response to angiotensin II and, to a lesser extent, in response to adrenocorticotropic hormone (ACTH). Aldosterone controls water and electrolyte balance mainly by stimulating sodium absorption by the renal distal tubules (19.II.B.4) but also by affecting the gastric mucosa and salivary glands.
   b. **Glucocorticoids,** mainly **cortisol** and **corticosterone,** are produced by the zona reticularis in response to ACTH and by the fasciculata after prolonged stimulation. Glucocorticoids control carbohydrate metabolism, especially by stimulating carbohydrate synthesis in the liver. They have the opposite effect in tissues that catabolize (degrade) carbohydrates to provide raw material for the liver. They also suppress the immune response by decreasing circulating lymphocytes and eosinophils.
   c. **Adrenal androgens,** mainly **dehydroepiandrosterone,** are secreted in response to ACTH by the zona reticularis and, after prolonged stimulation, by the fasciculata. Their masculinizing and anabolic effects are less potent than those of testosterone.
4. **Abnormal function.**
   a. **Hypersecretion. Cushing syndrome** is caused by hypersecretion of cortisol and often of androgens. Its symptoms include truncal obesity, a moon face, high blood sugar, dia-

betes mellitus, hirsutism, amenorrhea, acne, and emotional lability. Aldosterone hypersecretion (eg, **Conn syndrome**) causes sodium and water retention, which in turn causes increased blood pressure (hypertension).

    b. **Hyposecretion.** Chronic hypofunction of the adrenal cortex (eg, **Addison's disease**) causes low serum glucose, sodium, chloride, and bicarbonate and high serum potassium. It results in weakness, nausea, weight loss, and elevated ACTH levels (the last of which causes hyperpigmentation). Without testicular androgens to compensate, decreased adrenal androgen synthesis in women may cause the loss of pubic and axillary hair.

    5. **Fetal (or provisional) cortex.** The thickest adrenal layer before birth, the fetal cortex is located between the medulla and the immature thin permanent cortex. It produces sulfated androgens that are activated by the placenta and that enter the maternal circulation. After birth, the fetal cortex regresses and the permanent cortex develops the three layers described previously.

B. Adrenal Medulla:

    1. **Embryonic origin.** The adrenal medulla derives from the neural crest.

    2. **Structure.** It contains two major cell types: chromaffin and ganglion cells.

      a. **Chromaffin cells (pheochromocytes)** are the predominant medullary cells; they are postganglionic sympathetic neurons that have lost their axons and dendrites. They contain large nuclei, abundant electron-dense secretory granules filled with **catecholamines** (epinephrine or norepinephrine), a well-developed Golgi complex, a few profiles of RER, and many oval mitochondria. Their secretory granules have a strong affinity for chromium stains. Chromaffin cells synthesize and release their catecholamines in response to neural stimulation (especially stress) mediated by preganglionic sympathetic neurons.

      b. **Ganglion cells.** The few parasympathetic ganglion cells present exhibit typical morphologic characteristics of autonomic ganglion cells (9.V).

    3. **Normal function.** Chromaffin cells produce two types of catecholamines in response to preganglionic sympathetic stimulation (eg, stress). Both elevate blood glucose by stimulating glycogenolysis in the liver; they also increase blood flow to the heart. **Epinephrine** increases heart rate and dilates blood vessels to the organs needed to combat or escape stress, such as cardiac and skeletal muscle. It dilates bronchioles and constricts vessels in organs (eg, skin, digestive tract, kidneys) that are not essential in reacting to stress. **Norepinephrine** constricts blood vessels in nonessential organs. By increasing peripheral resistance, it increases blood pressure and blood flow to the heart, brain, and skeletal muscle.

    4. **Abnormal function.** Hypersecreting chromaffin cell tumors (**pheochromocytomas**) cause a sustained stress response (especially hypertension) even in the absence of stress. Ganglion cell tumors (**neuroblastomas** and **ganglioneuromas**) are more common, especially in children, but their clinical manifestations vary.

C. Adrenal Blood Supply:

    1. **Arteries.** Three main arteries supply each adrenal gland: the **superior suprarenal** from the inferior phrenic artery, the **middle suprarenal** artery from the aorta, and the **inferior suprarenal** from the renal artery. These arteries penetrate the capsule separately, and their branches anastomose to form a **subcapsular arterial plexus.** This plexus gives rise to three groups of arteries: the arteries of the capsule; the arteries of the cortex, which branch to form the **cortical capillaries** that pass between the secretory cells and drain into the medullary capillaries; and the arteries of the medulla, which traverse the cortex without branching until they reach the medulla, where they branch to form the medullary capillaries.

    2. **Medullary capillaries** receive a double blood supply from arteries of both the cortex and medulla and converge to form several medullary veins.

    3. **Medullary veins** converge to form a single large suprarenal vein.

    4. The **suprarenal vein** arises at the core of the medulla and drains into the renal vein or directly into the inferior vena cava.

III. ISLETS OF LANGERHANS

These small nests of endocrine cells distributed throughout the pancreas contain four major peptide hormone-secreting endocrine cell types:

A. **A Cells (Alpha Cells):** In response to low blood glucose, these cells secrete **glucagon,** whose effects are opposite those of insulin.

B. **B Cells (Beta Cells):** The most numerous islet cell type, these secrete **insulin** in response to high blood glucose. Before its release, insulin is stored (complexed with zinc) in cytoplasmic granules. Insulin enhances glucose uptake by most cells, glycogen synthesis by hepatocytes, and triglyceride synthesis by adipocytes. B-cell malfunction causes **diabetes mellitus,** a condition manifested by a great excess of blood glucose **(hyperglycemia)** that spills over into the urine **(glycosuria).** Hyperplasia and neoplasia of the B cells may result in **hyperinsulinism syndrome,** which is characterized by **hypoglycemia.**

C. **D Cells (Delta Cells):** These cells secrete **somatostatin,** which suppresses the release of insulin, glucagon, and growth hormone. They also may secrete **gastrin,** which stimulates secretion by gastric glands. **Zollinger-Ellison syndrome (gastrinoma)** causes peptic ulcers through excessive acid secretion by parietal cells in the gastric mucosa. Somatostatinomas are rare tumors with complex effects.

D. **F Cells (PP Cells):** These cells secrete **pancreatic polypeptide,** which inhibits pancreatic exocrine secretion of enzymes and bicarbonate. It also causes gallbladder relaxation and decreases bile secretion.

IV. THYROID GLAND

During week four of fetal development, the thyroid arises as an outpocketing of the endoderm lining the floor of the embryonic pharynx; it soon divides in two. In adults, the thyroid lies anterior to the larynx and has two lobes connected by an isthmus. Each lobe consists of many spherical follicles and is covered by a thin capsule penetrating the parenchyma to form septa.

A. **Thyroid Follicles:** Each follicle consists of an outer simple epithelium of follicular cells enclosing a central lumen filled with **colloid** (Fig. 21–1). The follicles, which vary in size, enlarge during stimulation.

B. **Thyroid Follicular Cells:**
  1. **Structure.** Thyroid follicular cells derive from endoderm and exhibit typical peptide hormone-secreting cell ultrastructure. Cell height ranges from squamous in inactive glands to columnar during stimulation.
  2. **Normal function.** Thyroid follicular cells differ from other endocrine cells in that they store an intermediate form of their secretory product **(thyroglobulin)** extracellularly in colloid

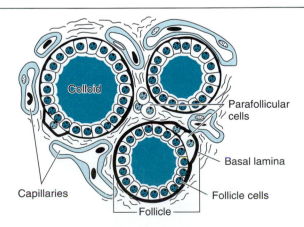

Figure 21–1. Schematic diagram of thyroid follicles and associated structures.

rather than internally in cytoplasmic granules. Stimulation by pituitary thyroid-stimulating hormone (TSH) (20.III.A.2.b), signaling an increased energy demand, increases synthesis and secretion.

a. **Synthesis and storage of thyroglobulin.** The steps required by this process (Fig. 21–2) are (1) synthesis of the tyrosine-rich protein thyroglobulin on the RER; (2) glycosylation in the ER and Golgi complex; (3) packaging in vesicles in the Golgi complex; and (4) fusion of the vesicles with the apical cell membrane, resulting in exocytosis of the thyroglobulin into the colloid in the lumen of the follicle.

b. **Uptake and oxidation of iodide.** A molecular pump in the follicular cell's basal plasma membrane transfers circulating iodide into the cytoplasm. It is oxidized by peroxidase and then transferred to the cell's apex. Iodide uptake is also stimulated by TSH.

c. **Iodination of thyroglobulin and formation of thyroid hormone.** Enzymes in the plasma membranes of the apical microvilli projecting into the colloid catalyze the iodination of tyrosine residues in the thyroglobulin, a reaction that occurs at the microvillus–colloid interface. One iodide molecule is added to tyrosine, forming **monoiodotyrosine (MIT)**. A second iodide molecule is subsequently added to some tyrosine residues, forming **diiodotyrosine (DIT)**. Coupling of the two iodinated tyrosines forms a thyronine molecule. The coupling of two DIT molecules forms **tetraiodothyronine (thyroxine, or T_4)**, whereas the coupling of one MIT molecule and one DIT molecule forms **triiodothyronine (T_3)**. Although T_4 makes up 90% of the thyroid hormone produced, it is less potent than T_3.

d. **Thyroid hormone secretion.** TSH causes the follicular cells to pinocytose portions of the colloid, forming vesicles containing iodinated thyroglobulin. These vesicles fuse with lysosomes carrying enzymes that cleave the thyroglobulin. The T_4 and T_3 released in this way diffuse out of the secondary lysosomes (2.III.F.2). They pass through the cytoplasm and cross the basolateral plasma membranes to reach the bloodstream.

e. **Targets and effects of thyroid hormones.** T_3 and T_4 act throughout the body to increase basal metabolic rate (ie, the rate at which cells use glucose), promote cell growth, increase heart rate, raise body temperature, and generally enhance all energy-requiring

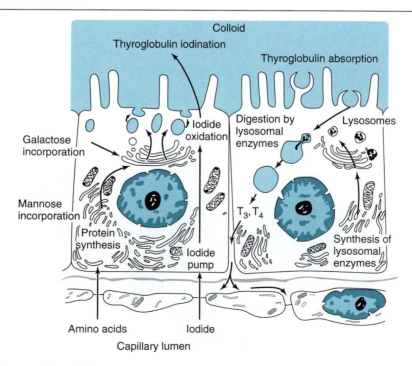

Figure 21–2. Diagram of thyroid follicle cells. The synthesis and iodination of thyroglobulin (*left*) and its absorption, digestion, and release (*right*) are shown. Both processes occur in the same cell. (Reproduced, with permission, from Junqueira LC, Carneiro J, Kelley RO: *Basic Histology,* 9th ed. Stamford, CT: Appleton & Lange, 1998.)

cell functions. They also act on the TRH-secreting cells of the hypothalamus (20.III.E.1) and the thyrotrophs in the adenohypophysis (20.III.A.2.b) to reduce TSH secretion (**negative feedback**).

3. **Abnormal function.**
   a. **Hyperthyroidism.** Overproduction of thyroid hormone (thyrotoxicosis) causes nervousness, palpitation, rapid pulse, muscular weakness, fatigue, weight loss with good appetite, excessive perspiration, heat intolerance, and emotional lability. Hyperactive thyroid follicles enlarge, owing to increased follicular epithelial height and increased thyroglobulin deposits, causing thyroid swelling known as **goiter.**
   b. **Hypothyroidism.** Termed **cretinism** in children and **myxedema** in adults, hypothyroidism causes poor glucose use. Its symptoms include lethargy, cold intolerance, slowed intellectual and motor skills, glycosaminoglycan accumulation in the dermis (with consequent bloating), and sometimes weight gain. Because iodine is required for normal thyroid function, iodine-deficient diets reduce functional thyroxine production and often underlie cretinism and myxedema. Because uniodinated thyroxine caused by iodine deficiency cannot provide negative feedback on TSH production, follicular enlargement and goiter may accompany this type of hypothyroidism.

C. **Parafollicular Cells (C Cells):** These lie interspersed among follicular cells or in clusters between follicles (see Fig. 21–1). Their cytoplasm stains poorly with standard stains and typically appears clear or white. Electron micrographs reveal many small secretory granules. These cells secrete the peptide hormone **calcitonin** in response to high blood calcium. Calcitonin lowers blood calcium by causing calcium uptake by cells and increased calcium deposition in bone.

V. PARATHYROID GLANDS

These four small glands on the posterior surface of the thyroid gland derive from the third and fourth pharyngeal pouches (endoderm). In adults, they consist of two major parenchymal cell types: **chief** and **oxyphil cells** (which may be different forms of the same cell type).

A. **Chief Cells:** These are the most numerous of the parenchymal cells.
   1. **Structure.** These small (4–8 μm in diameter) polygonal cells exhibit typical peptide secretory ultrastructure; they contain many small granules in a pale-staining cytoplasm.
   2. **Normal function.** Chief cells secrete **parathyroid hormone** (**PTH,** or **parathormone**) in response to low blood calcium. PTH, a peptide hormone, acts on three targets to increase blood calcium. In **bone,** PTH increases resorption. In the **kidneys,** it increases phosphate excretion and calcium reabsorption and causes activation of a vitamin D precursor. In the **intestines,** PTH (perhaps by activating vitamin D) increases calcium absorption by the intestinal mucosa. Increased blood calcium levels decrease PTH secretion.
   3. **Abnormal function.**
      a. **Hyperparathyroidism.** Excessive PTH secretion elevates serum calcium (**hypercalcemia**) and decreases serum phosphate (**hypophosphatemia**). It causes increased urine calcium, abnormal calcium deposits in the arteries and kidneys, and excessive calcium loss from bones, resulting in **osteomalacia** and **osteitis fibrosa cystica.**
      b. **Hypoparathyroidism.** Insufficient PTH secretion disrupts neuromuscular function. The resulting low blood calcium leads to spontaneous and uncontrolled firing of action potentials. In peripheral nerves, this may cause spastic muscle contraction, or **tetany.** The spontaneous firing of neurons in the brain may have behavioral effects.

B. **Oxyphil Cells:** These are larger and less numerous than chief cells; their abundant mitochondria make them intensely acidophilic. Oxyphil function is not clearly understood.

VI. PINEAL BODY (EPIPHYSIS CEREBRI)

This small (3–5 mm × 5–8 mm) conical organ attaches by a stalk to the roof of the diencephalon near the posterior aspect of the third ventricle. Pia mater penetrates the organ, carrying blood vessels and forming irregular septa. The pineal body contains clusters of globular, basophilic, calcified matrix known as **brain sand (corpora arenacea),** which increase in size, number, and calcification with age. The radiopacity of these bodies, in conjunction with the pineal body's central location in the skull, provides a useful landmark for radiologists. The two major pineal cell types are pinealocytes and astroglial cells.

A. Pinealocytes:

1. **Structure.** These cells have large irregular nuclei with prominent nucleoli and pale basophilic cytoplasm. With silver stain, they exhibit long cytoplasmic processes that terminate as swellings in the septa near blood vessels. The role of pineal innervation (both sympathetic and from the posterior commissure through the pineal stalk) is uncertain.

2. **Normal function.** Pinealocytes secrete the indolamine **melatonin.** Cyclic changes in plasma melatonin levels follow changes in environmental lighting, but the precise relationship between these phenomena remains unknown. Melatonin may help establish circadian rhythms and also may have antigonadotropic effects that delay the onset of sexual maturity until puberty. Other pineal secretions include arginine vasotocin and possibly some substances that exert an antigonadotropic effect by means of the hypothalamohypophysial axis.

3. **Abnormal function.** Pineal lesions occur most often in young males and may cause either precocious or delayed sexual maturity. Because of their location, pineal tumors may restrict the flow of cerebrospinal fluid through the sylvian aqueduct, causing hydrocephalus.

B. Astroglial Cells: Also known as **interstitial cells,** these glialike cells have elongated heterochromatic nuclei and long cytoplasmic processes that contain intermediate filaments. They lie near blood vessels and between clusters of pinealocytes.

MULTIPLE-CHOICE QUESTIONS

Select the single best answer.

21.1. The rounded group of cells indicated by an arrow in Figure 21–3 is a sample of which of the following?
(A) Acinus
(B) Adrenal medulla
(C) Glomerulus
(D) Pancreatic islet of Langerhans
(E) Parafollicular cells
(F) Pineal astroglial cells
(G) Thyroid follicle

21.2. Which of the following organs is shown in Figure 21–4?
(A) Adrenal gland
(B) Pancreas
(C) Parathyroid gland
(D) Pineal body
(E) Thyroid gland

21.3. Which of the following organs is shown in Figure 21–5?
(A) Adrenal gland
(B) Pancreas
(C) Parathyroid gland
(D) Pineal body
(E) Thyroid gland

21.4. The capillaries of the adrenal medulla receive their double blood supply directly from which of the following vessels?

(A) Cortical and medullary veins
(B) Cortical capillaries and medullary arteries
(C) Middle and inferior suprarenal arteries
(D) Subcapsular arterial plexus and suprarenal vein
(E) Superior and inferior suprarenal arteries

21.5. Which of the following is true of glucocorticoids?
(A) Include the steroid hormone aldosterone
(B) Are produced in response to stimulation by ACTH
(C) Are produced primarily by the zona glomerulosa
(D) Typically enhance the immune response
(E) Include the steroid hormone dehydroepiandrosterone

21.6. Which of the following is true of chromaffin cells?
(A) Comprise a minor cell type in the adrenal medulla
(B) Are modified postganglionic parasympathetic neurons
(C) Derive from neural crest
(D) Secrete the neurotransmitter acetylcholine
(E) Are important components of the fetal adrenal cortex.

21.7. Which of the following is true of oxyphil cells?

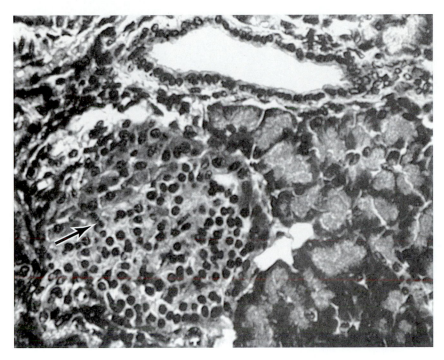

Figure 21–3.

(A) Are typically smaller than parathyroid chief cells
(B) Contain abundant mitochondria
(C) Cytoplasm is characteristically basophilic
(D) Are the predominant cell type in the parathyroid gland
(E) Secrete calcitonin

21.8. Which of the following is true of the colloid within a thyroid follicle?
(A) Contains the storage forms of both T_3 and T_4
(B) Is produced by the adenohypophysis and stored in the thyroid

(C) Consists of numerous thyroglobulin-containing membrane-limited vesicles
(D) Normally lacks significant concentrations of iodine
(E) Is pinocytosed and digested by the follicular cells in response to increased circulating levels of thyroxine

21.9. Which of the labels in Figure 21–6 corresponds to the site of aldosterone synthesis?

21.10. Which of the labels in Figure 21–6 corresponds to the zona fasciculata?

21.11. Which of the labels in Figure 21–6 corresponds to the layer that is least responsive to ACTH?

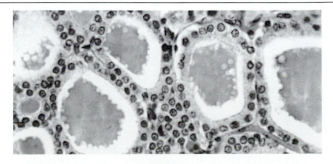

Figure 21–4.

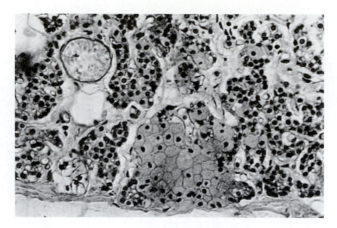

Figure 21–5.

21.12. Which of the following hormones is produced in response to angiotensin II?
(A) Aldosterone
(B) Corticosterone
(C) Dehydroepiandrosterone
(D) Epinephrine
(E) Norepinephrine

21.13. Which of the following hormones is produced in response to decreased blood calcium levels?
(A) Glucagon
(B) Insulin
(C) Somatostatin
(D) Pancreatic polypeptide

(E) Thyroxine
(F) Calcitonin
(G) Parathyroid hormone
(H) Melatonin

21.14. Which of the following hormones is a glucocorticoid?
(A) Aldosterone
(B) Corticosterone
(C) Dehydroepiandrosterone
(D) Epinephrine
(E) Glucagon
(F) Insulin
(G) Norepinephrine

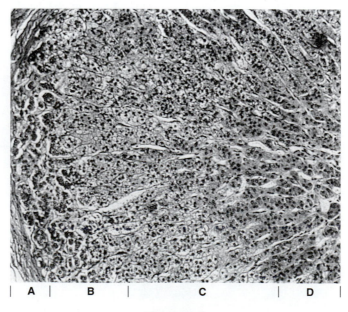

Figure 21–6.

21.15. Which of the following hormones is produced in response to decreased blood sugar levels?
(A) Glucagon
(B) Insulin
(C) Somatostatin
(D) Pancreatic polypeptide
(E) Thyroxine
(F) Calcitonin
(G) Parathyroid hormone
(H) Melatonin

21.16. The production of which of the following hormones is influenced by external visual stimuli?
(A) Glucagon
(B) Insulin
(C) Somatostatin
(D) Pancreatic polypeptide
(E) Thyroxine
(F) Calcitonin
(G) Parathyroid hormone
(H) Melatonin

21.17. Which of the following hormones characteristically raises the body's basal metabolic rate?
(A) Glucagon
(B) Insulin
(C) Somatostatin
(D) Pancreatic polypeptide
(E) Thyroxine
(F) Calcitonin
(G) Parathyroid hormone
(H) Melatonin

21.18. Which of the following cell types characteristically produces glucagon?
(A) A (alpha) cell
(B) B (beta) cell
(C) C (parafollicular) cell
(D) D (delta) cell
(E) F cell

21.19. Which of the following is true of the adrenal cortex?
(A) Derives from embryonic neural crest
(B) Consists mainly of chromaffin cells
(C) Contains cells with abundant SER
(D) Secretes catecholamines
(E) Excessive secretory activity results in Addison's disease

21.20. Which of the following is true of parathyroid hormone?
(A) Characteristically lowers blood calcium
(B) Has target cells in the kidney, bone, and small intestines
(C) Excessive circulating levels characteristically cause tetany
(D) Is synthesized mainly by the thyroid's parafollicular cells

21.21. Which of the following is true of brain sand?
(A) Is found only in diseased pineal bodies
(B) Becomes decalcified with increasing age
(C) Is typically intensely acidophilic
(D) May serve as a landmark for radiologists
(E) Serves as a storage depot for melatonin

21.22. Which of the following is true of the fetal (provisional) adrenal cortex?
(A) Rapidly increases in size shortly after birth
(B) Is located between the capsule and the adult cortex
(C) Produces sulfated androgens
(D) Forms the ganglion cells of the adult adrenal medulla
(E) Secretions typically are restricted to the placenta and prevented from reaching the maternal circulation

21.23. Patients with Zollinger-Ellison syndrome (gastrinoma) suffer from hypersecretion of HCl from cells in the gastric mucosa. Which cells are responsible for the hypersecretion of gastrin that may lead to the gastric and duodenal ulcers seen in these patients?
(A) A (alpha) cells
(B) B (beta) cells
(C) C (parafollicular) cells
(D) D (delta) cells
(E) F cells
(F) Parietal cells
(G) Paneth's cells

ANSWERS TO MULTIPLE-CHOICE QUESTIONS

21.1. D (note surrounding pancreatic acini and duct above)
21.2. E (IV.A; Fig. 21–2; note colloid-filled follicles)
21.3. C (V.A.1 and B; note predominance of chief cells and cluster of larger oxyphil cells in lower central area)
21.4. B (II.C.1 and 2)
21.5. B (II.A.2.a–c and 3.a,b)

21.6. C (II.B.1, 2.a,b and 3)
21.7. B (V.A.1,2 and B; Fig. 21–6)
21.8. A (IV.B.2.a–c)
21.9. B (II.A.2.a and 3.a)
21.10. C (II.A.2.b)
21.11. A (ie, the capsule)
21.12. A (II.A.3.a)
21.13. G (V.A.2)
21.14. B (II.A.3.b)

21.15. A (III.A)

21.16. H (VI.A.2)

21.17. E (IV.B.2.e)

21.18. A (III.A)

21.19. C (II.A.2; 4.VI.C.5)

21.20. B (V.A.2)

21.21. D (VI)

21.22. C (II.A.5)

21.23. D (III.D)

Male Reproductive System

22

OBJECTIVES

This chapter should help the student to:

- Name the male reproductive glands, ducts, and external genitalia and give their locations.
- Describe the general organization of the testis as it appears in a midsagittal section.
- Trace the life cycle of the male gametes (spermatozoa), beginning with their embryonic origin and ending with their path through the ducts of the male reproductive system.
- Describe the wall structure and functions of each duct of the male reproductive system.
- Distinguish between spermatogenesis, spermatocytogenesis, and spermiogenesis and describe the changes in the number of chromosomes and amount of DNA that occur during each process.
- Describe Sertoli (supporting) and Leydig (interstitial) cells in terms of their structure, function, location, and embryonic origin.
- Describe the roles of temperature, the pituitary gland, and the testes's cells in spermatogenesis.
- Compare the accessory glands of the male reproductive system in terms of general organization, epithelial lining, secretory products, and the point(s) at which their secretions enter the duct system.
- Describe the three erectile bodies of the penis in terms of their histologic structure and location.
- Describe the blood supply to the erectile tissue of the penis and the factors that control the transitions between flaccid and erect states.

MAX-Yield™ STUDY QUESTIONS

1. Sketch a testis in sagittal section (Fig. 22–1). Include and label the following components:
    a. Anterior surface
    b. Posterior surface
    c. Tunica albuginea
    d. Mediastinum testis
    e. Septa
    f. Tunica vaginalis
    g. Testicular lobule
    h. Seminiferous tubules
    i. Tubuli recti (straight tubules)
    j. Rete testis
    k. Ductuli efferentes
2. Name the three layers of the wall of the seminiferous tubule, from exterior to lumen (II.C.1[1]).
3. Name the two types of cells in the germinal (seminiferous) epithelium (II.C.2 and 3).
4. Name, in order, the three phases of spermatogenesis (II.D.1–3) and compare them in terms of the cell type present at the beginning and end of each phase.
5. List the spermatogenic cell types in order of appearance during spermatogenesis (II.C.2.a–e) and compare them in terms of location, size, nuclear morphology, chromosome number, and amount of DNA.
6. Describe the appearance, changes, and movement of the following during spermiogenesis (II.D.3.a–f):
    a. Acrosome
    b. Centrioles
    c. Nucleus
    d. Manchette
    e. Mitochondria
    f. Residual bodies
7. Describe the contents and function of the acrosome (acrosomal cap; II.D.3.a).
8. Describe the difference between spermatogonia A and spermatogonia B in terms of the cell type their progeny (daughter cells) become (II.D.1).

---

[1] See footnote on page 1.

9. Describe Sertoli (supporting) cells (II.C.3.a and b) in terms of location, shape and size, nuclear morphology, and functions.

10. Describe the blood–testis barrier (II.G) in terms of the cell type and cellular structures responsible for its formation, its location, the contents of the basal and adluminal compartments, and its function(s).

11. Describe interstitial (Leydig's) cells (II.F) in terms of location, shape, staining properties, primary function, and organelle(s) responsible for their primary function.

12. Name the cells chiefly responsible for producing the following: testosterone (II.F); androgen-binding protein (II.C.3.b); inhibin (II.C.3.b); FSH (II.C.3.b); and LH (Table 20–1). Indicate the major target site(s) of each secretion and describe the secretion's effects (in males).

13. Why must the testes be maintained at a temperature below core body temperature? How do the scrotum and pampiniform plexus maintain the optimal 35 °C (I.A.1.c)?

14. List, in order, the three intratesticular ducts through which spermatozoa pass after leaving the seminiferous tubules (III.A.1–3).

15. List, in order, the three ducts (III.B.1–3) through which spermatozoa pass as they travel from the intratesticular ducts to the prostatic urethra and compare these ducts in terms of epithelial lining, amount and organization of smooth muscle, and extent of coiling.

16. Compare stereocilia (III.B.1; 4.IV.A.4) with true cilia (2.III.J; 4.IV.A.1) in terms of the presence of basal bodies and microtubules.

17. What is secreted by the cells lining the ductus epididymis (III.B.1)?

18. Compare the ampulla with the rest of the ductus deferens in terms of diameter and mucosal folds (III.B.2).

19. Compare the seminal vesicles, prostate gland, and bulbourethral glands (IV.A–C) in terms of:
   a. Location and the point(s) at which their ducts enter the reproductive tract
   b. Secretory products
   c. Contribution to seminal fluid volume

20. Describe the corpora amylacea (IV.B) in terms of location, biochemical composition, and change in number with age.

21. Sketch the penis in cross-section (Fig. 22–2) and longitudinal (parasagittal) section (Fig. 22–1) and label the following:
   a. In both cross- and longitudinal sections
      (1) Corpus cavernosum
      (2) Corpus spongiosum
      (3) Tunica albuginea
      (4) Urethra
      (5) Erectile tissue
      (6) Loose connective tissue
      (7) Skin
      (8) Prepuce
   b. In cross-section only
      (1) Superficial and deep dorsal veins
      (2) Dorsal arteries
      (3) Deep arteries
   c. Longitudinal section only
      (1) Glans penis
      (2) Fossa navicularis
      (3) Meatus

22. Describe the role(s) of the following in penile erection (V.B and C):
   a. Parasympathetic innervation
   b. Cavernous spaces (lacunae)
   c. Helicine arteries
   d. Arteriovenous shunts

SYNOPSIS

I. GENERAL FEATURES OF THE MALE REPRODUCTIVE SYSTEM

This system (Fig. 22–1) consists of the external genitalia and the glands and ducts that produce and transport the male gametes (spermatozoa) and seminal fluid. Together, the seminal fluid and spermatozoa constitute the semen.

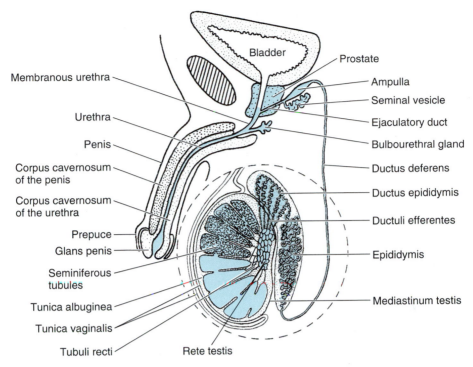

Figure 22–1. Schematic diagram of the male reproductive system. The testis and epididymis (*circled*) are magnified in comparison with the other organs. (Reproduced, with permission, from Junqueira LC, Carneiro J, Kelley RO: *Basic Histology,* 9th ed. Stamford, CT: Appleton & Lange, 1998.)

A. **Glands:** The glands of this system include the paired testes and several accessory glands.
   1. The **testes** are the male gonads and are located in the scrotum; they are the primary glands of the male reproductive system and have both exocrine and endocrine functions.
      a. The **exocrine component** comprises the seminiferous tubules, where spermatozoa are produced (spermatogenesis), and the intratesticular genital ducts, into which the seminiferous tubules deliver their products for transport to the excretory genital ducts (III.B).
      b. The **endocrine component** consists of nests of testosterone-secreting interstitial cells in the connective tissue between the seminiferous tubules.
      c. **Role of temperature.** Testosterone production can occur at the normal core body temperature (37°C), but normal sperm production occurs only at 35 °C. To maintain this lower temperature, the testes are held away from the body in the scrotum. In addition to the evaporation of sweat from the scrotal surface, cooling is aided by the pampiniform plexus of veins that surrounds each testicular artery. The plexus contains cooler blood returning from the scrotum, which in turn cools the testes' blood supply.
   2. **Accessory glands** are located along the excretory genital duct system (see Fig. 22–1).
      a. **Seminal vesicles** are paired glands whose secretions increase the seminal fluid's volume, raise its pH, and add fructose, providing nourishment and energy for the gametes.
      b. The **prostate gland** adds fluid rich in citric acid and acid phosphatase.
      c. **Bulbourethral glands** and **glands of Littre** are mucous glands that lubricate the duct system's distal segment.

B. **Ducts:** The ducts of the male reproductive system are described in terms of their location, number, length, diameter, wall structure, and function.
   1. **Intratesticular ducts.** Located within the testes, these are continuous with the seminiferous tubules and include the tubuli recti, rete testis, and ductuli efferentes.
   2. The **excretory genital ducts** are located outside the testes and include the ductus epididymis, ductus (or vas) deferens, ejaculatory duct, and urethra. In these larger ducts, the secretions of the testes and the accessory glands combine to form semen.

C. External Genitalia: These include the penis and scrotum. The penis contains the penile ure-thra, which is the most distal element of the duct system. The scrotum contains the testes. The histology and reproductive function of the penis are described in this chapter, with an emphasis on erectile tissue (V.A), blood flow (V.B), and autonomic innervation (V.C).

II. TESTES

A. Embryonic Origin: **Primordial germ cells,** originating from **yolk sac endoderm,** migrate into the abdominal cavity's dorsal wall and invade the mesoderm of the **genital ridge,** where they collect to form the **primitive sex cords.** The primordial germ cells form the spermatogo-nia, and the mesoderm gives rise to the Sertoli cells, interstitial cells, and connective tissue be-tween the cords. The solid sex cords form seminiferous tubules, developing central lumens and anastomosing with adjacent cords. Subsequently, the seminiferous tubules anastomose with remnants of the mesonephric tubules and the excretory duct system. The entire developing testis becomes encapsulated by connective tissue, separates from the dorsal wall, and descends into the scrotum.

B. General Organization: The **tunica vaginalis** is a double-layered mesothelial sac covering the anterior surface of each testis. This extension of the peritoneum is picked up during the testes' descent into the scrotum. The **tunica albuginea** is a dense fibrous connective tissue cap-sule that thickens along the posterior surface to form the **mediastinum testis.** The septa are ex-tensions of the tunica albuginea that penetrate each testis and divide it into approximately 250 compartments, or lobules. Each **lobule** comprises between one and four seminiferous tubules (the exocrine component) and loose vascular connective tissue between the tubules that con-tains clusters of testosterone-secreting interstitial cells of Leydig (the endocrine component).

C. Seminiferous Tubules:
1. **General structure.** Each long (40- to 70-cm), narrow (0.2-mm) tubule is highly coiled and packed into a small space. The walls of each tubule, from the exterior to the lumen, have three layers. The **tunica propria** is a thin sleeve of fibrous connective tissue with several layers of fibroblasts. The innermost layer includes contractile **myoid cells** that attach to the basal lamina. A well-defined **basal lamina** separates the tunica propria and the seminifer-ous epithelium. The stratified **seminiferous epithelium** consists of two cell lineages: sper-matogenic cells and supportive (Sertoli's) cells.
2. **Spermatogenic cells.** These cells derive from embryonic yolk sac endoderm and undergo spermatogenesis (II.D)—a multistep process of differentiation that begins with the cells closest to the basal lamina (spermatogonia). The process ends with the release of spermato-zoa into the tubule lumen. Cells at different steps in the process are identified according to their size, nuclear morphology, and position in the epithelium.
   a. **Spermatogonia** are small round cells near the basal lamina. They are the least differen-tiated and are the only spermatogenic cell type present before puberty. They have a round nucleus with patchy heterochromatin. Like most of the body's cells, they have a diploid chromosome number (46, 2n) and diploid DNA (2N) until stimulated to divide.
   b. **Primary spermatocytes,** which are closer to the lumen, are the largest germ cells pres-ent. Each has a large round nucleus with dark strands of heterochromatin that resemble tangled string. These cells are seen in prophase of meiosis I, the longest phase of meiosis (as long as 22 days). They have a diploid chromosome number (46, 2n) and tetraploid DNA (4N) in preparation for the first meiotic division.
   c. **Secondary spermatocytes.** Still closer to the lumen, these products of the first meiotic division are approximately half the size of the primary spermatocytes that divide to form them. They are rare in histologic section because they undergo the second meiotic divi-sion almost immediately after forming. Their chromosome number is haploid (23, n), and their DNA is diploid (2N).
   d. **Spermatids** are products of the second meiotic division of secondary spermatocytes and are located next to the lumen. Small cells with dark heterochromatic nuclei, they may exhibit a range of nuclear morphology, depending on the stage of spermiogenesis. Both their chromosome number (23, n) and DNA (N) are haploid.

    e. Spermatozoa are located in the lumen. They are the products of spermiogenesis (differentiation of spermatids; II.D.3) and are recognized by their long flagella (II.E). Both their chromosome number (23, n) and DNA (N) are haploid.

   3. Supporting (Sertoli's) cells derive from the embryonic genital ridge mesoderm.

    a. Structure. These elongated, branched, pyramidal epithelial cells extend from the basal lamina to the lumen of each seminiferous tubule. They exhibit deep cytoplasmic infoldings that embrace the developing spermatogenic cells. Their large pale nuclei are ovoid and indented and contain a prominent nucleolus. Sertoli cells have a well-developed SER and Golgi complex, numerous mitochondria, and some RER. The cells' margins are bound tightly to neighboring Sertoli cells by occluding junctions, forming a continuous sheath around the tubule lumen (II.G).

    b. Function. Sertoli cell functions include (1) **physical support** for the spermatogenic cells, which attach to one another by cytoplasmic bridges; (2) **nutritional regulation** of the developing spermatozoa, which are isolated from the blood supply by the occluding junctions between the supporting cells and therefore depend on these cells to mediate the exchange of nutrients and metabolites with the blood; (3) **protection** from autoimmune attack by immunoglobulins in the blood; (4) **phagocytosis** of residual bodies shed by the maturing spermatozoa; and (5) **secretion of fluid** for sperm transport, including **androgen-binding protein (ABP),** which combines with the testosterone produced by interstitial cells and is released into the tubule lumen (ABP secretion increases in response to increased levels of FSH and testosterone), and **inhibin,** which acts on the pituitary to decrease FSH production. Sertoli's cells also may secrete estrogen.

 D. Spermatogenesis: The entire multistep process that transforms spermatogonia into spermatozoa is testosterone-dependent and divisible into three phases: spermatocytogenesis, meiosis, and spermiogenesis.

   1. Spermatocytogenesis is the production of primary spermatocytes from spermatogonia through a series of mitotic divisions. Daughter cells of early divisions form two types of spermatogonia. **Spermatogonia A** remain undifferentiated stem cells that are able to produce more A and B cells. **Spermatogonia B** may undergo further mitoses to form more spermatogonia B or may enter meiosis to form primary spermatocytes.

   2. Meiosis involves two successive cell divisions that yield four haploid spermatids from one diploid primary spermatocyte.

    a. Meiosis I. During the first meiotic division, primary spermatocytes divide to yield secondary spermatocytes. During the preceding S phase (DNA synthesis), DNA doubles (as in mitosis), becoming tetraploid (4N) while the chromosome number remains diploid (2n). During an extended prophase, the 23 pairs of homologous chromosomes (22 pairs of autosomes + XY) thicken by coiling. They pair up point for point (synapsis) and form bridges (chiasma), allowing the trading of DNA between paired chromosomes ("crossing over"). The nuclear membrane disintegrates during late prophase. During metaphase, the 23 pairs line up at the equatorial plate. During anaphase, the pairs separate, with one member of each pair moving toward the opposite pole. During telophase, the nuclear membranes reform in the daughter cells. Each secondary spermatocyte is subsequently haploid (1n), containing only one member of each homologous chromosome pair (22 + X or 22 + Y), but has diploid DNA (2N). The equatorial constriction tightens but leaves the daughter cells connected by a narrow cytoplasmic bridge.

    b. Meiosis II. During the second meiotic division, the secondary spermatocytes divide to yield spermatids. During this division, the chromosome number in each cell remains the same (haploid), but the amount of DNA is halved (as during a standard mitosis), resulting in spermatids that are haploid in both chromosome number (1n) and DNA (1N).

   3. Spermiogenesis is the complex process of cytodifferentiation by which spermatids become spermatozoa. Many structural changes occur and spermatids exhibit stage-related variations in appearance. Spermiogenesis includes the following processes.

    a. Acrosome formation. Proacrosomal granules form in the Golgi complex and coalesce to form a large membrane-bound acrosomal vesicle that attaches to the nuclear envelope. The vesicle membrane spreads over the nucleus, covering its anterior two thirds and forming a head cap whose contents redistribute to form the mature acrosome (acrosomal cap). The contents of this large specialized lysosome are rich in carbohydrate and enzymes such as hyaluronidase, neuraminidase, acid phosphatase, and a trypsinlike protease. The enzymes aid in penetrating the egg's corona radiata and zona pellucida (23.II.B.2.a) during fertilization (23.V).

b. Centriole migration and **flagellum formation.** The centrioles migrate to the spermatid's posterior pole. A flagellum emerges from one, perpendicular to the cell surface, and forms the tail. The other centriole forms a collar around the flagellum base.

c. Shift of cytoplasm toward the flagellum. The anterior plasma membrane subsequently contacts the acrosome. Excess cytoplasm forms a **residual body.**

d. Migration of mitochondria. The mitochondria move toward the flagellum and form a spiral collar around the proximal part of the tail (the middle piece), concentrating at the future site of high energy consumption. Fructose and citric acid in semen aid in sperm motility, serving as mitochondrial metabolites.

e. Condensation of nuclear chromatin. The chromatin forms a dense mass with no visible substructure. A cylindrical band of microtubules (**manchette**) surrounds the nucleus, associating with the acrosome's posterior border and causing the nucleus to flatten and elongate.

f. Sloughing of residual bodies. Immediately after their release into the tubule lumen, the spermatozoa shed their residual cytoplasm. Much of this is phagocytosed by the Sertoli cells.

E. Structure of Mature Spermatozoa:

1. Head. In frontal view, the head has an oval outline. On its side, it appears as a 4- to 5-μm-long spearhead. It is mostly nucleus, with the anterior two thirds of the nucleus covered by the acrosome and the posterior covered by the manchette.

2. Tail. Approximately 55-μm long, the tail is enveloped by plasma membrane. It has four parts. The **neck** includes the proximal centriole, connecting piece, capitulum, flagellar base, and an occasional mitochondrion. The **middle piece** contains many mitochondria arranged end to end in a helical sheath around the flagellum. It is 5- to 7-μm long and 1-μm thick, with the annulus at its posterior end. In this region, the flagellum has nine outer dense fibers and a "9 + 2" core microtubule pattern. The flagellum in the 50-μm-long **principal piece** is surrounded by an outer fibrous sheath with dorsal and ventral longitudinal columns connected by circumferential ribs. The flagellum itself has seven dense outer fibers collected in two compartments and a 9 + 2 core microtubule pattern. The **end piece** lacks the fibrous sheath but otherwise has the same structure as the principal piece. As it tapers toward the tip, its nine doublets dissociate to form 18 single microtubules.

F. Interstitial (Leydig's) Cells: Derived from embryonic genital ridge mesoderm, these cells secrete **testosterone** in response to stimulation by pituitary LH (interstitial cell–stimulating hormone [ICSH]). They occur as vascular nests of pale acidophilic cells in the loose connective tissue between the seminiferous tubules. Their large pale nuclei contain one or two prominent nucleoli. Their cytoplasm contains the extensive SER typical of steroid-secreting cells, a well-developed Golgi complex, and lipid droplets.

G. Blood–Testis Barrier: Spermatogenesis involves the appearance of new sperm-specific proteins and glycoproteins on the differentiating spermatogenic cells. Because this process begins at puberty, well after immune system development, these surface molecules may be recognized as nonself antigens. The blood–testis barrier protects developing sperm from autoimmune damage. The barrier comprises a belt of junctional complexes joining the Sertoli cells at their lateral surfaces. It separates the seminiferous tubule lining into two functional compartments. The **basal compartment** houses the spermatogonia. It lies between the basal lamina and the junctional belt and is accessible to any blood-borne substance that can penetrate the basal lamina. The **adluminal compartment** extends from the junctional belt inward to the lumen. It is inaccessible to blood-borne substances except those that are taken up by the supporting cells and passed through their cytoplasm to this privileged space.

III. DUCT SYSTEM

Extending from the tubuli recti in the testis through the urethra, the duct system functions in the maturation, storage, and transport of spermatozoa. Its parts require testosterone to maintain normal function. The secretory epithelial lining provides spermatozoa with nutrients. Two sets of ducts (one per testis) empty into a common urethra.

A. Intratesticular Genital Ducts: The **tubuli recti** (straight tubules) connect the seminiferous tubules to the rete testis. The epithelium closest to the seminiferous tubules is similar to that of the seminiferous tubules. As the tubuli recti approach the rete testis, they gradually lose the spermatogenic cells until only Sertoli's cells remain. The main segment is lined by simple cuboidal epithelium supported by a dense connective tissue sheath. The **rete testis** is an anastomosing network of tubules lying in the mediastinum testis. It is lined by low cuboidal epithelium. The **ductuli efferentes (efferent ductules)** are 10 to 20 4- to 6-mm-long ducts connecting the rete testis with the epididymis. The walls contain smooth muscle, and the epithelium has alternating groups of simple cuboidal and ciliated columnar cells. The cuboidal cells absorb much of the fluid secreted by seminiferous tubules while the cilia sweep the spermatozoa toward the epididymis. Together, the ductules form the head of the epididymis, and they converge to form a single ductus epididymis.

B. Excretory Genital Ducts:
1. **Ductus epididymis.** This single, highly coiled 4- to 6-m-long tube forms the body and tail of the epididymis. It is an important site of sperm maturation. The cells of its pseudostratified columnar epithelium have abundant apical **stereocilia** (long, irregular, nonmotile microvilli) and secrete glycerophosphocholine (a possible capacitation inhibitor) and a spermatozoon-binding glycoprotein of unknown function. They also phagocytose and digest residual bodies sloughed during spermatogenesis. A sheath of circular smooth muscle underlies the basal lamina, gradually thickening along the length of the tube. Its peristaltic contractions propel sperm toward the ductus deferens. Sperm move slowly through this long coiled tube and are often seen in its lumen in tissue sections.
2. **Ductus deferens (vas deferens).** This single straight tube with thick muscular walls begins in the scrotum at the termination of the epididymis. It ascends within the spermatic cord through the inguinal canal into the abdomen, joining with the duct of the seminal vesicle in the pelvic cavity near the prostate. The lumen is narrowed by longitudinal mucosal folds. The pseudostratified columnar epithelial lining has fewer stereocilia than in the epididymis. The three layers of smooth muscle in the wall (inner and outer longitudinal, middle circular) are capable of powerful peristaltic contractions during ejaculation. The diameter increases near the termination to form the **ampulla,** which has a highly folded mucosa.
3. **Ejaculatory duct.** This short duct, lined by pseudostratified columnar epithelium, is formed by the junction of the ductus deferens and the duct of the seminal vesicle. It penetrates the prostate to empty into the prostatic urethra.
4. **Urethra.** The male urethra is a genitourinary passageway shared by the urinary and reproductive systems (see 19.VI.A for its divisions, structure, and epithelial lining). It contains small, mucus-secreting **glands of Littre** in its wall.

IV. ACCESSORY GENITAL GLANDS

A. Seminal Vesicles: The paired seminal vesicles each consist of two highly coiled 15-cm-long tubes that develop as outgrowths of the ductus deferens. Their mucosa is highly folded, with primary, secondary, and tertiary branching; it is rich in fructose and also contains citrate, inositol, prostaglandins, and several proteins. The smooth muscle underlying the lamina propria contracts during ejaculation. In each gland, the tubes converge to form a single duct that joins with the ductus deferens of each side to form the ejaculatory ducts. Seminal vesicle secretions make up approximately 70% of the human ejaculate.

B. Prostate Gland: The prostate surrounds the urethra at its origin below the bladder. It consists of 30 to 50 compound tubuloalveolar glands arranged in three concentric groups (mucosal, submucosal, and main) whose ducts empty independently into the urethra. The mucosa is folded, and the epithelium varies from tall cuboidal to pseudostratified columnar. The epithelium produces prostatic fluid, which is rich in citric acid and acid phosphatase and also contains amylase, fibrinolysin, and lipids. The entire gland is surrounded by a fibroelastic capsule containing smooth muscle that contracts during ejaculation, expelling the prostatic fluid into the urethra. Extensions of the capsule form septa that divide the gland into indistinct lobes and aid in expelling the prostatic fluid. A characteristic feature of the prostate is the presence of **corpora amylacea** in the gland's lumen. These small glycoprotein spheres become larger, more numerous, and calcified with age, but their significance is unknown. The prostate is a common site of disease in men older than 50 years of age.

C. **Bulbourethral Glands (Cowper's Glands):**   These paired, spherical tubuloalveolar glands are 3 to 5 mm in diameter and are lined by cuboidal epithelium. Their ducts empty clear lubricating mucus into the membranous urethra.

V. PENIS

A. **General Organization:**   The penis comprises three cylinders of spongy erectile tissue surrounded by a common connective tissue sheath and covered by hairless thin skin (Fig. 22–2).
   1. **Corpora cavernosa.** Each of these two dorsal erectile cylinders is penetrated by a deep artery and ensheathed by a thick, dense connective tissue called the **tunica albuginea.**
   2. **Corpus spongiosum (corpus cavernosum urethrae).** This single, smaller, ventral cylinder is surrounded by a thinner connective tissue sheath. Its expanded distal tip is the **glans penis.** The corpus spongiosum is penetrated along its length by the cavernous (penile) urethra, whose lumen communicates with the exterior through an opening **(meatus)** in the glans.
   3. **Erectile tissue.** Within each cylinder is an irregular network of fibrous connective tissue trabeculae containing smooth muscle fibers. The trabeculae form the supporting framework between the numerous endothelium-lined lacunae (vascular sinuses).

B. **Blood Supply:**   The blood supply of the penis depends on its functional state.
   1. **Flaccid.** The peripheral dorsal arteries in the loose connective tissue sheath supply much of the arterial blood, which is drained by the superficial veins. In the flaccid state, arteriovenous shunts between the deep arteries in the corpora cavernosa and the superficial veins are open, and the deep artery branches feeding the vascular spaces (the **helicine arteries**) are closed.
   2. **Erect.** In this state, the arteriovenous shunt closes down. The deep arteries in the corpora cavernosa force blood through the dilated helicine arteries into the vascular spaces in the erectile tissue. The sudden filling of the lacunae may block the veins draining them.

C. **Innervation:**   Parasympathetic stimulation (through the ventral roots of spinal nerves S2, S3, and S4) causes erection by affecting the arteriovenous shunts and helicine arteries. The sympathetic discharge accompanying ejaculation contributes to the subsequent decline of parasympathetic activity and return to the flaccid state.

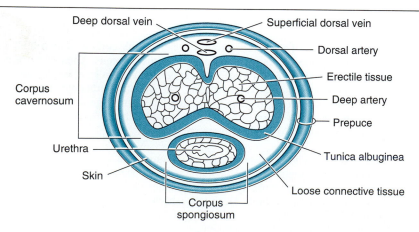

Figure 22–2. Schematic diagram of a cross-section through the penis.

MULTIPLE-CHOICE QUESTIONS

Select the single best answer.

22.1. Which of the following structures is (are) shown in Figure 22–3?
(A) Ductuli efferentes
(B) Ductus deferens
(C) Ductus epididymis
(D) Rete testis
(E) Seminiferous tubules
(F) Seminal vesicle
(G) Tubuli recti

22.2. Which of the following structures is (are) shown in Figure 22–4?
(A) Ductuli efferentes
(B) Ductus deferens
(C) Ductus epididymis
(D) Rete testis
(E) Seminiferous tubules
(F) Seminal vesicle
(G) Tubuli recti

22.3. Which of the following is the connective tissue sheath that surrounds each seminiferous tubule?

(A) Dartos tunic
(B) Mediastinum testis
(C) Rete testis
(D) Septum
(E) Tunica albuginea
(F) Tunica propria
(G) Tunica vaginalis

22.4. Which function of the Sertoli cells is most directly attributable to their zonula occludens junctions?
(A) Phagocytosis of residual bodies
(B) Physical support of the spermatogenic cells
(C) Protection of spermatozoa from autoimmune attack
(D) Synthesis and secretion of androgen-binding protein

22.5. Which of the following is true of interstitial (Leydig's) cells?
(A) Secrete testosterone
(B) Have basophilic cytoplasm
(C) Contain abundant RER
(D) Are components of the seminiferous epithelium
(E) Derive from yolk sac endoderm

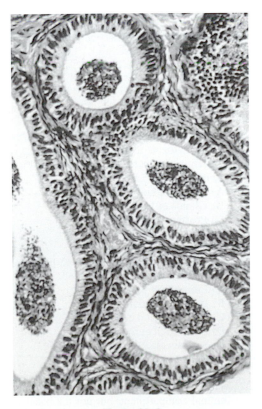

Figure 22–3.

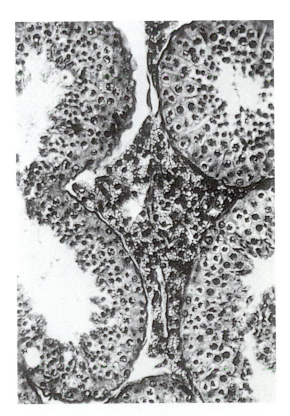

Figure 22–4.

22.6. Which of the labels in Figure 22–5 corresponds to the bulbourethral (Cowper's) gland?

22.7. Which of the labels in Figure 22–5 corresponds to the reproductive tract component that is characteristically surrounded by three layers of smooth muscle?

22.8. Which of the labels in Figure 22–5 corresponds to the rete testis?

22.9. Which of the labels in Figure 22–5 corresponds to the seminal vesicle?

22.10. Which of the labels in Figure 22–5 corresponds to a layer composed entirely of dense regular connective tissue?

22.11. Which of the labels in Figure 22–5 corresponds to the location of corpora amylacea?

22.12. Which of the labels in Figure 22–5 corresponds to the ejaculatory duct?

22.13. Which of the labels in Figure 22–5 corresponds to the site in which recently formed spermatozoa undergo further maturation?

22.14. Which of the labels in Figure 22–5 corresponds to the structure that is transected in a vasectomy?

22.15. Which of the labels in Figure 22–5 corresponds to the structure that is an extension of the peritoneum?

22.16. Which of the following structures has lining cells that exhibit apical stereocila?

(A) Ductuli efferentes
(B) Ductus epididymis
(C) Ejaculatory duct
(D) Membranous urethra
(E) Penile urethra
(F) Prostatic urethra
(G) Rete testis
(H) Tubuli recti

22.17. A portion of which of the following structures is lined exclusively with Sertoli cells?
(A) Ductuli efferentes
(B) Ductus epididymis
(C) Ejaculatory duct
(D) Membranous urethra
(E) Penile urethra
(F) Prostatic urethra
(G) Rete testis
(H) Tubuli recti

22.18. Which of the following structures is lined by transitional epithelium?
(A) Ductuli efferentes
(B) Ductus epididymis
(C) Ductus deferens
(D) Ejaculatory duct
(E) Membranous urethra
(F) Penile urethra
(G) Prostatic urethra
(H) Tubuli recti

22.19. Which of the following is the most common site of disease in older males?
(A) Ductuli efferentes
(B) Ductus epididymis
(C) Ductus deferens
(D) Ejaculatory duct
(E) Membranous urethra
(F) Penile urethra
(G) Prostatic urethra
(H) Tubuli recti

22.20. Fructose, a potential fuel for the mitochondria of spermatozoa, is present in the highest concentrations in secretions of which of the following glands?
(A) Bulbourethral gland
(B) Glands of Littre
(C) Prostate gland
(D) Seminal vesicle
(E) Testis

22.21. Which of the following produce both androgen-binding protein and inhibin?
(A) Fibroblasts
(B) Leydig (interstitial) cells
(C) Myoepithelial cells
(D) Primordial germ cells
(E) Sertoli cells
(F) Spermatids
(G) Spermatogonia

22.22. Which of the following is a feature of the adluminal compartment, but not the basal compartment, of the seminiferous tubules?

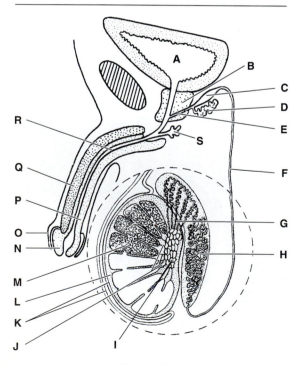

Figure 22–5.

(A) Is accessible to the immune system
(B) Contains Sertoli cells
(C) Contains spermatids
(D) Contains spermatogonia
(E) Contains testosterone-secreting cells

22.23. Which of the following is the spermatogonia's tissue of origin?
(A) Epithelium lining the mesonephric tubules
(B) Mesoderm in the genital ridge
(C) Neural ectoderm
(D) Neural crest mesenchyme
(E) Sertoli cells
(F) Yolk sac endoderm

22.24. Nondisjunction (failure of chromosomes to separate normally) during the first meiotic division in the seminiferous epithelium results in abnormal numbers of chromosomes in which of the following cell types?
(A) Primary spermatocyte
(B) Sertoli cell
(C) Secondary spermatocyte
(D) Spermatogonia

22.25. Which of the following is true of spermiogenesis?
(A) Occurs before puberty
(B) Involves cytodifferentiation of early spermatids
(C) Involves spermatocytogenesis, meiosis, and spermatogenesis
(D) Takes place in diploid cells
(E) Results in the production of primary spermatocytes

22.26. Which of the following cells bear the primary responsibility for forming the blood–testis barrier?
(A) Capillary endothelial cells
(B) Interstitial cells (of Leydig)
(C) Myoid cells
(D) Primary spermatocytes
(E) Sertoli cells
(F) Spermatogonia

22.27. Atherosclerosis, which is often attributable to excessive dietary cholesterol, has been known to restrict blood flow through the helicine arteries of the penis. Which of the following is most likely to be caused by atherosclerotic lesions at this site?
(A) Autoimmune infertility
(B) Birth defects
(C) Impotence (failure to achieve erection)
(D) Secondary priapism (persistent erection from obstructed vascular outflow)

22.28. A male patient with a pituitary gonadotropic tumor, which results in the hyposecretion of follicle-stimulating hormone (FSH), is most likely to exhibit which of the following?
(A) No symptoms (he has no ovarian follicles)
(B) Loss of libido (sex drive)
(C) Low serum testosterone levels
(D) Low sperm count
(E) Prostatic hypertrophy

ANSWERS TO MULTIPLE-CHOICE QUESTIONS

22.1. C (III.B.1; note pseudostratified columnar epithelium and abundant sperm in lumen; compare with Fig. 22–4)
22.2. E (II.C.1 and 2.b; note various nuclear shapes and sizes, and abundant primary spermatocytes)
22.3. F (II.C.1)
22.4. C (II.C.3.b)
22.5. A (II.F)
22.6. S (IV.C; Fig. 22–1)
22.7. F (III.B.2; Fig. 22–1)
22.8. I (III.A.2; Fig. 22–1)
22.9. D (IV.A; Fig. 22–1)
22.10. L (II.B; Fig. 22–1)
22.11. B (IV.B; Fig. 22–1)
22.12. E (III.B.3; Fig. 22–1)

22.13. H (III.B.1; Fig. 22–1)
22.14. F (III.B.2; Fig. 22–1)
22.15. K (II.B; Fig. 22–1)
22.16. B (III.B.1)
22.17. H (III.A.1)
22.18. G (III.B.4; 19.VI.A.1)
22.19. G (IV.B)
22.20. D (IV.A)
22.21. E (II.C.3.b)
22.22. C (II.G)
22.23. F (II.A)
22.24. C (II.D.2.a)
22.25. D (II.D and D.3)
22.26. E (II.C.3)
22.27. C (V.B.1 and 2)
22.28. D (II.C.3.b; Table 20–2)

Female Reproductive System

23

OBJECTIVES

This chapter should help the student to:

- Name the internal organs and external genitalia of the female reproductive system and give the structure, function, and location of each.
- Trace female gamete (ovum) development from embryonic origin through ovulation and implantation.
- Describe the structural changes and the roles of FSH and LH at each step of follicle maturation.
- Name the cells that produce estrogen and progesterone. Describe the conditions under which these hormones are produced and explain their effects on FSH and LH production by the pituitary gland.
- Identify each follicle's stage in the continuum of development and atresia, and the follicles' component cells and structures in a micrograph of an ovary.
- Describe the endometrium in terms of location, structure, blood supply, and changes accompanying the menstrual cycle phases; correlate the endometrial changes with events occurring in the ovary and with changing pituitary and ovarian hormone levels.
- Identify, in a micrograph, the uterine wall and its layers and components, and the uterine cervix and its characteristic structural features.
- Describe the changes in the structure and location of the conceptus between fertilization and implantation, and know the amount of time that typically elapses between these events.
- Describe implantation in terms of the structural changes in the blastocyst and endometrium.
- Describe the placenta in terms of fetal and maternal contributions, the steps in chorionic villus development, and the layers of the placental barrier.
- Identify, in a micrograph of an implanted blastocyst or embryo, the components of the developing embryo and placenta.
- Describe the vaginal wall in terms of its layers, response to estrogen, and innervation. Distinguish the vaginal wall from the esophagus in micrographs.
- Describe the vulva in terms of the structure and innervation of its components.
- Describe, and identify in a micrograph, the histologic structure of the mammary gland and its components in the prepubertal, resting adult, pregnant, and lactating states.
- Describe the synthesis and secretion of milk and name the hormones responsible for breast growth, activation and maintenance of lactation, and milk ejection.

MAX-Yield™ STUDY QUESTIONS

1. Diagram and label the organs of the female reproductive system (I.A[1]; Fig. 23–1).
2. Sketch a cross-section of an ovary (Fig. 23–2) and label the germinal epithelium, tunica albuginea, cortex, medulla, and ovarian follicles.
3. Name the stages—from primordial to mature (graafian)—of follicle development (II.B.1–3).
4. Distinguish between primordial and unilaminar primary follicles (II.B.1 and 2.a).
5. Distinguish between unilaminar and multilaminar primary follicles (II.B.2.a).
6. Distinguish between primary and secondary follicles (II.B.2.a and b).

---

[1] See footnote on page 1.

7. Sketch a mature (graafian) follicle (Fig. 23–3) and label the following (II.B.2.a,b and 3):
   a. Theca externa
   b. Theca interna
   c. Antrum
   d. Liquor folliculi
   e. Follicle (granulosa) cells
   f. Oocyte
   g. Cumulus oophorus
   h. Corona radiata
8. Explain the following aspects of ovulation (II.D):
   a. Assuming menses begin on day 1 of an idealized 28-day cycle, determine when ovulation occurs
   b. Name the pituitary hormone whose production surges just before ovulation
   c. Name the components of the mature follicle (II.B.1–3) that are carried with the ovum and those that are left behind after ovulation
9. Sketch a corpus luteum (Fig. 23–2; II.E) and indicate the location of the granulosa lutein cells, the theca lutein cells, and the surrounding stroma.
10. Compare the corpus luteum of menstruation with that of pregnancy (II.E.1 and 2) in terms of the length of time that they persist, the hormones they secrete, the hormones that stimulate their formation and maintain their function, and their size.
11. Name the hormone(s) that has the following functions (II.F):
    a. Stimulate(s) follicle growth, oocyte maturation, and estrogen production
    b. Stimulate(s) ovulation
    c. Stimulate(s) the formation of the corpus luteum
    d. Maintain(s) the corpus luteum of menstruation and stimulate(s) progesterone production
    e. Maintain(s) the corpus luteum of pregnancy
    f. Inhibit(s) FSH production and stimulates LH production
    g. Inhibit(s) LH production
12. Compare atretic follicles (II.B.4) with corpora albicans (II.E.3) in terms of which degenerating ovarian structures they represent, their relative size, their predominant tissue types, and their persistence.
13. Name the four segments of each oviduct (III.B) and compare the segments in terms of their location (in relation to the uterus and ovaries; label the segments on your diagram for question 1), their luminal diameter, and the size of their mucosal folds; also indicate the common site of fertilization.
14. Name the two epithelial cell types that line the lumen of the uterine tubes (III.C) and compare them in terms of their height, apical specializations, and secretory activity.
15. On the diagram for question 1, label the boundaries of the three parts of the uterus (IV).
16. Name the three basic layers in the uterine wall (IV).
17. Name the two layers of the endometrium (IV.A) and indicate which layer:
    a. Includes the uterine epithelium
    b. Undergoes cyclic thickening and shedding
    c. Remains unchanged during the menstrual cycle
    d. Is more responsive to ovarian hormones
    e. Provides, through parts of the endometrial glands, the epithelial cells that cover the uterine surface after menstruation
    f. Contains only straight arteries
    g. Contains coiled arteries
18. Describe the myometrium (IV.B) in terms of:
    a. The type of muscle fibers it contains
    b. The location(s) and names of its major vessels
    c. The two mechanisms responsible for its increased muscle mass during pregnancy
    d. Its response to oxytocin during copulation and childbirth (20.IV.A.1)
19. Compare the proliferative, secretory, and menstrual phases of the menstrual cycle (IV.D; Table 23–1) in terms of the:
    a. Relative thickness of the stratum functionale
    b. Appearance of the endometrial glands in histologic section
    c. Degree of coiling of the coiled arteries and their proximity to the epithelial surface
20. Draw a coronal section of the uterine cervix (Fig. 23–1) and indicate the location of the simple columnar and stratified squamous epithelia and the cervical glands (IV.E).
21. Describe fertilization (V) in terms of:
    a. Where it typically takes place
    b. Whether it occurs before or after the production of the second polar body

    c. The change in chromosome number in the ovum

    d. The name applied to the fertilized ovum

22. Compare the zygote, the morula, and the blastocyst (V; VI) in terms of:

    a. Relative size

    b. Relative number of component cells

    c. Size of component cells

    d. The presence of a fluid-filled cavity

    e. The stage at which the embryo reaches the uterus

    f. The stage at which the inner cell mass and trophoblast initially form

    g. The stage at which the zona pellucida disappears, allowing direct contact between the embryo and uterine wall and leading to implantation

23. Name the structures formed by the inner cell mass and by the trophoblast (VI.A).

24. Distinguish between the syncytiotrophoblast and the cytotrophoblast in terms of location and structure (VI.A). Which disappears during the later stages of development (VII.A)?

25. Describe decidual cells (VI.B) in terms of structure, function, location, and origin.

26. Distinguish among the decidua basalis, decidua capsularis, and decidua parietalis (VI.B).

27. Distinguish between the chorion frondosum and the chorion laeve (VI.A) in terms of location.

28. Describe the lacunae of the developing placenta (VII.A) in terms of:

    a. How these cavities in the endometrium originate

    b. How they become filled with blood

    c. The tissue that forms their lining

29. Name the types of chorionic villi (VII.A) and distinguish between them on the basis of the presence or absence of the syncytiotrophoblast, cytotrophoblast, extraembryonic mesenchyme, and fetal blood vessels.

30. Distinguish between the fetal and maternal parts of the placenta (VI.A and B; VII.A and B) in terms of the specific structures and regions from which they originate, the hormones they secrete, and the location of the fibrinoid.

31. Sketch a chorionic villus extending into a lacuna (Fig. 23–4) and label the six layers of the placental barrier (VII.B.1) present during the first trimester of pregnancy. Which layer is absent during the last trimester?

32. Name the three layers that form the vaginal wall and describe the tissue type(s) in each layer (VIII.A–C).

33. Explain the usually acidic pH of the vagina and the increase of fluid in the vagina during sexual arousal despite the lack of glands in the vaginal wall (VIII.A).

34. List the major components of the vulva and the structures that border the vestibule (IX.A–D).

35. Compare the glandulae vestibulares majores (Bartholin's glands) with the glandulae vestibulares minores (vestibular glands) in terms of their size, number, location, and secretory product and their developmental counterparts in the male (IX.B).

36. Compare the clitoris and the penis (IX.A) in terms of the presence of a glans and prepuce and the component number of erectile bodies.

37. Compare the labia majora and minora (IX.C and D) in terms of their location; the thickness of the stratum corneum on their external and internal surfaces; the presence of sweat glands, sebaceous glands, and coarse hairs; and their developmental counterparts in the male.

38. Compare the prepubertal, resting adult, pregnant, lactating, and senile mammary glands (X.B–G) in terms of the presence and number of:

    a. Lactiferous sinuses and ducts

    b. Plasma cells

    c. Adipose tissue

    d. Lobules

    e. Alveoli

    f. Fat droplets and secretory vesicles in alveolar cells

    g. Alveolar lumens distended with secretory product (milk)

39. Compare the mode of secretion of milk proteins with that of milk lipids (X.F).

40. List the steps in the milk ejection reflex (X.F.3).

41. Describe the role of plasma cells and colostrum in conferring passive immunity on newborns (X.E and F).

42. Make a time line of the idealized 28-day menstrual cycle. Designate day 1 as the first day of menstruation and indicate at which point, or during which days, the following events occur:

    a. Uterus (IV.A and D; Table 23–1)
       (1) Menstrual phase
       (2) Proliferative phase
       (3) Secretory phase
    b. Ovaries (II.B, D, E, and F; Table 23–1)
       (1) Follicular growth
       (2) Degeneration of the corpus luteum of menstruation
       (3) Estrogen production
       (4) Ovulation
       (5) Progesterone production
       (6) Corpus luteum formation
    c. Pregnancy
       (1) Fertilization (II.D)
       (2) Entry of the blastocyst into the uterus (V)
       (3) Implantation (V)
    d. Pituitary gland (II.F; Table 23–1)
       (1) FSH production
       (2) LH surge
43. Name the cells that produce the following:
    a. Relaxin (II.E.2)
    b. Estrogen (II.E and F)
    c. Follicle-stimulating hormone (II.F)
    d. Luteinizing hormone (II.F)
    e. Progesterone (II.E and F)
    f. Prolactin (VII.B.2; Table 20–2)
    g. Oxytocin (X.F.3)
    h. Androgens (II.B.4)
    i. Chorionic gonadotropin (VII.B.2)
    j. Human placental lactogen (VII.B.2)

SYNOPSIS

I. GENERAL FEATURES OF THE FEMALE REPRODUCTIVE SYSTEM

  A. Components of the System: The female reproductive system comprises the ovaries, uterine tubes (oviducts), uterus, vagina, and external genitalia (Fig. 23–1).
    1. The **ovaries** are the female **gonads.** They contain the **ovarian follicles,** which produce estrogen and harbor and promote the development of the **ova** (eggs). After ovulation, follicle remnants form a **corpus luteum,** which produces estrogen and progesterone.
    2. **Uterine tubes (oviducts)** capture the released ovum, serve as the primary site of fertilization, and convey the ovum (fertilized or not) to the uterus.
    3. The **uterus** is a hollow muscular organ lined by a mucosa (**endometrium**) that undergoes cyclic changes controlled by ovarian hormones. The changes prepare the uterus for the implantation and nourishment of the fertilized ovum.
    4. The **vagina** is a tubular organ that directs spermatozoa to the cervical canal. Vaginal fluid increases sperm motility. Estrogen causes the vaginal epithelium to thicken and its cells to accumulate glycogen, which is released into the lumen by means of exfoliation.
    5. The **external genitalia** include the **clitoris** and the **labia minora** and **majora;** they contain numerous nerve endings that play a role in sexual arousal.

  B. Cyclic Changes: Roughly every 28 days between **menarche** (first menses) and **menopause,** cyclic changes occur in the structure and activity of each organ, especially in the ovaries and uterus. The synchronization of these changes is crucial to normal reproductive function and is achieved mainly by pituitary FSH and LH. These hormones directly modulate follicle growth and development, as well as ovarian hormone production. Ovarian hormones (estrogen and progesterone) control the cyclic changes in the uterine lining and influence pituitary FSH and LH production through negative feedback.

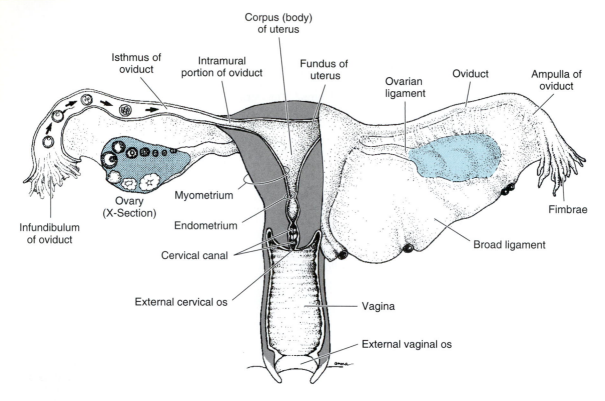

Figure 23–1. Internal organs of the female reproductive system. (Modified and reproduced, with permission, from Junqueira LC, Carneiro J, Kelley RO: *Basic Histology,* 9th ed. Stamford, CT: Appleton & Lange, 1998.)

C. Early Embryonic Development: Selected aspects of early development are described as they relate to reproductive structure and function. These include **fertilization** (V), **preimplantation embryonic development** (V), **implantation** (VI), and the formation and functions of the **placenta** (VII).

D. Mammary Glands: Mammary glands also undergo histologic changes related to the menstrual cycle, pregnancy, and pituitary and ovarian hormones (X).

II. OVARIES

A. General Organization: Located in the pelvic cavity, the ovaries are paired, almond-shaped organs ($3 \times 1.5 \times 1$ cm). Their outer covering, the **germinal epithelium,** does not form oocytes as the name suggests. It is a simple cuboidal epithelium derived from the peritoneum. The inner covering, the **tunica albuginea,** is a dense connective tissue capsule between the germinal epithelium and ovarian cortex. Each ovary (Fig. 23–2) has a peripheral **cortex** and a central **medulla.** The cortex harbors most of the oocyte-containing ovarian follicles embedded in connective tissue **(stroma).** The medulla consists of stroma containing a rich vascular bed.

B. Ovarian Follicles: Each follicle consists of a single **oocyte** surrounded by one or more layers of **follicle (granulosa) cells.** The cortex contains follicles at various stages of development.
   1. Primordial follicles are inactive and characterize the earliest stage of follicle development. They are the predominant follicles present before puberty, and they constitute the majority thereafter. Each consists of a **primary oocyte** (most in the diplotene stage of meiosis I prophase) surrounded by one layer of squamous follicle cells.

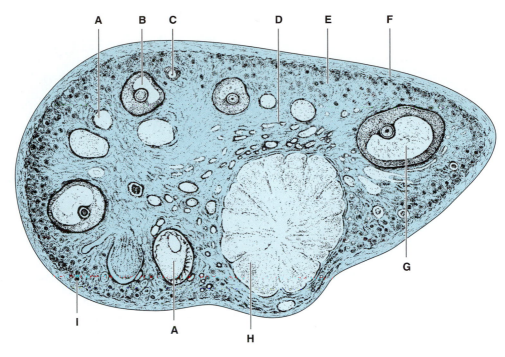

Figure 23–2. Schematic diagram of a cross-section of the ovary. Labeled structures include the atretic follicles (A), secondary follicle (B), multilaminar primary follicle (C), medulla (D), cortex (E), tunica albuginea (F), mature or graafian follicle (G), corpus luteum (H), and primordial follicle (I). (Modified and reproduced, with permission, from Copenhaver WM, Bunge RP, Bunge MTS: *Bailey's Textbook of Histology,* 16th ed. Baltimore, MD: Williams & Wilkins, 1972.)

2. **Growing follicles.** During follicle growth, which is stimulated by pituitary FSH, the oocyte enlarges to a diameter of 125 to 150 μm. The follicle epithelium becomes cuboidal and proliferates to become stratified (multilaminar). The stromal tissue immediately surrounding the follicle differentiates into the steroid hormone–producing **theca folliculi.**

   a. **Primary follicles** consist of a primary oocyte surrounded by single or multiple layers of cuboidal follicle cells. They have no antrum. **Unilaminar primary follicles** have a single layer of cuboidal follicle cells. At this stage, the glycoprotein-rich **zona pellucida** forms between the oocyte and the follicle cells. **Multilaminar primary follicles** have multiple layers of follicle cells. During this stage, the zona pellucida thickens and the theca folliculi forms.

   b. **Secondary follicles.** During this stage, cavities filled with fluid **(liquor folliculi)** appear between the follicle cells, gradually coalescing to form one large cavity, or **antrum.** The theca folliculi forms two layers: the **theca interna,** containing a rich vascular network and steroid-secreting cuboidal cells with abundant SER, and the **theca externa,** consisting mainly of vascular connective tissue.

3. **Mature (graafian) follicles** (Fig. 23–3) differ from late secondary follicles mainly by their large diameter (2.5 cm). During this stage, which immediately precedes ovulation, antrum size increases greatly. The oocyte is displaced to one side of the follicle, surrounded by a few layers of follicle cells **(corona radiata),** and rests on a pedestal of follicle cells **(cumulus oophorus).**

4. **Atretic follicles.** Of the 400,000 follicles present at birth, approximately 450 develop to maturity. More than 99% become atretic (ie, degenerate by autolysis) at various stages. Atresia of the primordial follicles leaves a space that is filled by stroma; as a result, no vestiges of atretic primordial follicles are seen in adult ovaries. Autolytic remnants of larger primary and secondary follicles are removed by macrophages and are replaced, by the stromal cells, with a wavy collagenous scar. The scar is gradually removed and remodeled into normal stromal tissue. Some thecal cells from the atretic follicles may remain, becoming interstitial cells that secrete steroids, especially androgens.

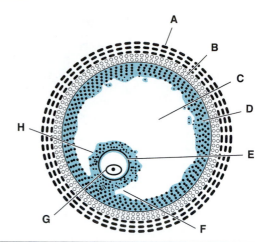

Figure 23–3. Schematic diagram of a mature (graafian) follicle. Labeled structures include the theca externa (A), theca interna (B), antrum (C), granulosa cells of the membrana granulosa (D), oocyte (E), cumulus oophorus (F), nucleus of the oocyte (G), and corona radiata (H). (Reproduced, with permission, from Junqueira LC, Carneiro J, Kelley RO: *Basic Histology,* 9th ed. Stamford, CT: Appleton & Lange, 1998.)

C. **Origin and Maturation of Oocytes:** In the embryo, yolk sac endoderm gives rise to **primordial germ cells,** which migrate to the genital ridges in the posterior wall of the abdominal cavity, from which the ovaries develop. These germ cells are surrounded by the flattened follicle cells of primordial follicles; subsequently, they enter the first meiotic division and arrest in prophase. At this point, they are **primary oocytes** (comparable to primary spermatocytes [22.II.C.2.b]). The first meiotic division is completed just before ovulation and involves equal division of the chromatin but unequal division of the cytoplasm between the resulting **secondary oocytes.** The secondary oocyte that retains almost all the cytoplasm is the **ovum;** the other is termed the **first polar body.** After it is formed, but before ovulation, the ovum begins the second meiotic division, which halts in metaphase until fertilization occurs. During fertilization, the second meiotic division is completed and the **second polar body** is formed. The fertilized ovum is called the **zygote.**

D. **Ovulation:** On day 14 of an idealized 28-day cycle, ovulation involves the rupture of a mature follicle and the release of its ovum. It is preceded and stimulated by a surge in pituitary LH production. As the amount of liquor folliculi in the antrum increases, the ovum and its surrounding zona pellucida and corona radiata detach from the cumulus oophorus and float in the antrum. Perhaps owing to collagenase activity, the stroma thins and becomes ischemic between the follicle and the ovary surface, indicating the site of imminent rupture, or **stigma.** During rupture, the ovum, with its corona intact, is expelled by the ovary and captured by the uterine tube. Unless it is fertilized within 24 hours, the ovum degenerates.

E. **Corpus Luteum:** This temporary endocrine gland is formed by the remnants of the follicle after ovulation. After ovulation, the follicle collapses and the granulosal lining becomes folded. Cells in the granulosa layer and theca interna enlarge and begin secreting steroids. The **granulosa lutein cells** are large, pale-staining, progesterone-secreting cells derived from the granulosa cells. The **theca lutein cells,** which secrete estrogen, are smaller, darker-staining cells derived from the cells of the theca interna.
   1. **Corpus luteum of menstruation.** If fertilization does not occur, the corpus luteum degenerates after approximately 14 days.
   2. **Corpus luteum of pregnancy.** If fertilization does occur, the corpus luteum enlarges. It is maintained for 6 months; although it gradually declines thereafter, it persists until the end of pregnancy. In addition to estrogen and progesterone, it produces **relaxin,** a polypeptide hormone that loosens the fibrocartilage attachment of the symphysis pubis, allowing the pelvic opening to enlarge during parturition.
   3. **Corpus albicans.** This dense connective tissue scar, which replaces a degenerated corpus luteum, is larger for a corpus luteum of pregnancy than for that of menstruation. Like atretic follicles, it is eventually removed by macrophages.

F. **Hormones and Ovarian Function:** Pituitary FSH stimulates follicle growth during the first half of the menstrual cycle. Growing follicles produce estrogen, whose high mid-cycle level exerts negative feedback on the pituitary gland, reducing FSH production. This causes the LH surge, which directs follicle maturation, stimulates ovulation, and controls the formation and maintenance of the corpus luteum. The corpus luteum produces both estrogen and progesterone. Progesterone inhibits LH production, causing the corpus luteum to degenerate after approximately 14 days unless fertilization occurs. If the ovum is fertilized and implants in the uterus, **chorionic gonadotropin** produced by the developing placenta (VII.B.2) maintains the corpus luteum in the absence of LH.

III. UTERINE TUBES (OVIDUCTS, FALLOPIAN TUBES)

These are paired 12-cm-long muscular tubes whose lumens are continuous proximally with the uterine cavity (see Fig. 23–1). Each tube's distal end opens to the peritoneal cavity near the ovary.

A. Function: The uterine tube moves close to the ovary before ovulation and captures the ovulated ovum. It provides a suitable environment for, and is the most common site of, fertilization. This tube subsequently transports the zygote to the uterus.

B. Uterine Tube Segments: Each uterine tube has four segments (see Fig. 23–1). The **pars interstitialis** (intramural segment) penetrates the uterine wall. It contains the fewest mucosal folds, and the myometrium contributes to its muscularis. The **isthmus** is the narrow segment adjacent to the uterine wall; it contains few mucosal folds. The **ampulla** is the widened segment near the distal end; it contains extensive branched mucosal folds and is the common site of fertilization. The **infundibulum,** which is the funnel-shaped distal segment, opens near the ovary. Fingerlike extensions of its mucosal folds, the **fimbriae,** project from the opening toward the ovary.

C. Wall Structure: The uterine tube wall has three layers: mucosa, muscularis, and serosa. It does not contain a definitive submucosa. The **mucosa** includes the lamina propria and the lining epithelium. The mucosal folds, which are largest and most numerous in the ampulla and infundibulum, decrease in size and number toward the uterus. The lining is simple columnar epithelium with two main cell types. The cilia on the surface of the abundant **ciliated columnar cells** beat in waves. Most beat toward the uterus and thus aid in egg transport. Shorter, mucus-secreting **peg cells** are interspersed among the ciliated cells. The film they produce, which is propelled toward the uterus by cilia, helps transport the ovum and hinders bacterial access to the peritoneal cavity. The **muscularis** has inner circular and outer longitudinal smooth muscle layers; its wavelike contractions move the ovum toward the uterus. The tubes' outer covering is a **serosa** of visceral peritoneum.

IV. UTERUS

A pear-shaped muscular organ in the pelvic cavity, the uterus (womb) is the site of embryonic implantation and development. It is grossly divided into three regions (see Fig. 23–1). The **body,** or **corpus,** is its large, round middle region. The **fundus** is the part of the body above the entry of the uterine tubes. The neck, or **cervix,** is the narrow, downward extension of the uterus into the vagina. The uterine wall consists of three layers: the endometrium, myometrium, and serosa or adventitia.

A. Endometrium: This layer (uterine mucosa) consists of simple columnar epithelium supported by a lamina propria. Simple tubular glands extend from the luminal surface into the lamina propria; their lining is continuous with the surface. The endometrium receives a dual blood supply and comprises two layers. The **stratum functionale (pars functionalis)** is the temporary layer at the luminal surface. It responds to ovarian hormones by undergoing cyclic thickening and shedding. It can be subdivided, based on the density of the lamina propria, into a **zona compacta** near the lumen and a deeper **zona spongiosa.** The **stratum basale (pars basalis)** is the thinner, deeper, permanent layer containing the basal portions of the endometrial glands and is retained during menstruation. The epithelial cells lining these glands divide and cover the raw surface exposed during menstruation. Paired **uterine arteries** branch to form the **arcuate arteries** in the middle of the myometrium. The arcuate arteries give rise to two sets of arteries:

straight arteries to the stratum basale and **coiled arteries** to the stratum functionale. The dual supply to the endometrium is important in the cyclic shedding of the stratum functionale, when the coiled arteries are lost and the straight arteries are retained.

B. **Myometrium:** The uterine muscularis is its thickest tunic, comprising four poorly defined smooth muscle layers. The middle layers contain the arcuate arteries. During pregnancy, the myometrium grows extensively by means of hypertrophy and hyperplasia. At birth, a pituitary oxytocin surge promotes the forceful myometrial contractions that expel the fetus.

C. **Serosa:** The fundus and most of the body are covered by serosa.

D. **Menstrual Cycle:** The endometrium undergoes cyclic changes controlled by the ovarian hormones estrogen and progesterone. Ovarian hormone production is in turn controlled by the pituitary hormones FSH and LH and is related to follicle growth, to ovulation, and to the formation and degeneration of the corpus luteum. The menstrual cycle is divided into three phases based on structural and functional changes in the endometrium: the **menstrual phase,** the **proliferative** (or **follicular**) **phase,** and the **secretory** (or **luteal**) **phase.** Table 23–1 describes an idealized 28-day menstrual cycle in terms of its three main phases, the part of the cycle each occupies, the endometrial changes that occur during each phase, and the correlated changes in ovarian function.

E. **Uterine Cervix:** The external surface of the cervix bulges into the vaginal canal. Its wall consists mainly of dense connective tissue, with a small amount of smooth muscle. The mucosa has a tall simple columnar epithelium that lines the cervical canal and branched cervical glands that penetrate into the lamina propria. Stratified squamous epithelium covers its external (vaginal) surface. The switch in epithelial type occurs just inside the cervical canal, where it opens into the vagina (external os of the cervix), the most common site of cervical cancer. The cervical mucosa is not shed during menstruation, but cyclic changes in the amount and viscosity of cervical secretions do occur. At ovulation, watery secretions permit access by sperm; in the luteal phase and during pregnancy, the secretions are abundant and more viscous. Cervical dilation preceding parturition occurs in response to intense collagenase activity in the cervical wall.

Table 23–1. The menstrual cycle.

Phase of Cycle	Endometrial Changes	Correlated Ovarian Changes
Menstrual phase: first day of menstrual bleeding through day 3–5 of cycle	Decline of ovarian progesterone production (owing to degeneration of corpus luteum) causes intermittent constriction of coiled arteries. Leads to ischemia, degeneration, and shedding of functionalis. Fragments of functionalis tissue along with blood and uterine fluid are discharged through vagina as menstrual fluid. Because straight arteries supplying basale do not react to hormonal changes, basale remains intact.	Without chorionic gonadotropin from an implanted embryo, corpus luteum degenerates and progesterone production ceases.
Proliferative phase (follicular phase): days 4–6 to day 14 of the cycle	Under influence of increasing estrogen (and in preparation for possible implantation), endometrium regenerates from basale. As functionalis thickens, glands lengthen, remaining relatively straight.	Under influence of pituitary FSH, follicles grow and produce estrogen. LH surge on day 14 induces ovulation and supports formation of corpus luteum.
Secretory phase (luteal phase): days 14–28 of the cycle	Progesterone from developing corpus luteum causes edema of lamina propria and endometrial thickening. Glands grow and become highly coiled, exhibiting sawtooth appearance in longitudinal section. Secretion of glycoproteins (nutrients for embryo before implantation) increases and dilates lumens. Coiled arteries elongate and grow closer to luminal surface. By day 20 of the cycle, endometrium is prepared to receive implanting embryo. Without implantation, cycle begins.	Pituitary LH supports corpus luteum development. Granulosa lutein cells begin producing progesterone, and theca lutein cells produce estrogen. Elevated progesterone inhibits LH production by negative feedback. At end of phase, without chorionic gonadotropin from implanted embryo, corpus luteum degenerates, progesterone production ceases, and menstruation occurs.

V. FERTILIZATION & PREIMPLANTATION DEVELOPMENT

Fertilization occurs at the **ampullaroisthmic junction** in the uterine tube. Sperm penetrate the corona radiata and subsequently the zona pellucida. Only one sperm head fuses with the plasma membrane of the ovum (**oolemma).** Fertilization induces completion of the ovum's second meiotic division and thereby the formation of the second polar body. Finally, the haploid male and female pronuclei fuse to form the diploid nucleus of the zygote. The zygote undergoes several rounds of mitosis, without cell growth between divisions, to become a solid ball of smaller cells, or **morula,** as it moves along the oviduct toward the uterus. As mitosis continues, a cavity forms at the center of the embryo, which is now called a **blastocyst.** By this stage (day 4 after fertilization), the embryo has entered the uterus. The blastomeres—the cells of the blastocyst—form two layers: a peripheral trophoblast, which will form the fetal part of the placenta, and a disk of cells (the inner cell mass), which will form the embryo, bulging into the cavity. After it enters uterus, the blastocyst floats free for 2 to 3 days before implantation. The zona pellucida dissipates at this time, allowing the trophoblast cells to contact the endometrium directly.

VI. IMPLANTATION

This is the penetration of the uterine epithelium by the blastocyst. It is the first step in placentation and involves important activities in the blastocyst and in the uterine linings (ie, the decidual reaction).

A. Blastocyst Activity: The **trophoblast** cells attach to the endometrium, divide rapidly, and differentiate into two layers. The **syncytiotrophoblast,** the highly invasive outer layer, consists of multiple nuclei in a single large cytoplasm. It is formed by the fusion of mononucleated cells from the underlying layer, the **cytotrophoblast.** The trophoblast erodes the uterine epithelium, allowing the embryo to invade the stroma. By day 9 after fertilization, the embryo is completely embedded in the endometrium and is surrounded by a trophoblastic shell. Implantation in which the embryo becomes completely embedded in the endometrium is termed **interstitial implantation.** The **inner cell mass** forms a bilaminar disk (**blastodisk),** which becomes the embryo itself, and a shell of **extraembryonic mesoderm,** which lines the inner surface of the cytotrophoblast. The blastodisk is separated from the **extraembryonic mesoderm** by a cavity called the **extraembryonic coelom.** The embryo is thus separated from the endometrium by a three-layered shell, or chorion. The **chorion** includes derivatives of both the trophoblast (syncytiotrophoblast and cytotrophoblast) and the inner cell mass (extraembryonic mesoderm). It has two regions. The **chorion frondosum** is the part that lies adjacent to the decidua basalis (VI.B) and forms the fetal part of the placenta. The **chorion laeve** is the part adjacent to the decidua capsularis (VI.B). Midway through pregnancy, this layer fuses with the decidua parietalis (VI.B) on the opposite side of the uterus, obliterating the uterine cavity.

B. Decidual Reaction: Immediately after implantation, the endometrium undergoes changes referred to as the decidual reaction (the pregnant endometrium is now termed the **decidua).** During this reaction, the endometrium thickens and its stromal cells enlarge to become **decidual cells,** which secrete prolactin. The decidual reaction helps prevent invasion of the trophoblast beyond the endometrium (a condition termed **decidua increta** or **decidua percreta).** The decidua has three parts. The **decidua basalis** is the part underlying the implantation site; it forms the maternal part of the placenta. The **decidua capsularis** is the part overlying the implanted embryo and separating it from the uterine cavity. The **decidua parietalis** is the remainder of the endometrium (ie, the portion not in direct contact with the embryo).

VII. PLACENTA

This is a temporary organ whose formation begins during implantation. It has both embryonic (chorion frondosum) and maternal (decidua basalis) components. The placenta transfers maternal nutrients and oxygen to the embryo, cleanses the fetal blood, and secretes hormones.

A. Steps in Placental Development (Placentation): The invading syncytiotrophoblast surrounds small islands of endometrium that contain blood vessels. Enzymes secreted by the syncytiotrophoblast lyse the maternal tissue, leaving spaces, or **lacunae,** and rupturing blood vessels. The ruptured vessels fill the syncytiotrophoblast-lined lacunae with maternal blood. Solid cords of chorionic tissue **(chorionic villi)** grow into these lacunae and develop; these cords bring the blood in the fetal vessels close enough to the maternal blood in the lacunae for exchange to occur and also form a selectively permeable **placental barrier** (VII.B; Fig. 23–4). **Primary villi** are tongues of syncytiotrophoblast and cytotrophoblast. The underlying extraembryonic mesenchyme invades the primary villi to form **secondary villi,** composed of syncytiotrophoblast, cytotrophoblast, and a core of extraembryonic mesenchyme. The extraembryonic mesenchyme differentiates into blood vessels that later establish connections with the umbilical vessels of the fetus. **Tertiary villi** are thus composed of syncytiotrophoblast, cytotrophoblast, and extraembryonic mesenchyme that contains blood vessels in its core. The cytotrophoblast gradually disappears as its cells fuse with the syncytiotrophoblast.

B. Placental Functions:
1. **Transfer of nutrients and wastes.** By day 23 of gestation, the fetal blood is circulating through the tertiary villi. Nutrients from the maternal blood in the lacunae reach the fetal circulation by passing successively through (1) the syncytiotrophoblast; (2) the cytotrophoblast, which later disappears; (3) the basal lamina of the trophoblast; (4) the extraembryonic mesenchyme; (5) the basal lamina of the vessels in the tertiary villi; and (6) the fetal vascular endothelial cells. These six layers constitute the placental barrier (see Fig. 23–4), which restricts the substances that cross between the maternal and fetal circulations. The maternal–fetal boundary is further marked by **fibrinoid,** a layer of the products of necrosis that may form a nonantigenic barrier and may explain maternal tolerance of fetal antigens.
2. **Placental hormones.** Many hormones are secreted by the syncytiotrophoblast of the chorion, and a few additional hormones are produced by the decidual cells. Placental hormones include chorionic gonadotropin, chorionic thyrotropin, chorionic corticotropin, estrogens, progesterone, prolactin, and placental lactogen.

VIII. VAGINA

This muscular tube extends from the cervix to the external genitalia. Its walls lack glands; thus, vaginal lubrication involves secretions produced by the cervical and Bartholin's glands and smaller mucous glands in the vestibule. The vaginal walls have three layers: mucosa, muscularis, and adventitia.

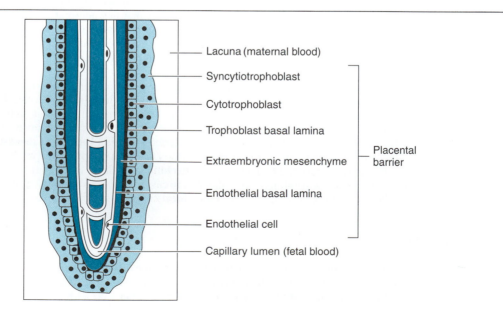

- Lacuna (maternal blood)
- Syncytiotrophoblast
- Cytotrophoblast
- Trophoblast basal lamina
- Extraembryonic mesenchyme — Placental barrier
- Endothelial basal lamina
- Endothelial cell
- Capillary lumen (fetal blood)

Figure 23–4. Schematic diagram of a tertiary chorionic villus showing the composition of the placental barrier between maternal and fetal blood.

A. Mucosa: The vaginal mucosa is a stratified squamous epithelium that is rich in glycogen and is supported by an elastic fiber–rich lamina propria. Bacterial metabolism of glycogen from the lining cells that are sloughed into the lumen results in lactic acid accumulation, which causes the vagina's low pH. The extensive capillary plexus in the lamina propria provides much of the fluid that seeps into the lumen during sexual arousal. The vaginal mucosa contains few sensory nerve endings.

B. Muscularis: The vaginal muscularis consists mainly of longitudinal smooth muscle but includes some circular fibers near the mucosa.

C. Adventitia: The vagina is surrounded by a sheath of dense connective tissue that is rich in elastic fibers. It contains an extensive venous plexus, bundles of nerve fibers, and clusters of neurons.

IX. EXTERNAL GENITALIA (VULVA)

This area is richly innervated with Meissner's and pacinian corpuscles and free nerve endings.

A. Clitoris: A homologue of the dorsal part of the penis, the clitoris consists of two erectile corpora cavernosa that end in a **glans clitoridis.** It is surrounded by a **prepuce** and covered by stratified squamous epithelium.

B. Vestibule: This is the area that receives the vaginal and urethral openings. It is covered by stratified squamous epithelium and includes two types of glands. **Bartholin's glands (glandulae vestibulares majores)** are two large tubuloalveolar mucous glands on opposite sides of the vestibule and are analogous to the male bulbourethral (Cowper's) glands. The **vestibular glands (glandulae vestibulares minores)** are smaller and more numerous mucous glands scattered around the vestibule. These are analogous to the glands of Littre in males; most lie near the urethra and clitoris.

C. Labia Minora: These are skin folds with a core of spongy (erectile) connective tissue that are covered by stratified squamous epithelium; they are analogous to the male corpus spongiosum. They have a thin keratinized layer on their surfaces and, although they contain sweat and sebaceous glands on both surfaces, they lack hairs.

D. Labia Majora: These folds of skin have a core of subcutaneous fat and a thin layer of muscle. The inner surface of each is similar to that of the labia minora; the outer surface has more keratin and contains coarse hairs. Both surfaces contain numerous sebaceous and sweat glands. Their developmental analogue in males is the scrotum.

X. MAMMARY GLANDS

These accessory glands of the skin are specialized to secrete milk. Each of these compound tubuloalveolar glands contains 15 to 25 lobes, separated by adipose tissue and bands of dense connective tissue. Each lobe empties through a **lactiferous duct,** which exhibits a terminal expansion, or **lactiferous sinus,** before opening independently on the surface of the richly innervated **nipple (papilla).** These glands undergo extensive changes correlated with age and the functional state of the reproductive system.

A. Embryonic Development: **Milk lines,** which are paired ventral epidermal thickenings running from forelimb to hindlimb, appear at 6 weeks. Their caudal portions regress early. During the second trimester, 15 to 25 epithelial invaginations develop along these lines on each side of the thorax; these are the future lactiferous ducts. The rest of the milk lines normally degenerate. Mammary secretion in newborns (stimulated by placental and maternal hormones) is common.

B. Prepubertal Mammary Gland: This nonfunctional gland consists of lactiferous ducts and sinuses. The small nipple is surrounded by a lightly pigmented **areola.**

C. **Changes During Puberty:** The female breasts enlarge, owing to the accumulation of adipose tissue and collagenous connective tissue, and the nipples enlarge and become more prominent. Increased ovarian estrogen production stimulates fat deposition and the proliferation (including increased length, diameter, and branching) of the lactiferous ducts.

D. **Resting Adult Gland:** The basic subunits of this gland, the lobules, develop as a result of growth during puberty. The lobules are separated by loose connective tissue, and few secretory alveoli are present. Most of the lobules consist of several blind-ended **intralobular ducts;** these are lined by cuboidal epithelium lying on a basal lamina and surrounded by myoepithelial cells. All the intralobular ducts from one lobule empty into a single terminal **interlobular duct,** which leads to a larger **lactiferous duct.** The lactiferous ducts are lined by cuboidal to columnar epithelium, overlying a layer of densely packed, spindle-shaped, longitudinally oriented **myoepithelial cells,** that separates the epithelium from its basal lamina. The lactiferous ducts empty through stratified squamous-lined **lactiferous sinuses.** Minor changes occur during the menstrual cycle. The estrogen peak at ovulation induces further proliferation of the ducts, which can cause premenstrual breast enlargement, attended by transient edema and tenderness.

E. **Pregnant Adult Gland:** The influence of several hormones, including estrogen, progesterone, prolactin, and human placental lactogen, causes an intense proliferation of the ducts and the growth of alveoli at their ends, thereby enlarging the breasts. The terminal epithelium of the intralobular ducts proliferates and differentiates into milk-secreting cells, forming many secretory alveoli within the lobules. During pregnancy, the mammary alveolar cells are characterized by basal nuclei that are surrounded by a supranuclear Golgi complex, scattered mitochondria, lysosomes, milk protein–containing secretory vesicles, and a few apical fat droplets. Myoepithelial cells intervene between the alveolar cells and their basal lamina. During the third trimester of pregnancy, the number of plasma cells in the interlobular connective tissue increases. These cells add secretory IgA and IgG to the mammary secretions (especially colostrum) and confer passive immunity on the newborn. Although the glands are well-developed during pregnancy, secretions are not found in their lumens until late in pregnancy, when they contain **colostrum** (protein-rich first milk), or during lactation, when they contain lipid-rich milk.

F. **Lactating Adult Gland:** With the loss of the placenta at birth, estrogen and progesterone decrease and prolactin increases. The major histologic change from pregnant glands involves the accumulation of milk in the alveolar lumens and their consequent dilation. The secretory cells drop in height from low columnar to low cuboidal, and many neutral-triglyceride–containing fat droplets collect in their cytoplasm. During secretion, the fat droplets acquire a membrane from the cell apex **(apocrine secretion).** Secretory vesicles containing milk proteins also collect in the cytoplasm and are released by means of **merocrine secretion.**
   1. **Milk composition.** Milk typically contains 4% lipids, 1.5% proteins (caseins, lactalbumin, IgA), 7% lactose (disaccharide of glucose and galactose), and 87.5% water.
   2. **Maintaining lactation.** The sensory stimuli of suckling inhibits hypothalamic dopamine (or prolactin-inhibiting hormone) secretion. This increases the release of prolactin from the anterior pituitary, which in turn stimulates milk production. With weaning (the cessation of suckling), prolactin levels fall and the alveoli degenerate.
   3. **Milk ejection reflex.** The sensory stimulus of suckling also causes oxytocin synthesis in the hypothalamus. Oxytocin release by the posterior pituitary stimulates myoepithelial cell contraction, ejecting milk from the alveoli into the lactiferous ducts.

G. **Senile Involution:** After menopause, the secretory components, ducts, and adipose and interlobular connective tissues in the breasts atrophy.

MULTIPLE-CHOICE QUESTIONS

Select the single best answer.

23.1. Which stage of ovarian follicle development immediately follows the multilaminar primary follicle stage?
(A) Corpus albicans
(B) Corpus luteum
(C) Graafian follicle
(D) Primordial follicle
(E) Secondary follicle
(F) Unilaminar primary follicle

23.2. Which of the following organs is characteristically lined by the epithelial type shown in Figure 23–5?
(A) Ovary
(B) Uterus
(C) Uterine tubes
(D) Vagina

23.3. Which of the following structures comprise the placenta?
(A) Decidua basalis and chorion frondosum
(B) Decidua basalis and chorion laeve
(C) Decidua capsularis and chorion laeve
(D) Decidua parietalis and chorion frondosum
(E) Decidua parietalis and chorion laeve

23.4. A cross-section of which of the following structures is shown in Figure 23–6?
(A) Ampulla of the uterine tube
(B) Body of the uterus
(C) Isthmus of the uterine tube
(D) Lactiferous duct
(E) Pars interstitialis of the uterine tube
(F) Placenta
(G) Uterine cervix

23.5. Which of the following structures is shown in Figure 23–7?
(A) Chorionic villus
(B) Fimbria
(C) Labium minorum
(D) Mucosal fold of the uterine tube
(E) Uterine cervix
(F) Uterine gland
(G) Vaginal gland

23.6. Which of the following is true of secondary chorionic villi but not of primary chorionic villi?
(A) Include the syncytiotrophoblast
(B) Include the cytotrophoblast
(C) Include the extraembryonic mesenchyme
(D) Do not include the cytotrophoblast

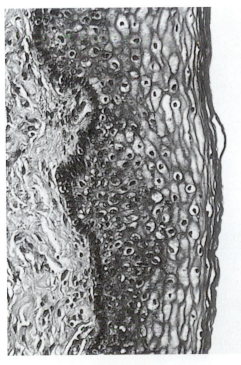

Figure 23–5.

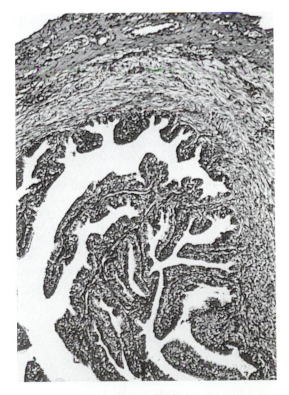

Figure 23–6.

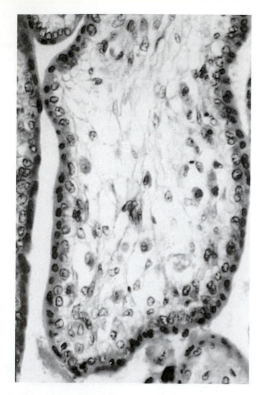

Figure 23–7.

(E) Do not include fetal blood vessels

(F) Do not include the syncytiotrophoblast

23.7. Which of the following structural changes accompanies follicle development?

(A) Reduction in size of the antrum

(B) Reduction in size of the follicle

(C) Reduction in size of the ovum

(D) Reduction in number of follicle (granulosa) cell layers

(E) Reduction in thickness of the zona pellucida

(F) Detachment of ovum and corona radiata from the cumulus oophorus

(G) Replacement of the theca interna by the theca externa

23.8. Which of the following structures is most likely to remain a part of the ovary after ovulation?

(A) Corona radiata

(B) First polar body

(C) Follicular liquor

(D) Theca interna

(E) Zona pellucida

23.9. Which of the following is true of the granulosa lutein cells?

(A) Are a minor cell type in the corpus luteum

(B) Derive from the theca interna

(C) Contain abundant RER

(D) Are small and dark-staining

(E) Secrete progesterone

23.10. Which of the following is true of fertilization?

(A) Typically occurs at the ampullaroisthmic junction of the uterine tube

(B) Typically occurs after the production of the second polar body

(C) Typically occurs after the ovum develops into a zygote

(D) Typically occurs 3 to 5 days after ovulation

(E) Typically occurs 3 to 5 days after insemination

23.11. Milk protein is produced by which of the following mechanisms?

(A) Apocrine secretion

(B) Autocrine secretion

(C) Cytocrine secretion

(D) Holocrine secretion

(E) Merocrine secretion

(F) Paracrine secretion

23.12. Which of the labels in Figure 23–8 corresponds to the site of the decidual reaction?

23.13. Which of the labels in Figure 23–8 corresponds to the location of most of the peg cells?

23.14. Which of the labels in Figure 23–8 corresponds to the site of normal transition from stratified squamous epithelium to simple columnar epithelium?

23.15. Which of the labels in Figure 23–9 corresponds to the location of cells that become theca lutein cells during the formation of the corpus luteum?

23.16. Which of the labels in Figure 23–9 corresponds to the location of the follicular liquor?

23.17. Which of the labels in Figure 23–9 corresponds to the location of the cells that accompany the ovum during ovulation?

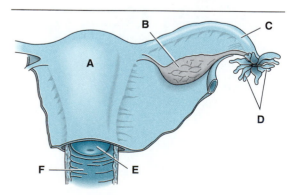

Figure 23–8.

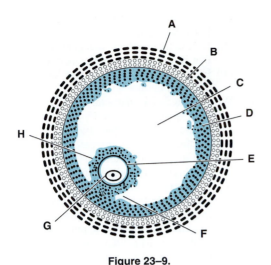

Figure 23–9.

23.21. Which of the structures in Figure 23–10 most closely resembles a graafian follicle?

23.22. Which of the structures in Figure 23–10 corresponds to the type of follicle that is most abundant before puberty?

23.23. At which of the following stages of embryonic development is the zona pellucida typically shed?
(**A**) Blastocyst
(**B**) Graafian follicle
(**C**) Fertilization
(**D**) Implantation
(**E**) Morula
(**F**) Ovulation
(**G**) Zygote formation

23.24. Which of the following is true of the corpora albicans but not of atretic follicles?
(**A**) May leave hormone-secreting interstitial cells after degeneration
(**B**) Characteristically resemble large collagenous scars
(**C**) Eventually may be removed by macrophages and replaced by stroma
(**D**) Are remnants of follicles that degenerate before ovulation
(**E**) May contain degenerating ova
(**F**) Often enlarge during pregnancy
(**G**) May contain degenerating granulosa cells floating in the remnants of the follicular liquor

23.18. Which of the labels in Figure 23–9 corresponds to the location of a cell whose progeny secrete chorionic gonadotropin after implantation?

23.19. Which of the labels in Figure 23–10 corresponds to the most active site of progesterone synthesis?

23.20. Which of the labels in Figure 23–10 indicates the location of an atretic follicle?

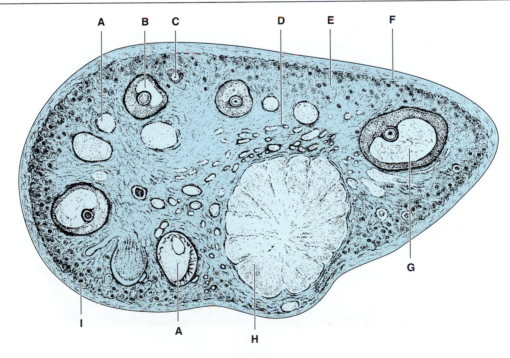

Figure 23–10.

23.25. Which of the following is true of the uterine tube's ampulla but not of its isthmus?
(A) Contains mucus-secreting peg cells
(B) Contains ciliated epithelial cells
(C) Is covered by serosa derived from the peritoneum
(D) Contains large mucosal folds that are extensively branched
(E) Contains smooth muscle in its walls

23.26. Which of the following is true of the stratum basale but not of the stratum functionale?
(A) Includes the uterine surface epithelium
(B) Includes connective tissue
(C) Contains cells that are highly responsive to ovarian hormones
(D) Is a layer that undergoes cyclic thickening and shedding
(E) Contains cells that replace the surface epithelium after menstruation
(F) Contains coiled arteries

23.27. Which of the following is true of the cytotrophoblast but not of the syncytiotrophoblast?
(A) Is a multinucleated mass formed by cell fusion
(B) Lines the inner walls of placental lacunae
(C) Contains cells that are derived from extraembryonic mesenchyme

(D) Contains cells that are derived from the inner cell mass
(E) Is a component of the placental barrier that is lost during late pregnancy
(F) Is an epithelial cell layer that lacks a typical basal lamina

23.28. Which stages of mammary gland development are shown in Figure 23–11?
(A) Senile (A) and prepubertal (B)
(B) Lactating (A) and pregnant (B)
(C) Senile (A) and lactating (B)
(D) Resting adult (A) and senile (B)
(E) Pregnant (A) and lactating (B)
(F) Resting adult (A) and prepubertal (B)
(G) Prepubertal (A) and resting adult (B)

23.29. Which of the following is the pituitary hormone responsible for maintaining progesterone secretion in the corpus luteum during menstruation?
(A) ADH
(B) Chorionic gonadotropin
(C) Follicle-stimulating hormone
(D) Luteinizing hormone
(E) Oxytocin
(F) Pregnenolone
(G) Prolactin

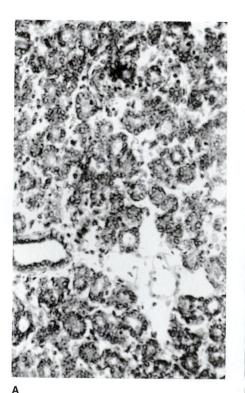
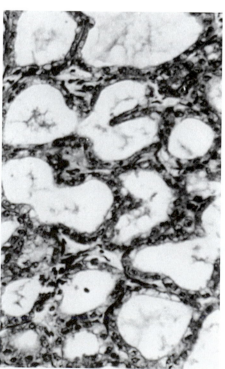

A
B

Figure 23–11.

23.30. Which of the following occurs in response to the neural stimulation that accompanies suckling of the mother's breast by a newborn infant?
(A) Hypothalamic inhibition of the release of oxytocin
(B) Relaxation of the myoepithelial cells surrounding mammary gland alveoli
(C) Contraction of smooth muscle in the myometrium
(D) Relaxation of smooth muscle in the nipple and areola
(E) Proliferation of plasma cells in the mammary interlobular septa

23.31. In which of the following sites are the FSH receptors located on the follicle (granulosa) cells of a primary follicle?
(A) Cytoplasm
(B) Nuclear envelope
(C) Nucleoplasm
(D) Plasma membrane
(E) SER

23.32. Which of the following is the primary cause of the normally acidic pH in the vaginal lumen?
(A) Bacterial conversion of mucus (secreted by vaginal glands) into lactic acid
(B) Bacterial conversion of glycogen (from exfoliated cells) into lactic acid
(C) Secretion of acidic mucopolysaccharides by glands in the vaginal walls
(D) Secretion of HCl by the lining epithelium of the vagina
(E) Acidic secretions from uterine glands

23.33. The endometrium shown in Figure 23–12 was taken from a uterus during which phase?
(A) Follicular
(B) Menstrual
(C) Proliferative
(D) Secretory

23.34. The blood that supplies the nutrients used to rebuild the endometrium after menstruation is contained in which of the following?
(A) Coiled arteries in the zona spongiosa
(B) Coiled arteries in the zona compacta
(C) Straight arteries in the stratum functionale
(D) Straight arteries in the stratum basale
(E) Menstrual fluid in the uterine lumen

23.35. Which of the following is true of the colostrum?
(A) Contains mainly lipid
(B) Is first released during weaning

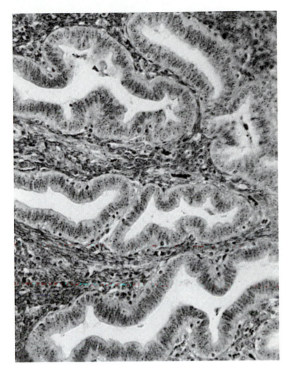

Figure 23–12.

(C) Is an incomplete form of milk produced by hormonally deficient mothers
(D) Provides a form of passive immunity to the infant
(E) Is produced mainly by apocrine secretion

23.36. Which letter in Figure 23–13 (an idealized time line of the human menstrual cycle) most closely corresponds to the period during which LH supports corpus luteum development and progesterone production by its granulosa lutein cells?

23.37. Ovulation is triggered by a surge in the release of which substance by the hypothalamic neurons?
(A) Dopamine
(B) FSH
(C) GnRH
(D) LH
(E) Oxytocin
(F) Somatostatin

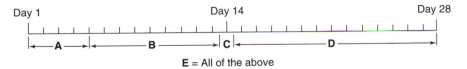

Day 1   Day 14   Day 28

A   B   C   D

E = All of the above

Figure 23–13.

ANSWERS TO MULTIPLE-CHOICE QUESTIONS

23.1. E (II.B.2.a and b)

23.2. D (VIII.A)

23.3. A (VII)

23.4. A (III.B, IV, IV.E, VII.A, and X)

23.5. A (VII.A; note syncytiotrophoblast and incomplete cytotrophoblast)

23.6. C (VII.A)

23.7. D (II.B.2.a,b and 3)

23.8. D (II.B.3, C, and D)

23.9. E (II.E)

23.10. A (V)

23.11. E (X.F)

23.12. A (VI.B; Fig. 23–1)

23.13. C (III.C; Fig. 23–1)

23.14. E (IV.E; Fig. 23–1)

23.15. B (II.E; Fig. 23–3)

23.16. C (II.B.2.b; Fig. 23–3)

23.17. H (II.B.3 and D; Fig. 23–3)

23.18. G (II.D, V, VI.A and B.2; Fig. 23–3)

23.19. H (II.E and F; Fig. 23–2)

23.20. A (II.B.4; Fig. 23–2)

23.21. G (II.B.1–3; Fig. 23–2)

23.22. I (II.B.1; Fig. 23–2)

23.23. A (V)

23.24. B (II.B.4 and E.3)

23.25. D (III.B and C)

23.26. E (IV.A)

23.27. E (VII.A)

23.28. E (X.B–G)

23.29. D (II.F)

23.30. C (X.F.3; 20.IV.A.1)

23.31. D (II.B.2; 20.I.D.1 and III.A.2.b)

23.32. B (VIII and VIII.A)

23.33. D (IV.D; Table 23–1; note "sawtooth" glands)

23.34. D (IV.A)

23.35. D (X.E and F; 14.II.B.1 and 2)

23.36. C (II.E and F; Table 23–1)

23.37. D (II.D; 20.IV.A and A.1)

INTEGRATIVE MULTIPLE-CHOICE QUESTIONS: THE ENDOCRINE SYSTEM

Select the single best answer.

ES.1. Peptide hormones and steroid hormones share which of the following characteristics?
(A) Bind to receptors on the surface of their target cells
(B) Stimulate the production of intracellular second messengers (eg, cyclic AMP)
(C) Have dramatic effects on their target cells at very low concentrations
(D) Affect gene transcription by binding directly to DNA
(E) Are typically secreted through ducts

ES.2. Which of the following hormones are produced by pituitary acidophils?
(A) Luteinizing hormone and thyroid-stimulating hormone
(B) Luteinizing hormone and follicle-stimulating hormone
(C) Adrenocorticotropin and thyroid-stimulating hormone
(D) Adrenocorticotropin and β-melanocyte-stimulating hormone
(E) Growth hormone and prolactin
(F) Growth hormone and thyroid-stimulating hormone

ES.3. Which of the following is true of both endocrine and exocrine glands?
(A) Are incapable of merocrine secretion
(B) Have secretory cells that derive from lining epithelia
(C) Contain abundant SER in peptide- or protein-secreting cells
(D) Nerves do not affect their secretory activity
(E) Release their secretions directly into the blood

ES.4. Which of the following is true of ACTH?
(A) Is a steroid hormone
(B) Inhibits the production of glucocorticoids by the adrenal cortex
(C) Enhances the production of mineralocorticoids by the adrenal cortex
(D) Enhances the production of corticotropin-releasing hormone by the hypothalamus
(E) Is produced by acidophils in the adenohypophysis
(F) Causes Addison's disease when produced in excess
(G) Its target cells include those that line the renal collecting ducts

ES.5. Which pituitary hormone stimulates the growth of granulosa cells and oocytes in the ovarian follicles?
(A) Chorionic gonadotropin
(B) Dehydroepiandrosterone
(C) Estrogen
(D) FSH
(E) LH
(F) Progesterone
(G) Prolactin
(H) Testosterone

ES.6. Which of the following hormones is secreted by the syncytiotrophoblast?
(A) Chorionic gonadotropin
(B) Dehydroepiandrosterone
(C) Estrogen
(D) FSH
(E) LH
(F) Progesterone
(G) Prolactin
(H) Testosterone

ES.7. Which ovarian hormone inhibits pituitary LH secretion?
(A) Chorionic gonadotropin
(B) Dehydroepiandrosterone
(C) Estrogen
(D) FSH
(E) LH
(F) Progesterone
(G) Prolactin
(H) Testosterone

ES.8. Pituitary FSH stimulates the production of which ovarian hormone?
(A) Chorionic gonadotropin
(B) Dehydroepiandrosterone
(C) Estrogen
(D) FSH
(E) LH
(F) Progesterone
(G) Prolactin
(H) Testosterone

ES.9. Which of the following cells increase their secretory activity when LH receptors in their plasma membranes are occupied?
(A) Chromaffin cells
(B) Ganglion cells of the adrenal medulla
(C) Hypothalamic neurons responsible for GnRH production
(D) Interstitial (Leydig's) cells
(E) Pituitary luteotrophs
(F) Secretory cells in the zona glomerulosa
(G) Sertoli cells

ANSWERS TO INTEGRATIVE MULTIPLE-CHOICE QUESTIONS

ES.1. C (20.I.D)

ES.2. E (20.III.A.2.a and b)

ES.3. B (4.II, V.A and VI.C.1–5; Table 4–3; 20.I.D)

ES.4. C (19.II.C.2; 20.III.A.2.b; Table 20–1; 21.II. A.3.a–c and 4.a,b)

ES.5. D (Table 20–2; 23.II.F)

ES.6. A (23.II.F and VII.B.2)

ES.7. F (23.II.F)

ES.8. C (Table 20–2; 23.II.B.2 and F)

ES.9. D (Table 20–2; 22.II.F)

Sense Organs

24

OBJECTIVES

This chapter should help the student to:

- Name the classes of sensory receptors and list the receptors in each class.
- Describe the structure, function, and location(s) of each type of sensory receptor.
- Describe the process by which each receptor type transduces a stimulus into a neuronal potential.
- Locate and identify the receptors in a micrograph of an organ or tissue.
- Identify the components of receptors in a diagram or micrograph.
- Relate the microscopic structure of each receptor to its "adequate stimulus."

MAX-Yield™ STUDY QUESTIONS

1. In which receptor types are generator and receptor potentials formed (I.A.1.a–c[1])?
2. List all the receptors that fall into the following categories (Table 24–1) and note the overlaps:
   a. Mechanoreceptors
   b. Chemoreceptors
   c. Thermoreceptors
   d. Photoreceptors
   e. Neuronal receptors (I.A.1.a)
   f. Epithelial receptor cells (I.A.1.b)
   g. Neuroepithelial receptors (I.A.1.c)
   h. Proprioceptors (III.A and B)
   i. Touch and pressure receptors
   j. Warmth receptors
   k. Pain receptors (nociceptors)
   l. Cutaneous (skin) receptors (II. A–G)
3. List the types of sensations detected by free (unencapsulated) nerve endings (II.A).
4. Compare Merkel's, Meissner's, Ruffini's, and pacinian corpuscles (II.B, D, E, and F) in terms of:
   a. Nerve fiber number and arrangement
   b. Capsule presence and structure
   c. Adequate stimulus
   d. Location
5. To what do the terms "intrafusal" and "extrafusal" refer (III.A.1)?
6. Name the two types of intrafusal muscle fibers (III.A.1) and compare them in terms of diameter and length, distribution of nuclei, and type of sensory nerve ending.
7. Compare muscle spindles with Golgi tendon organs in terms of function and the fibers (other than nerve) inside the capsule (III.A.1 and B).
8. List the receptors found in joint capsules and describe their function (Table 24–1).
9. Which sense organ in the head (other than the eye) contributes important information for proprioception (sense of equilibrium and position in space; VI.D; Table 24–1)?
10. Name the three types of lingual papillae that have taste buds and compare them in terms of taste bud distribution (IV.A).
11. Name the cell types that occur in taste buds (Table 24–2) and compare their structure, function, and location.
12. In which part of the nasal cavity is the olfactory epithelium located (IV.B)?
13. Name three cell types in olfactory epithelium (IV.B) and compare their shape, nuclear position, apical surface specializations (where applicable), and function.
14. Compare taste buds (IV.A; Table 24–2) and olfactory epithelium (IV.B) in terms of:
    a. Basic function
    b. Receptor cell type (I.A.1.b and c)

---

[1] See footnote on page 1.

    c. Receptor cell apical surface specializations

    d. Signal transmission (I.A.1.b and c)

15. List, in order, the major steps in the embryonic development of the eye (V.A).

16. Name the eye's three compartments and describe their boundaries (V.C.3 and I; Fig. 24–2).

17. Name, from outermost to innermost, the globe's three basic tunics (or layers) (V.A–D); list the major components of each from anterior to posterior.

18. List, from anterior to posterior, the cornea's five layers and describe the composition of each (V.B.1). Which layer is thickest?

19. Describe the sclera (V.B.3) in terms of its predominant tissue type, its vascularity, and the proportion of the eye it covers.

20. Compare the sclera and cornea (V.B.1 and 3) in terms of transparency, blood supply, source of nourishment (V.B.2,3 and C.1), and amount of sensory innervation (V.B.1).

21. Beginning with its outer surface (attached to the sclera), name the choroid's layers and compare them in terms of the number and size of their blood vessels (V.C.1).

22. Name the ringlike, triangular anterior thickening of the eye's vascular layer (V.C.2).

23. What is the function of the ciliary muscles (V.C.2)?

24. How is the ciliary body attached to the lens (V.C.2)?

25. Describe the site of production, composition, and circulatory route of the aqueous humor (V.B.2 and C.2).

26. Name the two muscles of the iris (V.C.3.b) and compare them in terms of:

    a. Cell type

    b. Fiber orientation

    c. Innervation

    d. Effect of contraction on light entering the eye

27. Name, from outside to inside, the layers of the lens (V.J.1–3).

28. Select the condition of the following when the eye is focusing on a near object (V.C.2):

    a. Ciliary muscles (contracted or relaxed)

    b. Lens shape (rounded or flattened)

29. Name the two principal layers of the retina (V.D.1 and 2) and compare them in terms of:

    a. Embryonic origin (V.A)

    b. Location (inner or outer; V.D.1 and 2)

    c. Tissue type (V.D.1 and E)

    d. Photosensitivity (V.D.1 and 2)

    e. Phagocytic capacity (V.D.1)

    f. Melanin content (V.D.1)

    g. How far and in what form they extend anteriorly (V.D)

30. Name the three layers of retinal nerve cells that receive and relay visual signals to the brain (V.E.1–3) and answer the following questions:

    a. Which cells serve as true photoreceptors?

    b. In which order do the cells relay a signal?

    c. Which layer is crossed first by incoming light?

    d. Which cells contribute axons to the optic nerve?

    e. Which cells can generate an action potential?

31. Compare rods and cones (V.E.1; Table 24–3) in terms of:

    a. Shape

    b. Photoreceptor function

    c. Relative number and location

    d. Presence of outer and inner segments

    e. Independence of flattened vesicles from the plasma membrane

    f. Phagocytosis of flattened vesicles by pigmented epithelium

    g. Visual pigment type and distribution

    h. Visual acuity in bright and low light

    i. Sensitivity to shades of gray and color

    j. Association with monosynaptic bipolar cells

32. Name the neuroglia in the neural retina (V.E.4).

33. Beginning with the tear layer on the cornea, list the components (eg, layers, fluids, compartments) through which light must pass to reach the outer segments of the rods and cones (V.L).

34. Beginning with the bleaching of the visual pigment by light, list the steps leading to the generation of an action potential by the ganglion cells (V.L).

35. Name the muscles that open and close the eyelids (palpebrae; V.K.2).
36. Name the three types of glands found in the eyelids and compare them in terms of their secretions and the locations of their duct openings (V.K.2).
37. Describe the lacrimal gland (V.K.3) in terms of its location, classification, and glandular epithelium.
38. Trace the flow of tears over the eye and through the components of the lacrimal apparatus (V.K.3). Why do our noses "run" when we cry?
39. List the major structural components of the external ear (VI.A), the middle ear (VI.B), and the inner ear (VI.C) and name the major function of each component.
40. Name the regions of the bony labyrinth and the parts of the membranous labyrinth each contains (VI.C.2).
41. Compare the perilymph and the endolymph (VI.C.2 and E.2) in terms of location and site of production.
42. List the divisions of the vestibular part of the membranous labyrinth (VI.D.1 and 2) and name the sensory organ located in each.
43. Name the auditory part of the membranous labyrinth (VI.E) and the sensory organ it contains (VI.E).
44. Make a schematic drawing of the contents of the cochlea as they would appear if they were removed and uncoiled (Fig. 24–4) and show the location of the following:
    a. Oval window
    b. Scala vestibuli
    c. Helicotrema
    d. Scala tympani
    e. Round window
    f. Scala media (cochlear duct)
    g. Perilymph
    h. Endolymph

    The drawing should resemble a narrow tube (cochlear duct) within a wider tube, with the scala vestibuli extending from the oval window to the helicotrema, and the scala tympani extending from the helicotrema to the round window.
45. In situ, the cochlear duct (scala media) is triangular (Fig. 24–5). Name the structure(s) forming its roof, lateral wall, and floor (VI.E.2).
46. Name the structures in the maculae (VI.D.1) and in the crista ampullaris (VI.D.2) that correspond to the following structures in the organ of Corti (V.E.2.b):
    a. Inner and outer hair cells
    b. Stereocilia
    c. Pillar cells
    d. Tectorial membrane
47. Compare the organ of Corti's apex and base in terms of the sound frequency to which each responds best (VI.E.3).
48. Beginning with the act of lying down, describe (naming all vestibular apparatus structures that play a part) the sequence of events that send a signal to the brain that the head's position has changed (VI.D.1).
49. Beginning with a sound entering the external ear, list the sequence of events (naming all related cavities, moving bones, membranes, and cells) that cause the bipolar spiral ganglion cells to generate an action potential (V.E.3).

SYNOPSIS

I. GENERAL FEATURES OF THE SENSE ORGANS

Sense organs respond to stimuli by generating action potentials in a sensory (**afferent**) nerve cell process. Signals travel to the central nervous system (CNS) for integration, enabling reflex or conscious reactions to environmental changes. Sensory fibers typically carry signals for only one sensory modality (eg, pain, touch, or temperature).

A. **Classification:** Receptors are classified by their relationship to the nervous system, their stimulus sensitivity, and the presence or absence of a capsule.
  1. **Relationship between the receptor and the nervous system.**
     a. **Neuronal receptors.** The sensory nerve is stimulated directly. Each stimulus partly depolarizes the nerve ending. Many such **generator potentials** are summed until threshold is achieved, which results in the firing of an **action potential.** Examples include skin receptors and proprioceptors.

     b. Epithelial receptors are epithelial cells that generate a **receptor potential** (depolarize) when stimulated. These receptor cells release neurotransmitters to stimulate (indirectly) nearby nerve endings. Common in special sense organs, these receptors include the eye's rods and cones, the ear's hair cells, and the taste buds' sensory cells.

     c. Neuroepithelial receptors. These peripherally located neurons receive a stimulus and transmit the signal to the CNS along their own axons (eg, olfactory receptors).

  2. Adequate stimulus is the stimulus to which receptors are most sensitive. Table 24–1 classifies the sensory receptors according to adequate stimulus. Receptors in one class also may respond to other stimuli at higher intensities.

  3. Presence or absence of a capsule. Encapsulated receptors are surrounded by a specialized connective tissue capsule. Free nerve endings lack a capsule.

B. Distribution: Sense organ distribution maximizes adequate stimulus detection. The **receptive field** is the body region in which a stimulus evokes a response. More receptors in a given area yield greater sensitivity. Touch receptors, for example, are abundant in the fingertips.

C. Adaptation: This term refers to how rapidly receptors recover and respond to repeated stimuli.

II. RECEPTORS FOR SUPERFICIAL & DEEP SENSATION

A. Free Nerve Endings: These are the numerous and widely distributed peripheral dendritic branches of sensory neurons, whose somata lie mainly in craniospinal ganglia. These unencapsulated receptors are usually branches of unmyelinated or lightly myelinated fibers that occur in bundles beneath epithelia. As they penetrate an epithelium, they lose their myelin and branch among the cells. Branches of one nerve may cover a wide area and overlap the territories of others. Different free nerve endings (Fig. 24–1) have different adequate stimuli (eg, pain, cold, warmth, touch, or pressure) but are structurally indistinguishable. Of the skin's many receptors, these are the most numerous. They also occur in the walls of hollow organs, where some monitor dilation and contraction. Others sense pain in the body's interior.

B. Merkel's Corpuscles: These unencapsulated touch receptors are located deep in the epidermis; they sense direct pressure and, by means of desmosomal attachments, indirect pressure. More abundant in thick skin, they have two main components: the DNES-like **Merkel's cell** (18.II.D) and a specialized nerve ending that loses its Schwann's cell sheath as it penetrates the basal lamina. The nerve's free end forms a flat **Merkel's disk** that contacts the Merkel's cell in synapselike junctions. Merkel's cell granules lack catecholamines.

C. Other Nerve Endings with Expanded Tips: These also occur in deep epidermis; some are cold receptors, which respond to local cooling.

D. Meissner's Corpuscles: These elongated ovoid mechanoreceptors, which register touch and superficial pressure, have a thin capsule and contain many layers of flat Schwann's cells and fibroblasts (see Fig. 24–1). Nerve endings enter from the base and zigzag through the stack. Commonly found in the skin's dermal papillae (18.III.A), they are more numerous in the fingertips, palms, soles, and nipples.

E. Pacinian Corpuscles: These pressure receptors occur in deep dermis, hypodermis, periosteum, joint capsules, and mesenteries (see Fig. 24–1). Their capsules consist of layers of flat fibroblastlike cells separated by narrow fluid-filled spaces. Larger and rounder than Meissner's corpuscles, they resemble onion slices in tissue sections. A nerve enters the capsule, loses its myelin, penetrates the core while covered by a few layers of flat Schwann's cells, and gives off several blunt branches near the pole opposite its entry.

F. Ruffini's Corpuscles: These slow-adapting mechanoreceptors are common in dermis, hypodermis, and joint capsules (see Fig. 24–1). Their thin capsules surround a fluid-filled cavity containing a collagen mesh that penetrates the capsule to anchor it in the surrounding tissue. A single nerve ending loses its Schwann's cell sheath as it enters. Its many branches weave around the collagen fibers and respond to movement of the surrounding tissue.

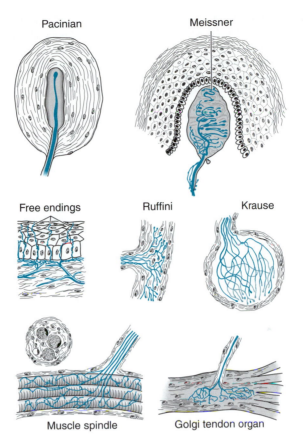

Figure 24–1. Several types of sensory nerve endings (not drawn to the same scale). (Based on a drawing in Ham AW: *Histology,* 6th ed. Philadelphia: JB Lippincott, 1969.)

Table 24–1. Classification of sensory receptors by adequate stimulus.

Receptor Class	Adequate Stimulus	Examples and Location
Mechanoreceptors	Touch	Free nerve endings, Merkel's and Meissner's corpuscles, encapsulated end-bulbs.
	Pressure	Free nerve endings; Meissner's, pacinian, and Ruffini's corpuscles; carotid sinus.
	Vibration	Touch and pressure receptors plus those of the inner ear.
Thermoreceptors	Temperature changes	Most are free nerve endings. Some with expanded tips have been identified as cold receptors. No specialized warmth receptors other than free nerve endings are known.
Nociceptors	Pain	Free nerve endings, usually near epithelial surfaces (eg, skin, gut, blood vessels). Abundant in cornea. None in brain except in blood vessels.
Chemoreceptors	Chemical changes	Taste buds, olfactory epithelium, and carotid and aortic bodies.
Proprioceptors	Body position changes	Muscle spindles, Golgi tendon organs, and vestibular apparatus of inner ear. Some mechanoreceptors in joint capsules (eg, free nerve endings, Ruffini's and pacinian corpuscles) respond to displacement and may contribute to proprioception.
Photoreceptors	Light	The retina of the eye (rods and cones).

G. End-bulbs: These fluid-filled bulbs with thin capsules (eg, Krause's end-bulbs) contain many nerve endings that enter at one pole and branch internally (see Fig. 24–1). They are relatively common and vary in size; most are mechanoreceptors. The largest are the genital corpuscles in the genital connective tissue; the smallest occur in the conjunctiva. Others occur in subepithelial connective tissue of the oral and nasal cavities, in the peritoneum, and in the connective tissue around joints and nerve trunks.

H. Carotid Sinus: This is a baroreceptor (a type of mechanoreceptor; see 11.II.F).

III. PROPRIOCEPTORS

A. Muscle Spindles: These are encapsulated fusiform (spindle-shaped, with wide equator and tapered poles) proprioceptors in striated muscles (see Fig. 24–1). They have both sensory and motor innervation. Layers of flattened fibroblasts make up the capsule. Muscles for delicate and precise movements (eg, eye muscles) require more spindles.

1. **Intrafusal fibers** comprise a bundle of 2 to 20 specialized muscle cells within the capsule. Oriented parallel to **extrafusal fibers** (typical striated muscle outside the capsule), the shorter, narrower intrafusal fibers cross the spindle capsule from pole to pole, attaching at the poles by their striated ends. Their nonstriated, dilated centers lack myofilaments, contain nuclei, and lie at the spindle's equator. The two types of intrafusal fibers include **nuclear chain fibers,** which are short and numerous and whose nuclei are arranged in rows, and **nuclear bag fibers,** which are less numerous and longer (may extend beyond the capsule) and whose nuclei occur in a cluster in their baglike dilated centers.

2. **Sensory innervation** is of two types. **Primary annulospiral endings** show dynamic sensitivity (ie, they are more sensitive to initial muscle stretching). A single, large myelinated sensory (afferent) fiber penetrates each spindle, losing its myelin sheath inside the capsule. The unmyelinated section branches to form spiral endings that embed in the sarcolemma around the intrafusal fibers' dilated centers. **Secondary flower spray** and **annulospiral endings** occur in most spindles. One or more small myelinated sensory nerves loses its myelin as it branches within the spindle. Branches may end in spirals, like primary endings, or in a flower spray pattern, where expanded tips of each branch contact the sarcolemma. Unlike primary endings, these terminate mainly on nuclear chain fibers and exhibit static sensitivity (ie, they are more sensitive to prolonged stretching).

3. **Motor innervation** reaches intrafusal fibers by means of small myelinated motor (efferent) fibers from cells in the ventral spinal cord gray matter. These **gamma motor neurons** contact the intrafusal fibers' striated polar regions in two ways. Most endings on nuclear bag fibers resemble motor end-plates (10.II.C), with small clusters of **boutons terminaux.** Those on nuclear chain fibers form multiple **boutons en passage.** Stimulation by these neurons contracts the intrafusal fibers' striated poles, stretching their nonstriated centers where the sensory endings are located. Thus, motor neurons can heighten a spindle's sensitivity to further stretching of the muscle in which it is embedded.

B. Golgi Tendon Organs: These occur mainly near muscle–tendon junctions (see Fig. 24–1). Each has a capsule composed of flat fibroblasts and is filled with collagen bundles that may extend beyond the capsule and insert in the tendon. Some extrafusal muscle fibers may attach to these bundles, allowing activation by muscle stretching or contraction. One large myelinated sensory fiber penetrates the capsule and gives off unmyelinated branches between the collagen bundles.

IV. CHEMORECEPTORS

A. Taste Buds: These ovoid chemoreceptors occur on the tongue's dorsal surface and on the soft palate and epiglottis. Lingual taste buds are embedded in epithelial projections called papillae. Taste buds occur on the apical surfaces of **fungiform papillae** and on the lateral surfaces of **foliate** and **circumvallate papillae.** Some papillae are more sensitive to certain tastes than others, but no exclusive structure–function relationships have been established. Chemicals in the oral cavity reach the taste buds through a small **taste pore** to stimulate the receptors. The

Table 24–2. Taste bud cell types.

Type	Structure	Function	Location
Type-I (supporting) cells	Dark-staining. Most numerous. Thin columnar cells. Apical microvilli. Long, dark nucleus. Abundant fine filaments, granules, ribosomes, and RER.	No sensory innervation, but may serve as developmental intermediates between basal and sensory cells. Granules release glycosaminoglycans into taste pore.	Between sensory cells
Type-II (sensory) cells	Light-staining. Second most numerous. Broad columnar cells. Apical microvilli. Pale, round, basal nucleus. Fewer filaments than type I. Abundant SER and mitochondria.	Chemical stimuli contact apical microvilli, are converted into electrical impulses, and are transmitted through electrical synapses to unmyelinated sensory nerve endings contacting cell's base. Perceived as taste.	Between supporting cells
Type-III (sensory) cells	Similar to type II. Several sensory nerve endings contact each cell's basal and lateral surfaces.	Transmit chemical stimuli by chemical synapse (neurotransmitter release) to sensory nerve endings.	Between supporting cells
Type-IV (basal) cells	Small, polygonal cells. Fewer, less-developed organelles.	Stem cells that may derive from type-V cells and give rise to sensory and supporting cells.	Base of taste bud
Type-V cells	Crescent-shaped cells with elongated nucleus, filament bundles, and few organelles.	May enclose some nerve endings and give rise to basal cells.	Around taste bud periphery (perigemmal)

five cell types that occur in taste buds (Table 24–2) may be different stages in the life of a single cell type undergoing continual turnover.

B. **Olfactory Epithelium:**   The main receptor for **olfaction** (smell), this pseudostratified columnar epithelium occurs mainly on the upper surface of the **superior concha (turbinate bone)** in the nasal cavity. The epithelium has three cell types, whose nuclei lie in three rows. **Olfactory cells** are bipolar neurons (broad middle, narrow apex, and base) derived from neural ectoderm. Their large round pale nuclei lie between the supporting and basal cell nuclei. From the cell apex, 6 to 20 long, nonmotile cilia extend into the nasal cavity, increasing the chemoreceptive surface. Chemicals interact with receptor sites on the cilia to elicit receptor potentials. The base of each cell tapers into an axon that carries the impulse to the brain. **Supporting cells** are columnar but are wider at their microvillus-covered apices. Their pale nuclei form the top row in the broad region between the surface and the olfactory cell nuclei. They contain RER, SER, lysosomes, lipid droplets, and a red–brown pigment that helps distinguish the olfactory region from the surrounding epithelium. **Basal cells** are small, conical cells at the base of the olfactory epithelium. Their nuclei form the deepest row, and their narrow apical processes extend between the other cells.

C. **Carotid and Aortic Bodies:**   See 11.II.E for a discussion of these chemoreceptors.

V. THE EYE

This complex organ (Fig. 24–2) refines and projects images onto its photosensitive retina, which subsequently generates the signals interpreted by the brain as vision. The basic structure of the **globe** (eyeball) comprises three principal layers: the tunica interna (retina; V.D), tunica vasculosa (V.C), and tunica fibrosa (V.B). This arrangement is better understood in light of its embryonic origin.

A. **Embryonic Development:**   The eye begins as the **optic bulb,** or **optic vesicle**—a hollow outgrowth of the embryonic brain. As this outgrowth contacts the overlying ectoderm, two crucial events occur: the bulb induces the ectoderm to form the **lens placode,** and the part of the bulb contacting the ectoderm invaginates to form the double-walled **optic cup.** The cup subsequently forms the **tunica interna (retina).** Its outer layer forms the retina's **pigmented epithelium,** and its inner layer forms the photosensitive **neural retina.** The stalk connecting the optic cup to the

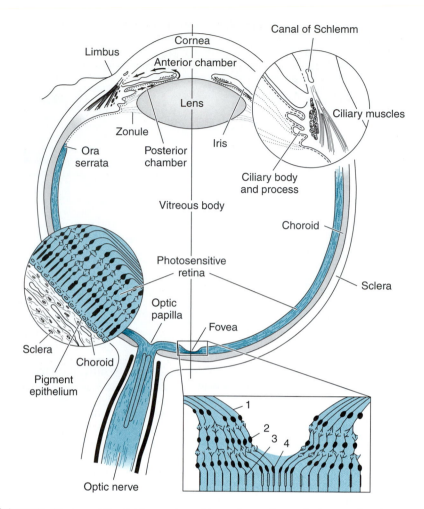

Figure 24–2. Schematic diagram of the right eye, viewed from above. Arrows in the anterior chamber show the flow of aqueous humor. *Lower right*: Enlarged diagram of the fovea shows 1, axons of the ganglion cells; 2, bipolar cells; 3, rods; and 4, cones. (Modified and reproduced, with permission, from Ham AW: *Histology,* 6th ed. Philadelphia: JB Lippincott, 1969.)

brain becomes the **optic nerve.** The lens placode thickens and invaginates to form the **lens vesicle,** which pinches off to form the **lens** and positions itself in the mouth of the optic cup. Mesenchyme condenses around the tunica interna to form the globe's two outer tunics: the outermost **tunica fibrosa** and the **tunica vasculosa** (which lies between the fibrosa and the interna).

B. Tunica Fibrosa: The eye's outermost tunic has two main components. The anterior sixth forms the transparent **cornea;** the posterior five sixths form the opaque (white) **sclera.** The junction between the cornea and sclera is the **limbus.**

1. **Cornea.** This transparent avascular disk bulging from the front of the eye has five layers. The **anterior epithelium** is outermost. This thin, nonkeratinized stratified squamous epithelium contains many free nerve endings. **Bowman's membrane** (anterior limiting lamina) is a cell-free, thick basement membrane composed of ground substance and reticular fibers. The **stroma** (substantia propria) forms the cornea's core and 90% of its thickness. It has many layers of collagen bundles oriented parallel to those in the same layer and perpendicular to those in adjacent layers. Fibroblasts lie between the layers. **Descemet's membrane** (posterior limiting membrane) is a thick basement membrane that differs from Bowman's

membrane in position and composition. It has elastin, but no elastic fibers. Its network of atypical collagen fibers is decorated with granules. The **corneal endothelium** (posterior epithelium) is a simple cuboidal epithelium lining the cornea's internal surface.

2. **Limbus.** The vessels of this highly vascular, ringlike junction between the cornea and sclera, together with the fluid in the anterior chamber, nourish the avascular cornea. Near the limbus, the stroma contains an endothelial channel called **Schlemm's canal,** which drains fluid from the anterior chamber toward veins in the limbus. Blockage of this canal raises intraocular pressure and causes **glaucoma.**

3. **Sclera.** This opaque white connective tissue covers the eye's posterior five sixths. It is anchored in the orbit by the dense connective tissue of **Tenon's capsule.** The sclera has three layers. The **episclera** is the sclera's outermost layer of fibroelastic tissue. The **substantia propria** is a dense mat of collagen bundles and fibroblasts that form the sclera's thick middle layer, where the ocular muscles insert. The **lamina fusca,** which consists of loose connective tissue, is the sclera's inner layer. It contains elastic fibers and melanocytes and is separated from the choroid by the narrow perichoroidal space.

C. **Tunica Vasculosa (Uvea):** This middle tunic of the eye has three major components: the choroid (posterior), ciliary body, and iris (anterior).

1. **Choroid.** The choroid lies between the sclera and the retina's pigmented epithelium and has four layers. The **suprachoroidal lamina** resembles the sclera's lamina fusca, from which it is separated by the perichoroidal space. The **vascular lamina** comprises loose connective tissue with many whorl-like veins that converge to form four larger vortex veins, which exit the back of the eye through the sclera. The **choriocapillary layer** (choriocapillaris) is a layer of fenestrated sinusoids embedded in loose connective tissue. **Bruch's membrane,** the choroid's inner layer, is the basement membrane of the retina's pigmented epithelium.

2. **Ciliary body.** The ciliary body extends forward from the choroid as a ringlike triangular thickening at the level of the lens. It has the same layers as the choroid, minus the choriocapillaris. Its primary structural components are the ciliary processes and the ciliary muscles. The **ciliary processes** are irregular epithelium-covered connective tissue outgrowths of the ciliary body that extend toward the lens. They serve as origins for the fibers of the circular **ligament of Zinn (zonule),** which insert in the edge of the lens to anchor it. The two layers of pigmented epithelium covering these processes derive from the layers of the optic cup. The inner **ciliary epithelium** borders the internal cavity of the eye. Its cells have the basolateral plasma membrane infoldings typical of ion- and water-transporting cells. They secrete **aqueous humor,** which flows through the pupil to the anterior chamber. From here, the fluid penetrates the tissue near the limbus to reach Schlemm's canal (V.B.2). The deeper, simple columnar epithelial layer derives from the optic cup's outer layer. The **ciliary muscles** comprise three smooth muscle bundles near the junction of the ciliary body and sclera. The contraction of all of these muscles pulls the ciliary body and choroid forward, releasing tension on the zonule and allowing the lens to become rounded for near vision. The relaxation of all groups increases tension on the zonule, flattening the lens to allow focus on distant objects. The adjustment of individual muscles enables the eye to focus on intermediate distances.

3. **Iris.** This structure controls the amount of light that reaches the retina and gives the eye its color. In front of the lens, it projects as a flat ring from the ciliary body, leaving a circular opening—the **pupil**—at its center. The iris includes the most anterior extensions of the tunica vasculosa and tunica interna, forming the border between the anterior and posterior chambers. The **anterior chamber** lies between the cornea and the iris; the **posterior chamber** lies between the iris and the lens–zonule complex.

a. **Layers.** The **anterior surface** of the iris is rough and contains pigment cells and fibroblasts. Its **stroma** is a poorly vascularized connective tissue with fibroblasts and melanocytes. The **vascular stratum,** between the stroma and posterior surface, has many blood vessels. The **posterior surface** is smooth, heavily pigmented, and continuous with the double-layered epithelium covering the ciliary processes.

b. **Involuntary muscles.** The **sphincter pupillae** is a ring of smooth muscle in the pupillary margin that contracts under parasympathetic control to partly close the pupil. The **dilator pupillae** fibers, which extend like spokes between the ciliary body and the pupillary margin, contract under sympathetic control to open the pupil.

D. Tunica Interna (Retina): This derivative of the optic cup (V.A) is considered an extension of the CNS. Its nonphotosensitive anterior section forms part of the ciliary body and iris. Its posterior section is a highly specialized photoreceptor. The junction between its anterior and posterior parts, the **ora serrata,** lies behind the ciliary body. The posterior part of the retina is further divisible into two layers, based on structure, function, and embryonic origin.

    1. Pigmented epithelium. This melanin-rich simple cuboidal epithelium derives from the optic cup's outer wall. It rests on a thick elastic basement membrane (Bruch's membrane; V.C.1) that separates it from the choroid. The cells' many apical microvilli embrace the outer segments of the rods and cones in the overlying neural retina. Although it is not a photoreceptor, this layer is crucial to vision. It absorbs light that has passed through the photosensitive layer, ensuring that light stimulates the rods and cones only on its first pass. Its cells have basal plasma membrane infoldings and mitochondria typical of ion- and water-transporting cells. They also phagocytose and degrade the vesicles shed by the rods' outer segments.

    2. Neural retina. This highly organized, sensitive photoreceptor derives from the optic cup's inner wall. Its structure and function are considered in detail in the sections that follow (V.E–H). Although the pigmented epithelium extends to the ciliary body and iris, the neural retina extends only to the ora serrata.

E. Cells of the Neural Retina: There are ten retinal layers (Fig. 24–3), but only three layers of retinal neurons receive, integrate, and transmit visual signals to the brain as nerve impulses. These are the **photoreceptor cells** (**rods** and **cones**), **bipolar cells,** and **ganglion cells.**

    1. Rods and **cones** are best understood in terms of their similarities and differences. Their basic similarities are discussed here, and their differences are detailed in Table 24–3. Both rod and cone cells are photoreceptors located deep in the retina, next to the pigmented epithelium. Light must penetrate the more superficial layers of the retina to reach and stimulate

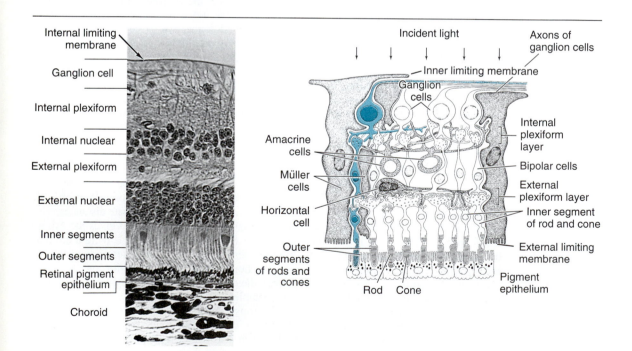

Figure 24–3. The layers of the retina. *Left:* Section from a monkey retina; light enters from the top (x 655). *Right:* Schematic drawing of the three layers of retinal neurons. Arrows show the direction of the light path. Stimulation of the rods and cones by incident light generates an impulse that is transmitted to the bipolar cells and subsequently the ganglion cells (ie, in a direction opposite the path of light). (Reconfigured and reproduced, with permission, from Junqueira LC, Carneiro J, Kelley RO: *Basic Histology,* 9th ed. Appleton & Lange, 1998; and Dowling JE, Boycott BB: *Proc R Soc Lond (Biol)* 166:80, 1966.)

Table 24–3. Important differences between rods and cones.

	Rods	Cones
Cell shape	Long, narrow cells with long cylindrical outer segments.	Plumper cells with shorter conical outer segments.
Vesicles	Flat vesicles are entirely separate from outer segment plasma membrane. Visual pigment is synthesized on polyribosomes and incorporated in small vesicles that move from inner to outer segment and fuse to form flat vesicles. These move toward outer segment apex to be shed and phagocytosed by pigmented epithelium.	Flat vesicles are invaginations of outer segment plasma membrane and turn over more slowly than in rods.
Cell number	Approximately 120 million per retina.	Approximately 6 million per retina.
Visual pigment	The globular protein **rhodopsin** is synthesized in the rod's inner segment and collects on the outer surface of the stacked vesicles. It is bleached by light, a photochemical reaction that initiates the visual process.	**Iodopsin** is evenly distributed in cones' outer segments and is most sensitive to red light.
Sensitivity	More active in low light (night vision). Respond to slight differences in light intensity and shades of gray.	Require more intense light (day vision), permit greater visual acuity, responsible for color vision.
Location	Concentrated peripherally.	Concentrated in fovea centralis (macula lutea).

them. Each cell has two structurally different segments. The **outer segment,** or specialized dendritic (receptive) part of each cell, contains a stack of flat membrane-limited vesicles and the visual pigment. This segment's distal tip is embraced by pigment epithelial cell microvilli. The **inner segment** is rich in polyribosomes and glycogen and is separated from the outer segment by a constriction. Basal bodies near the constriction anchor one or two intracytoplasmic cilia that extend through the constriction and into the outer segment. Mitochondria near the constriction provide energy for the visual process.

2. **Bipolar cells** lie in the middle of the neural retina and comprise two populations of interneurons that relay visual signals from the photoreceptors to the ganglion cells. Each **diffuse bipolar cell** synapses with ganglion cells and two or more photoreceptors (mostly rods). Each **monosynaptic bipolar cell** synapses with a single cone and a single ganglion cell, perhaps accounting for the cones' greater visual acuity.

3. **Ganglion cells** lie near the globe's inner surface and have large cell bodies and nuclei. Their dendrites make synaptic contact with the bipolar cells. Their axons form the layer of nerve fibers covering the retina's inner surface. They converge to exit the eye through the optic nerve, carrying visual signals to regions of the brain responsible for vision.

4. **Other cell types** in the neural retina include two minor populations of neurons and important glial cells. **Horizontal cells** and **amacrine cells** are neurons whose functions may include integrating visual signals before they reach the brain. Horizontal cell processes terminate near the synapses between photoreceptor and bipolar cells. Amacrine cell processes terminate near synapses between bipolar and ganglion cells. Glial cells include **astrocytes, microglia,** and the large, highly branched **Müller cells** that span the neural retina's entire width and embrace the processes of the retinal neurons.

F. **Fovea Centralis:** Directly opposite the center of the lens, the retina's fovea lies at the center of a small yellowish disk called the **macula lutea** (see Fig 24–2). Because it has the greatest concentration of cones, it is the retinal region with the greatest visual acuity. The lateral displacement of all retinal layers except the photoreceptors makes it the retina's thinnest region. It lacks blood vessels and is nourished by the underlying choriocapillaris.

G. **Optic Disk and Retinal Blood Supply:** Also known as the **papilla** (see Fig. 24–2), **nerve head,** and **blind spot,** the optic disk is the site where the ganglion cells' axons converge at the back of the eye and exit to form the optic nerve. Because it lacks photoreceptors, it is insensitive to light. Retinal vessels enter and exit the eye through the optic disk's center and branch over the retina's internal surface. Capillaries penetrate the retina's inner layers except near the fovea. The retina's outer layers are supplied by diffusion from vessels in the choriocapillaris.

H. Optic Nerve: This nerve consists of ganglion cell axons that converge to exit the eye at the optic disk. Within the nerve, the axons acquire myelin from oligodendrocytes. As it leaves the eye, the nerve acquires a dura mater sheath that is continuous with the sclera, as well as arachnoid and pia mater coverings. It contains the retinal artery and vein at its core.

I. Vitreous Body: This transparent, gel-like body (mostly water and hyaluronan) fills the large **vitreous space** between the lens and the retina (see Fig. 24–2). Some peripheral fibrils form its capsule. It contains a few macrophages and hyalocytes—stellate cells with oval nuclei that produce the fibrils and hyaluronan. During development, the central artery extends from the optic disk through the vitreous to the lens as the **hyaloid artery.** It subsequently degenerates, leaving the narrow **hyaloid canal.**

J. Lens: This is a transparent, elastic, biconvex structure of epithelial origin (see Fig. 24–2). Nourished by aqueous humor, it has neither blood nor nerve supply. It is suspended by the zonule of the ciliary body behind the pupil. Ciliary muscle contraction changes the curvature of the lens to enable focus on objects near or far, a process called **accommodation.** The lens has three main components:

1. The **lens capsule** is an elastic and transparent basal lamina that covers the entire lens and prevents wandering cells from penetrating it. It consists mainly of fine type-III and type-IV collagen fibrils embedded in a glycoprotein- and glycosaminoglycan-rich matrix.

2. **Subcapsular epithelium.** The height of this low cuboidal epithelium beneath the capsule on the anterior lens surface increases to columnar near the lens equator, where cell division occurs. Its cells contain few organelles and form the lens fibers.

3. **Lens fibers** are long, narrow, hexagonal, specialized epithelial cells that make up most of the lens. During differentiation, they lose their nuclei, fill with proteins called **crystallins,** and develop a variety of specialized plasma membrane components, including junctional complexes and ridgelike processes. Crystallins belong to the family of heat-shock proteins.

K. Accessory Structures of the Eye:

1. The **conjunctiva** comprises two parts. The **bulbar conjunctiva** is a thin, nonkeratinized stratified squamous epithelium covering the eye's anterior surface to the cornea. The **palpebral conjunctiva** is a stratified columnar epithelium covering the inner eyelid surface. The conjunctiva overlie a loose, vascular lamina propria containing many lymphocytes, plasma cells, and macrophages.

2. **Eyelids (palpebrae)** are five-layered skin folds that protect the eyes. The thin skin on the surface lacks hairs except at the free margin. It is underlain by loose connective tissue containing the **orbicularis oculi muscles** that close the eyes. The dense connective tissue core, or **tarsal plate,** provides flexible support and harbors the **tarsal (meibomian) glands.** These release oily secretions from openings in the free margins that prevent the opposing lids from sticking and also slow tear evaporation. A thin lamina propria and overlying palpebral conjunctiva cover the internal surface. The loosely coiled sweat **glands of Moll** and the sebaceous **glands of Zeis** accompanying the eyelashes also open on the free margins. Near the top of the upper lid, the **levator palpebrae superioris muscle** inserts into the skin and top of the tarsal plate to open the lid.

3. **Lacrimal apparatus.** This system of glands and ducts provides tears to lubricate and protect the eyes. The **lacrimal glands** are tear-secreting compound tubuloalveolar glands located superolaterally in the bony orbits. The secretory units are surrounded by myoepithelial cells and divided into lobes by connective tissue. The columnar secretory cells have pale granules and secrete antibacterial **lysozyme.** Tears are released behind the upper eyelid and flow over the eye's anterior surface. The excess drains through the **lacrimal puncta,** one small hole in the free margin of each lid near the medial palpebral angle. Tears enter by capillary attraction and, with the aid of a pumping action provided by the orbicularis oculi muscles, follow the **lacrimal canaliculi** into a short duct that empties into the **lacrimal sac.** This dilation of the lacrimal drainage system delivers tears to the **nasolacrimal duct** with the aid of gravity. This duct empties through a bony canal into the nasal cavity through the inferior meatus.

L. Brief Summary of Light Path and Vision: Light penetrates the tear layer and subsequently the transparent cornea (V.B.1). Crossing the anterior chamber (V.C.3), a limited amount of light passes through the pupil (V.C.3) and across the lens (V.J), which focuses the image and projects it through the vitreous body (V.I) onto the retina (V.D). The central part of the image focuses on the fovea of the macula lutea (V.F). Here, light penetrates and excites the photoreceptors (V.E.1) before it is absorbed by the pigment epithelium (V.D.1). The bleaching of the visual pigments in the excited rods and cones (see Table 24–3) generates a receptor potential (I.A.1.b) that is transmitted to the bipolar cells (V.E.2), which integrate and relay the signal to the ganglion cells (V.E.3), which subsequently transmit the signal to the brain by means of the optic nerve (V.H).

VI. THE EAR

This collection of structures for hearing and balance (Fig. 24–4) has three major components: the **external, middle,** and **internal ears.**

A. **External Ear:** The external ear's three main components are the auricle, external auditory meatus, and tympanic membrane. The **auricle (pinna)** is a funnel-like plate of elastic cartilage sandwiched between two layers of skin. Modified apocrine sweat glands in the skin, the **ceruminous glands,** secrete a waxy **cerumen.** The auricle collects and focuses sound waves toward the tubelike **external auditory meatus.** This canal is surrounded by elastic cartilage along its outer third and by bone along its inner two thirds. Sounds gathered by the auricle are carried inward by the meatus to vibrate the **tympanic membrane (eardrum)** covering its internal orifice. The membrane's three layers are the outer epidermis, middle dense connective tissue, and inner cuboidal epithelium.

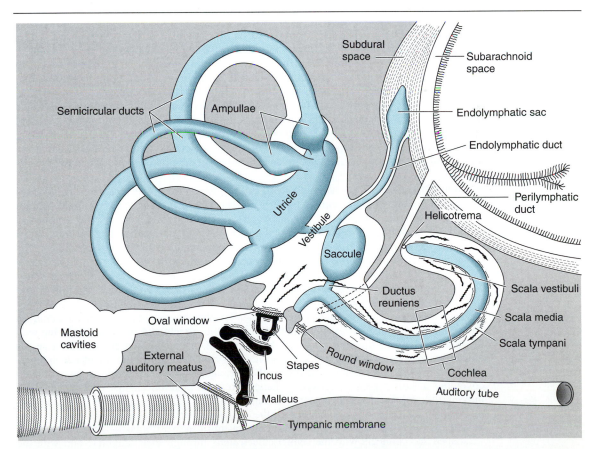

Figure 24–4. Schematic diagram of the vestibulocochlear apparatus. Sound waves shown in the external auditory meatus cause vibration of the tympanic membrane, ossicles, and perilymph. (Redrawn and reproduced, with permission, from Best CH, Taylor NB: *The Physiological Basis of Medical Practice,* 8th ed. Baltimore, MD: Williams & Wilkins, 1966.)

B. Middle Ear: This component lies in a cavity in the temporal bone. Its connection with the auditory meatus is closed by the tympanic membrane. It communicates with the nasopharynx through the **auditory (eustachian) tube** and also communicates with the **mastoid air cells.** The auditory tube is surrounded by elastic cartilage. Its walls are collapsed except during swallowing, when they separate to allow pressure in the middle ear cavity to equilibrate with the environment.

   1. **Windows.** The medial bony wall of the middle ear cavity has two membrane-covered openings, the **oval** and **round windows,** which lie between the middle and internal ears (see Fig. 24–4).
   2. **Auditory ossicles.** The main functional components of the middle ear are three uniquely shaped small bones that span the middle ear cavity from the tympanic membrane to the oval window membrane. The **malleus** (hammer), **incus** (anvil), and **stapes** (stirrup) transmit vibrations from the tympanic membrane to the fluid in the inner ear. Small muscles restrict ossicle movement to limit damage from loud noises.
   3. **Mucosa.** The simple squamous-to-cuboidal epithelium that lines the middle ear contains mucous or seromucous secretory cells. A thin lamina propria binds the lining to the periosteum of the walls and ossicles. Near the auditory canal, the lining changes to the pseudostratified columnar epithelium of the canal and nasopharynx.

C. Internal Ear (Labyrinth, Vestibulocochlear Apparatus): This component consists of two mechanoreceptors: the **cochlea** (for hearing; VI.E) and the **vestibule** (for equilibrium; VI.D).

   1. **Embryonic development** begins with a thickened ectodermal disk (**otic placode**) on each side of the head. Each placode invaginates to form an **otic vesicle,** which undergoes further outpocketing to form a series of interconnected chambers and canals called the **membranous labyrinth.** After the chambers form, their lining differentiates to form special cells and sensory organs (VI.D and E). Mesenchyme condenses around the membranous labyrinth to form the bony labyrinth.
   2. **General organization.** The internal ear consists of a complex of bony cavities and canals called the bony labyrinth, which houses the delicate membranous labyrinth and its organs of hearing and balance. The **bony labyrinth** consists of two interconnected compartments: the **vestibule** and the **cochlea.** The space between the membranous and bony labyrinths contains a fluid called **perilymph.** The **membranous labyrinth** lies within and conforms to the shape of the bony labyrinth; its interconnected chambers and canals contain a fluid called **endolymph.** Small sections of its simple squamous endothelial lining develop into a sensory epithelium in which the cells rise gradually to form a columnar shape. Despite their origin as outpocketings of the same otic vesicle, the cochlear and vestibular receptors are structurally and functionally distinct and are considered separately in the sections that follow.

D. Vestibular Organs: Bony labyrinth components associated with balance include the **vestibule** and the **semicircular canals** (see Fig. 24–4). The membranous part of each includes a special sensory organ composed of two main cell types: **hair cells** (receptors) and **supporting cells.** Each hair cell has several long stereocilia and one true cilium extending from its apical surface. The goblet-shaped type-I hair cells are surrounded by afferent nerve endings. The columnar type-II hair cells contact both afferent and efferent endings on their basal and lateral surfaces. The columnar supporting cells have basal nuclei, lie between the hair cells, and produce the glycoprotein-rich **gelatinous layer** that covers the sensory epithelium and bulges into the membranous labyrinth's lumen.

   1. The **vestibule** is an oblong cavity in the inner ear that houses two saclike, membranous labyrinth components concerned with equilibrium: the **utricle** and the **saccule.** In the wall of each is a sensory **macula**—an ovoid button of sensory epithelium covered by a gelatinous layer into which the hair cells' stereocilia and cilia extend. In both the utricle and saccule, tiny rocklike crystals of calcium carbonate and protein—the **otoliths** (statoliths)—cover the macula's gelatinous layer. Changes in head position change endolymph flow, moving the otoliths. The movements are transmitted through the gelatinous layer, displacing the hair cell processes and stimulating the associated nerve endings.
      a. The **utricle** is the largest membranous component of the vestibular system. This kidney-shaped sac connects with the semicircular canals through their ampullae and with the saccule by means of the narrow **utriculosaccular duct.**
      b. The **saccule** is spherical and smaller. It communicates with the cochlear duct through the short narrow **ductus reuniens** and with the utricle through the utriculosaccular duct.

    c. The **endolymphatic duct** is a tubular evagination of the utriculosaccular duct; it terminates as a blind expansion called the **endolymphatic sac.** The sac has a tall columnar epithelial lining and is surrounded by vascular connective tissue. Duct and sac functions may include producing endolymph and clearing debris from it.

  2. **Semicircular canals.** These three thin bony canals in the temporal bone are oriented at right angles to each other in three planes and communicate with the vestibule by small openings. They contain the three **semicircular ducts** of the membranous labyrinth, which communicate with the utricle. The superior, lateral, and posterior ducts leave the utricle as dilated **ampullae** and rejoin the utricle at their terminations. Each ampulla contains a sensory **crista ampullaris** whose hair cells and supporting cells resemble those in the maculae but are arranged in ridges (cristae) rather than buttons (maculae). The conical gelatinous layer of each crista is termed a **cupula** and lacks otoliths.

E.   **Cochlea:**   This snail shell–like spiral canal houses the **cochlear duct,** the part of the membranous labyrinth concerned with hearing. The cochlea's screwlike bony core, the **modiolus,** houses the auditory nerve cell bodies of the **spiral ganglion.** The thread of the screwlike modiolus is a spiral bony shelf, the **osseous spiral lamina,** which supports the auditory epithelium (**spiral organ of Corti**).

  1. **The cochlear duct uncoiled.** The auditory part of the membranous labyrinth is more easily understood if one pictures it removed from the bony cochlea and uncoiled (see Fig. 24–4). Removed and uncoiled, it appears as a tube within a tube. The outer tube has a simple squamous epithelial lining bound tightly to the cochlea's bony walls; the tube begins at the oval window, ends at the round window, and contains the perilymph. The inner tube is the cochlear duct. Its cavity, the **scala media,** is filled with **endolymph.**

  2. **The cochlear duct in situ.** Within the coiled cochlea, a section through a single turn (Fig. 24–5) reveals that the cochlear duct has a triangular shape, with a **roof** (the **vestibular,** or **Reissner's, membrane** separates the scala media from the scala vestibuli), a **lateral wall** (mainly the **stria vascularis,** an unusual epithelium covering many capillaries that together produce the endolymph that fills the duct), and a **floor,** which includes the spiral organ of Corti and the spiral lamina on which it rests.

    a. The **spiral lamina** has both bony (osseous) and membranous parts. The osseous spiral lamina is the thread of the screwlike modiolus. The membranous spiral lamina extends across the cochlear canal from the edge of the thread to the spiral crest on the lateral wall. At its core lies the thin, fibrous **basilar membrane.** The organ of Corti lies on the membranous portion of the spiral lamina.

    b. The **spiral organ of Corti** is a mechanoreceptor that is highly sensitive to vibration. It is anchored to the osseous spiral lamina by the **spiral limbus,** which consists of epithelium-covered connective tissue. The glycoprotein-rich **tectorial membrane** extends from the limbus to cover the sensory cells' apices as the gelatinous layer covers the vestibular maculae. The spiral organ of Corti has two main cell types: supporting and sensory cells.

      (1) The **supporting cells** occur in two groups and are termed **inner** and **outer pillar cells.** Their broad bases contain their nuclei, and their elongated apices contain tonofilament bundles. They underlie and support the sensory cells and form the walls of a channel between the two groups of sensory cells, which is called the inner tunnel.

      (2) The **sensory cells,** as in the vestibular system, are called hair cells. **Inner hair cells** form a single row between the inner tunnel and the internal spiral tunnel formed by the tectorial membrane as it bridges the space between the limbus and the organ of Corti. These goblet-shaped cells have apical stereocilia and many basal mitochondria. **Outer hair cells** form three parallel rows lateral to the inner tunnel and rest on columnar supporting cells. Outer hair cells have basal nuclei and mitochondria and are more columnar than inner hair cells. The 100 or so stereocilia on each cell are arranged in a V or W pattern and penetrate the cuticular plate, which is an expansion of the outer pillar cell. The tips of the stereocilia are covered by the tectorial membrane. Cochlear hair cells lack true cilia. They are innervated by the dendritic processes that pass from the bipolar neurons of the spiral ganglia in the modiolus through the spiral limbus and basilar membrane to reach them.

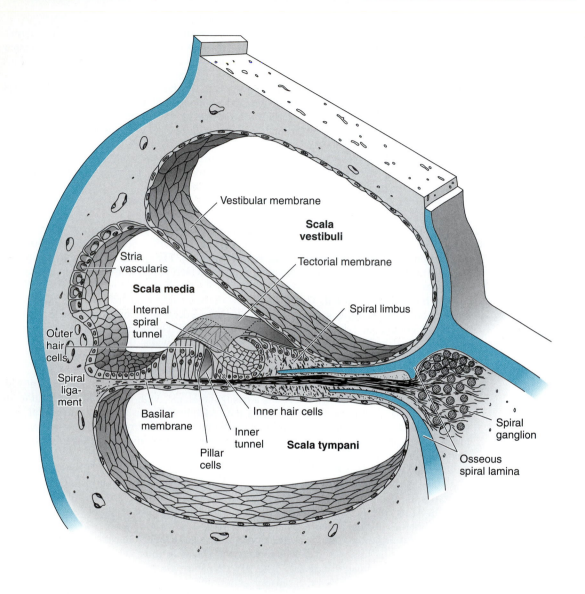

Figure 24–5. Schematic diagram of a section through a single turn of the cochlea. (Redrawn and reproduced, with permission, from Bloom W, Fawcett DW: *A Textbook of Histology,* 9th ed. Philadelphia: WB Saunders, 1968.)

3. **Hearing.** Sounds collected by the auricle traverse the auditory meatus and vibrate the tympanic membrane. This activity moves the ossicles, and movement of the stapes in the oval window transmits vibrations to the perilymph in the scala vestibuli. The perilymph carries these vibrations through the helicotrema and into the scala tympani. Vibrations in the perilymph cause the delicate membranous spiral lamina and associated organ of Corti to move in relation to the tectorial membrane, displacing the hair cell stereocilia. Movement of the stereocilia generates an action potential in the bipolar spiral ganglion cell processes. Neural signals generated in the cochlea are carried to the brain through the cochlear nerve (the collected axons of the bipolar cells). Cochlear sensitivity is tonotopic; sounds of different frequencies are perceived in specific locations. The organ of Corti in the basal cochlea responds best to high frequencies; its counterpart in the apex responds best to low frequencies.

MULTIPLE-CHOICE QUESTIONS

Select the single best answer.

24.1. Which of the following receptors are classified as unencapsulated?
(A) Free nerve endings
(B) Golgi tendon organs
(C) Krause's end-bulbs
(D) Meissner's corpuscles
(E) Muscle spindles
(F) Pacinian corpuscles
(G) Ruffini's corpuscles

24.2. Which of the following structures are located outside the muscle spindle capsule?
(A) Extrafusal fibers
(B) Flower spray endings
(C) Intrafusal fibers
(D) Nuclear bag fibers
(E) Nuclear chain fibers

24.3. Which of the following is the thickest layer of the cornea?
(A) Anterior epithelium
(B) Bowman's membrane
(C) Corneal endothelium
(D) Descemet's membrane
(E) Stroma

24.4. Which of the following is the junction between the cornea and the sclera?
(A) Ciliary body
(B) Episclera
(C) Iris
(D) Lamina fusca
(E) Limbus
(F) Macula lutea
(G) Ora serrata

24.5. Which of the following produces the aqueous humor of the eye?
(A) Ciliary epithelium
(B) Choriocapillary layer
(C) Corneal endothelium
(D) Endolymphatic sac
(E) Schlemm's canal
(F) Stria vascularis
(G) Substantia propria

24.6. Which of the following forms the border between the anterior and posterior chambers of the eye?
(A) Ciliary body
(B) Cornea
(C) Iris
(D) Lens
(E) Ora serrata
(F) Vitreous body
(G) Zonule

24.7. Which of the following is true of the fovea centralis of the retina?
(A) Is also known as the blind spot
(B) Is the region in which the axons of the ganglion cells converge and leave the eye as the optic nerve
(C) Is the region in which retinal veins converge before leaving the eye
(D) Is the thinnest part of the neural retina
(E) Does not contain photoreceptive cells

24.8. Which of the following is true of Golgi tendon organs?
(A) Are unencapsulated
(B) Are thermoreceptors
(C) Possess both sensory and motor innervation
(D) Occur at muscle–tendon junctions
(E) Contain muscle fiber bundles

24.9. Which of the following is true of type-III cells of taste buds?
(A) Possess apical cilia
(B) Phagocytose neurotransmitters in response to chemical stimuli
(C) Are innervated by several sensory fibers
(D) Stain darkly with standard dyes
(E) Give rise to basal cells

24.10. Which of the following is true of olfactory (sensory) cells?
(A) Each has a tuft of specialized stereocilia on its apical surface
(B) Are specialized neurons
(C) House the upper (superficial) row of nuclei in the olfactory epithelium
(D) Are distributed in the respiratory epithelium throughout the nasal cavity
(E) Possess small, dark, conical nuclei

24.11. Which of the following is true of the cornea?
(A) Constitutes the anterior half of the tunica fibrosa
(B) Is composed of 10 layers
(C) Is avascular
(D) Accounts for eye color
(E) Is attached to the ciliary body by the zonule

24.12. Which of the following is true of ciliary muscle contraction?
(A) Is under voluntary control
(B) Pulls the ciliary body forward
(C) Is needed to focus on distant objects
(D) Prevents visual accommodation
(E) Increases tension exerted on the zonule
(F) Causes the lens to flatten

24.13. Which of the following is true of the pigmented epithelium of the retina?
 (A) Rests on Descemet's membrane
 (B) Contributes anteriorly to the lens
 (C) Derives from the outer wall of the embryonic optic cup
 (D) Phagocytoses flattened vesicles shed by cones
 (E) Is classified as a photoreceptor

24.14. Which of the following is true of the lens fibers?
 (A) Are highly differentiated fibroblasts
 (B) Contain long heterochromatic nuclei
 (C) Contain abundant proteins called crystallins
 (D) Consist of type-III collagen

24.15. Which of the following is true of tears?
 (A) Are secreted by meibomian glands
 (B) Contain lysozyme
 (C) Are secreted by the glands of Moll
 (D) Are secreted by the glands of Zeiss
 (E) Enter the eye through the lacrimal puncta
 (F) Are secreted by the conjunctiva

24.16. Which of the following receptor classifications includes the olfactory epithelium?
 (A) Chemoreceptor
 (B) Mechanoreceptor
 (C) Nociceptor
 (D) Proprioceptor
 (E) Photoreceptor
 (F) Thermoreceptor

24.17. Which of the following receptor classifications includes free nerve endings for which the adequate stimulus is pain?
 (A) Chemoreceptor
 (B) Mechanoreceptor
 (C) Nociceptor
 (D) Proprioceptor
 (E) Photoreceptor
 (F) Thermoreceptor

24.18. Because muscle spindles and Golgi tendon organs contribute information about body position, they are included in which of the following receptor classifications?
 (A) Chemoreceptor
 (B) Mechanoreceptor
 (C) Nociceptor
 (D) Proprioceptor
 (E) Photoreceptor
 (F) Thermoreceptor

24.19. The organ of Corti's exceptional sensitivity to vibration places it in which of the following receptor classifications?
 (A) Chemoreceptor
 (B) Mechanoreceptor
 (C) Nociceptor
 (D) Proprioceptor
 (E) Photoreceptor
 (F) Thermoreceptor

24.20. Which of the following structures is a component of the eye's tunica fibrosa?
 (A) Choroid
 (B) Ciliary body
 (C) Cornea
 (D) Iris
 (E) Lens
 (F) Pigmented epithelium
 (G) Uvea
 (H) Zonule

24.21. Which of the following structures is a derivative of the embryonic optic vesicle?
 (A) Choroid
 (B) Ciliary body
 (C) Cornea
 (D) Iris
 (E) Lens
 (F) Pigmented epithelium
 (G) Uvea
 (H) Zonule

24.22. Which of the following structures is the most anterior extension of the eye's tunica vasculosa?
 (A) Anterior epithelium
 (B) Bruch's membrane
 (C) Ciliary body
 (D) Cornea
 (E) Iris
 (F) Lens
 (G) Pigmented epithelium
 (H) Zonule

24.23. Which of the following pairs of eye components are formed by embryonic mesenchyme?
 (A) Choroid and retina
 (B) Choroid and sclera
 (C) Cornea and choroid
 (D) Cornea and lens
 (E) Lens and choroid
 (F) Lens and retina
 (G) Lens and sclera
 (H) Sclera and retina

24.24. Which of the following cells transmit visual signals from the retina to the brain?
 (A) Amacrine cells
 (B) Bipolar cells
 (C) Cone cells
 (D) Ganglion cells
 (E) Horizontal cells
 (F) Müller cells
 (G) Rod cells

24.25. Which of the following cells span the entire width of the neural retina?
 (A) Amacrine cells
 (B) Bipolar cells
 (C) Cone cells
 (D) Ganglion cells
 (E) Horizontal cells
 (F) Müller cells
 (G) Rod cells

24.26. Which of the following is true of cone cells but not of rod cells?
  (A) Are derived from embryonic neural ectoderm
  (B) Contain flattened vesicles that are independent of the plasma membrane
  (C) Contain the visual pigment rhodopsin
  (D) Are highly responsive to small amounts of light
  (E) Comprise approximately 95% of the retina's photoreceptor cells
  (F) Are responsible for color vision
  (G) Are more abundant in the periphery of the retina

24.27. Which of the labels in Figure 24–6 corresponds to the structure responsible for forming the endolymph?

24.28. Which of the labels in Figure 24–6 corresponds to the location of the spiral ganglion in the modiolus?

24.29. Which of the labels in Figure 24–6 corresponds to the lumen of the cochlear duct?

24.30. Which structure in Figure 24–6 is located in the cochlea and is analogous in structure and function to the cupula of each semicircular canal?

24.31. Which of the labels in Figure 24–6 corresponds to the location of the outer hair cells?

24.32. Which of the following is true of the middle ear?
  (A) Contains the endolymph
  (B) Contains the labyrinth
  (C) Contains the modiolus
  (D) Contains the ossicles
  (E) Contains the perilymph
  (F) Contains the semicircular canals
  (G) Contains the vestibular apparatus

24.33. Which of the following is true of the internal ear?
  (A) Communicates with the pharynx through the eustachian tube
  (B) Communicates with the mastoid air cells
  (C) Contains the cochlea
  (D) Contains stapes
  (E) Includes the auricle
  (F) Includes the ceruminous glands
  (G) Includes the external auditory meatus

24.34. Which of the following is true of the smooth muscles of the iris?
  (A) Constrict the pupil in response to epinephrine
  (B) Are incapable of graded contractions
  (C) Are poorly innervated
  (D) Are of ectodermal origin
  (E) Release tension on the zonule when they contract

24.35. Which of the following lingual papillae have taste buds on their apical surfaces?
  (A) Circumvallate
  (B) Filiform
  (C) Foliate
  (D) Fungiform

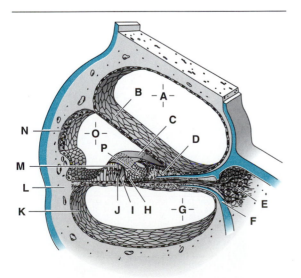

Figure 24–6.

ANSWERS TO MULTIPLE-CHOICE QUESTIONS

24.1. A (I.A.3; II.A–F)
24.2. A (III.A.1)
24.3. E (V.B.1)
24.4. E (V.B.2; Fig. 24–2)
24.5. A (V.C.2)
24.6. C (V.C.3; Fig. 24–2)
24.7. D (V.F; Fig. 24–2)
24.8. D (III.A and B)
24.9. C (Table 24–2)
24.10. B (IV.B)
24.11. C (V.B.1)
24.12. B (V.C.2)
24.13. C (V.A, B.1, C.1,2, and D.1)
24.14. C (V.J.3)
24.15. B (V.K.2 and 3)
24.16. A (IV.B; Table 24–1)
24.17. C (II.A; Table 24–1)
24.18. D (III.A and B; Table 24–1)

24.19. B (VI.C; Table 24–1)
24.20. C (V.B.1)
24.21. F (V.A)
24.22. E (V.C.3; Fig. 24–2)
24.23. B (V.A, B, and C)
24.24. (V.E.3)
24.25. F (V.E.4; Fig. 24–3)
24.26. F (V.E.1; Table 24–3)
24.27. N (VI.E.2; Fig. 24–5)
24.28. E (VI.E; Fig. 24–5)
24.29. O (VI.E.1; Figs. 24–4 and 24–5)
24.30. C (VI.D.2 and E.2.b; Fig. 24–5)
24.31. M (VI.E.2.b.[2]; Fig. 24–5)
24.32. D (VI.A, B, and C)
24.33. C (VI.A, B, and C)
24.34. D (V.C.3.b)
24.35. D (IV.A)

Diagnostic Final Examination

25

ANSWER SHEET FOR DIAGNOSTIC FINAL EXAMINATION

1. Take this test using the answer sheet provided. You may want to use a copy of the blank form to preserve the original.
2. Use the answer key to identify your correct and incorrect answers.
3. Total your correct answers for each row (rA through rK) and for each column (c1 through c5). Compare these totals to identify your relative strengths and weaknesses.
4. Use the key provided to determine the topic emphasized in each row or column.
5. Use your performance on this test to budget your time for examination preparation.

1. A B C D E F G	2. A B C D E F G	3. A B C D E F G	4. A B C D E F G	5. A B C D E F G	rA____
6. A B C D E F G	7. A B C D E F G	8. A B C D E F G	9. A B C D E F G	10. A B C D E F G	rB____
11. A B C D E F G	12. A B C D E F G	13. A B C D E F G	14. A B C D E F G	15. A B C D E F G	rC____
16. A B C D E F G	17. A B C D E F G	18. A B C D E F G	19. A B C D E F G	20. A B C D E F G	rD____
21. A B C D E F G	22. A B C D E F G	23. A B C D E F G	24. A B C D E F G	25. A B C D E F G	rE____
26. A B C D E F G	27. A B C D E F G	28. A B C D E F G	29. A B C D E F G	30. A B C D E F G	rF____
31. A B C D E F G	32. A B C D E F G	33. A B C D E F G	34. A B C D E F G	35. A B C D E F G	rG____
36. A B C D E F G	37. A B C D E F G	38. A B C D E F G	39. A B C D E F G	40. A B C D E F G	rH____
41. A B C D E F G	42. A B C D E F G	43. A B C D E F G	44. A B C D E F G	45. A B C D E F G	rI ____
46. A B C D E F G	47. A B C D E F G	48. A B C D E F G	49. A B C D E F G	50. A B C D E F G	rJ ____
51. A B C D E F G	52. A B C D E F G	53. A B C D E F G	54. A B C D E F G	55. A B C D E F G	rK____
c1____	c2____	c3____	c4____	c5____	

Key to Answer Sheet Topics
Columns:

> c1 = Cells and Tissues
> c2 = Growth, Development, Aging, and Repair
> c3 = Circulatory and Lymphoid Systems
> c4 = Respiratory, Digestive, and Urinary Systems
> c5 = Nervous System, Endocrine System, and Sense Organs

Rows:

> rA = Methods of Study
> rB = Cytology/Cell Biology
> rC = Integumentary System
> rD = Musculoskeletal System
> rE = Circulatory System
> rF = Lymphoid System
> rG = Respiratory System
> rH = Digestive System
> rI = Urinary System
> rJ = Endocrine System
> rK = Reproductive System

Select the single best answer.

1. A randomly and continuously dividing culture is established, with a cell type for which the S phase occupies 20% of the cell cycle. The culture undergoes a pulse/chase procedure. It is incubated in medium containing [^{3}H]-thymidine for 5 hours (the "pulse" component); subsequently, the radioactive medium is removed and replaced with medium containing nonradioactive thymidine for another 2 hours (the "chase" component). At the end of this procedure, the cells are fixed and subjected to autoradiography. The results show that 30% of the cells incorporated the radiolabel into their nuclei. Estimate the length of the S phase for these cells.
   (A) 2 hours
   (B) 5 hours
   (C) 10 hours
   (D) 15 hours
   (E) 20 hours
   (F) 25 hours
   (G) 30 hours

2. Which of the following cell types would be most heavily labeled by an antisense cyclin D riboprobe during *in situ* hybridization studies?
   (A) Multipolar neurons in the spinal cord
   (B) Skeletal muscle fibers in the tongue
   (C) Melanocytes in the epidermis
   (D) Keratinocytes of the stratum basale
   (E) Fibroblasts in the transversalis fascia

3. A 1-year-old African–American child is brought by her adoptive parents to a family physician's office with a history of increasingly frequent episodes of painful swelling in her hands and feet. A blood sample is obtained to screen for sickle cell anemia. Which of the following tests would provide the most reliable information for distinguishing among sickle cell disease, sickle cell trait, and other possible hemoglobinopathies?
   (A) Isoelectric focusing
   (B) SDS-PAGE
   (C) Blood smear with differential blood count
   (D) Standard hemoglobin and hematocrit
   (E) Density gradient centrifugation

4. To study the sequence of events in the synthesis and secretion of insulin, investigators fed 10 rats a high-sugar meal. Thirty minutes after the meal, the investigators administered a [^{3}H]-leucine pulse injection, which was followed 2 hours later with a chase injection of cold (nonradiolabeled) leucine at a concentration higher than that of the labeled leucine. Subsequently, five pairs of rats were sacrificed—one pair at 1, 2.5, 5, 10, and 15 hours after the chase. Next, the pancreas of each animal was processed for EM autoradiography. Which of the following sequences best reflects the pattern of [^{3}H]-leucine incorporation into the newly synthesized insulin seen in the EM autoradiographs?
   (A) Nucleus → cytoplasm → RER → Golgi complex
   (B) Cytoplasm → nucleus → RER → Golgi complex
   (C) Cytoplasm → nucleus → RER → secretory vesicles
   (D) Cytoplasm → RER → Golgi complex → secretory vesicles
   (E) Cytoplasm → RER → Golgi complex → lysosomes
   (F) Lysosomes → cytoplasm → RER → transfer vesicles
   (G) Lysosomes → cytoplasm → SER → transfer vesicles

5. An autopsy performed after an automobile accident reveals that a 22-year-old woman died of a cerebral hemorrhage, without other discernible trauma to her head. A large pituitary adenoma centered in the pars distalis is found near the site of the hemorrhage. Breast tissue histology indicates active mammary glands, although no physical evidence or documentation suggests a current or recent pregnancy. Which of the following histologic analyses of the pituitary tumor would confirm the existence of a mammotropic tumor?
   (A) H & E stain
   (B) TEM
   (C) PAS stain
   (D) Immunohistochemistry for prolactin
   (E) Enzyme histochemistry for lactase
   (F) Scanning electron microscopy

6. A 35-year-old man visits his physician after discovering a small lump (nodule) on his right testis. A biopsy of the site shows a rapidly dividing, metastatic germ-cell tumor. Which of the following cell-cycle phases is most likely to be absent among the randomly but continuously dividing cells of the tumor?
   (A) G_1
   (B) Prophase
   (C) G_0
   (D) Metaphase
   (E) S
   (F) Anaphase
   (G) G_2

7. In a 53-year-old man, the cytoplasm of a motor neuron in the lateral cell column (Clarke's column) of the spinal cord gray matter would be expected to contain more of which substance compared with the cytoplasm of a 15-year-old male?
   (A) Ribophorin
   (B) Lipofuscin
   (C) Epinephrine
   (D) Cyclin
   (E) Euchromatin
   (F) Norepinephrine
   (G) Heterochromatin

8. A 15-year-old male patient visits his family physician for the checkup required to join his high school football team. Giemsa staining of a standard smear of his blood reveals that most of his mature RBCs lack ribosomes. From this information, and from knowledge of ribosome function and distribution, the physician can deduce which of the following?
   (A) The patient's mature RBCs are incapable of protein synthesis
   (B) The patient's RBC precursors never possess a nucleolus during their development
   (C) The patient's mature RBCs contain RER
   (D) The patient's mature RBCs are incapable of oxidative phosphorylation
   (E) The patient's mature RBCs can synthesize only hemoglobin
   (F) The patient has a hemolytic anemia
   (G) The patient has a severe bacterial infection

9. Which of the following is a primary secretory product of the cell shown in Figure FE–1?
   (A) Pepsinogen
   (B) Lecithin
   (C) Mucin
   (D) Hydrochloric acid
   (E) Sulfated glycosaminoglycan
   (F) Sulfated androgen
   (G) Melatonin

10. A 12-year-old child had a cell-cycle mutation in a bone marrow stem cell that led to uncontrolled division of many of its derivatives. Which of the following cell types in this patient's brain most likely caused the brain tumor that ultimately led to her death?
   (A) Purkinje cell
   (B) Oligodendrocyte
   (C) Fibrous astrocyte
   (D) Microglial cell
   (E) Protoplasmic astrocyte
   (F) Pyramidal cell
   (G) Interneuron

11. A 4-year-old child is brought by his parents to a physician's office for the physical examination required for matriculation at a public kindergarten. The child is wearing thick glasses and has very pale skin and snow-white hair. Previously diagnosed with oculocutaneous albinism type 1 (OCA-1), the child possesses two different genetic mutations. One mutation is in each of two copies of the gene coding for which of the following enzymes?
   (A) Collagenase
   (B) Lysyl oxidase
   (C) Prolyl hydroxylase
   (D) Tyrosinase
   (E) Urate oxidase

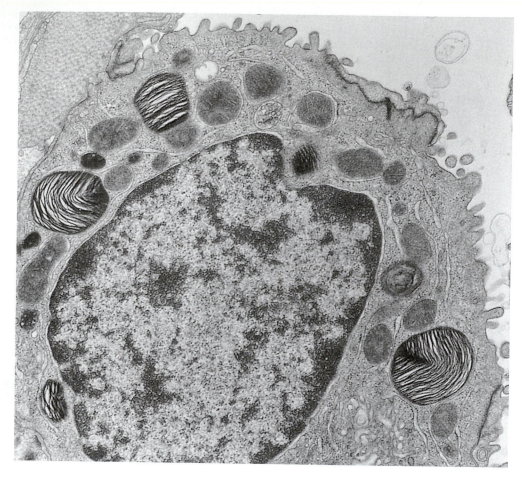

Figure FE–1.

12. A 26-year-old man visits his physician because of an extremely painful 10-cm lesion on his left thigh, which was obtained during a desperate slide to recover a ball in an amateur soccer game. Examination of the wound reveals an abrasion to the depth of the reticular dermis and only a few minor lacerations that penetrate deeper. Most likely, the abraded epithelium will be replaced by epidermal cells originating from which site(s) in and around the wound?
   (A) Stratum corneum
   (B) External root sheath
   (C) Stratum granulosum
   (D) Deep fascia (epimysium)
   (E) Papillary capillaries
   (F) Meissner's corpuscles
   (G) Sebaceous glands

13. A 38-year-old woman is brought by ambulance to the emergency room from the scene of an automobile accident. She is conscious but restless and apprehensive. Her skin is cool, moist, and pale. She struggles to sit up and immediately faints. Her blood pressure is 100/60 and her heart rate is rapid (110 beats per minute). Abrasions and a large contusion are evident in the right upper quadrant of her distended abdomen. Which tissue layer contains most of the structures responsible for the skin's pale, clammy condition?
   (A) Stratum corneum
   (B) Stratum spinosum
   (C) Papillary dermis
   (D) Reticular dermis
   (E) Hypodermis

14. An 85-year-old woman has been confined to a nursing home in a far northern region of the country for the past 6 years. Bone density scans reveal that she has developed osteoporosis. Although the amount of calcium in her diet is adequate, she is incapable of absorbing enough calcium to maintain bone mineral. One factor contributing to her condition most likely is insufficient production of which of the following substances by the dermis of her skin?
    (A) Parathormone
    (B) Vitamin A
    (C) Calcitonin
    (D) Vitamin D
    (E) Melanin

15. Which of the following is the most likely location of the vibration receptors shown in Figure FE–2?
    (A) Epidermis
    (B) Inner ear
    (C) Dermal papillae
    (D) Patellar tendon
    (E) Reticular dermis

16. A sample of mRNA isolated from predominantly slow (red) muscles is compared with that isolated from predominantly fast (white) muscles. The sample from the slow muscles would be expected to contain a higher proportion of mRNA for which of the following proteins?
    (A) Actin
    (B) Tropomyosin
    (C) α-Actinin
    (D) Troponin T
    (E) Myosin heavy chain type I
    (F) Troponin I
    (G) Glycogen

17. A 45-year-old woman visits her physician because of increasing pain and stiffness in her right knee during the past 2 years. She played competitive amateur softball in high school and continued until she was 35 years old. In college, she seriously injured her left knee in a collision with another ball player but made a full recovery. During the past 10 years, she has become more sedentary and has gained 35 pounds. Radiographs show cartilage erosion and bone degeneration around the knee joint. The degenerative joint disease can be attributed to normal effects of aging, previous acute trauma, and stress from excessive weight bearing on the knee joint. Such factors combine to cause abnormal sulfated glycosaminoglycan and type-II collagen secretion by which cells?

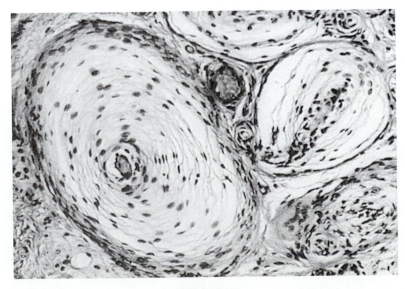

Figure FE–2.

(A) Chondrocytes
(B) Fibroblasts
(C) Osteoblasts
(D) Synovial lining cells
(E) Osteoclasts
(F) Bone marrow stem cells
(G) Muscle satellite cells

18. Potential bone marrow donors for a 53-year-old man undergoing treatment (chemotherapy and radiation) for acute leukemia include his 44-year-old brother and his 8-year-old grandson. Which of the following sites would be appropriate for obtaining a marrow sample from his grandson but not from his brother?
(A) Iliac crest
(B) Tibia
(C) Sternum
(D) Ribs
(E) Vertebrae

19. Which of the following body sites is the most likely source of the tissue section shown in Figure FE–3?
(A) Colon
(B) Tongue
(C) Parotid gland
(D) Pyloric stomach
(E) Liver
(F) Duodenum
(G) Trachea

20. An 18-year-old woman visits her physician because she tires easily and has experienced increasing weakness in her hands that no longer seems to subside completely, even after long intervals of rest. After a chest radiograph reveals an enlarged thymus, myasthenia gravis is suspected and is confirmed by a muscle biopsy. The biopsy indicates a marked reduction in the number of acetylcholine receptors (compared with normal) at which of the following sites?
(A) Presynaptic membranes of the myoneural junctions
(B) Primary synaptic clefts of the myoneural junctions
(C) Postsynaptic membranes of the myoneural junctions
(D) Nonjunctional sarcolemmas of myofibers
(E) Presynaptic membranes of Golgi tendon organs

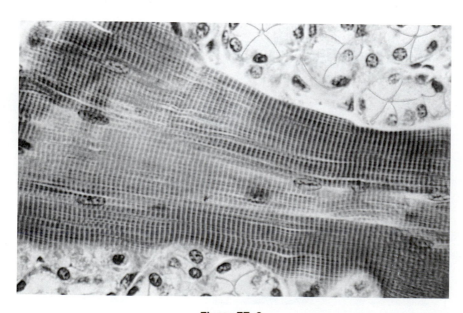

Figure FE–3.

(F) Transverse tubules of myofibers

(G) Sarcoplasmic reticulum of myofibers

21. Several members of an extended family visit a physician's office for treatment of diarrhea and nausea. All had attended a family reunion (a "luau") 2 days earlier, where members ate pit-roasted wild boar that was prepared by an inexperienced cook. The differential blood count of each affected family member reveals excessive numbers of the cell type shown in Figure FE–4. What is the likely cause of the family's symptoms?

   (A) Parasitic infection of blood monocytes and macrophages

   (B) Trichinosis (helminth infection)

   (C) Food poisoning (bacterial infection) from contaminated meat

   (D) Indigestion from overeating

   (E) Lead poisoning from ingested shotgun pellets

22. The cell whose nucleus is labeled "N" in Figure FE–5 derives from the same embryonic germ layer as which of the following cell types?

   (A) Keratinocyte

   (B) Erythrocyte

   (C) Type-I alveolar cell

   (D) Ependymal cell

   (E) Melanocyte

23. Patients with Marfan syndrome, a connective tissue disorder, are predisposed to dissecting aneurysms of the ascending aorta. These aneurysms are typically caused by a derangement of the elastic laminae that predominate at which site in this organ?

   (A) Intima

   (B) Serosa

   (C) Media

   (D) Valves

   (E) Adventitia

24. A 65-year-old man with a childhood history of rheumatic fever and a 45-year history of cigarette smoking arrives at the emergency room and complains that he has experienced severe shortness of breath for the past 18 hours. He reports that his chronic cough has notably worsened during the past several months. Sputum specimens are negative for bacterial pathogens but have a pink

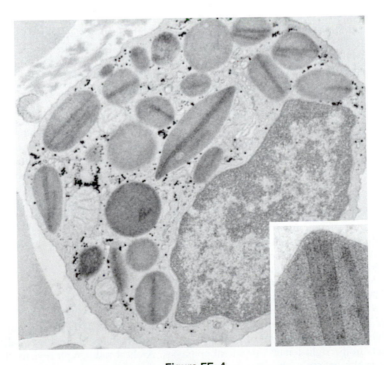

Figure FE–4.

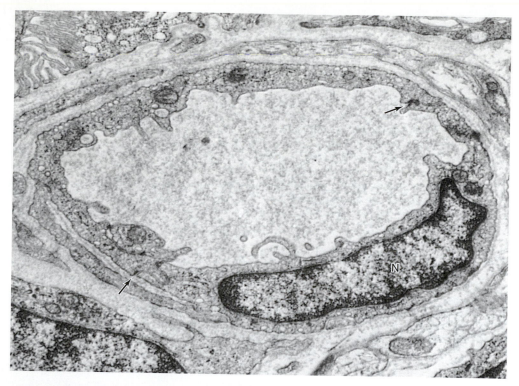

Figure FE–5.

tinge. Chest radiographs reveal pulmonary edema and an enlarged left atrium. The excess fluid and blood in this patient's alveoli can be most directly attributed to which of the following?

(A) The patient's worsening chronic cough
(B) Elevated hydrostatic pressure in alveolar septal capillaries
(C) Degradation of the blood–air barrier by an active pneumonia
(D) Paralysis of the cilia in the patient's small airways
(E) A normal age-dependent decline in pulmonary function
(F) Smoking-related metaplasia in the trachea and primary bronchi

25. A 53-year-old woman visits her physician and complains of a 2-week history of increasingly frequent headaches, fatigue, muscle spasms, weakness, urinary frequency, and thirst. Her blood pressure is 180/110 mmHg. Blood tests reveal hypernatremia and hypokalemia. CAT scans indicate normal adrenal glands. However, the patient's left kidney is reduced in size, apparently from stenosis of the left renal artery. A subsequent blood test reveals higher-than-normal aldosterone concentration in the blood. The hyperaldosteronism in this patient is most directly attributed to endocrine hypersecretion by cells in which of the following locations?

(A) Renal afferent arterioles
(B) Adrenal zona reticularis
(C) Adrenocortical adenoma in the zona glomerulosa
(D) Renal distal convoluted tubule
(E) Adrenal zona fasciculata
(F) Persistent fetal adrenal cortical tissue
(G) Renal papillary collecting ducts

26. After falling down several stairs, a 62-year-old man arrives at the emergency room in pain and unable to walk. Radiographs of both hips indicate no fractures but reveal small, round radiolucent defects in the right femur's proximal epiphysis. A chest radiograph reveals similar rounded defects in the patient's ribs. The findings are consistent with a diagnosis of multiple myeloma, a

lymphoproliferative disorder characterized by uncontrolled hyperplasia of functional B-cell derivatives. A significant increase in the amount of which plasma protein would be consistent with this diagnosis?

(A) Serum albumin
(B) Granulocyte–macrophage colony-stimulating factor (GM-CSF)
(C) Erythropoietin
(D) Fibrinogen
(E) Gamma globulin
(F) Complement
(G) Parathyroid hormone

27. Among the benefits provided to newborns by breast feeding is the passive immunity conferred by which of the following cells?
(A) Macrophages in the areolar papillary dermis
(B) CD8+ cells in the mammary lymphatic vessels
(C) Epithelial cells in the lactiferous duct
(D) Plasma cells in the mammary connective tissue
(E) Neutrophils in the colostrum
(F) Myoepithelial cells in the mammary alveoli

28. Most dietary fats absorbed by enterocytes first enter the peripheral blood at which of the following sites?
(A) Hepatic sinusoids
(B) Junction of the jugular and subclavian veins
(C) Blood capillaries in the intestinal villi
(D) High endothelial postcapillary venules of lymph nodes
(E) Hepatic portal vein

29. Appendicitis often is attributed to a fecalith lodged in the lumen of the appendix. A fecalith can trap antigens in the distal lumen and cause the appendix to enlarge in response to cell proliferation in which of the following sites?
(A) Bases of the crypts
(B) Tips of the villi
(C) Submucosal glands
(D) Surface epithelium
(E) Germinal centers
(F) Capillary lumen

30. A 26-year-old man reports a 6-month history of unsuccessful attempts to conceive a child with his 24-year-old wife. He suspects he is infertile because his wife has two children from a previous marriage. A sample of the patient's semen exhibits oligospermia, many nonmotile sperm, and a few neutrophils. Analysis of proteins in the semen reveals the presence of immunoglobulins, and subsequent immunohistochemistry reveals that many of the sperm are coated by immunoglobulins. A structural defect of which cell type in the seminiferous tubules is most directly related to the presence of immunoglobulins in the semen?
(A) Sertoli cell
(B) Leydig cell
(C) Myoid cell
(D) Spermatogonia
(E) Spermatid
(F) Endothelial cell
(G) Plasma cell

31. Hydrocortisone treatment of premature infants with respiratory distress syndrome relieves their symptoms by enhancing the synthesis and secretion of pulmonary surfactant by which of the following cells?
(A) Alveolar macrophages
(B) Goblet cells
(C) Clara's cells
(D) Brush cells
(E) Endothelial cells
(F) Type-I alveolar cells
(G) Type-II alveolar (septal) cells

32. A 33-year-old firefighter is brought to the emergency room with acute respiratory distress syndrome caused by smoke, heat, and chemical damage to her type-I alveolar cells. Her damaged type-I alveolar cells must be replaced primarily by which of the following healthy cells?
    (A) Vascular pericytes
    (B) Clara's cells
    (C) Endothelial cells
    (D) Type-II alveolar (septal) cells
    (E) Fibroblasts
    (F) Vascular smooth muscle
    (G) Small granule cells

33. A 12-year-old child complains of a 24-hour history of watery eyes, sneezing, and nasal congestion. The child's nasal mucosa is inflamed, but his throat is clear and he is afebrile. Which of the following cells in the child's nasal mucosa synthesized the protein that triggered his apparent allergic reaction?
    (A) Goblet cells
    (B) Olfactory cells
    (C) Mast cells
    (D) Bowman's gland alveolar cells
    (E) Plasma cells
    (F) Dust cells
    (G) Endothelial cells

34. A 24-year-old man visits his physician and complains that he has had difficulty breathing and has produced red-tinged sputum for 2 days. During the past 24 hours, he has also noticed a red tinge to his urine and has experienced pain in his side and back. Microorganisms are not found in his sputum, but his urine contains red blood cells and casts. During the following 2 days, the patient rapidly develops hematuria and renal failure. Immunohistochemical studies of lung and kidney biopsies reveal extracellular IgG accumulations around the pulmonary alveoli and within the renal glomeruli. All findings are consistent with an autoimmune attack. Which of the following extracellular proteins is a normal component of both the blood–air barrier and the glomerular filtration barrier and thus could be the target of this patient's impaired immune system?
    (A) Type-II collagen
    (B) Laminin
    (C) Heparan sulfate
    (D) Tropocollagen
    (E) Hyaluronan
    (F) Actin

35. An 18-year-old amateur boxer suffers a severe blow to the face during a street fight. Radiographs reveal fractures of the nasal septum and nasal bone, the nasal spine of the frontal bone, and parts of the ethmoid bone (including the cribriform plate). Two weeks later, much of the swelling has subsided, but the patient complains that he has lost his sense of smell. Histology of a needle biopsy from the nasal mucosa over the superior concha reveals chromatolysis in which of the following cells?
    (A) Goblet cells
    (B) Basal cells
    (C) Oligodendrocytes
    (D) Ciliated columnar epithelial cells
    (E) Olfactory cells
    (F) Schwann cells
    (G) Sustentacular cells

36. Which of the following is associated with hyposecretion by the cell shown in Figure FE–6?
    (A) Acromegaly
    (B) Myxedema
    (C) Diabetes mellitus
    (D) Duodenal ulcer
    (E) Osteoporosis
    (F) Pernicious anemia
    (G) Addison's disease

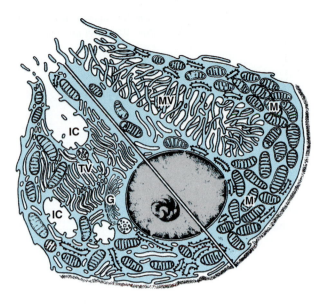

Figure FE–6.

37. The founder cells of a hereditary pancreatic exocrine adenoma arise from which embryonic cell type?
    (A) Extraembryonic mesoderm
    (B) Endoderm
    (C) Neural crest
    (D) Visceral (splanchnic) mesoderm
    (E) Neural ectoderm
    (F) Hypoblast
    (G) Cytotrophoblast

38. A 52-year-old man with a 20-year history of both alcohol abuse and hypertension visits his physician to renew his prescription for blood pressure medication. Although the patient has lost 12 pounds since his last visit 4 months ago, his abdomen has become round and swollen. He complains of hemorrhoids and that his feet swell after he has been standing or sitting for long periods of time. His liver is enlarged and hard, his spleen is palpable, and his skin and eyes have a yellowish tinge. A liver biopsy reveals advanced cirrhosis and extensive replacement of damaged hepatocytes with scar tissue. Which of the following best explains his large and swollen abdomen?
    (A) Hyperplasia of both the liver and the spleen
    (B) Overeating and severe constipation
    (C) Reduction in both hepatic blood flow and albumin synthesis
    (D) Leaky blood vessels and high concentration of serum bilirubin
    (E) Diuretic effects of both alcohol and blood pressure medications

39. Peripheral vasoconstriction caused by chronic stress and heavy smoking can cause localized ischemia in the gastric mucosa. Such ischemia may favor the development of a gastric ulcer by impairing the nourishment, and therefore the normal secretory function, of which epithelial cell type?
    (A) Surface mucous cell
    (B) Parietal (oxyntic) cell
    (C) Enteroendocrine cell
    (D) Gastric chief cell
    (E) Goblet cell
    (F) Pancreatic acinar cell
    (G) Vascular endothelial cell

40. Duodenal ulcers experienced by patients with Zollinger–Ellison syndrome are mainly caused by the hyperstimulation of which cells in the digestive tract mucosa?
 (A) Gastric chief cells
 (B) Paneth's cells
 (C) Goblet cells
 (D) Parietal (oxyntic) cells
 (E) Enterocytes
 (F) Pancreatic δ cells
 (G) Pancreatic acinar cells

41. Acquired Fanconi syndrome is a renal disease that involves relatively selective damage to the proximal convoluted tubule. The condition interferes with the proximal tubule's normal structure and function but typically spares the glomerulus. Which of the following symptoms would be consistent with this condition?
 (A) Hyperglycemia
 (B) Hypocalciuria
 (C) Hyperphosphatemia
 (D) Aminoaciduria
 (E) Hematuria
 (F) Hypertension
 (G) Proteinuria

42. Among children younger than 5 years of age, Wilms tumor (nephroblastoma) is a relatively common renal malignancy. In affected individuals, cells that should form the lining of the renal tubules instead retain the structural and functional characteristics of the embryonic cells from which they derive. Thus, it can be said that these cells fail to undergo what normal process?
 (A) Anaplasia
 (B) Proliferation
 (C) Differentiation
 (D) Homeostasis
 (E) Metaplasia
 (F) Hypertrophy
 (G) Hyperplasia

43. Nephrotic syndrome is a renal disease characterized by proteinuria, hypoalbuminemia, and generalized edema. A major cause of this condition is membranous glomerulonephritis, which is characterized by abnormal accumulations on the glomerular basement membrane of proteins secreted by which cells?
 (A) Juxtaglomerular cells
 (B) Plasma cells
 (C) Kupffer cells
 (D) Glomerular capillary endothelial cells
 (E) Mesangial cells
 (F) Hepatocytes
 (G) Podocytes

44. Furosemide is a diuretic that increases urine volume through its effects on the $Na^+/K^+/Cl^-$ pump (symporter) located in which component(s) of the nephron?
 (A) Macula densa
 (B) Vasa recta
 (C) Glomerular filtration barrier
 (D) Ascending limb of the loop of Henle
 (E) Medullary collecting ducts
 (F) Proximal convoluted tubules
 (G) Peritubular capillaries

45. A 22-year-old man visits his physician because of a 2-week history of frequent urination (every 1 to 2 hours) and excessive thirst. Two months earlier, he had been diagnosed with bipolar affective disorder and had begun lithium therapy. The patient's urine is dilute and pale. Circulating levels of the antidiuretic hormone arginine vasopressin (AVP) range from normal to slightly elevated. CT scans of the pituitary gland are normal, and no other evidence of endocrine malfunction is discovered. A 5-hour water fast increases serum AVP levels but fails to increase urine solute concentration. Likewise, treatment with a synthetic AVP (DAVP) has no effect on urine concentration. What is the most likely location of the lesion causing this patient's symptoms?

    (A) Glomerular filtration barrier
    (B) Hypothalamus
    (C) Proximal convoluted tubules
    (D) Primary capillary plexus
    (E) Renal collecting ducts
    (F) Pars nervosa
    (G) Distal convoluted tubule

46. A genetic defect leading to a disruption of G protein–coupled receptor-mediated transmembrane signal transduction would most likely affect the expression of a gene normally induced by which hormone?
    (A) Thyroid hormone
    (B) Estradiol
    (C) Testosterone
    (D) Epinephrine
    (E) Insulin
    (F) Hydrocortisone
    (G) Progesterone

47. A 4-year-old child is brought to a pediatrician by his parents because of his short stature (64 cm). Blood analyses reveal abnormally low levels of somatotropin. Cells normally responsible for producing this hormone derive from which embryonic cell population?
    (A) Neural crest
    (B) Neural ectoderm
    (C) Sclerotome
    (D) Endoderm
    (E) Extraembryonic mesenchyme
    (F) Oral ectoderm
    (G) Dermamyotome

48. In a patient with lupus erythematosus, a clone of B lymphocytes developed that produced an antibody to the patient's own TSH receptors. This particular antibody bound and blocked the TSH receptors without activating them. Consequently, T_3 and T_4 production were substantially reduced, preventing negative feedback to the hypothalamus and pituitary. After blood levels of TSH were sustained at higher-than-normal levels, they dropped precipitously to abnormally low levels. A subsequent biopsy of this patient's adenohypophysis revealed an unusually high percentage of chromophobes (62%). The abnormally large numbers of these cells most likely can be attributed to:
    (A) A malignant chromophoboma
    (B) Excess thyrotropic degranulation
    (C) Excess colloid in Rathke's cysts
    (D) Hypophyseal lymphocyte infiltration
    (E) A pheochromocytoma
    (F) Adenohypophyseal scarring
    (G) Pituicyte hyperplasia

49. A dramatic reduction in the oxygen-carrying capacity of a patient's blood after extensive erythrocyte lysis during a sickle cell crisis triggers the secretion of which substance from the patient's kidneys?
    (A) Angiotensinogen
    (B) Renin
    (C) Erythropoietin
    (D) Aldosterone
    (E) ADH
    (F) Hemoglobin
    (G) Fibroblast growth factor

50. A morbidly obese patient has abnormally low plasma levels of both thyroxin and thyrotropin. MRI reveals a possible tumor in the anterior pituitary, suggesting a malfunction of the thyrotrophs. Which of the following most likely contributes directly to this patient's obesity?
    (A) Poor use of glucose throughout the body
    (B) Hyperabsorption of nutrients by enterocytes
    (C) Bloating from excess ADH secretion
    (D) Hyperactivity associated with hypothyroidism
    (E) Hypersecretion by the adrenal medulla

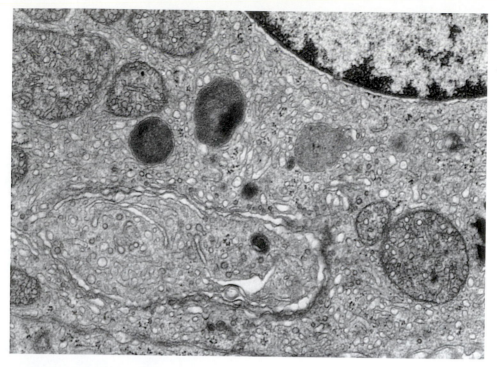

Figure FE–7.

51. A 33-year-old woman visits her physician during her 24th week of pregnancy and complains of hirsutism. She states that she noticed the problem 2 months earlier but that it seems to have worsened during the past couple of weeks. Where would the cell shown in Figure FE–7 most likely be located in this patient?
    (A) Pancreatic islets of Langerhans
    (B) Hair follicles
    (C) Adrenal medulla
    (D) Fetal brain
    (E) Corpus luteum
    (F) Thyroid gland
    (G) Placenta

52. Mitotic figures in which site reflect the process of cell division that replaces the endometrial surface epithelium lost during menstruation?
    (A) Glands in the stratum basale
    (B) Coiled arteries in the zona spongiosa
    (C) Stroma in the stratum functionalis
    (D) Cytotrophoblast in the primary villi
    (E) Arcuate arteries in the myometrium

53. A 23-year-old employee at a chemical storage facility was exposed briefly to several liquid and aerosolized toxic chemicals during an explosion at his work site. Two hours after the accident, he had detectable blood levels of several toxins. Three months later, he was referred to a fertility clinic to assess long-term damage to his reproductive function. His semen specimen indicated normal volume, sperm count, and sperm mobility. Which cells contributed most to protecting his spermatozoa from damage by toxic chemicals in his blood stream?
    (A) Myoid cells
    (B) Spermatogonia
    (C) Endothelial cells
    (D) Sertoli cells
    (E) Interstitial cells

     (**F**) Fibroblasts in the tunica propria
     (**G**) Fibroblasts in the tunica albuginea

54. Androgen deprivation can be used to slow tumor growth in many cases of prostatic adenocarcinoma. There are several approaches to this type of therapy, including surgical castration. One less-invasive approach involves the use of synthetic analogs of luteinizing hormone–releasing hormone (LHRH) that bind and block normal LHRH receptor activity and thereby indirectly block testosterone synthesis. What is the main target of such direct LHRH analog binding?
     (**A**) Hypothalamus
     (**B**) Interstitial cells (of Leydig)
     (**C**) Adenohypophysis
     (**D**) Sertoli cells
     (**E**) Prostatic adenocarcinoma cells

55. A 28-year-old woman complains to her physician of a 6-month history of increasingly frequent headaches and blurred vision. A CT scan reveals an intracranial tumor, which subsequently is surgically removed. After surgery, the patient's headaches stop and her vision improves. However, she becomes extremely lethargic and stops menstruating. Blood analyses reveal very low circulating levels of estrogens, thyroid hormone, and adrenal steroids. The patient's postoperative symptoms are the result of damage to, or removal of, which structure(s) during surgery?
     (**A**) Epiphysis cerebri
     (**B**) Neurohypophysis
     (**C**) Rathke's cysts
     (**D**) Adenohypophysis
     (**E**) Internal carotid artery
     (**F**) Herring bodies
     (**G**) Internal jugular vein

DIAGNOSTIC FINAL EXAMINATION: *ANSWERS WITH EXPLANATIONS*

1. *C.* Nuclear labeling with [^{3}H]-thymidine occurs only during the S phase, when the label is incorporated into newly synthesized DNA (3.VI.B.2; Fig. 3–2). In a randomly and continuously dividing cell population, the percentage of cells in any phase of the cycle is equal to the percentage of the cycle that that phase occupies. Thus, in this example, 20% of the cells are in S phase at all times. The number of cells labeled equals the number of cells in S phase when the label is added plus any that enter S phase during the 5-hour incubation period. Because 30% of the cells incorporated the radiolabel into their nuclei during the 5-hour incubation, an additional 10% of the cell population had to enter S phase during incubation. Because 10% of the continuously cycling population crossed a single point in the cycle in 5 hours, 5 hours constitutes 10% of the duration of the full cell cycle. (Thus, the length of the full cell cycle is 50 hours.) Because S phase occupies 20% of the 50-hour cell cycle, the duration of S phase is 10 hours.

2. *D.* In situ hybridization allows localization of specific mRNA transcripts in cells through the annealing of labeled complementary (antisense) RNA sequences (riboprobes) to the transcript (1.X). Cyclins are so named because their expression (transcription and translation) levels rise and fall during the cell cycle (3.VI.B.4). Cyclin D expression, and its association with cyclin-dependent kinases (Cdk4 and Cdk6), signals the transition from the cell cycle's G_1 checkpoint (Fig. 3–2). Thus, at any point in time, a continuously dividing cell population contains more cells with detectable levels of cyclin D mRNA expression than a nondividing or slowly dividing population. The neurons, skeletal muscle fibers, melanocytes, and fibroblasts in the sites mentioned all represent terminally differentiated cells and most likely would not contain detectable levels of cyclin D transcripts. Cells in the epidermal stratum basale divide continuously to replace cells sloughed from the stratum corneum and thus would be more likely to contain detectable levels of cyclin D mRNA.

3. *A.* The amino acid sequence for the hemoglobin S (HbS) molecule of sickle cell anemia differs from the normal hemoglobin A (HbA) molecule in the substitution of a valine (neutral) for a glutamic acid (acidic) in the sixth position on the β chain (12.III.A.3). The difference in net charge is sufficient to allow separation of HbA and HbS by isoelectric focusing (IEF; 1.XIV.B). Indeed, IEF is commonly used to screen the umbilical cord blood of newborns for hemoglobinopathies.

HbF, whose amino acid sequence provides it with an additional isoelectric point, occurs in both normal individuals and in those with sickle cell anemia. Sickle cell *disease* is characterized by homozygosity for HbS (SS), whereas sickle cell *trait* corresponds to the heterozygotic condition (SA). Thus, IEF reveals two hemoglobin bands for normal hemoglobin (HbA, HbF), two for sickle cell disease (HbS, HbF), and three for sickle cell trait (HbA, HbS, HbF). SDS-PAGE separates proteins based on molecular weight and cannot clearly resolve the effect of changing a single amino acid in a protein as large as hemoglobin (1.XIV.A). A blood smear can reveal the presence of sickled cells but cannot be used reliably to distinguish between the disease and the trait (12.I.E and III.A.3). The complete blood count, together with the standard hemoglobin and hematocrit, may reveal the reduction of red cells caused by hemolysis associated with sickling, but these tests cannot identify the cause of the anemia or distinguish between the disease and the trait (12.I.D and E). Density gradient centrifugation is even less sensitive than SDS-PAGE as a method for separating molecules and particles based on weight (1.XII).

4. **D.** The labeled leucine is transported across the plasma membrane by means of facilitated transport, without any requirement for endocytosis (2.II.C.1.b). Labeled leucine is incorporated into insulin during protein synthesis, which begins in the cytoplasm on free polyribosomes (2.III.B.2). Because insulin is secreted, it has an initial signal sequence that, once formed, binds to a signal recognition particle that halts further translation until the polyribosome docks with the RER (2.III.C.1.b). Additional label is incorporated as the rest of the protein is synthesized and deposited in the RER cisterna; subsequently, it is carried by transfer vesicles to the Golgi apparatus for further processing and packaging into secretory vesicles (2.III.D.2.d). These vesicles are shuttled to the plasma membrane by molecular motors attached to microtubule arrays (2.III.I.1.b). True secretion would require a second stimulus (another elevation of blood sugar).

5. **D.** Only immunohistochemistry using primary antibodies to human prolactin and labeled secondary antibodies would provide definitive proof of a selective increase in the number of mammotrophs (prolactin-secreting cells) in the pituitary tumor (1.IX; 20.III.A.2.a). Hematoxylin and eosin can indicate whether tumor cells are acidophils or basophils but cannot distinguish between mammotrophs and somatotrophs. A PAS stain would be even less useful because it would stain only the pituitary basophils (20.III.A.2.b). Enzyme histochemistry (1.VIII) for lactase would be fruitless because the disaccharidase required to digest milk sugar is primarily associated with the microvilli of the small intestine's enterocytes (15.I.C.1 and VII.B.3.a). Transmission and scanning microscopy are both very powerful methods but would not provide the low-power overview of the tumor that is possible with light microscopy; further, without special staining (eg, immunogold), neither of these methods would be suitable for identifying mammotrophs.

6. **C.** In the randomly and continuously dividing cell population in this malignant tumor, all stages of the cell cycle are likely to be represented except G_0 (3.VI.B.1). G_0 cells must reenter the cell cycle at the G_1–S boundary to begin dividing again (3.VI.B.5.a).

7. **B.** Because they are terminally differentiated, nondividing cells, neurons tend to accumulate lipofuscin pigment (indigestible products of endocytosis) in their cytoplasm with advancing age (2.III.F.3; 9.I.F). Ribophorin, a component of the RER (2.III.C.1.a), is less likely to change in amount in cells synthesizing mainly acetylcholine for neurosecretion, although some increase with maturation is possible. Neurons may become more heterochromatic with age; however, chromatin is located not in the cytoplasm but in the nucleus. Although these motor axons are indeed associated with the sympathetic nervous system, they are *pre*synaptic, and only *post*synaptic neurons of this system characteristically release epinephrine or norepinephrine (9.I.D.2; Fig. 9–1; Table 9–2).

8. **A.** The absence of ribosomes is completely normal for mature circulating red blood cells. Although some ribosomes may be present at the reticulocyte stage, they are extruded during final maturation (13.IV.B.5). Ribosome components are synthesized in the nucleolus (3.IV.A and B). One or more nucleoli are present in the precursors of all red blood cells at the proerythroblast and basophilic erythroblast stages (13.IV.B.1 and 2); however, they are extruded with the nucleus at the end of the normoblast stage of development (13.IV.B.4). The ribosomes synthesized from components made in the nucleolus are responsible for the cytoplasmic basophilia of the basophilic erythroblasts. Ribosomes are the molecular machines required for protein synthesis (2.III.B) and thus are critical to hemoglobin synthesis. Synthesis of protein (ie, hemoglobin) is impossible in their absence. In patients with hemolytic anemia, increased numbers of immature red blood cells (eg, reticulocytes) are often released to the circulation from the bone marrow to help replace erythrocytes lost to cell lysis (12.III.A.2). Thus, in such patients, increased numbers

of erythrocytes containing ribosomes in peripheral blood would be expected (although the mature erythrocytes would lack ribosomes). Red blood cells are indeed incapable of oxidative phosphorylation and obtain the energy needed to maintain their structure and function from glycolysis; however, it is the lack of mitochondria, rather than the lack of ribosomes, that accounts for this functional property (12.III.A.1). Larger-than-normal numbers of white blood cells, particularly neutrophils, would provide evidence of a severe bacterial infection (12.III.B.2.a; 13.V.A.2.e).

9. **B.** The large multilammellar bodies in this cell are characteristic of the alveolar type-II (septal) cell. They contain pulmonary surfactant, which septal cells release into the lumens of the alveoli to reduce surface tension and help keep the alveoli open during expiration (17.VIII.B.2). This process facilitates reinflation of the lungs during inspiration. Surfactant is composed chiefly of dipalmitoyl phosphatidylcholine, also known as dipalmitoyl lecithin (17.VIII.C). Pepsinogen is a proenzyme secreted by gastric chief cells, which contain abundant RER (15.VI.B.5). Mucin-secreting cells are characterized by abundant, large open granules that fill their apical cytoplasm (4.VI.C.3). Cells specialized to release HCl (eg, gastric parietal cells [15.VI.B.4] and osteoclasts [8.III.A.1.d]) pump H+ ions into the extracellular space (15.VI.B.4.b); they are characterized by highly folded plasma membranes at the site where the ions are released. Connective tissue cells (eg, fibroblasts and chondrocytes) secrete sulfated glycosaminoglycans (5.II.D; 7.II.A.1.b), which they synthesize in their prominent Golgi complexes (2.III.D.2.b). These cells typically are completely surrounded by abundant extracellular matrix of their own making, including collagen fibers synthesized on the abundant RER in their cytoplasm. Sulfated androgens are products of the fetal, and to a lesser extent the adult, adrenal cortex (21.II.A.3.c and 5). Steroid-secreting cells are characterized by abundant SER and mitochondria containing tubular cristae (4.VI.C.5). The melatonin-secreting cells of the pineal parenchyma (21.VI.A) do not have specific ultrastructural characteristics that would distinguish them from other cells without additional information.

10. **D.** The only bone marrow–derived cell among this group is the microglial cell (9.III.A.3), which derives from the CFU-GM stem cells that form macrophages and other components of the mononuclear phagocyte system (5.II.E.2.b; 13.V.B.1 and D). All of the other neurons and glial cells listed derive from neural ectoderm. Bone marrow and all of its derivatives arise from mesoderm (13.I.A).

11. **D.** The diagnosis of OCA-1 specifies a defect in the genes encoding the enzyme tyrosinase, which is the rate-limiting enzyme in the production of the pigment melanin (18.II.B.6). The result of this defect is a substantial reduction in skin, hair, and eye pigmentation. The condition is often accompanied by reduced visual acuity secondary to misrouting of optic nerve fibers. The child described in this example is a compound heterozygote, having received two different tyrosinase mutations—one from each parent. Collagenase, proline hydroxylase, and lysyl oxidase (5.II.A.1.a and b; IV.E) all occur normally in connective tissues (and thus in the skin and eyes) but have roles in the intracellular and extracellular processing of collagen and no direct impact on pigmentation. Urate oxidase is an enzyme found in peroxisomes (2.III.G); although its activity releases hydrogen peroxide, the small amount released is incapable of bleaching the skin, hair, and eyes.

12. **B.** Epidermis damaged or removed by burns, frostbite, abrasion, or the collection of skin grafts can be replaced by cells in the external root sheaths that line the hair follicles (18.V.3.b). Because the follicles of terminal hairs, such as those on the legs of adult males, often penetrate beyond the reticular dermis and into the hypodermis, shallower wounds are more likely to heal with an epidermal covering and are less likely to result in scarring. The external root sheath is continuous with, and composed of essentially the same cells as, the epidermal stratum malpighi, which also contributes new epidermal cells at the margins of wounds. Epidermal cells in the stratum corneum and stratum lucidum are enucleate and incapable of dividing to replace lost cells (18.II.A.4 and 5). Because the abrasion is to the depth of the reticular dermis in this example, the papillary capillaries in the site have been destroyed by the injury (18.I.B.3). Cells of capillaries and the deep fascia are mesodermally derived and cannot replace epidermis. Meissner's corpuscles consist of fibroblasts and unmyelinated dendritic processes of sensory nerves and are located in the dermal papillae (18.III.A; 24.II.D), which also was lost during this injury. Sebaceous glands are associated with hairs (18.VII.A) but are typically found close to the epidermis in the papillary dermis; most of these also were likely to have been damaged or lost in the injury. Even if they had not been lost or damaged, most cells of sebaceous glands are highly specialized for holocrine secretion (18.VII.B), terminally differentiated, and incapable of replacing the epidermis.

13. **D.** Hypovolemic shock caused by internal bleeding from liver trauma is strongly suggested in this case. The pale clammy skin reflects the reflex response to reduced venous return to the heart. This response involves the shunting of blood from peripheral vascular beds (eg, skin) to vital internal organs to protect oxygen supply to the brain and heart (18.IV.D). The skin's coolness and paleness is caused by the lack of blood in the papillary capillaries. Its moistness is caused by the consequent reduction in the evaporation of sweat (18.VIII.A.3) and by increased sweating in response to some sympathetic stimulation of the sweat glands. Sympathetic stimulation also causes contraction of smooth muscle in the metarterioles that lead to the dermal capillaries and thus prevents blood from entering the capillary beds. This process increases the blood volume flowing directly from the arteries to the veins through arteriovenous anastomoses (shunts). The skin's main arteriovenous anastomoses are located in the superficial and deep vascular plexuses at the borders between the reticular dermis and the papillary dermis and at the border between the reticular dermis and the hypodermis, respectively (Fig. 18–1). The coiled secretory components of the sweat glands, and much of the length of their ducts, are located in the reticular dermis (18.VIII.A.2). Thus, the reticular dermis contains important components of the body's reflex response to massive blood and fluid loss, which is especially important in cases of severe burns, when substantial areas of skin are burned beyond the depth of the reticular dermis. In such cases, the body's loss of fluid, its loss of a barrier to further water loss, and its inability to regulate the volume of venous return to the heart can be life threatening.

14. **D.** Bone is a metabolically active tissue that serves as an important depot for calcium (8.III.D.1). Its net amount of mineral results from the balance between bone deposition and resorption. When insufficient calcium is available, bone resorption is required to maintain circulating calcium at levels sufficient to support life. Intestinal absorption of dietary calcium, which is also critical to maintaining calcium balance in the body, depends largely on the availability of adequate active vitamin D. When skin is exposed to sunlight for just a few hours a day, the dermis produces sufficient vitamin D to allow adequate absorption of dietary calcium (8.III.D.3.c; 18.I.A). However, in elderly individuals, the efficiency of this process is often reduced, creating an even greater need for the vitamin. Moreover, the reduced mobility and reduced tolerance of cold that accompany aging often reduce the amount of time elderly individuals are able to spend in the sun, further hampering vitamin D production. Parathormone is produced in the parathyroid gland in response to low serum calcium levels (21.V.A.2); it also increases bone turnover and, under conditions of inadequate absorption of dietary calcium, contributes to bone demineralization. Calcitonin enhances the deposition of calcium in bone when blood calcium levels become excessive, but this hormone is produced in the thyroid and not in the dermis (21.IV.C). Melanin is produced in the skin, but by cells in the epidermis (18.II.B.2); its effect on vitamin D production is very indirect.

15. **E.** Pacinian corpuscles such as those shown in Figure FE–2 are often described as resembling a sliced onion (24.II.E; Fig. 24–1). These peripheral vibration and deep pressure receptors are most abundant in the deep reticular dermis, hypodermis, joint capsules, and mesenteries. The sensation of vibration appears to involve the integration of deep pressure signals from these receptors with signals from more-superficial cutaneous touch receptors (eg, free nerve endings and Merkel's disks in the epidermis [24.II.B], and Meissner's corpuscles in the dermal papillae [24.II.D]). For this reason, damage to a single peripheral nerve rarely causes the loss of vibratory sensation (tested with a tuning fork); such a loss often points to a more central lesion. The vestibular and cochlear vibration receptors in the inner ear, with their characteristic rows of sensory hair cells (24.VI.D and E), are structurally very different from Pacinian corpuscles. The Golgi tendon organs have a less unique structure, comprising sensory dendritic arborizations among parallel collagen fiber bundles (24.III.B); because these often occur at muscle–tendon junctions, some skeletal muscle is typically seen in micrographs of these receptors.

16. **E.** All muscle fibers must undergo turnover of their cytoplasmic proteins (including contractile proteins) to maintain function. Thus continual protein synthesis is necessary. Slow (red) muscles are characterized not only by greater amounts of myoglobin, but also by a predominance of fibers containing mainly myosin heavy chain (MHC) type I in their contractile apparatus (10.II.B.3.a). MHC type II is characteristic of fast (white) muscles and exists in three main forms: IIA, IIB, and IIX (10.II.B.3 and 3.a). Because continual protein synthesis is necessary, the proportion of mRNA in all tissues reflects the proportion of proteins present. Proportions of other protein components of the contractile apparatus (actin, troponins, tropomyosin) may differ depending on whether a muscle is slow or fast, but only slightly because their regular arrangement is necessary for the maintenance of strict stoichiometry. The Z-disk protein α-actinin

(10.II.B.1.c.[2]) may vary somewhat because of differences in slow and fast myofiber diameters but does not vary as much as MHC-type protein. Although glycogen is a normal component of the sarcoplasm (10.II.B), it is not a protein, but rather a polysaccharide (for which no mRNA exists).

17. A. The key to answering this question is the knowledge that only chondrocytes and chondroblasts secrete type-II collagen, which is the cartilage-specific collagen (5.II.A.2; 7.I.C and II.A.1.a). Fibroblasts, osteoblasts, and bone marrow stem cells all secrete collagen and glycosaminoglycan matrix materials, but they secrete type-I or type-III (and not type-II) collagen (5.II.A.2 and E.1.b and c). Glycosaminoglycans secreted by synovial lining cells include mainly hyaluronan and thus are less likely to be sulfated (8.IV.B). Osteoclasts degrade rather than produce matrix materials (8.III.A.1.d). Muscle satellite cells (10.V.A) can, under unusual conditions, be stimulated to produce skeletal connective tissue matrix materials but are not found in the knee joint. A diagnosis of osteoarthritis (degenerative joint disease) would be most appropriate for the patient described in this case. The unilateral joint disease, obesity, and earlier trauma are among the determining clues. The effects of osteoarthritis on the bone most likely are secondary to the compromised structure and impaired function of the articular cartilage (7.II.A.6; 8.IV.B), whose maintenance requires a balanced turnover of the cartilage matrix materials by the articular chondrocytes.

18. B. During the natural aging process, much of the active (red) marrow is gradually replaced by inactive (yellow) marrow (13.I.C and III.A). As a general rule, the change proceeds in a distal-to-proximal direction in the long bones of the extremities. Thus, although the tibia is an accessible site for bone marrow aspiration in children, it normally contains only yellow marrow in middle-aged adults. The other sites listed typically contain red marrow in both children and adults. The iliac crests, sternum, and ribs are used as resources for grafts and biopsies and are listed according to the relative size of the samples that can be obtained from a single aspiration. Although vertebral bodies contain red marrow, they are rarely used as a biopsy source because they are relatively inaccessible by needle.

19. B. The cross-striations and mainly peripheral nuclei, without evidence of intercalated disks, indicate that the central tissue is skeletal muscle in longitudinal section (10.II.B.1.c.[2] and d). The surrounding tissue is epithelial. The acinar organization of the cells indicates that these comprise an exocrine gland (16.I.E.4). The predominantly basal location of the nuclei and the light-staining apical cytoplasm indicate that these are mucous acini (16.II.B.1). Among the body sites containing both skeletal muscle and mucous acini are the tongue (15.II.C), the pharynx (15.IV), the soft palate (15.II.A), and the larynx (17.V and V.C). The parotid gland may have acinar glands and some associated skeletal muscle (facial muscle or masseter), but its acini are exclusively serous (16.II.C). The pyloric stomach (15.VI.C.3), duodenum (15.VII.C.1), and trachea (17.VI) may contain some mucous acini and muscles, but the muscle is exclusively smooth rather than skeletal. The colon contains both muscle and mucous glands, but the mucus comes from goblet cells, and the muscle is not skeletal but smooth (15.VIII.C).

20. C. The patient complaints and test results described here are consistent with a diagnosis of myasthenia gravis. This disease is often accompanied by thymic hypertrophy and appears to stem from autoimmune effects that significantly and progressively reduce the number of acetylcholine receptors in myoneural junctions. As with all synapses, the receptors for the neurotransmitters are concentrated in the postsynaptic membrane (9.IV.C; 10.II.C.3). Their reduced numbers result in a functional disconnection between the motor nerve and the target muscle, causing the muscle weakness.

21. B. Figure FE–4 reveals the characteristic EM appearance of a normal eosinophil, and the ability to recognize this morphology is crucial to determining the correct answer. The specific granules are elongated and have an electron-dense internum (12.III.B.2.b; Table 12–1); their appearance is unlike that of any other type of cell. Increased numbers of circulating eosinophils are often seen during the infectious stage of parasitic infections by helminths such as *Trichinella spiralis* (trichinosis; 12.III.B.2.b). Trichinosis can be acquired by eating undercooked pork, especially from wild (nondomesticated) animals. Parasitic infection is among a very limited number of conditions that elevate eosinophil counts in peripheral blood. Bacterial food poisoning can cause diarrhea and nausea but would not result in elevated circulating numbers of any granulocyte type other than neutrophils (12.III.B.2.a), whose granules are much smaller, rounder, and homogeneous (ie, lack an internum; Table 12–1).

22. B. Figure FE–5 shows a labeled endothelial cell in an EM of a capillary. Although all three embryonic germ layers, as well as the neural crest, may give rise to epithelial cells (4.I.H; Table

4–1), virtually all of the cells that comprise the circulatory system, including those comprising the vessel walls and blood cells, derive from mesoderm (11.II.A.2.a; 13.II.A.1). The keratinocytes, like most of the epidermis, derive from surface ectoderm (18.I.B.1). Type-I alveolar cells, like all epithelia lining the digestive tract and respiratory passageways (except DNES/APUD cells), derive from endoderm (17.I.B). Ependymal cells, like all cells of the CNS (except microglia), derive from neural ectoderm (9.I.E). Melanocytes are classic examples of neural crest derivatives (18.II.B.2).

23. *C.* Perhaps the most characteristic histologic feature of the aorta is the overwhelming abundance of elastic laminae in the tunica media (11.II.B; Table 11–1). Most arteries have a collection of elastin at the borders between the intima and media and between the media and adventitia (11.II.B; Table 11–1). However, in the aorta, these internal and external elastic laminae cannot be distinguished, owing to the fact that elastic laminae occur throughout the entire media. In patients with Marfan syndrome, derangements in collagen and elastin production are common and may lead to areas of fragmentation or complete absence of elastic laminae in the aorta. The resulting loss of flexibility may lead to mechanical damage to the associated intima, enabling the blood in the lumen (under very high pressure) to begin to dissect into the aorta's wall. Although vasa vasora (11.II.H) are present, they normally contain less elastin, and the blood in their lumens is under less pressure. The aorta, like other arteries, has no valves (except the aortic valve, which is considered a component of the heart).

24. *B.* There are several keys to the correct answer. The history of rheumatic fever should suggest rheumatic heart disease. Associated damage to the cardiac valves during active disease often leads over time to stenosis (narrowing) of the valve opening. Although all valves may be affected, the mitral valve, between the left atrium and ventricle, is most commonly affected. Damage to this valve increases the hydrostatic pressure in the left atrium, which in turn increases pressure in the pulmonary veins and their tributaries (including the capillary beds in the interalveolar septa). This pressure increases the effusion of fluid into the interstitium, and when the amount of this fluid exceeds that which the lymphatics can drain, the excess spills into the lumen (5.IV.B). In cases of chronic disease, some blood cells reach the alveolar lumen. The patient's smoking may have contributed significantly to his pulmonary hypertension, chronic cough, metaplasia, ciliary paralysis, and the increased pressure in the pulmonary capillaries, which is the proximate cause of the pulmonary edema. The cough, metaplasia, and ciliary paralysis are all better viewed as symptoms rather than causes of the pulmonary edema. This patient's negative sputum cultures indicate that he does not have pneumonia, which could occur and could exacerbate his illness.

25. *A.* This patient's symptoms are consistent with hyperaldosteronism (21.II.A.4.a). Excess aldosterone enhances potassium excretion by the distal convoluted tubule (leading to hypokalemia), enhances sodium retention (leading to hypernatremia), and expands blood volume (leading to hypertension; 19.II.B.4). Primary hyperaldosteronism (Conn syndrome) typically involves an adrenocortical adenoma, which is characterized by tumor cells that secrete excess aldosterone; because of the small size of the adrenals, such tumors are relatively easy to detect by means of a CAT scan. However, this patient's hyperaldosteronism is secondary to the reduced blood flow to the left kidney through the stenosed left renal artery (19.I.A.1.b and II.E). Such reduced flow causes the juxtaglomerular cells in the wall of the afferent arteriole to secrete excess renin in response to the perceived reduction in blood pressure (19.II.D). The enzyme renin cleaves angiotensin I to form the angiotensin II that upregulates aldosterone secretion by the adrenal zona glomerulosa (19.II.D). Thus, secondary hyperaldosteronism is the result of a normal endocrine response to an abnormal vascular condition (rather than the result of an abnormal endocrine condition). Although cells of the adrenal zona fasciculata and reticularis also secrete steroids, they mainly secrete glucocorticoids (eg, corticosterone) and adrenal androgens (21.II.A.2.a–c). The cells of the macula densa in the wall of the distal convoluted tubule are a component of the juxtaglomerular apparatus and are assumed to participate in blood volume and electrolyte regulation, but no endocrine secretory product with the regulatory capabilities of renin has yet been associated with these cells (19.II.D). The main secretory products of the fetal (provisional) adrenal cortex are sulfated androgens (eg, dehydroepiandrosterone sulfate) rather than mineralocorticoids such as aldosterone (21.II.A.5). The renal papillary collecting ducts participate in sodium and water balance by reabsorbing sodium in the presence of antidiuretic hormone (ADH) from the posterior pituitary and are not known to produce an endocrine secretory product (19.II.C.2).

26. *E.* Multiple myeloma involves uncontrolled division of the functional B-cell derivatives known as plasma cells (5.II.E.2.c; 14.III.C). Because these cells secrete immunoglobulins, an increase in their number increases circulating concentrations of immunoglobulin. Typically, a single class of immunoglobulin (ie, IgG, IgM, IgA, and, less frequently, IgE) shows a spike in plasma concentration. All circulating immunoglobulins belong to the class of plasma proteins called gamma globulins (12.II.B.1). Serum albumin is produced by the liver (16.IV.B) and is more likely to decrease than to increase in amount in response to increasing concentrations of another large plasma protein. Erythropoietin and GM-CSF regulate the proliferation of other blood cell precursors; alterations in their quantity are not diagnostic of multiple myeloma (13.I.E). Because certain immunoglobulins fix (bind) complement (14.II.B.1,2 and C.2), circulating complement levels are more likely to decrease than to increase in patients with multiple myeloma. Under certain conditions, elevated parathyroid hormone levels can lead to decreased bone density (increased radiolucency; 21.V.A.3.a), but in such cases, radiographs reveal a more general pattern of bone loss rather than the focal defects described here.

27. **D.** Plasma cells are differentiated effector cells of the B-lymphocyte lineage that secrete humoral antibodies—specifically IgA and IgG (5.II.E.2.c; 14.III.C). These cells increase in number in the interlobular connective tissue of the breast during the last trimester of pregnancy (23.X.E). The immunoglobulins they secrete are added to the protein-rich first milk (colostrum) and also to the lipid-rich milk during later lactation. The IgA helps precipitate antigens in the milk, on the breast surface, and in the infant's digestive tract (14.II.B.2). In newborns, the IgG in milk can cross the thin mucosa of the digestive tract and enter the infant's bloodstream (14.II.B.1). Macrophages may indirectly contribute to the activity of mammary plasma cells by helping to program their response to specific antigens (14.III.B) but are not directly involved in conferring passive immunity. CD8+ (cytotoxic T) lymphocytes must contact antigens directly to carry out their immune function (14.III.A.2); thus, those in the mammary lymphatic capillaries have no direct role in passive immunity. Secretory cells in the mammary glands produce a transport component that facilitates the formation and release of IgA dimers into the colostrum and milk (14.II.B.2). The ductal cells have no direct involvement in conferring passive immunity. Substantial numbers of neutrophils in the colostrum indicate a bacterial infection of the mother's breast, which can harm rather than protect the infant. Myoepithelial cells indirectly assist in conferring passive immunity by causing the secretions in the alveoli to reach the lactiferous ducts (23.X.F.3).

28. **B.** Dietary fats are absorbed by enterocytes and assembled into chylomicrons, which are released at the basolateral membranes of absorptive cells (15.VII.B.3.c). Most of these chylomicrons enter the lacteals in the lamina propria of the villi (15.VII.B.1). Lacteals are lymphatic capillaries that drain through mesenteric lymphatics to the cisterna chyli and subsequently flow through the thoracic duct to enter the venous blood at the junction between the left subclavian and jugular veins in the neck (11.I.B.2). Some (but not most) of these chylomicrons enter the blood in the capillaries of the intestinal lamina propria and pass through the portal circulation to the hepatic sinusoids. The high endothelial postcapillary venules of lymph nodes are selective sites of diapedesis for T lymphocytes leaving the blood to enter the paracortical zone of lymph nodes (14.VII.A.3).

29. **E.** Among the sites listed, it is only in the germinal centers of the lymphoid nodules (characteristic of the appendix walls) that cell proliferation is likely in response to trapped antigenic material in the appendix lumen (14.IV and V; 15.IX). Proliferation of epithelial cells at the bases of the crypts occurs normally and serves to replace sloughed enterocytes (15.VII.B.8). The tips of villi and the submucosal glands are not viable choices because neither are found in the appendix. Surface epithelial cells rarely divide, and although some new capillaries may form as an organ enlarges, this growth is secondary to the enlargement itself. Moreover, blood cells do not divide in the lumen of peripheral capillaries, and lymphocytes must leave the lumen and enter the interstitium to divide. It is worth noting here that the initial enlargement of the lymphoid nodules alone does not constitute appendicitis; however, the increased pressure it causes in the wall, in addition to the pressure it exerts against the fecalith, may obstruct blood flow to the organ. This obstruction can sufficiently restrict access to the organ by certain blood-borne components of the immune system (eg, antibodies and immunocompetent cells), allowing a bacterial infection to gain a foothold. This may lead to a vicious cycle of further enlargement and further restriction of blood flow.

30. **A.** The tight junctions between the Sertoli cells in the seminiferous tubules form a barrier that prevents blood-borne components of the immune system from gaining access to the tubule lumen (blood–testis barrier; 22.II.G). Because novel proteins may be expressed by developing spermatogenic cells as a result of genetic recombinations during synapsis, this barrier is critical to prevent-

ing autoimmune attack of the spermatozoa and their precursors. If this barrier is damaged, immunoglobulins may reach the developing cells or even the tubule lumen, bind to the surface of the developing gametes, and impair their development or function. Although Leydig cell secretions are important in maintaining the structure and function of Sertoli cells, they are not located in the seminiferous tubules (22.II.F). Myoid cells are located in the seminiferous tubules but have no barrier function per se (22.II.C.1). Spermatogonia have no barrier function and, because they are premeiotic, have no novel surface antigens (22.II.C.2.a). Spermatids are protected by the blood–testis barrier but do not contribute to its function (22.II.C.2.d). Endothelial cells and plasma cells must be excluded because they are not located in the seminiferous tubules. Defective endothelial cells may leak excess immunoglobulin from the plasma into the interstitium (11.II.A.4), and overactive plasma cells may secrete excess immunoglobulins in the interstitium (14.III.C); however, if the blood–testis barrier remains intact, such activity does not affect semen composition.

31. G. Respiratory distress syndrome in premature infants results when birth occurs before full differentiation of the pulmonary epithelium (17.VIII.C). In such cases, the type-II alveolar cells (also called alveolar septal cells; 17.VIII.B.2) are premature and unable to secrete sufficient surfactant to coat the alveolar lumens. The surfactant normally acts to reduce the surface tension in the alveoli, helping to flatten the type-I cells and facilitating full inflation (17.VIII.C). Goblet cells, Clara's cells, and brush cells (17.I.B.1.a; VII.E) have some secretory function but do not specifically secrete surfactant. The alveolar macrophages clear debris from the alveolar lumen. The endothelial cells of the alveolar capillary beds are critical to successful gaseous exchange but do not contact the alveolar lumen and do not produce surfactant (17.VIII.A.1).

32. D. Type-I alveolar cells that are lost because of normal turnover, trauma, or pathologic damage are replaced by derivatives of type-II alveolar (septal) cells (17.VIII.D). Vascular pericytes, smooth muscle, and endothelium are important in vascular regeneration but do not contribute to the replacement of type-I cells. Clara's cells (17.VII.E) and small granule cells (17.I.B.1.a) are too distant to contribute to the replacement of cells lining the alveoli. Fibroblasts (17.VIII.A), important in connective tissue repair, do not participate directly in replacing damaged epithelial cells. When they do assist in replacement, it is usually through scar tissue formation, which results in the loss of normal epithelial function.

33. E. IgE secreted by plasma cells (14.III.C) during previous contact with an allergen is bound by its Fc component to the surface of local mast cells and basophils in the lamina propria (14.II.B.4). In individuals with allergic hypersensitivity, the allergen binds and cross-links the Fab components of these cell-surface IgEs, thereby triggering degranulation (5.II.E.2.a). This process releases heparin and histamine (neither of which are proteins) into the interstitium. Histamine enhances local secretion of mucus, increases blood flow to the area (by relaxing vascular smooth muscle), and causes the capillaries to leak fluid into the tissue (11.II.A.4). The allergen-specific IgE synthesized by the plasma cells is the protein on which this response depends and is thus the best answer. The term mucosa, with regard to both the digestive system and the respiratory system, refers to the lining epithelium and its underlying lamina propria (17.II.B). Plasma cells and mast cells are components of loose connective tissue that are important defensive elements in all mucosal barriers (5.II.E.2.a,c, III.A.1, and IV.2.b). Goblet cells and Bowman's glands (17.II.B) may contribute to nasal congestion, but their secretions are responses to and not causes of the events that follow the cross-linking of IgE molecules. Olfactory cells (24.IV.B) synthesize and secrete neurotransmitters but play only a very indirect role, if any, in producing the symptoms of allergic reaction. Dust cells are alveolar macrophages (17.VIII.B.3) and are too distant from the inflamed tissue to have a direct role in the patient's reaction. Endothelial cells (11.II.A.2.a), like secretory cells, respond to but do not directly cause the release of histamine from nearby mast cells and basophils.

34. B. The findings in this case are consistent with Goodpasture syndrome, which involves a devastating, and almost simultaneous, autoimmune attack on both the glomerular and alveolar basal laminae (4.IV.C.1.a and b; 17.VIII.A.1; 19.II.B.1.c). Most autoimmune diseases are poorly understood, but strong evidence suggests that the immunoglobulins bound to the basal laminae in Goodpasture syndrome activate complement (14.II.C.2) locally and lead to serious damage to the cells that normally contribute to the blood–air barrier and the glomerular filtration barrier. As these barriers break down progressively, blood cells reach the alveolar lumen and the urinary space, and urine gains access to the blood. Determining the correct answer to this question requires knowledge of the important roles that basal laminae play as selective barriers at these sites, and also understanding their composition. Laminin, heparan sulfate, and type-IV collagen are common components of all basal laminae (4.IV.C.1.a). Another key to answering correctly is the phrase "extracellular protein." This descriptor eliminates actin (an *intracellular* protein; 2.III.I.2.a) and both heparan sulfate and hyaluronan (glycosaminoglycans rather than proteins; 5.II.D.1). Type-II collagen is easily ruled out because it occurs only in cartilage matrix (5.II.A.2). The more difficult answer to rule out is tropocollagen. Tropocollagen is produced when registration peptides are cleaved from procollagen, the secreted form of this extracellular protein. Procollagen peptidase must cleave procollagen to tropocollagen to allow incorporation of its triple-helical structure into collagen fibers (5.II.A.1.b). However, type-IV collagen is unusual in that its registration peptides are not cleaved; thus, it never forms tropocollagen or fibers (5.II.A.2).

35. **E.** The olfactory cells are bipolar neurons (Table 9–3) whose cilia are specialized dendrites that extend into the nasal cavity and whose axons penetrate the cribriform plate to terminate in the olfactory bulb (24.IV.B). Fracture of the cribriform plate can be expected to sever many olfactory axons and can instigate retrograde degeneration, which would include chromatolysis in the olfactory cell bodies (9.VIII.B.1). In description of cellular responses to injury, the term chromatolysis is applied almost exclusively to neurons and signifies dispersion of the Nissl substance that accompanies enlargement of the cell body and movement of the nucleus to an eccentric position in the cytoplasm. The olfactory epithelium, which covers part of the superior conchae, has few goblet cells compared with the standard pseudostratified, ciliated columnar epithelium that lines most of the nasal fossa. The presence of both Schwann cells and oligodendrocytes among the choices may have falsely indicated a need to distinguish between central and peripheral myelination (9.III.A.2 and B.1). However, olfactory axons are unmyelinated, which rules out any role for the oligodendrocytes (unmyelinated axons in the CNS are not invested by glia). These unmyelinated fibers are part of the PNS and are invested (but not myelinated) by Schwann cells. The Schwann cells participate in anterograde and retrograde degeneration (9.VIII.B.1 and 2), but because they are not neurons, they do not undergo chromatolysis. Basal and sustentacular cells are the epithelial components of the olfactory mucosa (24.IV.B) but also do not undergo chromatolysis per se.

36. **F.** Figure FE–6 shows the structure of an inactive (*left*) and active (*right*) gastric parietal cell (15.VI.B.4.a and b). Note the abundant mitochondria (*M*), tubulovesicles (*TV*), and large intracellular canaliculi (*IC*) on the left and the abundant microvilli (*MV*) lining the extended intracellular canaliculus (formed by fusion of the TV with the IC) on the right. The two secretory products currently associated with this cell are gastric HCl and intrinsic factor (15.VI.B.4). It is the insufficient secretion of intrinsic factor (owing to its importance in vitamin B_{12} absorption) that causes pernicious anemia (15.VI.B.4.b). Excessive secretion of HCl by this cell type, sometimes in association with excess gastrin secretion by gastric enteroendocrine (G) cells or pancreatic delta (D) cells (Zollinger–Ellison syndrome; 21.III.C), is associated with duodenal ulcers. Acromegaly (8.III.D.4.b) is another disease caused by *hyper*secretion (by pituitary somatotrophs; 20.III.A.2.a) that can be ruled out immediately. Because maintenance of bone mineralization involves a balance between bone deposition and resorption, osteoporosis (8.III.D.2) can be caused by hyposecretion of matrix by osteoblasts or hypersecretion of acid and enzymes by osteoclasts, neither of which is depicted in Figure FE–6. Myxedema (21.IV.B.3.b), diabetes mellitus (21.III.B), and Addison's disease (21.II.A.4.b) are caused by the hyposecretion of thyroid follicle cells, pancreatic β cells, and cells of the adrenal cortex, respectively—none of which resemble gastric parietal cells.

37. **B.** The pancreas is both an endocrine and an exocrine organ (16.III.A). The enzyme-secreting acinar cells comprise the exocrine component (16.III.A and B). These arise by the proliferation and invagination of epithelial cells lining the lumen of the embryonic duodenum (16.I.B). The entire digestive tract lining derives from embryonic endoderm, which in turn derives entirely from the epiblast (as do the mesoderm and ectoderm). The cells in the islets of Langerhans comprise the endocrine pancreas (21.III). The insulin-secreting β cells and glucagon-secreting α cells derive from endoderm (21.III and III.B). Gastrin-secreting and somatostatin-secreting δ cells are of neural crest origin (4.VI.C.2).

38. **C.** To correctly answer this question, it is necessary to understand the mechanisms that maintain fluid balance between the blood and the tissues (5.IV.B). Such balance is maintained by (1) hydrostatic pressure in the vessels; (2) plasma colloid osmotic pressure (amount of solute); (3) the permeability of vessel walls; and (4) lymphatic drainage. Cirrhosis is a scarring of the liver from chronic damage—in this case from alcohol toxicity. Progressive scarring reduces blood flow through the liver, causing increased pressure in the portal vein and its tributaries (16.IV.C.1–5). This increased hydrostatic pressure, which is called **portal hypertension,** causes fluid in the

mesenteric veins and their tributaries to back up in the tissues they normally drain (including the intestines and spleen). When tissue fluid accumulation exceeds that which the lymphatics can remove, it spills into the tissues (**edema**) and abdominal cavity (**ascites**) (5.IV.B). Ascites swells the abdomen as fluid accumulates. The patient's preexisting systemic hypertension and cumulative damage to the hepatocytes further increases the hydrostatic pressure. The damage to the hepatocytes also compromises the synthesis of albumin and its secretion into the blood (16.IV.D.1.b). Serum albumin maintains the plasma colloid osmotic pressure that normally holds fluid in the vessels (12.II.B.1). Decreased serum albumin allows more fluid to flow out of the vessels and into the tissues, exacerbating both edema and ascites. Cirrhotic livers do undergo some connective tissue hyperplasia (scarring), but the associated swelling of the spleen is caused by portal hypertension rather than hyperplasia; moreover, hepatosplenomegaly is usually insufficient to cause major abdominal swelling. Overeating does not typically accompany alcoholism; indeed, there is usually a reduction in food intake, which can further compound the decrease in serum albumin synthesis. Although the high levels of bilirubin in the blood that cause jaundice can result from hepatocyte damage and can cause leakage from the bile canaliculi (16.IV.D.1.a), they are not a direct cause of ascites. The diuretic effects of alcohol and of diuretics used to help control systemic hypertension, such as furosemide (19.II.C.2), limit ascites by increasing urinary volume.

39. **A.** Because epithelia are avascular, normal blood flow to capillary beds in the underlying lamina propria is required for normal epithelial cell function (4.II.F). Because they are the cells farthest from the lamina propria capillaries, the stomach's surface mucous cells would be the first and most seriously affected by a local reduction in blood flow (15.VI.B.1). The mucus secreted by these cells provides an important barrier to autodigestion because of the combined actions of pepsin secreted by the gastric chief cells and HCl secreted by the parietal cells (15.VI.B.1). A reduction in mucus secretion would naturally reduce its protective effect. Reduced protective mucus secretion is most often associated with gastric ulcers, and excess acid secretion is most often associated with duodenal ulcers. A reduction in the secretions of the chief and parietal cells (15.VI.B.4 and 5) would more likely reduce than increase the likelihood of gastric ulceration. Goblet cells first appear in the GI in the small intestine (15.VII.B.4) and thus have no protective effect on the stomach. The enzymatic secretions of pancreatic acinar cells may participate in ulceration owing to autodigestion in the duodenum but are less likely to do so in the stomach unless an impairment of pyloric sphincter function exists. Moreover, a reduction in the secretion of these cells would more likely reduce than increase the likelihood of ulceration. Certain secretions of vascular endothelial cells can be affected by localized ischemia; thus, these cells can be indirectly tied to ulceration (11.II.A.2.a). However, among the available choices, the surface mucous cells are most strongly linked to the development of a gastric ulcer.

40. **D.** Zollinger–Ellison syndrome (gastrinoma) involves a hypersecreting tumor of the δ cells in the pancreatic islets of Langerhans (21.III.C). The tumor cells secrete excess gastrin, which causes excess HCl release by the parietal cells in the gastric mucosa. Excess HCl secretion is a common cause of duodenal ulcers. Gastric chief cells are less directly responsive to gastrin secretion; moreover, without adequate HCl, the pepsinogen they secrete cannot be converted to the active proteolytic enzyme known as pepsin (15.VI.B.5.b). Paneth's cells secrete lysozyme, an antibacterial enzyme, and occur mainly in the ileum (15.VII.B.6). Excess secretion by duodenal goblet cells is more likely to protect against ulceration than to cause it (15.VII.B.4). Disaccharidase and dipeptidase secretions by enterocytes are not known to be triggered by gastrin (15.VII.B.3). Pancreatic acinar cells (16.III.B.1) and δ cells (21.III.C) are not components of the digestive tract mucosa.

41. **D.** The proximal tubule is characteristically lined by simple cuboidal epithelium with abundant apical microvilli (19.II.B.2). This morphology is characteristic of an epithelium whose main function is absorption. The proximal tubule's main function is reabsorption of most of the water and solutes that pass through the glomerular filtration barrier, including most of the inorganic salts, glucose, and amino acids (19.II.B.2; Fig. 19–3). A malfunctioning proximal tubule would cause higher-than-normal concentrations of these materials in the urine and lower-than-normal concentrations in the blood. Hyperglycemia is not likely to occur in individuals with Fanconi syndrome because glucose is not reabsorbed and is lost in excess in the urine. Because salts are lost at higher-than-normal rates, hypocalciuria and hyperphosphatemia also would not be expected. Because insufficient salt and water reabsorption occurs, blood volume decreases; thus, hypertension would not be likely to occur. However, aminoaciduria is a likely symptom of this disorder because the reabsorption of amino acids is an important function of the proximal tubules (19.II.B.2).

42. *C.* The process by which cells make a normal transition from the embryonic state to their mature state through alterations in the type of genes they express is termed differentiation (2.IV.A; 3.VI.C). One characteristic of embryonic cells that is typically lost during differentiation is continuous proliferation. Cells that form tumor masses such as those seen in Wilms tumors have not failed to undergo proliferation; indeed, their uncontrolled proliferation can be termed *hyperplasia.* The term *anaplasia* specifically refers to the failure of embryonic cells to differentiate, and *metaplasia* refers to a shift in the cell or tissue phenotype *after* differentiation has already taken place (4.II.B). Hypertrophy refers to growth in the size of individual cells, as opposed to an increase in cell numbers in a particular population (hyperplasia). Hypertrophy can be a part of normal growth and differentiation for some cell types (eg, skeletal muscle and cartilage). Homeostasis refers to the maintenance of a normal steady state and requires further qualification for use in this context; for example, continuously dividing Wilms tumor cells can be said to be maintaining homeostasis at the cellular level despite their failure to respond to normal homeostatic signals at tissue, organ, and organism levels.

43. *B.* Membranous glomerulonephritis is characterized by the deposition of excess immunoglobulin (mainly IgG) on the glomerular basement membrane. IgG is a secretory product of plasma cells, which are B-lymphocyte effectors (14.I.F.4 and III.C). This condition is typically accompanied by an increase in the "leakiness" of the glomerular basement membrane, which involves a mechanism that remains unclear. Surprisingly, there is no accompanying invasion of blood-borne phagocytic cells (neutrophils and monocytes) that could damage the glomerular filtration barrier (19.II.B.1.c) in the process of trying to remove the excess immunoglobulin. Thus, the leakiness may involve increased phagocytic activity of the mesangial cells, which most likely function to maintain the filtration barrier (19.II.B.1.b and f). Regardless of the mechanism, the usual size selectivity of the barrier is lost and plasma protein, including serum albumin, enters the filtrate in amounts too large to be reabsorbed by the proximal tubule or to be replaced with compensatory serum albumin secretion by the hepatocytes. Secretions of the renal cells themselves are insignificant compared with the large amounts of IgG trapped on the filtration barrier.

44. *D.* Furosemide selectively inactivates the $Na^+/K^+/Cl^-$ pump (symporter) in the cells lining the thick ascending limb of the loop of Henle (19.II.B.3.b.[2] and C.2). It is the action of this pump that normally increases the salt concentration of the medullary interstitium. This effect is caused by the inability of water in this part of the nephron to follow the salt into the interstitium (19.II.B.3.b.[2]). Without this pump, the production of hypertonic urine would be impossible. When the pump is inactivated by furosemide, the medulla approaches an isotonic state. When the medullary interstitium is isotonic, water in the medullary collecting ducts cannot diffuse into the interstitium—even in the presence of ADH—owing to the absence of a concentration gradient. Thus, a higher-than-normal volume of urine at lower-than-normal tonicity is released. The medullary collecting ducts, vasa recta, and peritubular capillaries must be ruled out as potential answers because they are not parts of the nephron (19.II.B.1–5). Unlike pumps, which are sites of active transport, the glomerulus and medullary collecting ducts are sites of passive diffusion. The precise function of the macula densa (19.II.B.4 and D) is not yet clear; although it may sense the tonicity of the filtrate leaving the medulla through the ascending thick limb, the latter is a better choice for the location of furosemide's diuretic effects by virtue of its presence in the hypertonic medulla.

45. *E.* This patient's frequent voiding of dilute urine (polyuria) and excessive thirst (polydipsia) are classic symptoms of diabetes insipidus (DI). The most common cause of DI is a chronic or temporary loss of (or reduction in) the pars nervosa's ability to produce arginine vasopressin, also known as antidiuretic hormone (AVP or ADH; 20.IV.A.1), in response to a hypovolemic or hyperosmotic challenge. The normal-to-elevated levels of serum AVP, and the evidence of normal pituitary structure and function, indicate that factors regulating AVP production (including the pituitary and hypothalamus) are not affected in this case. Thus, the hormone's target cells become candidates for the location of the lesion causing this patient's symptoms. Because normal and even excess levels of AVP fail to elicit the production of hypertonic urine, AVP receptors on the target cells may be absent or malfunctioning. The main target cells for ADH are those lining the renal medullary collecting ducts, which become permeable to water in the presence of the hormone (19.II.C.2). This permeability allows water to diffuse into the hypertonic medullary interstitium, increasing the concentration of the filtrate that remains in the collecting ducts. Recent studies indicate that a number of drugs, including lithium, can downregulate the ADH-responsive water channels in the medullary collecting ducts and produce symptoms of DI. DI caused by lesions in the kidney is referred to as nephrogenic DI.

46. *D.* Adrenergic receptors, such as those that mediate epinephrine and norepinephrine signaling, are perhaps the best understood G protein–coupled receptors (2.II.C.2.c). The receptor for glucagon also belongs to this group. Steroid hormones, such as the sex steroids (eg, estradiol, testosterone, progesterone) and glucocorticoids (eg, hydrocortisone), pass more freely through membranes and bind to intracellular receptors, which subsequently bind directly to DNA to affect transcription (2.II.C.2.d). Interestingly, thyroid hormone is a peptide hormone whose receptors belong to the same intracellular receptor superfamily as do the receptors for steroids. The insulin receptor is a transmembrane protein with an intracellular tyrosine kinase domain (2.II.C.2.b). G proteins may be among the targets indirectly phosphorylated by an activated insulin receptor, but they are not the only targets; thus, insulin receptors are not considered G protein–coupled receptors per se.

47. *F.* Somatotrophs of the adenohypophysis are the primary cells responsible for the secretion of somatotrophic hormone, also known as growth hormone (STH or GH; 20.III.A.2.a). The entire adenohypophysis, which contains the somatotrophs (a type of pituitary acidophil), forms from an invagination and upgrowth of the oral ectoderm (20.II; Fig. 20–1; Table 20–1). The neural crest forms the follicular cells (most chromophobes) and the pia mater and skeletal tissue surrounding the pituitary (sella turcica; 9.I.E). The neurohypophysis forms from neural ectoderm (20.II; Fig. 20–1; Table 20–1). The dermamyotome and sclerotome are somitic derivatives that contribute to the axial and appendicular skeleton but not to the pituitary gland. The extraembryonic mesenchyme makes a significant contribution only to the hematopoietic tissues (5.I.E; 13.II.A.1) and the placenta (23.VI.A).

48. *B.* Chromophobes normally comprise 50% of the cells in the adenohypophysis. They include three main types: (1) the follicular cells that form a stromal network (the majority of chromophobes); (2) undifferentiated nonsecretory stem cells that lack granules; and (3) degranulated chromophils containing very few granules (20.III.A.1). Hyperstimulation of the thyrotrophs (20.III.A.2.b) can result in degranulation that occurs faster than the granules can be replaced, resulting in increased numbers of the third type of chromophobes. Chromophobomas can occur but would not cause the increase and subsequent precipitous drop of circulating TSH described in this case. Colloid in Rathke's cysts (20.III.C) is unrelated to the symptoms described. Pheochromocytomas are tumors of the adrenal medulla (21.II.B.4) and are unlikely to directly affect thyroid function. Adenohypophyseal scarring is possible in response to an infarction but would result in many additional symptoms. The pituicytes are the stromal cells of the neurohypophysis (20.IV.C), and in this case, it is adenohypophyseal structure and function that are clearly affected.

49. *C.* Any reduction in oxygen-carrying capacity triggers the production of erythropoietin by the kidney (13.VII.A). This hormone increases the rate of red blood cell production from hematopoietic stem cells in the bone marrow (the only cells that normally synthesize hemoglobin). Renin is not a hormone but rather an enzyme that converts the liver prohormone angiotensinogen to angiotensin I, which is subsequently converted to the hypertensive agent angiotensin II in the lungs (19.II.D). Fibroblast growth factor is an angiogenic agent that is produced locally in hypoxic tissues and that stimulates angioneogenesis to increase the local blood supply.

50. *A.* The two main factors associated with hypothyroidism that contribute to obesity are poor glucose utilization throughout the body and inactivity (21.IV.B.2.e and 3.b). Poor glucose utilization favors glucose storage as lipid, glycogen, and even glycosaminoglycan. Inactivity, possibly related to inefficient energy metabolism, only compounds the problem of decreased utilization and increased storage of glucose. Unfortunately, nutrient absorption by the enterocytes is unaffected by these factors; thus, normal uptake results in excess storage. Bloating may occur but more likely in response to the deposition of excess glycosaminoglycan (ie, myxedema; 21.IV.B.3.b). Hypersecretion of the adrenal medulla (epinephrine and norepinephrine) is more likely to cause hyperactivity than lethargy and is likely to promote fat mobilization; thus, this condition is likely to reduce the tendency toward obesity (21.II.B.3 and 4).

51. *G.* The cell shown in Figure FE–7 contains abundant SER and mitochondria with tubular cristae. This is classic morphology for a steroid-secreting cell (4.VI.C.5). Of the sites listed, the placenta contains the most active steroid-secreting cells during the last trimester of pregnancy (23.VII.B.2). Beginning at approximately 12 weeks' gestation, the corpus luteum's role in steroid secretion (23.II.E.2) is gradually supplanted by the placenta. The corpus luteum eventually degenerates into a corpus albicans, and the placenta assumes its role in steroid secretion. Although hirsutism is sometimes caused by excess secretion of adrenal androgens, these derive from the adrenal cortex rather than the medulla (21.II.A.3.c); thus, hirsutism is irrelevant. Secretions from the pancreatic islets and thyroid glands are polypeptides and thus require RER (4.VI.C.1 and 2). Cells in the brain, fetal or mature, do not typically exhibit the ultrastructure associated with steroid-secreting cells (9.II.A and III.A).

52. **A.** Just before menstruation, the endometrium is in the secretory phase and comprises a stratum functionale, which is subsequently shed, and a stratum basale, which remains after menstruation has taken place (23.IV.A; Table 23–1). Epithelial cells that line the remnants of the endometrial glands in the stratum basale proliferate to provide the cells that eventually cover the exposed stromal connective tissue. The entire stratum functionale is shed (including its zona spongiosa and zona compacta) and thus cannot contribute to endometrial regeneration. The cytotrophoblast and primary villi are components of the placenta (23.VII.A) that are not present after menstruation. Arteries in the myometrium feed but do not replace the growing endometrium (23.IV.B).

53. **D.** The structure that provides the most significant protection against blood-borne toxins and autoantibodies for the developing spermatozoa is the blood–testis barrier (22.II.G). This barrier consists of a series of tight junctions between the Sertoli cells in the seminiferous tubules. Together, these tight junctions separate the adluminal compartment (which contains the developing spermatozoa) from the basal compartment. The basal compartment contains the blood vessels, the interstitial cells, the myoid cells, and the spermatogonia. Because chemicals reached the bloodstream through this individual's skin and through his inhalation of fumes, endothelial cells lining the vasculature most likely have a limited ability to block the passage of these chemicals. The tunica albuginea (22.II.B) surrounding the entire testis is penetrated by the testis' blood supply, and the tunica propria (22.II.C.1) surrounding each seminiferous tubule is penetrated by capillaries that nourish the avascular seminiferous epithelium. Hepatocytes, through their highly developed capacity for chemical detoxification (16.IV.B), are an important component of systemic protection in this case but do not appear among the answers. Thus, the Sertoli cells are the best choice provided.

54. **C.** The main receptors for LHRH are those on the LH-secreting gonadotrophs in the adenohypophysis (20.III.A.2.b and E.1). When hypothalamic neurons release LHRH at nerve termini around the primary capillary plexus in the median eminence (20.III.D.1), the hormone passes through the hypophyseal portal system to the secondary capillary plexus in the adenohypophysis, where it binds to the LHRH receptors on the gonadotrophs, causing LH (ICSH) release into the bloodstream (20.III.F.6). Circulating LH stimulates testosterone secretion by the interstitial cells (of Leydig) in the testes (22.II.F). In this case, the testosterone stimulates the growth of the prostatic adenocarcinoma. A synthetic LHRH analog can be used to block the function of the LHRH receptors on the gonadotrophs, preventing the normal cascade that leads to testosterone synthesis and thus removing one signal that normally promotes the growth of the tumor.

55. **D.** This patient had an anterior pituitary (adenohypophyseal) adenoma that was treated by removal of most, if not all, of the adenohypophysis. The circulation was spared during removal of the adenoma. The headaches and blurred vision are typical in patients with this type of tumor. Headaches occur because of pressure from the tumor on the internal carotid artery and other components of the circle of Willis surrounding the infundibulum, which restricts blood flow to the brain (20.III.D.1). Blurred vision is caused by pressure on the optic chiasm, which lies directly anterior to the adenohypophysis (anterior pituitary) (20.II). A tumor in the epiphysis cerebri (pineal gland) would not result in the preoperative symptoms experienced by this patient and would not cause the postoperative effects she experienced if damaged or removed (21.VI). The neurohypophysis releases ADH and oxytocin (20.IV.A.1); this portion of the pituitary and the Herring bodies it contains (20.IV.A), have no direct involvement in maintaining thyroid and ovarian function. Rathke's cysts (20.III.C) have no known function. The cessation of the patient's headaches suggests that the internal carotids were spared during surgery. The internal jugulars lie outside the cranial cavity, and although they indirectly drain adenohypophyseal vessels (20.III.D.3), they generally are not subject to damage during a hypophysectomy. Owing to the location and normal function of the adenohypophysis, surgical removal of the tumor and damage to associated tissue provide the best explanation for both preoperative and postoperative symptoms in this case. The loss of adenohypophyseal function also can cause a loss of thyroid, adrenocortical, and ovarian function (20.III.A.1.a and b), as can severe damage to the hypothalamus.

Index

NOTE: Page numbers followed by the letter *f* indicate figures; page numbers followed by the letter *t* indicate tables.